The Freshwater Mussels of Tennessee

The Freshwater Mussels of Tennessee

Paul W. Parmalee and Arthur E. Bogan

Sponsored by
American Pearl Farms
Environmental Protection Agency
Shell Exporters of America, Inc.
Tennessee Department of Agriculture
Tennessee Wildlife Resources Agency
U.S. Department of Agriculture Forest Service
U.S. Fish and Wildlife Service

THE UNIVERSITY OF TENNESSEE PRESS / KNOXVILLE

Many State of Tennessee programs receive federal aid for
such wildlife programs as fish and/or wildlife. Policies of
the State and regulations of the U.S. Department of the
Interior prohibit discrimination on the basis of race, color,
religion, national origin, age, sex or handicap. If you believe that you have been discriminated against in any program, activity, or facility as described above, or if you desire more information, please write to the Office of Equal
Opportunity, U.S. Department of the Interior, Washington,
D.C. 20240.

Cover design by Sheila Hart, photographs by W. Miles
Wright. Unless otherwise noted, all photographs in this
volume are by W. Miles Wright.

Library of Congress Cataloging-in-Publication Data

Parmalee, Paul W.
    The freshwater mussels of Tennessee / Paul W. Parmalee
and Arthur E. Bogan. — 1st ed.
        p.   cm.
    Includes bibliographical references (p.   ) and index.
    ISBN 978-1-57233-013-9 (cloth)
    1. Freshwater mussels. I. Bogan, Arthur E. II. Title.
QL430.6.P37   1998
594' .4—dc21   97-45371

To Barbara and Cindy

Contents

Preface

There is no way of predicting which species will be important in the future, which will produce a miracle drug or which will prove to be a sensitive indicator of pollutants or nutrients. Every species plays a role in the complex calculus of the environment. To lose one deducts from the incomparable library of life on earth.

—E. E. Spamer, "Academy Helps Save Habitat," 1996

Of nearly 300 recorded species of freshwater mussels in the United States, approximately 130 are or were known to occur within the political boundaries of Tennessee. The mussel fauna of North America exhibits the greatest variety of species in the world and is concentrated mainly in the Southeast. Except for Alabama, the lakes, streams, and rivers of Tennessee once harbored the most diverse and abundant assemblage of these mollusks known in historic times. But with the settlement of land by European explorers and pioneers came lumbering of the forests, clearing and intensive farming activities, strip mining, industry, and construction of power dams. All of these factors, along with other related practices such as the channelization of numerous rivers and the commercial exploitation of mussel shell, brought about major reductions in species distribution and abundance, local extirpation, and, in at least a dozen cases, extinction.

Vast numbers of chalky mussel shells composing the so-called "shell mounds," plus lenses and scatterings of valves along the beaches and eroding banks of the Cumberland and Tennessee rivers and many of their major tributaries, attest to the aboriginal intensive use of these mollusks as a food resource. In spite of almost continual harvesting of shells from many of the most extensive mussel beds by these early peoples over a period of many millennia, there is no evidence to suggest that such gathering efforts resulted in the local extirpation or extinction of any species. The greatest overall detrimental impact on mussel populations probably can be attributed, directly or indirectly, to dam construction—especially those built in the 1930s, 1940s, and 1950s. Certainly there is no denying that biological and chemical effluents from certain agricultural practices, water-treatment plants, industry, and mining that are continually being discharged into our lakes and rivers also have had a devastating effect on mussels and other aquatic life. Although some species have invaded or expanded their ranges in the impounded stretches of rivers, such as in the Tennessee River reservoirs, the gain has been far less than the loss of mussel diversity that existed prior to dam construction. By 1998, among the approximately 300 species of freshwater mussels known to have occurred in the United States, over half were listed federally or by states as either threatened, endangered, or of special concern because of dwindling numbers, or they were listed as extinct.

The commercial value of shell notwithstanding, to the casual observer freshwater mussels generally appear to be valueless, rather dull, useless, and totally lacking in the aesthetic appeal generated by animals such as the bald eagle, spotted owl, or black-footed ferret. In reality, however, mussels are a valuable group of animals, not only for the pearls they produce, for their shell in the cultured pearl industry, and for their soft parts as food for a variety of birds, mammals, and fish, but also as indicators of water quality. As they filter water through their gills, dissolved heavy metals and other chemical and biological toxins are often absorbed and become concentrated in the shell and soft tissues. Because many species are extremely sensitive to pollution, analyses of these animals provide a means of detecting water contamination.

Since the beginning of the twentieth century hundreds of millions of dollars have been realized from the sale of freshwater mussel shells, first from the production of pearl shell buttons and later from the cultured pearl industry. The discovery, description, and naming of species new to science have persisted since about the beginning of the nineteenth century. However, only commercial shellers and a relatively few freshwater malacologists were interested in knowing the differences between one mussel and another. The basic and generally accepted classification of the unionoids involves structure of the soft parts (higher taxa) with the species level based on shell morphology. Much of the taxonomic confusion that exists today resulted from early workers assigning species names to what were only variations in shell shape.

But even with the generally accepted classification of freshwater mussels based on soft tissues, as this work goes to press the shell is still the basis for most identifications used by biologists. In this volume we have attempted to provide the reader with a concise treatment of each taxon, including a written description of the shell with accompanying color photograph of two "typical" specimens of each species, that will enable its identification. We also provide most of the various name combinations used at one time or another for a species and now in synonymy, plus general and Tennessee distribution records, notes on life history, ecology, host fish when known, and present status.

Acknowledgments

During the last two decades a great number of individuals have shown their interest in and enthusiasm for our efforts to establish a research and teaching collection of freshwater mussels at the University of Tennessee, Knoxville. Their collection efforts— either directly or incidentally during river and stream surveys and other specific aquatic studies or through just plain beachcombing— and their donations and exchanges of specimens have enabled us to obtain more comprehensive ecological data and distribution records. The generosity of many commercial shellers should certainly be recognized as well. In addition, several major buyers of raw shell, including George Borden, Crump, Tennessee; Lonnie Garner, U.S. Shell Co., Hollywood, Alabama; John Latendresse, Tennessee Shell Co. (now American Pearl Farms), Camden, Tennessee; and James Peach, American Shell Co., Nashville, Tennessee, are acknowledged for sharing their expert knowledge of mussels and the commercial shelling industry and for contributing specimens.

To all of the following people we express our gratitude and sincere appreciation for their contributions. If we have missed a few names, our apologies to those individuals, but to all who helped make this volume possible, we thank you: Patricia Adams, the late Paul Adams, Steve Ahlstedt, Bob Anderson, Herb Athearn, Don Ball, Bruce Bauer, Cindy Bogan, George Borden, Emanuel Breitburg, Doug Brewer, Clyde Brown, Dick Bryant, Bob Butler, Ken Cannon, Jim Chatters, Gary Coleman, the late Cliff Coney, Cailup Curren, Bill Dickinson, Gerald Dinkins, David Etnier, Mike Etnier, Glenn Fallo, Terry Ferguson, Patrice and Michael Fox, Sam Fuller, Jay Griswold, Andy Haines, John Harris, Marion Havlik, Willard Henning, Randy Hoeh, Rob Hoffman, Vic Hood, Don Hubbs, Mark Hughes, Ed Hunt, John Hurd, Eugene Keferl, Jim Kenon, Jim King, Walter Klippel, Martin Kohl, James Layzer, Forrest Loomis, Jerry Louton, Don Manning, Bruce Manzano, Harold Mathiak, Paula Mikkelsen, Pat Munson, Harold Murray, Baxter Napier, David Nieland, Ronald Oesch, the late Barbara Parmalee, David Parmalee, Evan Peacock, Wendell Pennington, Gregory Perino, Richard Polhemus, the late Burt Riedl, Bret Riggs, Neil Robison, John Schmidt, Guenter Schuster, Doug Shelton, Randy and Peggy Shute, Chris Skelton, David Stansbery, Lynn and Wayne Starnes, Carol Stein, Paul Stodola, Ed Styles, the late Don Tanner, James Theler, Terry Tune, Bill Turner, the late Henry van der Schalie, Kent Vickery, Malcolm Vidrine, G. Thomas Watters, Tom Whyte, Jenny Williams, Jim Williams, Bob Winters, Kirk Wright, and Miles Wright.

The loan of certain specimens for identification and photographic purposes has been indispensable in our efforts to assure as closely as possible accurate determinations and excellent examples for the illustrations. For their assistance in the identification and loan of specimens we gratefully acknowledge the following individuals: Steven A. Ahlstedt, U.S. Geological Survey, Knoxville, Tennessee; Douglas J. Brewer, Museum of Natural History, University of Illinois, Champaign-Urbana, Illinois; John B. Burch, Museum of Zoology, The University of Michigan, Ann Arbor, Michigan; George M. Davis, Academy of Natural Science of Philadelphia, Philadelphia, Pennsylvania; David J. Heath, Wisconsin Department of Natural Resources, Rhinelander, Wisconsin; Richard I. Johnson, Museum of Comparative Zoology, Harvard University, Cambridge, Massachusetts; Paula Mikkelsen, formerly of the Delaware Museum of Natural History, Wilmington, Delaware; and David H. Stansbery, Museum of Zoology, The Ohio State University, Columbus, Ohio.

We would like to recognize the time, effort, and talent that Richard Kirk, Environmental Services Division, Tennessee Wildlife Resources Agency, Nashville, devoted to the preparation of range maps and fig. 10. Robert M. Hatcher, coordinator of Nongame and Endangered Species, Tennessee Wildlife Resources Agency, Nashville, and Richard G. Biggins, U.S. Fish and Wildlife Service, Asheville, North Carolina, are gratefully acknowledged for their encouragement, funding, and support in the preparation of this volume. Jefferson Chapman, director, Frank H. McClung Museum, University of Tennessee, Knoxville, was most supportive with logistical and other forms of assistance during preparation of the manuscript. Betsy Bennett, director, North Carolina State Museum of Natural Sciences, Raleigh, was very supportive during the final stages of the preparation of the manuscript.

We wish to express our appreciation to R. Bruce McMillan, director, Illinois State Museum, Springfield, Illinois, for permission to use excerpts from *The Freshwater Mussels of Illinois* by Paul W. Parmalee, Illinois State Museum Popular Science Series, Vol. 8, 1967, and to Ronald B. Toll, Wesleyan College, Macon, Georgia, for allowing us to cite sections from *The Mussels (Mollusca: Bivalvia: Unionidae) of Tennessee* by Lynn B. Starnes and Arthur E. Bogan, *American Malacological Bulletin,* Vol. 6, 1988. We are especially grateful to the following outside reviewers of the manuscript who offered constructive criticisms, significant molluscan records and other data, and editorial suggestions for improving the text: Steven A. Ahlstedt, U.S. Geological Survey, Knoxville, Tennessee; Robert E. Warren, Illinois State Museum, Springfield, Illinois; G. Thomas Watters, the Ohio Biological Survey, Columbus, Ohio.

To Cindy Bogan, Betty Creech, Maria Richardson, and Misty Tilson we owe a special debt of gratitude for their tireless efforts in typing various sections of the manuscript.

If any one aspect of this book can be singled out as the most useful in assisting the reader to recognize and identify the various species of mussels, it is the quality of the colored photographs. We feel Miles Wright, Frank H. McClung Museum Photographer, did an exceptional job in capturing the true color and accurate detail of each species, so he and his able assistant Alice King are acknowledged with special appreciation for their time and effort invested and the quality product achieved.

MUSSELS AND MALACOLOGY IN TENNESSEE

Introduction

Tennessee is drained by three major rivers, the Tennessee, Cumberland, and Mississippi, as well as by several tributaries of the Ohio River and the Conasauga River, which is a part of the Mobile (Coosa) River system (figs. 1 and 2). These vast drainage systems and their water quality have been modified by the construction of dams, channelization of streams, domestic and agricultural pollution, and effluents from strip mining. All of these factors have served to modify extensively the aquatic environment which once supported vast numbers and varieties of molluscan species. Many of these mollusks were unable to adjust to such radical changes. One need only glance at figs. 3 and 4 to appreciate the major modifications of the Tennessee and Cumberland River systems that have resulted from 50 years of major dam construction. In light of continued or increasing channelization projects and dam construction, the Tennessee–Tombigbee project, and the Columbia (unfinished as of this writing) and Normandy dams on the Duck River, to cite just three examples, there is little doubt that additional species of aquatic mollusks will become extinct by the twenty-first century while others will swell the Endangered or Threatened lists.

The once rich and varied molluscan fauna of Tennessee provided a wealth of undescribed and unnamed species when it was first encountered by scientists about 1800. The taxonomic studies of C. S. Rafinesque, I. Lea, T. A. Conrad, G. W. Tryon, W. G. Binney, C. T. Simpson, L. S. Frierson, H. A. Pilsbry, and A. E. Ortmann left an impressive chronicle (albeit a confusing one at times) of the molluscan species found in the state. Numerous other malacologists contributed to our knowledge of this interesting group of animals as well. In addition to attempting to resolve taxonomic relationships, several workers (e.g., A. E. Ortmann) also provided valuable data on the life history and habitat requirements of many species and their regional significance. It was Binney (1885), for example, who observed a distinct assemblage of land snails living in the central portion of Tennessee. He then designated this area as the Cumberland Subregion, a part of the Interior Region of the Eastern Province. Later Ortmann (1924a) reported on the unique unionoid fauna which inhabited the Duck, Cumberland, and Tennessee rivers and which was unknown from the Interior Basin or outside of the Cumberland and Tennessee River drainages. He noted 45 species and forms which he considered endemic to these river systems. He also found 30 of these 45 "Cumberlandian" species living in the Tennessee River below Walden Gorge (immediately downstream from Chattanooga) and was able to more carefully define those areas inhabited by the Cumberlandian fauna (see fig. 10).

Ortmann (1924a:55–56) also observed that the Cumberlandian fauna inhabited those areas which included

CONASAUGA R. SYSTEM (MOBILE BASIN)

Fig. 1. Counties and major drainage basins of Tennessee. From David A. Etnier and Wayne C. Starnes, *The Fishes of Tennessee* (Knoxville: Univ. of Tennessee Press, 1993).

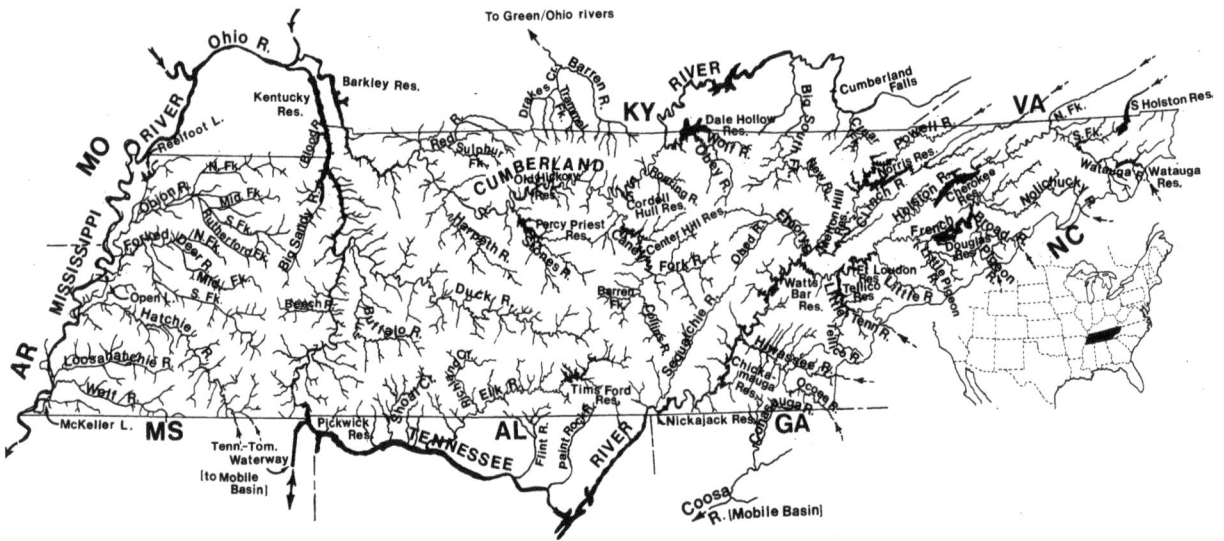

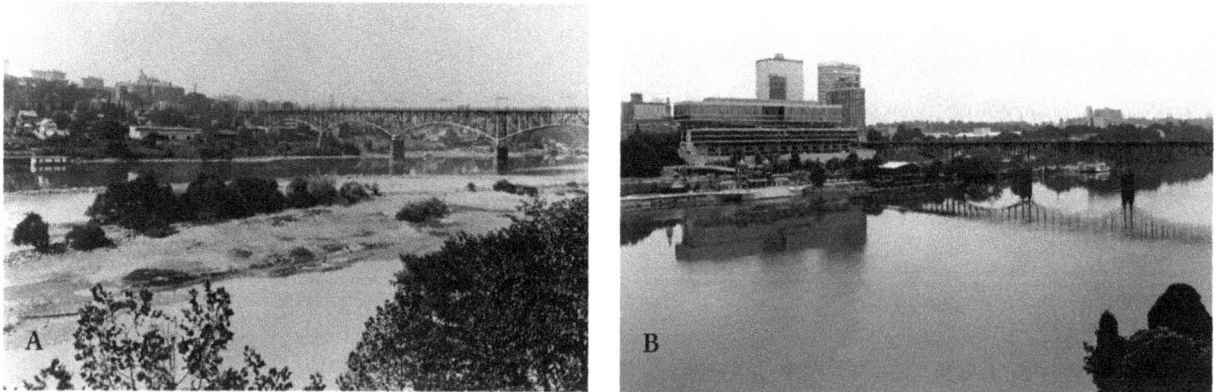

Fig. 2. Drainages of Tennessee denoting major streams, lakes, and reservoirs. From David A. Etnier and Wayne C. Starnes, *The Fishes of Tennessee* (Knoxville: Univ. of Tennessee Press, 1993).

Fig. 3. The Tennessee River at Knoxville, Tennessee. *A.* pre-impoundment view (circa 1925) compared with *B.* how the river (Fort Loudoun Lake) appears today. Left photo courtesy McClung Historical Collection.

4   *Mussels and Malacology in Tennessee*

Fig. 4. Views of Norris Lake. *A.* the lake at full pool (summer); *B.* at drawdown (winter); *C.* the Clinch River below the dam at normal flow; and *D.* the river when power is being generated as water is passing through the turbines.

"the drainages of the Cumberland and Tennessee rivers only, from the headwaters down, but leaving out the lower-most sections of the two rivers. It ends, in the Cumberland, about in the vicinity of Clarksville, Montgomery Co., Tenn." In addition, he noted that the Cumberlandian species were found in the upper Duck River and in the Tennessee River above Muscle Shoals, Alabama, but that they were absent at Dixie, Tennessee, in the lower Tennessee River. "Thus the lower-most Tennessee faunistically belongs to the Interior Basin" (Ortmann 1925:368). This regional faunal assemblage was further illustrated by van der Schalie and van der Schalie (1950) in their work dealing with the unionoids of the Mississippi River.

As of 1998, 129 species and 26 subspecies or forms of native freshwater mussels have been identified from Tennessee waters. Fourteen species of fingernail clams (Sphaeriidae) and three introduced species, the Asian clam *(Corbicula fluminea)* (Corbiculidae), Zebra Mussel *(Dreissena polymorpha),* and Dark Falsemussel *(Mytilopsis leucophaeata)* (Dreissenidae), bring the total to 171 freshwater bivalve taxa.

Structure, Development, and Growth

Freshwater mussels belong to a large and diversified group of animals classified by biologists as belonging to the Phylum Mollusca, mollusks, because they all possess certain similar structures or characteristics. Included in this major phylum are the chitons, tusk shells, snails, slugs, conchs, nautili, squids, octopi, clams, oysters, and mussels. The last three types are included in the Class Bivalvia (within the Phylum Mollusca) because they all possess a soft body that is enclosed by a shell of two parts. Because of these two separate shells, or valves, the animals are commonly referred to as bivalves.

A detailed description of the structure and function of the soft parts (the body), although often of significance in determining relationships among the different species, will not be presented here. This aspect of the mussels' structure is of special interest to the anatomist and taxonomist, but most individuals who collect and study the shells are concerned only with their structure and appearance. Usually similarities observed in the structure of shells of closely related species are present in their soft parts as well. However, two closely related species classified on the basis of similar internal structure may produce shells altogether different in appearance.

The body of a freshwater mussel consists of a thickened, central mass that is attached to the dorsal surface of the paired valves and contains various organs.

The anterior, ventral part of the body forms the muscular "foot." On each side of these structures hangs a thin double gill and, outside of these, thin sheets of tissue (the mantle) that attach to the inner surfaces of the shells. Compared with some of the more "advanced" groups of animals, such as fish and amphibians, the digestive, circulatory, reproductive, and nervous systems found in the mussel are fairly simple. At the posterior end of the body, two apertures or siphons are present through which water is brought in (incurrent siphon) and expelled (excurrent siphon). Oxygen is removed from the water by the gills. Food, which consists of microorganisms (protozoans, bacteria, diatoms) and organic particles suspended in the water, is moved across the surface of the gills by ciliary action, between the labial palps, then to the mouth, esophagus, and finally to the stomach.

Adult freshwater mussels live their entire lives partly embedded in the bottom of some body of permanent water; they are active mostly during the warm months. Movement is accomplished by extending and contracting the foot between the valves; extension of the foot enables the mussel to wedge itself into the substratum. During periods of rest or inactivity, the mussel remains partially embedded—the depth often depends on the particular species, water temperature, current, and other conditions—with the valves slightly spread and the posterior end of the shell (siphons) exposed. Scars on the inner surface of each valve indicate points of muscle attachment. The largest muscles are the anterior and posterior adductor muscles, which draw the valves together. Anterior and posterior retractor muscles draw the foot into the shell, while the anterior protractor muscle helps to extend the foot. Additional dorsal muscles help maintain the animal in the dorsal portion of the shell.

In freshwater mussels the sexes are usually separate. However, van der Schalie (1970) found that four species, the Paper Pondshell *(Anodonta [=Utterbackia] imbecillis),* Creek Heelsplitter *(Lasmigona compressa),* Green Floater *(L. subviridis),* and the Lilliput *(Toxolasma parvus)* were hermaphroditic, a condition in which an individual possesses both male and female sex organs. Some individuals of a few other species, such as the Spectaclecase *(Cumberlandia monodonta),* in which the sexes are normally separate, occasionally exhibit this condition. When the reproductive systems are distinct, the shells of the male and female of some taxa within several genera, e.g. *Epioblasma, Lampsilis,* and *Villosa* (Lampsilinae), exhibit structural differences. The posterior section of the valves of females are often more inflated and rounded. This is a compensation, at least in part, for the gills which become enlarged and distended when filled with developing eggs and glochidia. Sperm, the male sex cells, are released into the water; these are taken into the female's body through the incurrent aperture and then carried to tubes in the gills where the eggs, having been previously discharged from the ovaries, are apparently fertilized. The gills or portions of the gills then serve as brood pouches, called marsupia, as well as respiratory organs (see Tankersley and Dimock, 1992; Tankersley, 1996).

Development of a freshwater mussel from the fertilized egg is unique, since one stage of growth of the young, called glochidia, must usually take place on the skin, gills, or fins of a fish (or, in one species, *Simpsonaias ambigua,* on an aquatic salamander). Four distinct stages can be recognized in the growth of a mussel: (1) the fertilized egg; (2) the young or glochidium in the brood pouch (gill) of the female mussel; (3) the parasitic stage on a fish or salamander; and (4) the free-living stage with a completely formed shell (fig. 5).

Within the brood pouch or gill, the fertilized eggs develop into small larvae or young, the glochidia, which are typically rounded, oval, or triangular in outline. They

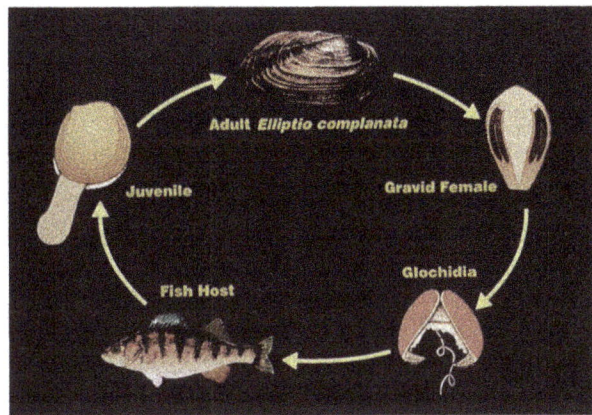

Fig. 5. A diagrammatic illustration of the life cycle of a freshwater mussel. Courtesy of John Christmas, Maryland Department of Natural Resources, Annapolis, Md.

possess only the embryonic stages of a mouth, intestines, heart, and foot when they are discharged through the excurrent siphon into the water (e.g., Pekkarinen and Valovirta, 1996). If these glochidia do not encounter a passing fish and attach to its gills or fins, they fall to the bottom. Many may be scattered considerable distances by the current, but all must attach themselves to a host or they will die in a short time. Glochidia of some species of mussels possess a hook or multiple hooks on the ventral margin of each valve which are assumed to assist in firmly attaching the glochidia on the host fish (see Hoggarth and Gaunt, 1988).

In some species this process is far more complicated and involves more sophisticated methods of insuring that the glochidia reach a fish host. One of the most intriguing aspects of this phase of the mussels' life cycle may be seen in the specialized development of the mantle in the Pocketbook, *Lampsilis ovata* (Kraemer, 1970; Barnhart and Roberts, 1996; and others). A portion of the mantle has been modified to appear as a minnow—eyespot, tail, and all. A similar structure may be seen in the mantle of the Wavyrayed Lampmussel, *Lampsilis fasciola* (fig. 6). By a waving action of this minnow-like mantle extension, fish are lured to the mussel where they receive a concentration of expelled glochidia as they attempt to devour the "fish." Other mussel taxa (e.g., some species of *Ptychobranchus*) produce glochidia in packets (conglutinates) that resemble tiny fish or larval insects. "A sticky thread trails behind and attaches the lure to a rock or stick, where it twirls invitingly in the current. When the 'lure' is bitten by a would-be predator, the eyespots rupture and releases the glochidia onto the fish's gills" (Barnhart and Roberts, 1996:25). Hartfield and Hartfield (1996) describe these conglutinates of the Triangular Kidneyshell, *Ptychobranchus greeni* (Conrad, 1834), as resembling aquatic dipteran (fly) larvae in shape, size, and coloration. Luo (1993) noted that conglutinates of the Fluted Kidneyshell, *P. subtentum* (Say, 1825), also resembled larval aquatic insects. Probably the most bizarre adaptation by freshwater mussels occurs in three species of *Lampsilis*. They develop a three- to six-foot-long clear, gelatinous tube with all of the glochidia placed in a structure at the end of this tube. The end containing the glochidia looks like a minnow, floating and moving in the current, appearing as a wounded or sick fish (Haag et al., 1995).

During the parasitic stage, each glochidium remains embedded in the tissues of the host fish, changing little in size but developing many of the adult organs and structures. The length of time the glochidia remain embedded depends on the species of host fish, place of attachment, and temperature of the water; this stage may last slightly over one week or as long as six weeks or more. Most unionoids appear to produce a single brood each year, but *Cumberlandia* has been documented as producing two broods per year (Howard, 1915; Gordon and Smith, 1990). Although each female mussel bears a very large number of glochidia, thus insuring continuation of the species, a large proportion fail to pass through the parasitic stage and, consequently, die.

Studies have shown that glochidia from some species of mussels react to, or will parasitize, only one species of fish, while certain kinds of fish may serve as host for several species of mussels. Normally, most infestations of glochidia are light and apparently do not

Fig. 6. The minnow-shaped mantle section of the Wavyrayed Lampmussel, *Lampsilis fasciola,* which serves as a lure to passing fish. Courtesy of William Bradley, Virginia Polytechnic Institute, Blacksburg, Va.

harm or cause injury to the host fish. It should be emphasized that the importance of knowing the host fish species for glochidia of freshwater mussels is essential to their effective propagation, conservation, and recovery. Upon completion of its parasitic stage, the young mussel breaks through the tissue of the fish and falls to the bottom where it begins an independent life. With adult structures being formed, including a juvenile shell, the animal begins its growth to adulthood.

The shape, size, thickness, and color of the shell vary enormously among species of freshwater mussels, but basic structures (fig. 7) and method of shell formation are the same for all. The shell has two parts, a right and left valve held together by a dorsal ligament. Each valve is made up of three layers. The external, or outside, horny covering is a thin epidermis or periostracum that protects the calcareous parts beneath it from injurious abrasion or leaching action of acidic water. Should this periostracum layer become thin or wear completely away in spots, which often happens in the dorsal beak area (umbo) of old individuals, the shell in these exposed areas becomes pitted or corroded. Variations in the color of this outer covering range from light yellow to green or brown or black; shells of many species are patterned with irregular-shaped markings or rays. The color and patterns are important characteristics used in identifying many species. Occurring beneath the epidermis or periostracum is a thin, calcareous layer composed of vertical prisms of calcium carbonate. The third layer, constituting the major portion of the shell, is referred to as the nacreous or mother-of-pearl layer. It consists of a large series of thin, calcium carbonate plates that lie one upon another and parallel to the surface of the shell. Some workers recognize a fourth inner layer called the hypostracum; in many species this inner layer is often iridescent. The nacre or mother-of-pearl varies from a pure silvery white, through shades of pink to dark purple; in a few species, it is a light salmon. The two outer shell layers are secreted by glands in the edge of the mantle, while the nacre, the inner layer, is produced by the entire surface of the mantle. Growth of the shell *in area* is effected by the addition of material around the margin and *in thickness* by successive deposits of nacre over the shell's entire inner surface.

The outer surface of the shell of many kinds of mussels is completely smooth; shells of some groups or species, on the other hand, possess knobs, pustules, spines,

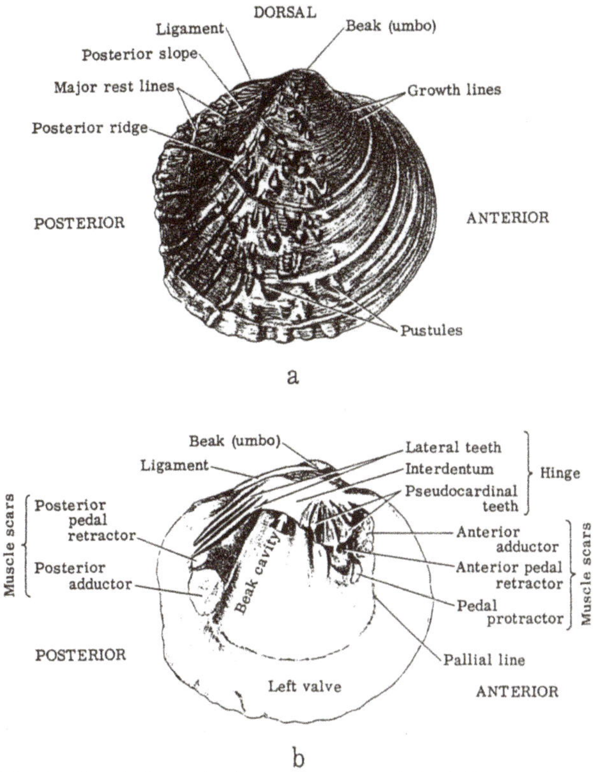

Fig. 7. Morphology of a freshwater mussel shell *(Cyclonaias tuberculata)* illustrating shell terminology. *A.* exterior of right valve; *B.* interior of left valve. From *Freshwater Unionacean Clams (Mollusca: Pelecypoda) of North America* (1975). Courtesy of John B. Burch, University of Michigan, Ann Arbor.

wrinkles, folds, or other configurations. Often the size, combination, or arrangement of these structures is diagnostic of a particular species, as are the ridges or loops on the beaks. The prominent, often raised, and/or darker concentric lines or rings on the surface of a valve indicate rest periods or stages when little or no growth took place (fig. 8). These lines or rings often result during the resting stage in winter, in periods of low water, or from some physical disturbance or other adverse condition. Occasionally deformed or pathologic (diseased) valves are encountered (fig. 9), most of which are largely the result of mechanical injuries. Mud frequently gets in between the shell and mantle; the mantle then secretes a thin film of nacreous material over the mud, thus forming a blister on the inner surface of the valve. More rarely, a shell may be so severely injured that the parts of the valve, or both valves, will continue to enlarge but produce a distortion in the typical shape of the mussel. These shell deformities are occasionally quite bizarre and may pose identification

problems. Generally, however, they are simply curiosities; F. C. Baker appears to have been the first malacologist to take note in print of such deformities of freshwater mussel valves in his 1901 article "Some Interesting Molluscan Monstrosities."

Generally, species characteristic of medium-sized and large rivers, or those living in swift current in small streams, develop thick shells, a heavier type of hinge teeth, and have the muscle scar areas deeply impressed. On the other hand, typical pond, lake, or quiet water species are thin-shelled, muscle scars are shallow, and often the valves "gape" or do not completely close at one or both ends (for example, species in the genera *Potamilus* and *Anodonta*). Identification of freshwater mussel shells relies not only on the presence or absence of external sculpture and raying patterns but also on the internal hinge structure of the shell. Different species display various combinations and levels of development of the hinge structure. Floaters *(Anodonta, Pyganodon)* completely lack any evidence of the hinge, while the pseudocardinal teeth in *Strophitus* are greatly reduced and the lateral teeth are little more than a rounded ridge. The size, shape, depth, and sculpture of the adductor muscle scars differ among species, e.g., they are flat and sculptured in *Amblema* and deep and smooth in *Fusconaia*. The size, shape, and orientation of the pseudocardinal teeth are also important, varying from large, massive, and heavily sculptured in *Amblema* and *Cyclonaias*, to thin and bladelike in *Villosa*. The presence and/or width of the interdentum is an important landmark in identification. It is closely tied to the depth of the beak cavity. Some species have no beak cavity (e.g., *Elliptio dilatata, Ptychobranchus fasciolaris*), while others have a deep, open beak cavity *(Quadrula* spp.) or a deep and compressed beak cavity *(Cyclonaias tuberculata)*.

Fig. 8. External surface of freshwater mussel valves showing dark concentric lines or rings indicative of rest periods.

Fig. 9. Mussel shell deformities. *A–F.* examples of freshwater mussel valves exhibiting various types of deformities, most often the result of trauma or physical damage to the mantle.

Taxonomy, Classification, and Common Names

The higher classification of freshwater bivalves used in this volume is presented to the generic level in table 1. The common names used for all of the freshwater bivalves follow Turgeon et al. (1988) except as modified or corrected for the second edition (Turgeon et al., n.d.). Table 2 lists all of the freshwater bivalve taxa known from Tennessee, their common names, a compilation of the federal and state level status of all of the Unionoida, and a listing of which species are presumed to be extinct (Williams et al., n.d.).

Sphaeriidae

Table 2 presents a list of 14 species of Sphaeriidae reported to occur in Tennessee. This list is based on Bickel (1968) and Burch (1975b). Taxonomy of the family follows Burch (1975b) and common names follow Turgeon et al. (1988). Although this family is not treated in this volume, individuals interested in the identification and ecology of the fingernail clams are referred to Burch (1975b) and the literature cited therein for further information.

Corbiculidae

The introduced Asian Clam is the only representative of this family in Tennessee.

Dreissenidae

The Zebra Mussel and the Dark Falsemussel are the only two representatives of this family introduced into Tennessee. The Dark Falsemussel is known from only a very few specimens apparently introduced into the state by commercial barge traffic.

Unionoidea

Higher-level classification of freshwater bivalves today is confusing to say the least. Various classifications have been proposed since that of Simpson (1900a, 1914) (see Modell, 1942, 1949, 1964; Haas, 1969a, b; Heard and Guckert, 1970; Starobogatov, 1970; Davis and Fuller, 1981; Lydeard et al. 1996), but it was Ortmann (1910, 1912a, 1916, 1919) who refined and developed a generally acceptable classification of North American Unionoidea. In 1919 Ortmann recognized two families, the Unionidae, with three subfamilies, and the Margaritiferidae. We have chosen here to follow the clear subfamily divisions provided by Ortmann's classification as used in his monograph of the naiads of Pennsylvania (Ortmann, 1919) and have followed Ortmann (1912a, 1919) and Smith and Wall (1984) and Lydeard et al. (1996) in recognizing the family Margaritiferidae. We feel that the conservative approach is best and the least confusing until a broad based phylogeny for the superfamily can be constructed.

Table 1

Higher Taxonomy of the Freshwater Bivalves of Tennessee

CLASS BIVALVIA	Subfamily Lampsilinae
Subclass Paleoheterodonta	*Actinonaias* Crosse and Fischer, 1893
Order Unionoida	*Cyprogenia* Agassiz, 1852
Superfamily Unionoidea	*Dromus* Simpson, 1900
Family Margaritiferidae	*Ellipsaria* Rafinesque, 1820
Cumberlandia Ortmann, 1912	*Epioblasma* Rafinesque, 1831
Family Unionidae	*Lampsilis* Rafinesque, 1820
Subfamily Unioninae	*Lemiox* Rafinesque, 1831
Amblema Rafinesque, 1819	*Leptodea* Rafinesque, 1820
Cyclonaias Pilsbry, 1922	*Ligumia* Swainson, 1840
Elliptio Rafinesque, 1820	*Medionidus* Simpson, 1900
Fusconaia Simpson, 1900	*Obliquaria* Rafinesque, 1820
Hemistena Rafinesque, 1820	*Obovaria* Rafinesque, 1820
Lexingtonia Ortmann, 1914	*Potamilus* Rafinesque, 1818
Megalonaias Utterback, 1915	*Ptychobranchus* Simpson, 1900
Plectomerus Conrad, 1853	*Toxolasma* Rafinesque, 1831
Plethobasus Simpson, 1900	*Truncilla* Rafinesque, 1819
Pleurobema Rafinesque, 1820	*Villosa* Frierson, 1927
Quadrula Rafinesque, 1820	Subclass Heterodonta
Tritogonia Agassiz, 1852	Order Veneroida
Uniomerus Conrad, 1853	Superfamily Corbiculoidea
Subfamily Anodontinae	Family Corbiculidae
Alasmidonta Say, 1818	*Corbicula* Megerle, 1811 [I]
Anodonta Lamarck, 1799	Family Sphaeriidae
Anodontoides Simpson in Baker, 1898	*Musculium* Link, 1807
Arcidens Simpson, 1900	*Pisidium* Pfeiffer, 1821
Lasmigona Rafinesque, 1831	*Sphaerium* Scopoli, 1777
Pegias Simpson, 1900	Superfamily Dreissenoidea
Pyganodon Crosse and Fischer, 1894	Family Dreissenidae
Simpsonaias Frierson, 1914	*Dreissena* Beneden, 1835 [I]
Strophitus Rafinesque, 1820	*Mytilopsis* Conrad, 1857 [I]
Utterbackia Baker, 1927	

I=Introduced

Table 2

List of Freshwater Bivalves from Tennessee with Common Names, and State and Federal Status

Taxa	Common Name	Federal[1]	TN	Williams et al. (1993)	Extinct[2]
Margaritiferidae					
Cumberlandia monodonta (Say, 1829)	Spectaclecase			T	
Unionidae					
Actinonaias ligamentina (Lamarck, 1819)	Mucket			CS	
Actinonaias pectorosa (Conrad, 1834)	Pheasantshell		SC	SC	
Alasmidonta atropurpurea (Rafinesque, 1831)	Cumberland Elktoe	E	E	E	
Alasmidonta marginata Say, 1818	Elktoe		SC	SC	
Alasmidonta raveneliana (Lea, 1834)	Appalachian Elktoe	E	E	E	
Alasmidonta viridis (Rafinesque, 1820)	Slippershell Mussel			SC	
Amblema plicata (Say, 1817)	Threeridge		CS	CS	
Anodonta suborbiculata Say, 1831	Flat Floater		CS	CS	
Anodontoides ferussacianus (Lea, 1834)	Cylindrical Papershell		E	CS	
Arcidens confragosus (Say, 1829)	Rock Pocketbook			CS	
Cyclonaias tuberculata (Rafinesque, 1820)	Purple Wartyback		SC	SC	
Cyprogenia stegaria (Rafinesque, 1820)	Fanshell	E	E	E	
Dromus dromas (Lea, 1834)	Dromedary Pearlymussel	E	E	E	
Ellipsaria lineolata (Rafinesque, 1820)	Butterfly			SC	
Elliptio arca (Conrad, 1834)	Alabama Spike		T	T	

Continued on next page

Table 2—*Continued*

Taxa	Common Name	Federal[1]	TN	Williams et al. (1993)	Extinct[2]
Elliptio arctata (Conrad, 1834)	Delicate Spike		E	SC	
Elliptio crassidens (Lamarck, 1819)	Elephantear		CS	CS	
Elliptio dilatata (Rafinesque, 1820)	Spike		CS	CS	
Epioblasma arcaeformis (Lea, 1831)	Sugarspoon		X	EX	X
Epioblasma biemarginata (Lea, 1857)	Angled Riffleshell		X	EX	X
Epioblasma brevidens (Lea, 1831)	Cumberlandian Combshell	E	E	E	
Epioblasma capsaeformis (Lea, 1834)	Oyster Mussel	E	E	E	
Epioblasma flexuosa (Rafinesque, 1820)	Leafshell		X	EX	X
Epioblasma florentina florentina (Lea, 1857)	Yellow Blossom	E	X	EX	X
Epioblasma florentina walkeri (Wilson & Clark, 1914)	Tan Riffleshell	E	E	E	
Epioblasma haysiana (Lea, 1834)	Acornshell		X	EX	X
Epioblasma lenior (Lea, 1842)	Narrow Catspaw		X	EX	X
Epioblasma lewisii (Walker, 1910)	Forkshell		X	EX	X
Epioblasma metastriata (Conrad, 1838)	Upland Combshell	E	E	E	
Epioblasma obliquata (Rafinesque, 1820)	Catspaw	E	E	E	
Epioblasma othcaloogensis (Lea, 1857)	Southern Acornshell	E	E	E	
Epioblasma personata (Say, 1829)	Round Combshell		X	EX	X
Epioblasma propinqua (Lea, 1857)	Tennessee Riffleshell		X	EX	X
Epioblasma stewardsonii (Lea, 1852)	Cumberland Leafshell		X	EX	X
Epioblasma torulosa torulosa (Rafinesque, 1820)	Tubercled Blossom E	X	EX	X	
Epioblasma torulosa gubernaculum (Reeve, 1865)	Green Blossom	E	X	EX	X
Epioblasma triquetra (Rafinesque, 1820)	Snuffbox		T	T	
Epioblasma turgidula (Lea, 1858)	Turgid Blossom	E	X	EX	X
Fusconaia barnesiana (Lea, 1838)	Tennessee Pigtoe		SC	SC	
Fusconaia cor (Conrad, 1834)	Shiny Pigtoe	E	E	E	
Fusconaia cuneolus (Lea, 1840)	Finerayed Pigtoe	E	E	E	
Fusconaia ebena (Lea, 1831)	Ebonyshell		CS	CS	
Fusconaia flava (Rafinesque, 1820)	Wabash Pigtoe		T	CS	
Fusconaia subrotunda (Lea, 1831)	Longsolid		T	SC	
Hemistena lata (Rafinesque, 1820)	Cracking Pearlymussel	E	E	E	
Lampsilis abrupta (Say, 1831)	Pink Mucket	E	E	E	
Lampsilis altilis (Conrad, 1834)	Finelined Pocketbook	T	T	T	
Lampsilis cardium Rafinesque, 1820	Plain Pocketbook		SC	SC	
Lampsilis fasciola Rafinesque, 1820	Wavyrayed Lampmussel		SC	CS	
Lampsilis ornata (Conrad, 1835)	Southern Pocketbook		E	SC	
Lampsilis ovata (Say, 1817)	Pocketbook		SC	SC	
Lampsilis siliquoidea (Barnes, 1823)	Fatmucket		CS	CS	
Lampsilis straminea claibornensis (Lea, 1838)	Southern Fatmucket		X	CS	
Lampsilis teres (Rafinesque, 1820)	Yellow Sandshell		CS	CS	
Lampsilis virescens (Lea, 1858)	Alabama Lampmussel	E	E	E	
Lasmigona complanata (Barnes, 1823)	White Heelsplitter		CS	CS	
Lasmigona costata (Rafinesque, 1820)	Flutedshell		CS	CS	
Lasmigona holstonia (Lea, 1838)	Tennessee Heelsplitter		T	SC	
Lasmigona subviridis (Conrad, 1835)	Green Floater		E?	T	
Lemiox rimosus (Rafinesque, 1831)	Birdwing Pearlymussel	E	E	E	
Leptodea fragilis (Rafinesque, 1820)	Fragile Papershell		CS	CS	
Leptodea leptodon (Rafinesque, 1820)	Scaleshell		E	E	
Lexingtonia dolabelloides (Lea, 1840)	Slabside Pearlymussel		T	T	
Ligumia recta (Lamarck, 1819)	Black Sandshell		SC	SC	
Ligumia subrostrata (Say, 1831)	Pondmussel		SC	CS	
Medionidus acutissimus (Lea, 1831)	Alabama Moccasinshell	T	T	T	
Medionidus conradicus (Lea, 1834)	Cumberland Moccasinshell	SC	SC		
Medionidus parvulus (Lea, 1860)	Coosa Moccasinshell	E	E	E	
Megalonaias nervosa (Rafinesque, 1820)	Washboard		CS	CS	
Obliquaria reflexa Rafinesque, 1820	Threehorn Wartyback		CS	CS	
Obovaria jacksoniana (Frierson, 1912)	Southern Hickorynut		T	SC	

Continued on next page

Table 2—*Continued*

Taxa	Common Name	Federal[1]	TN	Williams et al. (1993)	Extinct[2]
Obovaria olivaria (Rafinesque, 1820)	Hickorynut		SC	CS	
Obovaria retusa (Lamarck, 1819)	Ringpink	E	E	E	
Obovaria subrotunda (Rafinesque, 1820)	Round Hickorynut		T	SC	
Pegias fabula (Lea, 1838)	Littlewing Pearlymussel	E	E	E	
Plectomerus dombeyanus (Valenciennes, 1827)	Bankclimber		CS	CS	
Plethobasus cicatricosus (Say, 1829)	White Wartyback	E	E	E	
Plethobasus cooperianus (Lea, 1834)	Orangefoot Pimpleback	E	E	E	
Plethobasus cyphyus (Rafinesque, 1820)	Sheepnose		T	T	
Pleurobema chattanoogaense (Lea, 1858)	Painted Clubshell		E	E	
Pleurobema clava (Lamarck, 1819)	Clubshell	E	E	E	
Pleurobema cordatum (Rafinesque, 1820)	Ohio Pigtoe		T	SC	
Pleurobema georgianum (Lea, 1841)	Southern Pigtoe	E	E	E	
Pleurobema gibberum (Lea, 1838)	Cumberland Pigtoe	E	E	E	
Pleurobema hanleyanum (Lea, 1852)	Georgia Pigtoe		X	E	X
Pleurobema johannis (Lea, 1859)	Alabama Pigtoe		X	U	X
Pleurobema oviforme (Conrad, 1834)	Tennessee Clubshell		T/E?	SC	
Pleurobema perovatum (Conrad, 1834)	Ovate Clubshell	E	E	E	
Pleurobema plenum (Lea, 1840)	Rough Pigtoe	E	E	E	
Pleurobema rubellum (Conrad, 1834)	Warrior Pigtoe		X	E	X
Pleurobema rubrum (Rafinesque, 1820)	Pyramid Pigtoe		T	T	
Pleurobema sintoxia (Rafinesque, 1820)	Round Pigtoe		T	CS	
Pleurobema troschelianum (Lea, 1852)	Alabama Clubshell		X	E	X
Potamilus alatus (Say, 1817)	Pink Heelsplitter		CS	CS	
Potamilus ohiensis (Rafinesque, 1820)	Pink Papershell		T	CS	
Potamilus purpuratus (Lamarck, 1819)	Bleufer		CS	CS	
Ptychobranchus fasciolaris (Rafinesque, 1820)	Kidneyshell		CS	CS	
Ptychobranchus greeni (Conrad, 1834)	Triangular Kidneyshell	E	E	E	
Ptychobranchus subtentum (Say, 1825)	Fluted Kidneyshell		T	SC	
Pyganodon grandis (Say, 1829)	Giant Floater		CS	CS	
Quadrula apiculata (Say, 1829)	Southern Mapleleaf		CS	CS	
Quadrula cylindrica (Say, 1817)	Rabbitsfoot		T	T	
Quadrula cylindrica strigillata (Wright, 1898)	Rough Rabbitsfoot	E	E	E	
Quadrula fragosa (Conrad, 1835)	Winged Mapleleaf	E	E	E	
Quadrula intermedia (Conrad, 1836)	Cumberland Monkeyface	E	E	E	
Quadrula metanevra (Rafinesque, 1820)	Monkeyface		CS	CS	
Quadrula nodulata (Rafinesque, 1820)	Wartyback		CS	CS	
Quadrula pustulosa (Lea, 1831)	Pimpleback		CS	CS	
Quadrula quadrula (Rafinesque, 1820)	Mapleleaf		CS	CS	
Quadrula sparsa (Lea, 1841)	Appalachian Monkeyface	E	E	E	
Simpsonaias ambigua (Say, 1825)	Salamander Mussel		X	SC	
Strophitus connasaugaensis (Lea, 1858)	Alabama Creekmussel		E	SC	
Strophitus undulatus (Say, 1817)	Creeper		SC	CS	
Toxolasma cylindrellus (Lea, 1868)	Pale Lilliput	E	E	E	
Toxolasma lividus Rafinesque, 1831	Purple Lilliput		T	SC	
Toxolasma parvus (Barnes, 1823)	Lilliput		SC	CS	
Toxolasma texasensis (Lea, 1857)	Texas Lilliput		CS	CS	
Tritogonia verrucosa (Rafinesque, 1820)	Pistolgrip		CS	CS	
Truncilla donaciformis (Lea, 1828)	Fawnsfoot		CS	CS	
Truncilla truncata Rafinesque, 1820	Deertoe		CS	CS	
Uniomerus declivis (Say, 1831)	Tapered Pondhorn		CS	CS	
Uniomerus tetralasmus (Say, 1831)	Pondhorn		CS	CS	
Utterbackia imbecillis (Say, 1829)	Paper Pondshell		CS	CS	
Villosa fabalis (Lea, 1831)	Rayed Bean		T	SC	
Villosa iris (Lea, 1829)	Rainbow		SC	CS	
Villosa lienosa (Conrad, 1834)	Little Spectaclecase		T/E?	CS	
Villosa perpurpurea (Lea, 1861)	Purple Bean	E	E	E	

Continued on next page

Table 2—*Continued*

Taxa	Common Name	Federal[1]	TN	Williams et al. (1993)	Extinct[2]
Villosa taeniata (Conrad, 1834)	Painted Creekshell		T	CS	
Villosa trabalis (Conrad, 1834)	Cumberland Bean	E	E	E	
Villosa vanuxemensis (Lea, 1838)	Mountain Creekshell		SC	SC	
Villosa vanuxemensis umbrans (Lea, 1857)	Coosa Creekshell		E	SC	
Villosa vibex (Conrad, 1834)	Southern Rainbow		E	CS	
Corbiculidae					
Corbicula fluminea (Müller, 1774)	Asian Clam				
Dreissenidae					
Dreissena polymorpha (Pallas, 1771)	Zebra Mussel				
Mytilopsis leucophaeata (Conrad, 1831)	Dark Falsemussel				
Sphaeriidae					
Musculium lacustre (Müller, 1774)	Lake Fingernailclam				
Musculium partumeium (Say, 1822)	Swamp Fingernailclam				
Musculium securis (Prime, 1852)	Pond Fingernailclam				
Musculium transversum (Say, 1829)	Long Fingernailclam				
Pisidium adamsi Stimpson, 1851	Adam Peaclam				
Pisidium casertanum (Poli, 1791)	Ubiquitous Peaclam				
Pisidium compressum Prime, 1852	Ridgedbeak Peaclam				
Pisidium dubium (Say, 1817)	Greater Eastern Peaclam				
Pisidium nitidum Jenyns, 1832	Shiny Peaclam				
Pisidium punctatum Sterki, 1895	Perforated Peaclam				
Pisidium variabile Prime, 1852	Triangular Peaclam				
Sphaerium fabale (Prime, 1852)	River Fingernailclam				
Sphaerium occidentale (Lewis, 1856)	Herrington Fingernailclam				
Sphaerium striatinum (Lamarck, 1818)	Striated Fingernailclam				

[1]Based on U.S. Fish and Wildlife (1995) and subsequent notices in the Federal register.
[2]Based on Williams et al. (n.d.).
CS—Currently Stable (Williams et al., 1993)
E—Endangered
EX—Endangered, Possibly Extinct (Williams et al., 1993)
SC—Special Concern (Williams et al., 1993)
T—Threatened
U—Undetermined (Williams et al., 1993)
X—Extinct (Williams et al., n.d.)

A Brief History
of the Classification
# of North American
Freshwater Mussels

Since the late 1600s, the description, classification, and accounts of freshwater mussels have undergone constant modification and revision. With an even greater enthusiasm today for the study of these animals, plus more sophisticated analytical techniques such as DNA sequencing, taxonomic revision continues at a rapid pace.

The earliest known illustrations of freshwater bivalves from North America were of specimens from Virginia, figured and described by Martin Lister (1685). This work was reprinted twice. The description of unionoids from North America by Europeans really began in the latter eighteenth century with such works as Lightfoot (1786) and Spengler (1793). The early nineteenth century began with contributions by Bosc (1801, 1804).

The first paper on North American freshwater bivalves by an American writer was that of Thomas Say (1817). His chapter in the American edition of Nicholson's *Encyclopedia* (Say, 1817) set the stage for the early study and description of new terrestrial and freshwater mollusks of North America. This chapter was revised twice (Say, 1818a, 1819). Say introduced the new genus *Alasmidonta* (Say, 1818b) in the first volume of the *Journal of the Academy of Natural Sciences of Philadelphia* to replace *Monodonta* (Say, 1818a).

An important early naturalist, Jean Baptiste Pierre Antoine de Monet Chevalier de Lamarck (1819), contributed a major paper describing a series of new fresh-

water bivalve species from the New World. Many of Lamarck's taxa are still recognized today, for example *Actinonaias ligamentina* (Lamarck, 1819), *Elliptio crassidens* (Lamarck, 1819), and *Obovaria retusa* (Lamarck, 1819).

At this same time an immigrant French naturalist, Samuel Constantine Schmaltz Rafinesque, was beginning work on the fauna and flora of eastern North America. He published a series of short papers mostly listing taxa to be described, thereby introducing a series of generic and specific names without descriptions (Rafinesque, 1818a, b, 1819a, b).

Rafinesque (1820) introduced the world to the amazing diversity of freshwater bivalves and at the same time departed from the common European practice of placing all species in one of about four genera: *Unio, Margaritana, Alasmidonta,* or *Anodonta*. He introduced four new unionoid and one sphaeriid subfamilies, 20 new generic level taxa, and 68 new species with numerous subspecies and varieties. Part of the problem with this paper, however, was that it was published in Brussels in a short-lived journal in French, not easily accessible to most early naturalists in North America. The format of the paper was confusing to those who finally did obtain access to it. Rafinesque only listed those characters which would diagnose or distinguish the animals at the subfamily, generic, subgeneric, and species level. It is for this reason that his species descriptions appear to be so short, since he only included the characters needed to separate the species within the genus. The rest of the characters of a particular species could be found under the higher level taxa. The figures accompanying this paper were crudely executed and it is only with difficulty that some can be recognized.

Rafinesque (1831) continued his monographic work on the unionoids of the Ohio region, erecting 16 new genera and 46 new species. He noted that he had sent specimens to a variety of conchologists both in the United States and in Europe. C. A. Poulson (1832) provided an English translation of Rafinesque's 1820 monograph of the unionoids of the Ohio River and included a colored frontispiece of the Pistolgrip, *Tritogonia verrucosa* (Rafinesque, 1820). Rafinesque (1832) added another genus to the 36 genera of freshwater bivalves already erected. The two parts of Rafinesque's monograph of the unionoids of the Ohio have been variously reprinted and translated (see Bogan, 1988 for an unraveling of the history of these papers). Rafinesque compounded the problem in recognizing his taxa by not illustrating them or by using extremely poor figures and by not depositing identified specimens in recognized museum collections. Recognized specimens of some of Rafinesque's unionoid bivalve taxa, however, have subsequently been deposited in the Academy of Natural Sciences of Philadelphia (Charles A. Poulson and Samuel S. Haldeman collections) and the Museum National d'Histoire Naturelle, Paris (Férussac and Brongiart collections). The Museum of Natural History in Vienna has records of six species represented by seven specimens that were received from Rafinesque in 1827, but the specimens have not been located by modern researchers.

John Fleming (1828) has long been credited with the introduction of the family name Unionidae. However, Rafinesque (1820) first used the subfamily Unioninae, so the family name Unionidae must date from Rafinesque, 1820. The Western Academy of Natural Sciences of Cincinnati (1849) published a list of species with proposed synonymies. This was apparently in response to the difficulties naturalists had in trying to deal with the identification of the numerous taxa proposed by Rafinesque.

In 1823 Daniel H. Barnes published a very important and influential paper describing a series of new unionoid species and, especially significant, established a standard for the description, measurement, and illustration of new taxa. Thomas Say moved to the rural utopian settlement of New Harmony, Indiana, where he continued to publish malacological papers, describing many new taxa, in the *Disseminator of Useful Information* (Say, 1829–1831b). He also produced his impressive volume *American Conchology* in New Harmony (Say, 1830a–1834).

During this same time a series of other descriptive papers appeared by various authors: Daniel H. Barnes (1828), Jacob Green (1827), Samuel P. Hildreth (1828), and Jared P. Kirtland (1834); Augustus A. Gould (1841) on the mollusks of Massachusetts; James E. De Kay (1843) on the mollusks of New York; and Gerard Troost (1846) on Tennessee freshwater mollusks.

Timothy A. Conrad described a number of freshwater mollusks in his well-illustrated book of 1834. He caused a great deal of ink to be spilled by attempting to provide a synonymy and by taking the date of publication as the date when a name became available, not the date on which the species description was read at a meeting, as was Lea's policy (Conrad, 1853). Isaac Lea (1854, 1872a) had several rather vitriolic comments about Conrad's attempts to provide a synonymy of freshwater bivalves.

Baron André E. de Férussac (1835) reviewed the early descriptive work of Rafinesque, Say, and Lamarck, putting together synonymies and commenting on some of the species.

Isaac Lea published 279 papers between 1827 and 1876, covering primarily mollusks with an emphasis on freshwater bivalves, describing 899 new species of freshwater bivalves out of the 1,842 new species he erected. Lea provided anatomical diagnoses for 254 species and notes on the glochidia of 38 unionoid species (Scudder, 1885:lii–liii). His works were indexed by N. P. Scudder (1885). Lea summarized his views of the classification of freshwater bivalves using an artificial system based on the presence or absence of a dorsal wing and on the form and sculpture of the shell in a series of four editions of his *Synopsis* (Lea, 1836, 1838a, 1852c, 1870). These synopses formed the basis for much of the understanding of unionoids in North America. Lea's classification was challenged by Conrad and by Agassiz, among others, who felt that the genus *Unio* was too large and unnatural and should be split into smaller, more recognizable genera.

William Swainson (1840) proposed a classification of freshwater bivalves based on the quinary system, a scheme that organized suites of related species in circles that, in turn, touched other circles of related species. John E. Gray (1847) proposed a classification of the freshwater bivalves using the families Unionidae, Mutelidae, and Mycetopodidae. F. H. Troschel (1847) proposed a classification system based on anatomical characters, but due to lack of sufficient specimens he made some serious errors; however, this work was still very important to the further understanding of freshwater bivalves. Louis Agassiz (1852) published a classification of North American freshwater bivalves based on anatomical analyses, but apparently due to hasty work his groups were often heterogeneous. He recognized genera proposed by Rafinesque, Say, Swainson, and Schumacher and proposed five new genera himself, two of which are still recognized today.

Hermann von Ihering (1893) reported that some freshwater bivalves begin by hatching from an egg and then develop into a glochidium with a bivalved shell, while other species developed into a lasidium, a structure with three sections, only one with a shell. He placed those genera he believed to have glochidial larvae in the Unionidae and those with lasidia in the Mutelidae. This was a major step in attempting to clarify the relationships within the freshwater bivalves.

R. Ellsworth Call (1895) produced a well-illustrated volume of the freshwater bivalves of Arkansas. He followed this volume with a well-illustrated paper on the mollusks of Indiana (Call, 1900).

Charles T. Simpson (1896) first published a classification of freshwater bivalves based on shell characters. In his *Synopsis,* Simpson (1900a) recognized 533 species and 55 varieties in 25 genera. Simpson followed the work of von Ihering (1893) and recognized that the larval stage of the Unionidae was the glochidium and that the Mutelidae had a very different larval stage, the lasidium. Simpson (1900a) developed his classification of the Unionidae based upon the differences of the sexual characters of freshwater bivalves: the size, shape, and number of gills used as the marsupium. This work was expanded and was originally supposed to appear with illustrations, but it was finally produced in 1914 (Simpson, 1914) without them. Simpson (1914) continued his earlier support of the division of the freshwater bivalves into Unionidae with glochidia and Mutelidae with lasidia and still relied on the marsupial characters as the basis of his classification. This volume is still a very important tool in understanding the classification and shell characteristics of freshwater bivalves.

Edward G. Vanatta (1915) listed the "type specimens" bought from the Poulson collection which were listed as the types Poulson purchased from Rafinesque in 1831. Vanatta pointed out the priority of Rafinesque's unionoid taxa represented by specimens in the collection. Two lots of Rafinesque specimens came to the Academy of Natural Sciences of Philadelphia from the collection of S. S. Haldeman and are labeled in pencil on the inside of the valves in Haldeman's handwriting.

Bryant Walker (1916) responded to the review of the "types" of Rafinesque in the C. A. Poulson collection, noting that workers should use caution in accepting the priority of Rafinesque's taxa, especially those based on Rafinesque's presumed inadequate descriptions. Walker noted that this was only a small taste of the changes to be made when a serious and complete analysis of the confusion of available unionoid names was made using the rules of nomenclature. Walker (1918a, b) provided keys to the genera of freshwater mollusks of North America supplemented by figures of representatives of each genus.

Frank C. Baker (1898, 1928a) published two volumes treating the freshwater bivalve fauna of the Chicago area and Wisconsin and a series of papers on the Pleistocene molluscan fauna of the Midwest. Baker described a few new species and numerous new subspecies from the central portion of the United States.

Harold Hannibal (1912) summarized the information on Recent and Tertiary freshwater mollusks of California and provided information on the classification and evolution of the Unionidae. William I. Utterback (1915–1916a, 1916b) published "The Naiades of Missouri," bringing together the then current knowledge of these animals in that state and their taxonomy. This work summarized available information on each species in the fauna and also provided the common names by which each was known locally.

Lorraine S. Frierson (1914, 1927) published his own ideas on the classification and relationships of the freshwater bivalves of North America. He was a strong supporter of Rafinesque and campaigned for the long overdue recognition of those unionoid species described by Rafinesque (1820, 1831). His classified and annotated checklist was the last overview of the freshwater bivalves of North America until the major treatise by Fritz Haas (1969a) appeared. However, the classification used by Haas (1969a) is basically that of Frierson (1927) with some minor changes. Haas (1969b) provided the Unionoida section of the *Treatise of Invertebrate Paleontology*. The classification used in the *Treatise* differs somewhat from that used in his monograph (Haas 1969a).

Arnold E. Ortmann (1909a, 1918) was one of the first to recognize and report the detrimental effects of pollution and damming on the freshwater bivalve fauna. His early work on comparative anatomy of freshwater mussels resulted in the construction of a classification of North American bivalves that has withstood the test of time (see Davis and Fuller, 1981). Ortmann (1918, 1924a) introduced the concept of the Cumberlandian Province for unionoid bivalves, following the concept of regional endemism pointed out by Binney (1885) for land snails. Ortmann and Walker (1922), with the outside arbitration of Henry A. Pilsbry, attempted to settle a number of vexing taxonomic problems. Their efforts to stabilize nomenclature were quite successful and brought a certain level of stability to the names used for North American freshwater mussels.

Calvin Goodrich and Henry van der Schalie (1944) updated and expanded the earlier work of Call (1900) on the land and freshwater mollusks of Indiana. Henry van der Schalie and Annette van der Schalie (1950) summarized the unionoid work on the Mississippi River proper and provided an updated picture of the unionoid faunal provinces of North America. Henry van der Schalie also published numerous studies, some of which—for example the paper on the Tombigbee River unionoids (van der Schalie, 1939)—included descriptions of new species.

Joseph P. E. Morrison (1942) provided an early example of the importance of archaeological molluscan assemblages for understanding unionoid distribution prior to European colonization. He also was a proponent of using the earliest available names and a supporter of Rafinesque. However, Morrison did not explain his arguments for the resurrection of any of the unused senior synonyms he resurrected (Morrison, 1969).

William J. Clench (1959) revised the chapter on mollusks originally written by Walker (1918b) for the second edition of the classic volume *Freshwater Biology* by Henry B. Ward and George C. Whipple (1918). This chapter provided a key to the families of freshwater mollusks and discussions of the genera in each family with an illustration of a representative species of each genus.

Fritz Haas (1969a) published a compendium of his ideas on the classification of the freshwater bivalves

of the order Unionoida. In a companion article he presented the classification of Unionoida down to the generic level, listing type species (Haas, 1969b). Unfortunately, the original work was written early in his career and was not adequately updated before being published. Haas relied on Frierson (1927), the last comprehensive treatment of North American Unionidae up to that time.

Barry D. Valentine and David H. Stansbery (1971) provided some insight into unionoid taxonomic problems and presented a good overview of the unionoid fauna of the Red River in Oklahoma. William Heard and Richard H. Guckert (1970) developed a classification of the freshwater bivalves of North America based on anatomical data, with an emphasis on a portion of the demibranch, which demibranch or demibranches were used as marsupia, and the length of the breeding season.

Richard I. Johnson (e.g., 1970, 1972, 1977, 1978, 1980) has published a variety of papers on unionoid systematics and zoogeography. One of his approaches was to combine numerous very similarly shaped species together (e.g., *Elliptio complanata* (Lightfoot, 1786) and *Elliptio icterina* (Conrad, 1834)). His 1970 paper is the single most comprehensive volume on the unionoid fauna of the Southern Atlantic Slope region.

Arthur H. Clarke (1973) reviewed the freshwater mollusks of the Interior Basin of Canada, providing a popular volume covering the freshwater mollusks of Canada (Clarke, 1981a). He followed up his work in Canada with a detailed taxonomic evaluation of the tribe Alasmidontini (Clarke, 1981b, 1985).

George M. Davis and Samuel L. H. Fuller (1981) took the ideas of Heard and Guckert (1970) as a testable hypothesis and produced a new classification for some of the genera of North American freshwater bivalves, combining anatomical characters with electrophoretic studies. They concluded that their classification was very close to that originally proposed by Ortmann (1910, 1916), but failed to discuss the changes Ortmann included in his later works (Ortmann, 1912a, 1919). Davis and colleagues (e.g., Davis, 1983a, b; Davis and Mulvey, 1993; Davis et al., 1981) have continued their electrophoretic analyses of unionoids, especially on the species problems within the genus *Elliptio*.

Walter R. Hoeh (1990), using biochemical techniques, has presented the phylogenetic relationships within the eastern North American *Anodonta*. He elevated *Utterbackia* F. C. Baker, 1927, and *Pyganodon*

Crosse and Fischer, 1894, from subgeneric to generic rank based on the evidence for three highly differentiated clades within what was previously known as *Anodonta*.

Numerous state and regional field guides have been produced, including those by Robertson and Blakeslee (1948, for the Niagara Frontier Region), Murray and Leonard (1962, for Kansas), Parmalee (1967, for Illinois), Mathiak (1979, for Wisconsin), Oesch (1984, for Missouri), Harris and Gordon (1990, for Arkansas), Cummings and Mayer (1992, for the Midwest), Vidrine (1993, for Louisiana), Howells et al. (1996, for Texas), and others. All of these used the taxonomy current at the time of their publication, and all reflect some of the variations and changes in classification that have occurred over the years.

The American Fisheries Society's Committee on Names of Aquatic Invertebrates published the first edition of the *Common and Scientific Names of Aquatic Invertebrates from the United States and Canada: Mollusks* (Turgeon et al., 1988). This volume was produced by a committee from the Council of Systematic Malacologists and supported by the American Malacological Union. The freshwater bivalve portion of the checklist was prepared by Arthur E. Bogan and reviewed and commented on by numerous interested persons before the volume was published. This checklist recognized approximately 300 species of North American unionoid bivalves. The draft of the second edition, containing some revisions in scientific and/or common names, is now in press. Such references serve to standardize scientific and common names acceptable to the majority of malacologists. They reflect the "state of the art" in the science of classification (taxonomy), but some names and relationships will change as advances in anatomical and shell morphology research continue to be made.

Ecology of Freshwater Mussels

Mussels generally attain their greatest abundance and diversity in rivers and streams with good currents. Rivers are generally considered to be long-lived, geologically speaking, compared with lakes. Rivers were probably mussels' original habitat, since the center of diversity appears to have been in the South, where natural lakes may have been relatively few in number. Many strictly lake-dwelling mussels appear to have expanded or become established during or after the last glacial periods, since it was then that the majority of lakes were formed in the North. Riverine species are usually larger and heavier than those characteristic of natural lake environments; some scientists believe that this condition may be the result of a more abundant and continuing food supply which is made constantly available by river current. It has also been suggested that river mussels develop a heavier, stronger type of hinge teeth in order to withstand flow pressure and to facilitate movement in the current (e.g., Savazzi and Peiyi, 1992; Watters, 1994b).

Mussels spend their entire juvenile and adult lives partially or completely buried in the bottom of some permanent body of water. Usually each mussel will position itself so that the posterior portion of the valves protrudes from the substrate and is directed upstream. In this way, materials brought in through the incurrent siphon are partially forced into the mantle cavity by the current, and the waste products that are expelled through the excurrent siphon are quickly swept away. Although the mussel is able to change its position and location to some extent, an individual will rarely move more than a radial distance of a few hundred yards during a lifetime unless it is dislodged and carried elsewhere by current or some animal. The lives of these mussels are, therefore, directly subjected to effects exerted by the local bottom area which they inhabit and by the water passing over them.

In prehistoric times, as evidenced by the quantity and variety of shells found at aboriginal village and camp sites, the rivers of Tennessee were clear, with sand and gravel beds ideally suited to these mollusks. Shoals and riffles in big rivers, such as the once extensive stretch of the Tennessee River at Muscle (Mussel) Shoals in northwestern Alabama, provided ideal habitat for a diverse and abundant mussel fauna. Intensive agricultural practices have involved the clearing and plowing of more and more land; during spring run-off or at flood periods, tremendous quantities of topsoil are washed into streams and rivers, causing layers of fine sediment (silt and clay) to settle and cover the former sand and gravel bottoms. Species adapted to a sand and gravel bottom environment cannot long survive in one composed of fine sediment and are quickly destroyed by silt that clogs the gills, smothering the animal. Industrial and organic pollution have been the other major factors in reducing or completely destroying mussel populations in many sections of the rivers of Tennessee. Dam construction has turned the entire length of the Tennessee River into a series of reservoir impoundments. Although numerous species still thrive in this river-lake habitat, many others, because of the resulting change in water depth and temperature, loss of host fish, rates of flow, and substrate composition have been either extirpated or reduced to populations composed of nonreproducing relics.

Reduction in the diversity and abundance of mussels may well be attributed primarily to impoundment. Changes in mussel faunas as a result of impoundment are especially well documented for the Tennessee River, which now consists of a series of reservoirs formed by the construction of 36 multipurpose dams on the mainstem and on major tributaries. Changes in water depth and temperature, dissolved oxygen deprivation, increased sedimentation, and alteration or loss of resident fish host assemblages detrimentally affect the survival of mussels and their reproductive potential. As Neves et al. (1997) have pointed out, "Because mussels are thought to be the longest-lived freshwater invertebrates, with a longevity of more than 100 years for some species, population declines may continue for decades. Thus, the extirpation of species is a prolonged event, lagging decades behind the directly responsible factors of attrition of the fauna."

Compared with damming, pollution, and silting, the two major kinds of natural enemies, parasites and predators, generally have little overall effect on any given population of naiads. Freshwater mussels are infested with a variety of parasites and commensals; this latter term refers to any animal living in or on another, but one that usually does not harm or injure its host. The most common of these are the water mites (Unionicolidae) which infest the gills and, occasionally, the whole mantle cavity. The eggs and larvae are generally imbedded in the gills, foot, and mantle of the mussel, while the adult mites are active and move about over the gills and body. Occasionally mite larvae migrate into the area between the mantle and shell, become covered with pearly matter, and appear as small blisters or pimples on the inside of the shell. Vidrine (e.g., 1980, 1996) has undertaken extensive studies dealing with these parasitic mites. Of the 637 mussels examined from the Duck and Stones rivers in Middle Tennessee by Vidrine and Wilson (1991), approximately 60% were infested with one or more mite species. These animals apparently cause little if any damage to the mussel.

Occasionally leeches attach themselves to the mantle or, more rarely, to parts within the body cavity, but generally they are uncommon and do not remain attached permanently. There are several types of true parasites (an animal living in or on another at whose expense it obtains food and shelter) known to infest freshwater mussels. One of these is the fluke, a parasitic flatworm, several species of which have been reported parasitizing the body cavities of mussels. Several kinds of distomids, another type of trematode or parasitic flatworm, are known to occur in the muscular parts and in some organs of mussels. The majority of these parasites are apparently harmless.

Probably the most common mammal predator is the muskrat; piles of shells around muskrat houses, on feeding stations, and in entrances to their burrows in mud banks attest to the gathering activities of these rodents. The agility exercised by muskrats in opening mussels without damaging the valves in any fashion is still not completely understood. Apgar (1887:58–59) provided the following explanation:

When the Unio is traveling along, its foot projects a half inch or more from the lower side of the shell. If, while the foot is in this, its usual condition, the two valves be pinched, the foot will be caught between the closing shells; if the pinching be continued for a half or three-quarters of a minute, the animal, probably from the pain produced, becomes paralyzed and unable to make use of the adductor muscles. Now, if the shell be released, it will fly open about one-half inch, and can easily be torn entirely open. The strength needed to keep the foot from being drawn into the shell is not great, being far less than that of the jaws of the Musk Rat. So all that is necessary for Fiber [Ondatra] to do when he wants his dinner is to swim along until he sees a Unio at the bottom, dive, and quickly seize the animal, then swim leisurely to his hole or the bank. By the time he has reached a good place for eating his meal the Unio will be ready to open far enough for the insertion of paw or nose, and the luscious bivalve can be devoured from the whole shell. In my own experiments I was usually, though not always, successful. The failures I think were always due to the fact that not quite enough of the foot was caught by the closing shell; this was caused by my disturbing the animal before taking hold of it. If the Musk Rat be not more supple than I, he must occasionally miss his meal.

The effectiveness of this method is yet to be proven. Quite possibly the muskrat, upon dislodging the mussel from the substrate, is able to cut one or both adductor muscles with its incisors before the animal can close its valves. Raccoons, mink, and otters, as well as some species of birds (especially waterfowl) and turtles, also utilize mussels for food. The freshwater drum *(Aplodinotus grunniens)* feeds almost exclusively on them.

Modern extensive use of the freshwater mussel has centered around the shells as a source of raw material for the manufacture of pearl buttons and beads in connection with the cultured pearl industry. In this capacity humans have proved to be a major predator, and, by over-harvesting, they have depleted mussel beds in many localities. The cooked mussel bodies removed during the processing of shells were sometimes used as hog feed, and anglers occasionally use the soft parts for bait.

Distribution of mussels in a river is rarely uniform since the depth of water, rapidity of current, composition of the bottom, and other physical factors may be quite unsuited for mussel development and maintenance in certain areas. A constantly changing or shifting bottom will limit or often prevent the establishment of mussel beds. Dams or high falls, as documented at Cumberland Falls, Kentucky (Stansbery, 1969), may serve as barriers to their distribution by preventing host fish infested with glochidia from ascending the stream or river (Watters, 1996c). In fast-flowing, big rivers like the Mississippi, Ohio, and Tennessee (reservoirs), mussels can be found at a depth of 35 feet or more; however, the depth at which they occur is usually directly dependent upon temperature (as affected by the seasons; light and dark), amounts of dissolved oxygen, and quantity of suspended food materials. Each species has evolved its own combination of optimum habitat requirements, and these differ considerably among the various taxa.

Lakes without current usually have a different and more limited mussel fauna than that found in rivers and streams. Studies have shown that dams constructed across rivers, which greatly reduce or eliminate the current, will produce an environment no longer suitable for most riverine mussel species. Specific bottom conditions are important for establishment and survival of certain kinds of mussels, but the majority are tolerant of a bottom composed of various combinations of sand, gravel, and mud. Floaters (Anodontinae) and papershells and heelsplitters *(Potamilus* spp.) often occur in a bottom composed almost entirely of mud, but most mussels attain their greatest abundance and optimum development in bottom areas composed of a mixture of stable gravel, sand, and silt deposits.

The Unionoid Faunal Provinces of North America

Recognition of molluscan faunal regions within North America began with the terrestrial gastropods and was subsequently modified for the unionoid bivalves. The Atlantic slope fauna was long recognized as distinct from that of the Interior Basin (Conrad, 1834a). Simpson (1896, 1900a) divided the unionoid fauna of North America into three major sections: (1) the Atlantic Region, extending from Labrador to Florida, (2) the Mississippi Region, extending from the Rio Grande to the Chattahoochee River Basin, including all of the Mississippi River Basin, (3) the area west of the Rocky Mountains from Baja California to the Yukon was included as part of the Palaearctic Region. Hannibal (1912) defined the Californian Region and discussed the molluscan fauna, and Ortmann (1913a) examined the effects of the Alleghenian Divide on the distribution and relationships of unionoids in eastern North America. Van der Schalie and van der Schalie (1950) developed a map of North America with six major unionoid faunal regions delineated. Roback et al. (1980) and Vidrine (1993) provided maps of unionoid faunal zones and provinces, but theirs differed from the one of van der Schalie and van der Schalie (1950) by recognizing four additional faunal areas: Great Lakes–St. Lawrence Area, Mobile Basin, Central Gulf Coast, and Western Gulf Coast. Neck (1982) reviewed the faunal areas proposed by others and developed a scheme of four

zones for the Texas unionoid fauna: Texoma, Sabine, Central Texas, and Rio Grande faunal zones. In our interpretation (fig. 10) we recognize the Rio Grande, Central Texas and Sabine Provinces, but have included Neck's (1982) Texoma province as part of the Interior Basin faunal province since the fauna is quite similar to that of the lower Interior Basin.

Atlantic. Based on the works of Johnson (1970) and Sepkoski and Rex (1974) for the Atlantic Slope we have chosen to recognize only a northern and southern portion of this area realizing that there is a broad transition zone in the mid-Atlantic area. The southern margin of the south Atlantic province is somewhat nebulous, most often stopping at the Altamaha River Basin. The positions of the Satilla, St. Marys, and St. Johns rivers have been variously placed with the Eastern Gulf Coast area or in the Peninsular Florida area. We have chosen to place the fauna of these rivers in the Peninsular Florida area.

Pacific. The Pacific faunal area defined by Hannibal (1912) extends from the Yukon Basin in the north to the San Sebastian Vixcaino Bay, Baja California, in the south, and includes the Great Basin and Colorado River Basin. This is the same area Simpson (1896, 1900a) included as part of his Palaearctic Region. We have chosen to include the whole area west of the continental divide in the Pacific Province. The concept of Hannibal

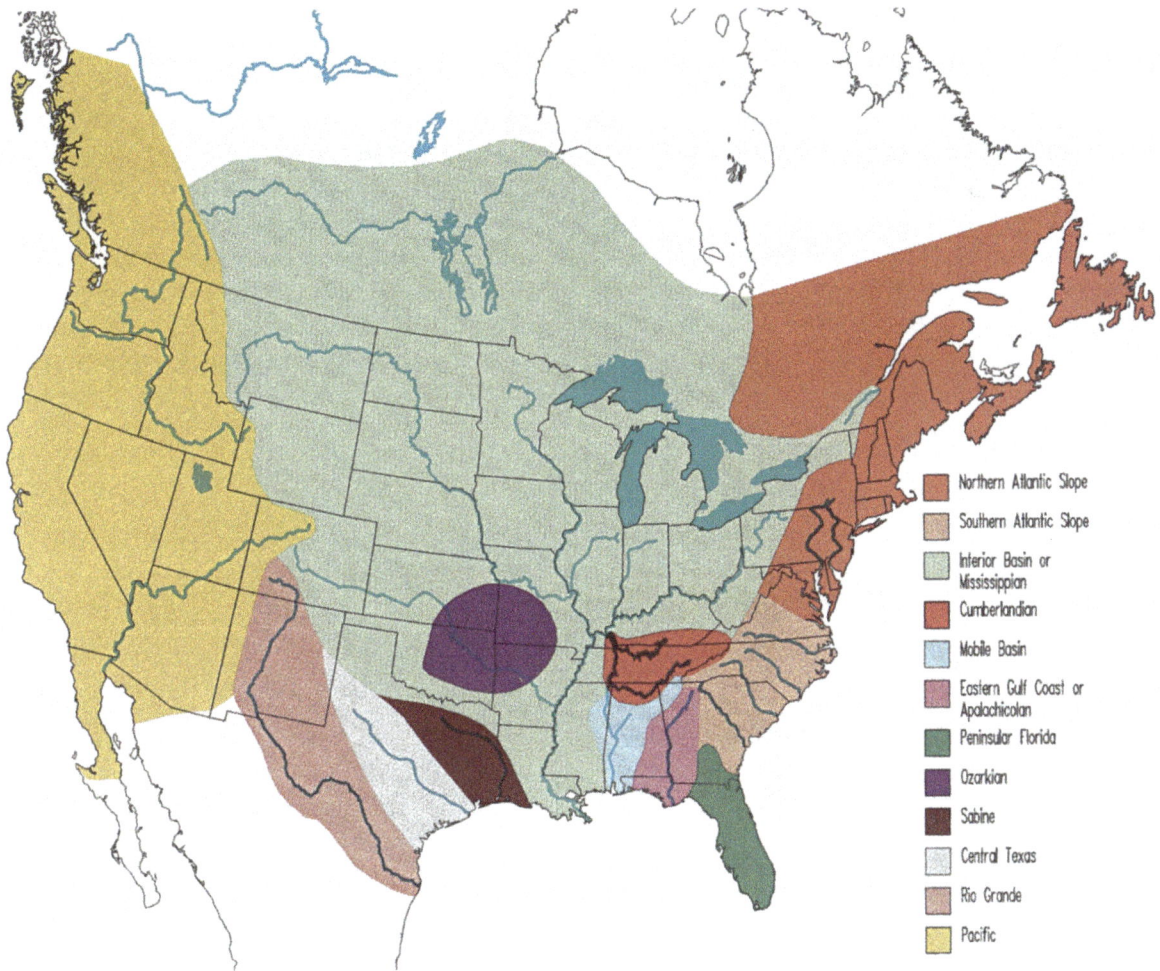

Fig. 10. The unionoid faunal provinces of North America.

(1912) and the one used here differs from that presented by van der Schalie and van der Schalie (1950) in being more inclusive. Van der Schalie and van der Schalie excluded southern California, the Great Basin, and the Colorado River Basin because the only unionoid in this area was *Anodonta californiensis*. The southern end of their Pacific province corresponds to the southern end of the distribution of *Margaritifera* and *Gonidea*.

Interior Basin. The Interior Basin faunal area includes the greatest geographic area of any of the provinces, encompassing the whole Mississippi River Basin with the exception of the Ozarkian and Cumberlandian faunal areas. Some would argue that this large province should be subdivided into an upper and a lower region based on the faunal differences, especially the area of the Mississippi River Embayment.

Ozarkian. Van der Schalie and van der Schalie (1950)

recognized this region as distinct from the surrounding Interior Basin province based on the number of endemic unionoid taxa.

Cumberlandian. Ortmann (1918, 1924a) defined a Cumberlandian region based on the high level of endemic freshwater bivalves in the Tennessee and Cumberland River basins. We have extended the downstream boundaries of the Cumberlandian Province in the Tennessee and Cumberland rivers to their mouths based on archaeological assemblages.

Mobile Bay Basin. Roback et al. (1980) and Vidrine (1993) recognized the Mobile Bay basin as a distinct faunal area based on the large number of endemic species.

Eastern Gulf Coast. The Apalachicolan faunal region was defined by Clench and Turner (1956) as those Gulf Coast drainages occurring from the Escambia River to the Suwannee River.

Peninsular Florida. Peninsular Florida was described as a faunal province by Clench and Turner (1956) and later by Johnson (1972). Butler (1989) redefined the area to include the St. Marys and St. Johns rivers. The Satilla River should also be included in this province based on close similarities among the unionoid assemblages.

Sabine. Neck (1982) defined this area of Texas as including the Sabine, Neches, and Trinity rivers.

Central Texas. Neck (1982) created the Central Texas faunal area, including the river basins extending from the Brazos River southwest to the Nueces River.

Rio Grande. Neck (1982) recognized the Rio Grande Basin as a distinct and separate faunal province.

Today the freshwater mussel fauna of Tennessee reflects a blending of taxa characteristic of the Interior Basin and Cumberlandian provinces. Species from the latter province continue to be more or less restricted to the upper Tennessee River drainage, while many from the Interior Basin have invaded and spread throughout much of the Cumberlandian Province, apparently adapting to the lakes and reservoirs that resulted from the damming of the Tennessee and Cumberland rivers and their major tributaries. Based on the study of aboriginal assemblages from the middle and lower stretches of these rivers, many Cumberlandian Province taxa had a much wider range in late prehistoric times.

Diversity of River Systems and Their Naiad Faunas

The State of Tennessee contains 129 unionoid species with only two being found in all four river basins: *Tritogonia verrucosa* and *Utterbackia imbecillis* (table 3).

The Mississippi River System

The nature and size of the Mississippi River forming the western border of Tennessee virtually precludes a diverse molluscan fauna. The river elevation annually fluctuates an average of 18 feet between winter highs and summer lows, and the substratum in shoal areas is composed of sand and gravel, where in pools it consists of shifting sand and silt. With few species recorded from the Mississippi River proper, most have come from oxbow lakes and tributary confluences.

Mississippi River tributaries in West Tennessee, with fishes providing the mechanism for dispersal, would be expected to support a relatively diverse fauna. Unfortunately, agricultural development of deep soils formed in loess and the resulting deposition of sediments in riverbeds led to channelization of these tributary rivers (the Forked Deer, Obion, Wolf, and Loosahatchie rivers) prior to thorough documentation of their mussel fauna. A total of 37 unionoid taxa have been reported for the Mississippi River Basin in Tennessee, 6 species are found only in this basin, while the remaining 31 taxa are also found in the Tennessee and/or Cumberland River basins (table 3).

The Hatchie River appears to contain the most varied unionoid fauna of all Mississippi River tributaries in Tennessee. Due to its relatively uniform sand and silt substratum, species diversity is relatively low; the limitation of habitat diversity is typical of direct Mississippi River tributaries in Tennessee. Approximately 75% of the species recorded from the Hatchie River occur in the Tennessee and Cumberland River systems. In his study of the naiad fauna of the Hatchie River, Manning (1989) recorded 33 native species, three of which *(Uniomerus declivis, Obovaria jacksoniana, Villosa vibex)* had not been previously reported as occurring in Tennessee.

Numerous backwater sloughs, cypress swamps, and oxbow lakes may be found in the floodplain belt of the Mississippi River. The largest of these, Reelfoot Lake, located in Obion and Lake counties in extreme northwestern Tennessee, now covers about 13,000 acres with an average depth of three to six feet (Petit, 1984). It has been greatly reduced in surface area and depth, estimated originally to have been 50,000 acres in extent with depths up to 36 feet after its formation following the New Madrid Earthquake of 1811–1812. All of the 11 species reported by Pilsbry and Rhoads (1896) and Ortmann (1926a) are either typical of lakes and river embayments (e.g., *Anodonta, Toxolasma*) or are adaptable to diverse conditions such as swift current

Table 3
List of Unionoidea from Tennessee by Major Drainage Basin within the State

	Tennessee River	Cumberland River	Mississippi River	Conasauga River
Margaritiferidae				
Cumberlandia monodonta (Say, 1829)	X	X		
Unionidae				
Actinonaias ligamentina (Lamarck, 1819)	X	X		
Actinonaias pectorosa (Conrad, 1834)	X	X		
Alasmidonta atropurpurea (Rafinesque, 1831)		X		
Alasmidonta marginata Say, 1818	X	X		
Alasmidonta raveneliana (Lea, 1834)	X			
Alasmidonta viridis (Rafinesque, 1820)	X	X		
Amblema plicata (Say, 1817)	X	X	X	
Anodonta suborbiculata Say, 1831	X	X	X	
Anodontoides ferussacianus (Lea, 1834)		X		
Arcidens confragosus (Say, 1829)	X	X	X	
Cyclonaias tuberculata (Rafinesque, 1820)	X	X		
Cyprogenia stegaria (Rafinesque, 1820)	X	X		
Dromus dromas (Lea, 1834)	X	X		
Ellipsaria lineolata (Rafinesque, 1820)	X	X		
Elliptio arca (Conrad, 1834)				X
Elliptio arctata (Conrad, 1834)				X
Elliptio crassidens (Lamarck, 1819)	X	X	X	
Elliptio dilatata (Rafinesque, 1820)	X	X		
Epioblasma arcaeformis (Lea, 1831)	X	X		
Epioblasma biemarginata (Lea, 1857)	X			
Epioblasma brevidens (Lea, 1831)	X	X		
Epioblasma capsaeformis (Lea, 1834)	X	X		
Epioblasma flexuosa (Rafinesque, 1820)	X	X		
Epioblasma florentina florentina (Lea, 1857)	X	X		
Epioblasma florentina walkeri (Wilson & Clark, 1914)	X	X		
Epioblasma haysiana (Lea, 1834)	X	X		
Epioblasma lenior (Lea, 1842)	X	X		
Epioblasma lewisii (Walker, 1910)	X	X		
Epioblasma metastriata (Conrad, 1838)				X
Epioblasma obliquata (Rafinesque, 1820)	X	X		
Epioblasma othcaloogensis (Lea, 1857)				X
Epioblasma personata (Say, 1829)	X	X		
Epioblasma propinqua (Lea, 1857)	X	X		
Epioblasma stewardsonii (Lea, 1852)	X	X		
Epioblasma torulosa torulosa (Rafinesque, 1820)	X	X		
Epioblasma torulosa gubernaculum (Reeve, 1865)	X			
Epioblasma triquetra (Rafinesque, 1820)	X	X		
Epioblasma turgidula (Lea, 1858)	X	X		
Fusconaia barnesiana (Lea, 1838)	X			
Fusconaia cor (Conrad, 1834)	X			
Fusconaia cuneolus (Lea, 1840)	X			
Fusconaia ebena (Lea, 1831)	X	X	X	
Fusconaia flava (Rafinesque, 1820)	X	X	X	
Fusconaia subrotunda (Lea, 1831)	X	X		
Hemistena lata (Rafinesque, 1820)	X	X		
Lampsilis abrupta (Say, 1831)	X	X		
Lampsilis altilis (Conrad, 1834)				X
Lampsilis cardium Rafinesque, 1820	X	X	X	
Lampsilis fasciola Rafinesque, 1820	X	X		
Lampsilis ornata (Conrad, 1835)				X
Lampsilis ovata (Say, 1817)	X	X	X	
Lampsilis siliquoidea (Barnes, 1823)			X	
Lampsilis straminea claibornensis (Lea, 1838)				X

Continued on next page

Table 3—*Continued*

	Tennessee River	Cumberland River	Mississippi River	Conasauga River
Lampsilis teres (Rafinesque, 1820)	X	X	X	
Lampsilis virescens (Lea, 1858)	X			
Lasmigona complanata (Barnes, 1823)	X	X	X	
Lasmigona costata (Rafinesque, 1820)	X	X		
Lasmigona holstonia (Lea, 1838)	X			X
Lasmigona subviridis (Conrad, 1835)	X			
Lemiox rimosus (Rafinesque, 1831)	X			
Leptodea fragilis (Rafinesque, 1820)	X	X	X	
Leptodea leptodon (Rafinesque, 1820)	X	X		
Lexingtonia dolabelloides (Lea, 1840)	X	X		
Ligumia recta (Lamarck, 1819)	X	X		
Ligumia subrostrata (Say, 1831)		X	X	
Medionidus acutissimus (Lea, 1831)				X
Medionidus conradicus (Lea, 1834)	X			
Medionidus parvulus (Lea, 1860)				X
Megalonaias nervosa (Rafinesque, 1820)	X	X	X	
Obliquaria reflexa Rafinesque, 1820	X	X	X	
Obovaria jacksoniana (Frierson, 1912)			X	
Obovaria olivaria (Rafinesque, 1820)	X	X	X	
Obovaria retusa (Lamarck, 1819)	X	X		
Obovaria subrotunda (Rafinesque, 1820)	X	X		
Pegias fabula (Lea, 1838)	X	X		
Plectomerus dombeyanus (Valenciennes, 1827)	X		X	
Plethobasus cicatricosus (Say, 1829)	X	X		
Plethobasus cooperianus (Lea, 1834)	X	X		
Plethobasus cyphyus (Rafinesque, 1820)	X	X	X	
Pleurobema chattanoogaense (Lea, 1858)				X
Pleurobema clava (Lamarck, 1819)	X	X		
Pleurobema cordatum (Rafinesque, 1820)	X	X	X	
Pleurobema georgianum (Lea, 1841)				X
Pleurobema gibberum (Lea, 1838)	X	X		
Pleurobema hanleyanum (Lea, 1852)				X
Pleurobema johannis (Lea, 1859)				X
Pleurobema oviforme (Conrad, 1834)	X	X		
Pleurobema perovatum (Conrad, 1834)				X
Pleurobema plenum (Lea, 1840)	X	X	X	
Pleurobema rubellum (Conrad, 1834)				X
Pleurobema rubrum (Rafinesque, 1820)	X	X		
Pleurobema sintoxia (Rafinesque, 1820)	X	X		
Pleurobema troschelianum (Lea, 1852)				X
Potamilus alatus (Say, 1817)	X	X		
Potamilus ohiensis (Rafinesque, 1820)	X	X	X	
Potamilus purpuratus (Lamarck, 1819)			X	
Ptychobranchus fasciolaris (Rafinesque, 1820)	X	X		
Ptychobranchus greeni (Conrad, 1834)				X
Ptychobranchus subtentum (Say, 1825)	X	X		
Pyganodon grandis (Say, 1829)	X	X	X	
Quadrula apiculata (Say, 1829)	X			
Quadrula cylindrica (Say, 1817)	X	X		
Quadrula cylindrica strigillata (Wright, 1898)	X			
Quadrula fragosa (Conrad, 1835)	X	X		
Quadrula intermedia (Conrad, 1836)	X			
Quadrula metanevra (Rafinesque, 1820)	X	X		
Quadrula nodulata (Rafinesque, 1820)	X		X	
Quadrula pustulosa (Lea, 1831)	X	X	X	
Quadrula quadrula (Rafinesque, 1820)	X	X	X	
Quadrula sparsa (Lea, 1841)	X			

Continued on next page

River Systems and Their Naiad Faunas   29

Table 3—*Continued*

	Tennessee River	Cumberland River	Mississippi River	Conasauga River
Simpsonaias ambigua (Say, 1825)		X		
Strophitus connasaugaensis (Lea, 1858)				X
Strophitus undulatus (Say, 1817)	X	X	X	
Toxolasma cylindrellus (Lea, 1868)	X			
Toxolasma lividus Rafinesque, 1831	X	X		
Toxolasma parvus (Barnes, 1823)	X	X	X	
Toxolasma texasensis (Lea, 1857)			X	
Tritogonia verrucosa (Rafinesque, 1820)	X	X	X	X
Truncilla donaciformis (Lea, 1828)	X	X	X	
Truncilla truncata Rafinesque, 1820	X	X	X	
Uniomerus declivis (Say, 1831)			X	
Uniomerus tetralasmus (Say, 1831)			X	
Utterbackia imbecillis (Say, 1829)	X	X	X	X
Villosa fabalis (Lea, 1831)	X			
Villosa iris (Lea, 1829)	X	X		
Villosa lienosa (Conrad, 1834)		X	X	
Villosa nebulosa (Conrad, 1834)				X
Villosa perpurpurea (Lea, 1861)	X			
Villosa taeniata (Conrad, 1834)	X	X		
Villosa trabalis (Conrad, 1834)	X	X		
Villosa vanuxemensis (Lea, 1838)	X	X		
Villosa vanuxemensis umbrans (Lea, 1857)				X
Villosa vibex (Conrad, 1834)				X
Total Taxa	102	87	37	24

or quiet water ponds and lakes (e.g., *Quadrula quadrula, Truncilla truncata*). None of the mussel taxa, with the possible exception of those of the *Anodonta* complex, can be considered numerous in Reelfoot Lake. Reduced oxygen levels, coupled with rapid filling of nutrient-rich soils that has resulted in the tremendous growth and spread of aquatic weeds (Petit, 1984), will undoubtedly bring about the extirpation of most mussel taxa in Reelfoot Lake.

The Tennessee River System

The Tennessee River and its tributaries, encompassing a watershed of over 40,540 mi^2, possess—or did possess at one time—the most diverse and abundant mussel fauna of any river in North America. This diversity is related to the geology of the area where the headwater tributaries of the river originate. The limestone-enriched provinces of the headwater drainages provide an ideal scenario for an expanded "Cumberlandian" mussel fauna: habitat diversity, abundant nutrients, and calcium-enriched ("hard") water. In addition, mussels of the Mississippi River or Interior Basin drainage, also once diverse and abundant, contributed to and made up most of the assemblages found in West

and Middle Tennessee, upstream in the Tennessee River to Muscle Shoals, Alabama, and in the Cumberland River to about Clarksville, Tennessee. Due to habitat modification resulting from human activity (e.g., pollution and impoundments), the greatest numbers and variety of naiad taxa in the state are now largely restricted to four Tennessee River tributaries: the Duck, Elk, Clinch, and Powell rivers.

The French Broad and Holston rivers join to form the Tennessee River at Knoxville, Knox County. The Clinch and Powell rivers, originating in the Ridge and Valley Province in southwestern Virginia, flow into the Tennessee River. The underlying geology is folded and faulted Paleozoic limestone lying in parallel northeast and southwest ridges. Stream substrata are gravel, rubble, and bedrock of primarily limestone (Fenneman, 1938). Water is hard and there are abundant nutrients (USEPA [United States Environmental Protection Agency] STORET Database). The 45 taxa that Ortmann (1924a) considered "Cumberlandian" have been recorded in this physiographic province.

The eastern headwater tributaries of the Tennessee River arise in the Blue Ridge Province. The Watauga, Nolichucky, French Broad, Pigeon, Little, Little Tennessee, and Hiwassee rivers originate along the western crest of the Blue Ridge physiographic province (1,800–2,400 feet). Except in lower reaches, streams are precipitous with soft water and low amounts of nutrients. Geologically, the area is composed of metamorphosed sedimentary rocks, gneisses, and schists (Fenneman, 1938). Boulders, cobbles, and siliceous rocks are typical substrata. While there are endemic fish species such as brook trout (*Salvelinus fontinalis* Mitchell) in the Blue Ridge Province, "Cumberlandian" unionoid species are rare or absent. Molluscan diversity and density, with few exceptions, increase after these streams enter the Ridge and Valley Province, lose gradient, and change water chemistry (Bogan and Starnes, 1983).

The Emory River, a tributary to the lower Clinch River, is a major stream draining the eastern portion of the Cumberland Plateau; it crosses geological strata that are characterized by Pennsylvanian sandstone, shale, and coal. The substratum is sandy with boulders, bedrock, and shale. The water is soft, slightly acidic, and nutrient limited. Twenty-two taxa, including 11 Cumberlandian endemics, have been recorded in this drainage, but most occurred in the lower reaches where the river enters the Ridge and Valley Province and the gradient decreases. The Sequatchie River, a southward-flowing tributary of the Tennessee River, drains the southern Cumberland Plateau. Twenty unionoid species are listed from the Sequatchie River.

The Highland Rim Province dominates Middle Tennessee and encompasses several major tributaries of the Tennessee River. Tributaries draining the crest of the Highland Rim from the south, elevations of 820–984 feet, include the Elk, Flint, and the Paint Rock rivers (the latter two do not contribute taxa to the state of Tennessee fauna). The Buffalo River drains the interior of the southwestern Highland Rim and the Duck River drains the eastern and western rim as well as the southern Nashville Basin. These rivers are moderate in gradient, nutrient enriched, and have hard water. Substrata consist of loose gravel and sand with limestone bedrock. Typically, these rivers contain a diverse unionoid fauna with the Duck River having 69 taxa, including 25 Cumberlandian species in the upper stretches. The Elk River has 61 taxa recorded from its waters. The Buffalo River, a tributary of the Duck River, is problematic; historically 27 taxa have been recorded from this river (van der Schalie, 1973), but few species

have been recently collected in the drainage (Ahlstedt, 1991b). This is despite the fact that water quality appears acceptable and faunal exchange could have occurred with the Tennessee or Duck rivers, since the substratum appears very similar to those rivers.

With the fairly recent discovery of local populations of the Bankclimber, *Plectomerus dombeyanus,* and the Southern Mapleleaf, *Quadrula apiculata,* in lower stretches of Kentucky Lake, a total of 102 native freshwater mussel taxa have now been recorded for the Tennessee River and its tributaries (table 3). The distribution of one species, *Lasmigona holstonia,* is restricted to the Tennessee and Mobile Bay basins (Upper Coosa River Drainage). *Plectomerus dombeyanus* and *Quadrula nodulata* are found only in the Tennessee and Mississippi river basins. A single species, *Quadrula apiculata,* appears to have been introduced into the lower Tennessee River. Two species on the southern edge of their range are restricted to the Tennessee River Basin: *Villosa fabalis* and *Lasmigona subviridis.* The Tennessee River Basin has 15 endemic taxa, and the Tennessee and Cumberland River basins share an additional 21 endemic taxa (table 3). Eleven species, all in the genus *Epioblasma,* which inhabited the shoal and riffles areas of the Tennessee River and its larger tributaries are now extinct, directly or indirectly as the result of impoundment. In many instances remaining populations of mussels in several of the Tennessee River reservoirs are no longer viable, their composition consisting of old nonreproducing individuals (Ahlstedt and McDonough, 1994).

Although some groups or individual species, such as the pimplebacks (*Quadrula* spp.), the Washboard (*Megalonaias nervosa*), and Ebonyshell (*Fusconaia ebena*), have adapted to reservoir conditions such as Kentucky Lake, the majority of taxa present in preimpoundment times have fared poorly. With dam construction and resulting impoundments throughout the main river course, the once free-flowing Tennessee River has experienced an upstream invasion of Interior drainage species such as the Washboard and Threehorn Wartyback, *Obliquaria reflexa.* Mussels characteristic of shallow, quiet water and a mud/silt substrate, such as the floaters (Anodontinae), have done exceptionally well. These additions to and changes in the present Tennessee River mussel assemblage can never compensate for the loss of its once rich and diverse fauna.

The Cumberland River System

The Cumberland River originates in the Cumberland Mountain Subprovince of the Cumberland Plateau in southeastern Kentucky. It extends 1,778 mi and has a drainage of 18,533 mi^2. The Cumberland Plateau is underlain by Pennsylvanian strata consisting of alternating layers of shale, sandstone, and coal. Water is soft and low in dissolved nutrients. While the upper Cumberland River is confined to Kentucky, the Big South Fork Cumberland River, a major tributary, drains the western Cumberland Plateau in Tennessee. Tributaries of the upper Cumberland River (Little South Fork Cumberland, Rockcastle, and Laurel rivers) flow through Pennsylvanian-age strata through most of their drainages. The Big South Fork Cumberland River has eroded through Pennsylvanian into Mississippian strata (limestone). Twenty-five unionoid species have been recorded from the Big South Fork Cumberland River drainage in Tennessee.

As the Cumberland River enters Tennessee from Kentucky it is joined by the Wolf, Obey, and Roaring rivers. These drain the eastern Highland Rim and possess substrata and water chemistry similar to the Duck and Buffalo rivers. Thirty unionoid species have been reported from the Obey River, but only seven from the Roaring River.

As the Cumberland River enters the Nashville Basin, it has a reduced gradient and meanders westward across the Basin until it re-enters the western Highland Rim. From the south, the Cumberland River receives drainage from the Stones River (in the central Nashville Basin) (Schmidt, 1982). The fauna of the Caney Fork River is substantially reduced due to a waterfall below the confluence of the Collins and Rocky rivers. Fourteen unionoid taxa have been recorded for the Caney Fork River and 49 for the Stones River.

After re-entering the Highland Rim, the Cumberland River flows westward through a deep alluvial floodplain. It receives several major tributaries draining the surrounding Highland Rim including the Harpeth and Red rivers and Yellow Creek. These tributaries have upland characteristics with predominately chert-gravel substrate. Prior to 1960 the Harpeth and Red rivers reportedly had 25 and 22 taxa, respectively (Starnes and Bogan, 1988), but more recent collections, albeit not exhaustive, suggest there has been a loss of several species from both rivers.

Eighty-seven species of naiads have been recorded from the Cumberland River and its tributaries in Tennessee (table 3). With 102 native species known from the Tennessee River system, it is apparent that the Cumberland River, in spite of providing excellent pre-impoundment mussel habitat, supported about 15 fewer species than did the Tennessee River. Several of the taxa absent from Cumberland River assemblage include Cumberlandian species, such as the Appalachian Monkeyface, *Quadrula sparsa*, Birdwing Pearlymussel, *Lemiox rimosus*, and the Slabside Pearlymussel, *Lexingtonia dolabelloides*. It is of interest to note that two species found only in the Cumberland River system are more northern Ohioan species with the Cumberland River basin being the southern edge of their range. *Alasmidonta atropurpurea* is the only Cumberland River basin endemic species. Two species occur only in the Cumberland and Mississippi River basins: *Ligumia subrostrata* and *Villosa lienosa* (table 3). The rest of the mussel species recorded from the Cumberland River also occur in the Tennessee River system.

The cause for this difference in total number of species is probably related to geology. The Cumberland River headwaters are in the nutrient-poor Pennsylvanian strata of the Cumberland Plateau. These tributaries have relatively depauperate faunas. It is only when streams cut through Pennsylvanian strata into limestone that diversity increases (Starnes and Bogan, 1982). A comparison of fauna in the Tennessee and Cumberland rivers reveals that primarily the headwater mussel species are absent from the Cumberland River (table 3). Thus, while these two rivers seem similar physiographically, they are discretely different and this translates into a slightly different mussel fauna.

The Coosa River System

One of the most northern tributaries of the Coosa River, the Conasauga River makes about a seven-mile-long loop into extreme southeastern Tennessee east of Chattanooga in Polk and Bradley counties. It originates in the Blue Ridge Province of northern Georgia and southern Tennessee. The geology of the area is dominated by granite, gneisses, schists, and metamorphic rocks (Fenneman, 1938) that produce soft water low in nutrients. Mussels are absent from this headwater area. After the river enters the Coosa Valley (Ridge and Valley) Province, water becomes hard, nutrients increase, and bivalves begin to appear. The Conasauga River in Tennessee contains 24 taxa, and of these *Utterbackia imbecillis* (Say), *Lasmigona holstonia* (Lea), *Tritogonia verrucosa* (Rafinesque), and *Villosa vanuxemensis* (Lea) also occur in the Tennessee or Cumberland rivers and/or their tributaries. The remaining 18 taxa are additions to the state species list that are typical of the Coosa River system and Gulf Coast streams (table 3).

Near the Tennessee/Georgia border unionoid species diversity increases. An additional 15 species were collected by Hurd (1974) immediately below the border in Georgia but have not been collected in the Tennessee stretch of river. These additional species may be limited by habitat diversity or stream size from expanding farther upstream in the Conasauga River.

Aboriginal Exploitation
of Freshwater Mussels

A Food Resource

Huge accumulations of chalky shells occurring above and in the banks bordering Tennessee's waterways, primarily along the large rivers and their major tributaries, attest to the aboriginal use of this once great food resource. The more obvious deposits, appearing as mounds or lenses and referred to by archaeologists as "middens," may contain literally hundreds of thousands of mussel valves and aquatic gastropods. Former human occupation levels within less obvious cave and rockshelter settings may also contain vast quantities of molluscan remains. Because of the high calcium carbonate content of the rock and soil fill in these natural caves and fissures, shell is often found especially well preserved, including nacre color and even remnants of the periostracum after being discarded several millennia ago.

It has been only within the last few decades that the identification and analysis of molluscan remains from aboriginal shell middens has become an especially important pursuit in enabling the archaeologist to arrive at a more accurate understanding of early subsistence patterns. This is not to say that there was a total lack of interest in the subject much earlier. A. E. Ortmann (1909c), for example, the malacologist who contributed most significantly to our knowledge of freshwater naiad taxonomy and ecology, published a short paper in 1909 titled "Unionidae from an Indian Garbage Heap."

In most of the early archaeological reports dealing with the excavation of particular sites, however, recovered mussel valves were discussed or illustrated only if they had been fashioned or modified into some form of ornament, tool, or utensil. In an early article titled "The Use of Molluscan Shells by the Cahokia [Illinois] Mound Builders," F. C. Baker (1923b:331), in addition to listing the mollusks recovered and the number of each, made the following general comment: "The first mentioned species [mussels] doubtless were used largely as food, for the ancient aborigine, like his more modern descendant, probably esteemed this bivalve as a valuable part of his menu." In a later study, Baker (1941:51) provided an in-depth study of vertebrates and mollusks represented in archaeological context in the Illinois River Valley. He noted that "[i]n some places unused clambakes have been found where the mussels have been carefully nested and covered with clay to be baked."

At about this same time, Morrison (1942) presented a detailed analysis of the molluscan taxa recovered in the Pickwick Basin mounds (Tennessee River, northwestern Alabama) during archaeological salvage operations prior to completion of Pickwick Landing Dam in 1938. He concluded that "[f]ifty-six kinds of mussels were eaten by the inhabitants," and that "[t]he shells were not used for any purpose (with the exception of 6 ornaments out of 100,000 specimens)." As for the cooking technique used, Morrison (1942:381) explained: "The mussels were steamed open for eating, as proven by the quantities of water-cracked rock fragments that were present in the shell deposits. These rocks were river cobbles, brought in by the Indians, and heated by them in a fire, before the mussels were placed over the rocks for cooking. They must have been used over and over again, as the pieces remaining are small, having been split apart many times by the action of the water and juices coming from the mussels."

These studies by Arnold E. Ortmann, Frank C. Baker, and Joseph P. E. Morrison were the forerunners of numerous others that have been undertaken in the decades that followed. In addition to attempting to determine the role or significance of aquatic mollusks in the diet of aboriginal peoples using this resource, these shell remains from archaeological context also provided a means of determining past naiad distribution and abundance. But an understanding of their potential value as a food resource remains paramount to the archaeologist. The vast numbers of shells forming aboriginal middens certainly attest to major exploitation of these animals by site occupants. With the possible exception of short periods of extreme cold and ice during some winters, mussels would have provided an abundant food resource that could be harvested throughout the year in Tennessee. But how nutritious and preferred were mussels compared to other meat sources, such as white-tailed deer, squirrels, and turkey? Were they relied upon only as an ever-present, abundant, and easily obtainable food but not as a preferred one, and were mussels consumed immediately upon harvesting or were some processed by smoking or drying and stored for later use?

Although answers to most or even all of these questions may never be known, some inferences may be drawn from studies of coastal shell mounds and shell heaps composed of marine shellfish. Black and Whitehead (1988) provide a concise review of ethnographic accounts from the Northeast Coast related to early harvesting and preservation of oysters and other saltwater mollusks, but those describing cooking and preservation methods are rare and somewhat vague. As one would expect, however, some of the shellfish were reported to have been shucked and eaten while fresh, while others were partially roasted or smoke dried. It can only be assumed that meat of freshwater mussels was treated in a like manner. In their effort to evalu-

ate mussels as a prehistoric food resource, Parmalee and Klippel (1974:432–433) obtained tissue analyses of two species of mussels (Potamilus alatus, Actinonaias ligamentina). The nutritional value proved to be low with an average of 68 calories per 100 grams of meat. They concluded that "on the basis of the analysis of fresh specimens it is submitted that when archaeologically derived samples are analyzed and compared to other faunal resources, freshwater mussels will prove to have played a less significant role in the total animal nutrition ingested—except perhaps on a short-term basis—than is frequently surmised from casual observations of total numbers of valves or minimum numbers of mussels from archaeological sites."

Accounts of the use of freshwater mussels as a food resource by early explorers and settlers in North America—if indeed these mollusks formed even a minor or occasional dietary item—appear to be lacking. Mussels inhabiting our rivers and lakes today are certainly edible, biological and chemical pollutants in their tissues notwithstanding, but the desirability of these animals as a food based on taste and texture is a matter of debate. Along these lines, the following is an interesting and rather amusing recounting of an experience by Pvt. Sam R. Watkins, Company H, First Tennessee Regiment, a Confederate soldier who survived all four years of the Civil War:

> Reader, did you ever eat a mussel? Well, we did, at Shelbyville. We were camped right upon the bank of Duck river, and one day Fred Dornin, Ed Voss, Andy Wilson and I went in the river mussel hunting. Every one of us had a meal sack. We would feel down with our feet until we felt a mussel and then dive for it. We soon filled our sacks with mussels in their shells. When we got to camp we cracked the shells and took out the mussels. We tried frying them, but the longer they fried the tougher they got. They were a little too large to swallow whole. Then we stewed them, and after a while we boiled them, and then we baked them, but every flank movement we would make on those mussels the more invulnerable they would get. We tried cutting them up with a hatchet, but they were so slick and tough the hatchet would not cut them. Well, we cooked them, and buttered them, and salted them, and peppered them, and battered them. They looked good, and smelt good, and tasted good; at least the fixings we put on them did, and we ate the mussels. I went to sleep that night. I dreamed that my stomach was four grindstones, and that they

turned in four directions, according to the four corners of the earth. I awoke to hear four men yell out, "O, save, O, save me from eating any more mussels!" (Sam R. Watkins, *"Co. Aytch,"* 83; reprinted by permission of Simon & Schuster, Inc., New York, 1962).

Tools, Utensils, and Ornaments from Mussel Shell

Considering the vast numbers of freshwater mussels harvested by Native Americans in eastern North America from about 8000 B.C. to historic times, relatively little use was made of the shells themselves. Despite the diversity of and often striking variation in the size, shape, thickness, and color of shells among freshwater species, nearly all finely worked or modified pieces were fashioned from marine mollusks obtained from the south Atlantic Coast and Gulf of Mexico. The majority of disc beads, pendants, dippers, gorgets, and engraved masks, typically found as burial accouterments, were cut from the marine whelk *Busycon* spp. For whatever reasons, valves of freshwater mussels seem to have held little aesthetic appeal to aboriginal peoples for use in the manufacture of decorative or symbolic objects.

In the case of utilitarian objects fashioned from mussel shell, however, the "spoon" or "dipper" was a type of utensil that appeared during the Archaic period (about 8500 to 1500 B.C.) and lasted into early historic times (fig. 11). Almost without exception the greatly inflated, deeply concave valves of the pocketbooks, *Lampsilis ovata* and *L. cardium,* were the ones used for this purpose. During the Archaic and later Woodland (about 1500 B.C.–A.D. 900) periods, the valves were used with little or no modification, but in the Mississippian cultural period (about A.D. 900–1400) many were altered by grinding away the lateral and pseudocardinal teeth and often portions of the shell margin. Some were modified further by notching the posterior end to form a handle; such utensils have been found not uncommonly with food bowls and other grave offerings.

With the onset of the Early Mississippian cultural period around A.D. 900 and the expansion of agricultural practices, a shell artifact appeared that is usually classified as a hoe. Typically, large thick valves of species such as the Mucket *(Actinonaias ligamentina),* Threeridge *(Amblema plicata),* and Washboard *(Megalonaias nervosa)* were used; a stick (wood handle) was

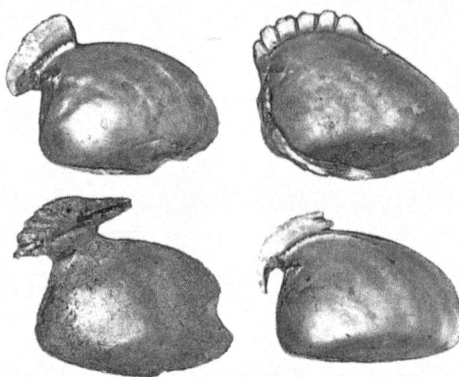

Fig. 11. Valves of the Pocketbook, *Lampsilis* cf. *ovata,* recovered from aboriginal sites which have been modified and probably used as spoons. Woodpecker head effigy spoon courtesy Tennessee State Museum, Gates P. Thurston Collection of Vanderbilt University.

hafted through a hole drilled near the center or close to the beak cavity area of the valve (fig. 12). With the posterior margin of the valve serving as the cutting or chopping edge, this implement was thought to have served as a type of cultivating tool. Mussel shells were also employed as scrapers, a handheld multipurpose tool that was used, for example, to remove excess tissue from hides or to shape and smooth wood objects and clay vessels before they were fired. Parmalee (1988) recovered a minimum of 175 valves of the Fluted Kidneyshell *(Ptychobranchus subtentum)* from a Late Mississippian site in Sevier County, Tennessee, that had served as scrapers. The ventral margin, sharp and strong when the valve is fresh, becomes with use worn down at an angle toward the posterior end.

Ornamental or decorative objects cut from freshwater mussel shells, as mentioned before, are relatively rare. In Baker's (1923b:329) report on shells from the Cahokia site, a major Mississippian period mound and village complex near East St. Louis, Madison/St. Clair counties, Illinois, he described a human head effigy and a gorget cut from shells of the Pink Heelsplitter *(Potamilus alatus)* and a valve of the Pocketbook *(Lampsilis ovata)* that "had been used as an ornament, several holes being drilled in the side." Beads cut from mussel valves are rare and only occasionally are shells encountered in archaeological context that have been drilled and otherwise modified for the purpose of suspension as pendants. Examples of these types of artifacts were recently recovered (Parmalee, 1994) in Dust Cave, an

Archaic rockshelter site along Pickwick Lake, Colbert County, Alabama: one valve of the Sugarspoon *(Epioblasma arcaeformis)* had most of the margins ground away in circular fashion with a centrally located drilled hole (bead?); one valve of the Fluted Kidneyshell had been smoothed by scraping, and a small hole drilled in the posterior end for suspension (pendant).

There is limited evidence suggesting that freshwater mussel shells were of certain symbolic significance to a few prehistoric aboriginal groups. Three examples of mussel shell effigy bowls were found with burials in stone box graves (Mississippian cultural period) at the Noel Site, approximately five miles south of Nashville, Davidson County, Tennessee. Two of these bowls were fashioned as paired valves while the third, a single left valve, almost certainly was modeled after the pocketbook, *Lampsilis ovata* or *L. cardium* (fig. 13). The artisan's attention to detail is evident by the distinct formation of the pseudocardinal teeth and the use of incised lines to depict lateral teeth.

The beauty and desirability of the freshwater pearl was not lost on prehistoric Native Americans. Their appreciation for pearls is reflected by the various decorative uses to which they were put. One of the most unusual uses was the inserting (inlaying) of a pearl in shallow holes drilled in the roots of bear canine teeth (Parmalee, 1959). The majority of evidence, however, indicates that drilled pearls were either sewn onto garments or strung as a necklace. One of the most impressive examples of the latter was a necklace composed of approximately 1,300 graduated pearl beads recovered in 1936 with a burial at the Hixon Site, Hamilton County, Tennessee (fig. 14).

Fig. 13. Freshwater mussel shell effigy bowls from the Noel site, Davidson County, Tennessee. Bowl at bottom depicts a left valve of the Pocketbook, *Lampsilis ovata* or *L. cardium*. Tennessee State Museum, Gates P. Thurston Collection of Vanderbilt University.

Fig. 12. Drilled valves of the Threeridge, Mucket, and Washboard. Large, thick valves like these were hafted and thought to have been used as hoes.

Fig. 14. A necklace of freshwater mussel pearls recovered with an aboriginal burial at the Hixon site, Hamilton County, Tennessee.

Historic Exploitation
of Freshwater Mussels

The Freshwater Mussel Pearl Button Industry
John F. Boepple (1854–1912), a button worker from
Ottensen, Germany, who immigrated to America in
1887, is generally acknowledged as the "founding fa-
ther" of the pearl button industry in this country. He
eventually reached Petersburg, Illinois, and, while bath-
ing in the Sangamon River, found what he had been
looking for—a source of shell from which quality buttons could be manufactured. Upon hearing about the
vast numbers of freshwater mussels that inhabited the
Mississippi River and its major tributaries, Boepple
moved to Rock Island, Illinois (adjacent to the Rock
River), and finally to Muscatine, Iowa, where he settled
and started the first freshwater pearl button company
in the world (O'Hara, 1980:5). This rather humble beginning in 1891 became a multimillion-dollar industry that flourished for the next 50 years.

Prior to this commercial use of the shells themselves,
it was the desire for pearls that often made the collection of freshwater mussels a lucrative, although terri-
bly wasteful, part-time or seasonal activity. O'Hara
(1980:4) noted that with the fortuitous finding of a
pearl in a freshwater mussel taken from a stream near
Paterson, New Jersey, in 1857, the local citizens "flocked
to the stream. They extracted an estimated $115,000
worth of pearls, but within two years there wasn't a
single mussel left in Notch Brook." It didn't take long

for the news to spread that there was great wealth to
be had from pearls that could be obtained from freshwater mussels that occurred in unlimited numbers in
nearly every river and stream. Finds were made in such
diverse regions of the country as Vermont, Washington, Tennessee, and Florida; it was about 1890 that
the pearling craze became rampant in the upper Mississippi River valley, especially in the Mississippi River
and the major rivers in Wisconsin, Iowa, and Illinois.
In Tennessee prior to 1894 the Caney Fork or Stones
River was the principal pearling center; the Calfkiller,
Elk, Duck, Cumberland, and Tennessee rivers were
also reported to have major sources of pearls, with
some pearling taking place in the Powell, French Broad,
and Obey rivers. After 1894, the Clinch River was the
major pearling river (Claassen, 1994). A historical
marker in Clinton, Tennessee, reports that Clinton was
one of the major pearl-buying cities in Tennessee.

Today, pearls are recovered usually as a by-product
when steaming open mussels harvested for their shells
which are to be used in the cultured pearl industry.
According to McGregor and Gordon (1992), "[i]t is
extremely rare to find a commercially valuable pearl
(1 in 10,000 mussels)."

From the beginning, it was realized that the seem-
ingly endless supply of shells, whose beautiful luster
and durability could withstand even the severest laun-

dering, would provide the raw material for a growing new industry in America. At the start of 1897 there were 13 button or blank-cutting factories in four cities along the Mississippi River, and by 1898 there were 49 plants in 13 cities on the Mississippi River with 12 plants on other rivers (O'Hara, 1980). Automatic machines were invented—one in 1903 did both the facing and drilling of blanks—that speeded up the production process. By 1912 there were nearly 200 plants in the United States utilizing valves of freshwater mussels for the manufacture of buttons, the sale of which amounted to almost $6,200,000. The annual sale of pearls and baroques ("slugs" of irregular shape but of high luster and prized for costume jewelry) averaged $300,000. Within the next 15 years, the industry grew tremendously; between 40,000 and 60,000 tons of raw shell were being taken yearly, selling for an estimated value of nearly $1,000,000. The rivers of Tennessee—the Clinch, Holston, Tennessee, French Broad, Elk, Duck, and Cumberland, among others—supplied a significant portion of mussel shell used in the manufacture of pearl buttons. Coker (1919), for example, recorded 98 tons of Holston River shells being sold in 1912, and from 1912 to 1914 a total of 50 tons from the Clinch River and 206 tons from the Tennessee River were sold. During the years 1929–1944, 61,476 tons of freshwater mussel shells harvested from Tennessee rivers were sold for the production of pearl buttons (Claassen, 1994:table 3).

A variety of tools and techniques, including diving, wading, and "grubbing" (picking up by hand), clam tongs, coke forks, rakes, and dredges, have been employed during the past century for harvesting mussels. Until the advent of diving as a standard means of harvesting mussels, with oxygen being supplied to the diver by either air tanks or a portable compressor on the boat, about 30 years ago probably 95% of the mussels collected in most large rivers were taken with an apparatus known as a crowfoot dredge or brail. A typical dredge consists of a 10- to 18-foot-long piece of one-inch pipe or a wood two-by-four with a large number of four-pronged hooks suspended from it by heavy cord or chain. Most mussel collectors use a flat-bottomed "john" boat, and as the boat drifts with the current across the mussel beds, the dredge is dragged along the bottom (figs. 15 and 16). When a prong of a hook touches or lodges in the soft parts of the mussel, which is partially imbedded in the bottom with the shell open

Fig. 15. Examples of mussel hooks attached to crowfoot dredges or brail bars.

Fig. 16. A johnboat with brail bar typical of those used by commercial shellers. Courtesy of Don Hubbs.

slightly, the valves instantly clamp shut on the prong. Since the hooks—sometimes as many as 100–200 making up a single brail—serve to anchor the boat, a "mule" is often employed as an underwater sail. Most of these "mules" are simply a piece of canvas stretched loosely on a wooden frame and submerged beneath the boat; the pushing of the current against the mule serves to propel the boat downstream.

Campsites operated by mussel fishermen normally have one to several "cookers" for killing and relaxing the clams (fig. 17). These are usually about five feet long, two feet wide, 12 to 18 inches deep, with wooden sides and a metal bottom. The unit is built over a shallow pit or elevated on blocks, where a fire is maintained under it to heat the bottom of the cooker. A small quantity of water is placed in the cooker with the mussels, and the top is covered with burlap or a wooden lid; the clams are steamed open, a process which takes only 20 or 30 minutes. The soft parts of the animal are then removed, usually by hand (and searched for pearls) or by passing the shells through a long, rotating metal cylinder perforated with holes through which the mussel meat falls after being shaken loose from the shells (fig. 18). Shells are then stacked in piles according to species, size, thickness, quality, and/or color. This type of operation was standard during the decades mussels were harvested for the manufacture of pearl buttons, but at the time of this writing, most of a particular day's catch may be sold "green," that is, uncooked with the animal still in the shell, depending upon the market price and the buyer's preference. One reason for this is that the buyer not only purchases the shell but also may increase the resale profit by recovering pearls. Buyers may drive to the sheller's camp or home where shells are stockpiled and buy them there, or the sheller, more often than not, trucks his catch to the buyer's warehouse or processing and shipping facility (fig. 19).

In the days when the pearl button industry was thriving, the shells arrived at the factory, were sorted

Fig. 18. Rotating metal cylinder (shaker) perforated with holes used in sorting out shells too small for commercial sale and removing excess mussel meat that falls through the holes. Courtesy of Freddy Couch.

Fig. 19. Weighing and bagging shell for shipment to Japan. U.S. Shell Co., Hollywood, Alabama. Courtesy of S. A. Ahlstedt.

first according to species, then to size. They were then sent to the cutting machines where the disks or "blanks" were cut out by a hollow, cylinder-type bit or hole saw (fig. 20). Waste and scrap shell was sometimes crushed and sold for poultry grit and lime, or the drilled valves were used for road fill. The blanks were placed in slowly revolving tumblers—oval, cast-iron drums—and the friction of the blanks rubbing on one another made the rough edges smooth. The blanks were then ground (on special grinding machines) to a uniform thickness; this process was followed by several days of soaking in water to prepare them for shaping, patterning, and drilling the eyes. The final process included polishing in wooden bucket-tumblers in which the buttons were placed first with pumice, then with a weak acid, and finally with a dry hardwood sawdust.

Fig. 17. Modern vats in which mussels are steamed and the meat removed. Courtesy of Freddy Couch.

Fig. 20. Examples of mussel shells from which button blanks were cut, and a selection of blanks.

As Claassen (1994:83) points out, "Literally hundreds of businesses came and went during the life of the freshwater shell button industry: finishing plants, saw works [blank cutting], and equipment suppliers." There were perhaps no more than eight or ten pearl button companies that operated in Tennessee: Claassen (1994) lists one for Knoxville, two for Nashville, one for Memphis and one for Clarksville. Development and refinement of plastics following World War II spelled the beginning of the end for this unique and profitable industry. By the mid-1960s, the last pearl button factory at Muscatine, Iowa, closed, thus representing the finish to a multimillion-dollar industry that flourished for almost 75 years. The pearl button industry could no longer compete with the quality and the low cost of production of plastic buttons.

The Cultured Pearl Industry

With the collapse of the pearl button industry in the early 1950s and the resulting loss of a market for shells, harvesting of freshwater mussels ceased, but not for long. Prior to 1900 all pearls, saltwater or freshwater, were natural; that is, they were formed under natural conditions as a result of fortuitous lodgment of a foreign object in the soft tissue. Because of the presence of this object, a pearl formed under natural conditions is not pearl throughout. During the 1920s there were over 300 natural pearl dealers in the United States; by the 1950s there were only six, and now there are none (Ward, 1985). Rather than harvesting freshwater mussels and marine bivalves strictly in search of natural pearls, those of commercial value being extremely rare, the vast majority of pearls sold today are "cultured." With the impetus of the cultured pearl industry coinciding with the end of the pearl button industry, harvesting of freshwater mussels continues to be profitable business.

A cultured pearl is one composed of a nucleus, typically cut in the form of a bead that has been strategically placed in the mantle or soft tissue of a bivalve, and the layers of nacre secreted by the clam or mussel around the nucleus. The Japanese are credited with developing the technique of inserting a nucleus (nucleation) into a marine oyster that resulted in the formation of a "cultured" pearl. By experimenting with numerous types of introduced materials, they found that a small mother-of-pearl pellet was the most successful stimulant to pearl production. It possessed an added virtue in that the resulting pearl would be composed entirely of nacreous material. One of the major factors in making this procedure successful has been the utilization of nuclei cut from valves of freshwater mussels. The quantity of mussels found in rivers from Wisconsin to Alabama, and the degree of hardness of their shells, is equaled almost nowhere else in the world. So shell from numerous rivers in eastern North America, including the Cumberland and Tennessee rivers in Tennessee, has been the source of nuclei material since the mid-1950s. It has been estimated that 95% of the world's round cultured pearls were nucleated with a bead cut and shaped from valves of freshwater mussels harvested from rivers in the United States. Because of their size, thickness, and hardness, about a dozen species have proven to be commercially valuable over the years, but the Threeridge *(Amblema plicata)*, Washboard *(Megalonaias nervosa)*, Ebonyshell *(Fusconaia ebena)*, and pimplebacks *(Quadrula* spp.) have remained the most preferred.

Beginning in the 1960s experiments were undertaken by the American Pearl Farms (then the Tennessee Shell Company, Inc.) in Camden, Tennessee, to develop a methodology for culturing pearls using freshwater mussels instead of pearl oysters (fig. 21). By following and modifying techniques developed by the Japanese, these experimenters have produced pearls in freshwater mussels that match the quality of those produced in marine bivalves. After the mussel is nucleated—the Washboard being the species most often used—it is placed in a mesh rack and suspended from a plastic frame. The mussels are held in these suspended nets for three to five years, after which time the pearls are removed. As time goes on, cultured pearls from freshwater mussels will be even greater competitors with those produced by marine bivalves. The market for raw shell from which to cut nuclei will continue and very possibly will expand as a result of this new source of cultured pearls, thus putting even greater pressure on remaining freshwater mussel resources.

The demand for North American freshwater mussel shells for use in the cultured pearl industry, whether marine pearl oysters or Tennessee River Washboards, continues unabated. The 1994 Tennessee mussel harvest totaled 2,707 tons. Hubbs (1995:7) reported that "Kentucky Reservoir continues to produce the majority of Tennessee's mussel harvest (83.2% during 1992, 95%

Fig. 21. A view of American Pearl Farms operation along the Tennessee River showing holding racks from which mussels, after being nucleated, are suspended while the cultured pearls are developing. Fred Ward © National Geographic Society.

during 1993, 97.3% during 1994). The wholesale value of 1994 mussel harvest was estimated at $8,492,089.78. This is close to the estimated $8,808,581 paid for 9,520,00 pounds harvested from Kentucky Reservoir alone during 1990." In addition to these sums, Hubbs (1995) further noted that the sale of commercial musseling licenses and mussel tax revenue associated with the commercial mussel program realized approximately $199,200.00 during 1994. These figures vary from year-to-year depending upon market demands, prices paid per pound (which varies depending upon the species), length of harvest season, shell size, and so forth. However, there is no doubt that the cultured pearl industry and the market it has created for raw shell continues to be a lucrative business for a large number of individuals.

Factors in Determining Distribution Records

The ability to determine and accurately plot exact early historic as well as modern range locations is plagued by several factors. Taxonomic confusion and differences of opinion as to species status as expressed in the literature by malacologists make accurate records impossible at times. To cite only one of numerous existing classification problems, the Pocketbooks *Lampsilis cardium* and *Lampsilis ovata* are presently considered distinct species. However, Arnold E. Ortmann, perhaps the most insightful unionoid systematist to have studied this group of animals, made the following observations (Ortmann, 1918:583–584): "All along its *[Lampsilis ovata]* range, and chiefly *above* Knoxville, it is accompanied by the var. *ventricosa [=L. cardium],* and intergrades with it. *L. ovata ventricosa* is found associated with the normal *L. ovata* in the larger rivers, but is less frequent there. It goes, however, beyond the upper limit of *L. ovata* in the headwaters, where it is found in its best development and as a pure race." He further noted (Ortmann, 1924a:48–49), with reference to *Lampsilis ovata ventricosa,* that its "[d]istribution [is] similar to that of the preceding form *[L. ovata],* often associated and integrading with it, and gradually taking the place of it in the headwaters. Also in Duck River, *ventricosa* gradually displaces *ovata* in the upstream direction. Intergrades between the two are frequent." In such cases as the *Lampsilis cardium–L.*

ovata complex, pinpointing exact locations where a species—or subspecies, form, or variety—occurs or did occur becomes somewhat problematical. In spite of the taxonomic problems with the Pocketbooks and other debatable "good" taxa, such as the Cumberland Bean, *Villosa trabalis,* and the Purple Bean, *Villosa perpurea,* we have accepted the 1998 status of species listed by Turgeon et al. (1988). One notable exception is our decision to follow Hoeh (1990) and his revision of the genus *Anodonta,* in which *A. imbecillis* is now placed in the genus *Utterbackia,* and *A. grandis* in the genus *Pyganodon.* The other change involves two species in the genus *Pleurobema: P. rubrum* (Rafinesque, 1820) is used in place of *P. pyramidatum* (Lea, 1840) and *P. sintoxia* (Rafinesque, 1820) for *P. coccineum* (Conrad, 1834), these species names now being used have priority (Rafinesque, 1820). Placement of dots on the maps indicating the known distribution of a mussel in Tennessee is based on published records, specimens in museum and private collections, and records obtained from aboriginal assemblages (fig. 22).

Numerous mussel species occurring in Tennessee are common and widely distributed, and it would have been impractical if not impossible to plot every known occurrence. For example, Ortmann (1918:556) commented that the Spike, *Elliptio dilatata,* was "[c]ommon, in large rivers as well as in small creeks, possibly the

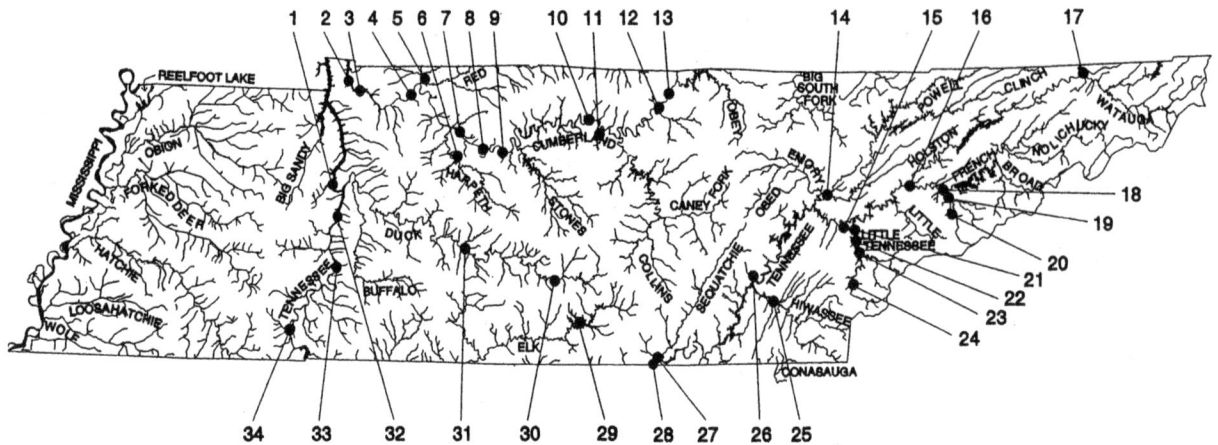

Fig. 22. Aboriginal site locations containing shell deposits that provided supplemental species distribution records.

1. Eva Site (40BN12), Benton County, Tennessee (Parmalee and Bogan, n.d.)
2. Stone Site (40SW23), Stewart County, Tennessee (Breitburg, n.d.a).
3. Hogan Site (40SW24), Stewart County, Tennessee (Parmalee and Hughes, n.d.a).
4. Meeks Site (40MT37), Montgomery County, Tennessee (Breitburg, n.d.b).
5. Dunbar Cave (40MT43), Cumberland/Red Rivers at Clarksville, Montgomery County, Tennessee (Breitburg, n.d.c).
6. Anderson Site (40WM9), Williamson County, Tennessee (Parmalee and O'Hare, 1989).
7. Late Archaic shell midden (40CH73), Cheatham County, Tennessee (Parmalee, n.d.a).
8. Cockrill Bend Site (40DV68), Cumberland River, Davidson County, Tennessee (Breitburg, n.d.d).
9. Jefferson Street Bridge (FAU-3258); The East Nashville Mounds (40DV4) and French Lick/Sulphur Dell (40DV5) sites, Davidson County, Tennessee (Walling et al., 1993).
10. Duncan Tract Site (40TR27), Trousdale County, Tennessee (Breitburg, 1983a).
11. Middle Cumberland River, Smith County, Tennessee (Parmalee et al., 1980).
12. Rockshelter midden, Cumberland River (ca. CRM 335.5), Jackson County, Tennessee (Parmalee and Klippel, n.d).
13. Penitentiary Branch (40JK73), Jackson County, Tennessee (Breitburg, 1983b).
14. Clinch River Breeder Reactor Plant Site (40RE108), Roane County, Tennessee (Parmalee and Bogan, 1986).
15. 40LD208, 40LD207 Site, Loudon County, Tennessee (Parmalee, 1990).
16. Lyons Site (40KN36) and Looney Island Site, (40KN4), Knoxville, Knox County, Tennessee (Parmalee, n.d.b).

17. Eastman Rockshelter (40SL34), Sullivan County, Tennessee (Manzano, 1986).
18. McCrosky Island Site (40SV43), Sevier County, Tennessee (Parmalee, n.d.c).
19. Dallas component, McCrosky Site (40SV9), Sevier County, Tennessee (Parmalee, n.d.d).
20. Little Pigeon River, Sevier County, Tennessee (Parmalee, 1988).
21. Citico (40MR7), Monroe County, Tennessee (Bogan, 1983).
22. Toqua Site (40MR6) (Bogan, 1987) and Tomotely Site (40MR5) (Robison 1978), Monroe County, Tennessee.
23. Martin Farm (40MR20), Monroe County, Tennessee (Bogan and Bogan, 1985).
24. Tellico River, Monroe County, Tennessee (Parmalee and Klippel, 1984).
25. Hiwassee River, Polk County, Tennessee (Parmalee and Hughes, 1994).
26. Sites along Chickamauga Reservoir, Tennessee (Parmalee et al., 1982; Bogan and Parmalee, 1977).
27. Site FN-3 Marion County, Tennessee (Bogan, 1977).
28. 40MI69, Marion County, Tennessee, and 1JA331, Jackson County, Alabama (Robison and Bogan, n.d.).
29/30. Owl Hollow Phase sites in Coffee, Franklin, and Bedford Counties, Tennessee (Robison, 1986).
31. 40MU261, Maury County, Tennessee (Parmalee and Klippel, 1986); Wiser-Stephens I Site (40CF81), Coffee County, Tennessee (Bogan, 1978).
32. Daniels Landing Site (40PY4), Tennessee River, Perry County, Tennessee (Parmalee and Hughes, n.d.b).
33. 40DR305, Decatur County, Tennessee (Parmalee and Layzer, n.d.).
34. Diamond Island (TRM196.0), Hardin County, Tennessee. (Parmalee, n.d.e).

most widely distributed species in the upper Tennessee region, so that it is hardly required to name special localities." In discussing the Mountain Creekshell, *Villosa vanuxemensis,* Ortmann (1918:581) stated that "[a]lso here it is unnecessary to give a list of localities, and it suffices to state that it is found practically over the whole upper Tennessee region." As with these common and widely distributed mussels, the total range of certain others, such as the Paper Pondshell, *Utterbackia imbecillis,* is not depicted by dots on the distri-

bution maps. Between comments in the discussion of each species in the "Tennessee Distribution" subheadings and the maps, however, the reader should be able to determine the generally known distribution of each. Finally, in preparing the distribution maps, a large format was used. The dots were purposely enlarged so they would be distinct when the maps were greatly reduced. We attempted to place the dots at specific, known locations, but because of the enlarged size, each dot may cover a mile or two of river length.

Translocation: An Answer to Species Survival?

Of the approximately 130 species of freshwater mussels known to occur or to have once occurred in Tennessee, the population status of only about one-third has been classified by federal standards as being Currently Stable (Williams et al., 1993). Of the remaining two thirds, nearly 20% are of Special Concern, 7% are Threatened, and 38% are Endangered (Williams et al., 1993). Of the 55 species listed as Endangered, 12 (all in the genus *Epioblasma*) are undoubtedly extinct based on extensive collections and other data obtained since the 1920s. If chemical and biological pollution, sedimentation, and all the other detrimental factors that contribute to the degradation of Tennessee's waterways continue unabated, many or even most of the 61 species classified as of Special Concern, Threatened, or Endangered will lose the battle for survival.

Most biologists today are of the opinion that the best approach to saving a species from extinction is to preserve the total habitat in which the animal occurs. In the case of mussels this is often impossible since much or all of a specific habitat, such as the former shoals and riffles of the Tennessee and Cumberland rivers, has been lost as the result of dam construction and other human activities or misuse. But when relict populations or individuals are still surviving under unfavorable conditions or when a stretch of river habitat supporting viable mussel populations is going to be adversely affected, relocating the population to a similar but stable and protected habitat is an option to be considered.

Such an approach has been tried in various regions of the United States, but with limited success in most instances. As one example, a total of six specimens of the Endangered Higgins Eye, *Lampsilis higginsi* (Lea, 1857), and numerous specimens of other locally common taxa were removed from bridge construction sites at Sylvan Slough (Oblad, 1980) and the Mississippi River (Nelson, 1982) in the Moline, Illinois–Davenport, Iowa, area. The mussels were relocated to what was considered comparable habitat from which they were removed; all six specimens of Higgins Eye and a large number of the other taxa were found alive one year later: the translocation was considered a success. Nelson (1982:106), however, concluded that "relocation of endangered mussels is a viable alternative but it only serves to save the individuals and does not save the habitat which is irreplaceable."

The most extensive and informative experiments in mussel translocation in the Midsouth were carried out by Sheehan et al. (1989) using 3,872 adult individuals of seven species. The naiads used in this study were transplanted to seven locations in the Clinch and North Fork Holston rivers in southwestern Virginia where they had previously been eliminated by water pollution. All seven species, common in numerous streams of the region, were obtained from viable populations in two tributary streams; the translocated populations were monitored up to four years. The researchers found that the annual rate of population decline was 10% or more, depending on the species, and identified three major factors that apparently influenced survival: changes in habitat, water quality, natural mortality. Sheehan et al. (1989:148) concluded that "[t]ranslocated mussels can persist for extended periods of time, but even under optimal conditions, populations may have to be supplemented periodically until recruitment through reproduction is sufficient to replace the loss of breeding adults due to natural mortality."

Perhaps "supplement" is the key word. If translocation is attempted with the remaining few individuals of relict populations that are already bordering on extinction, taxa such as the Catspaw *(Epioblasma obliquata obliquata)* of the Cumberland River and the Ringpink *(Obovaria retusa)* of the Cumberland and Tennessee (Kentucky Lake) rivers, it may of necessity be limited to a one-time effort. Although Ahlstedt (1991b:141) was successful in transplanting and establishing a breeding population of the Spiny Riversnail, *Io fluvialis* (Say, 1825), in sections of the North Fork Holston River in Virginia where it had been previously extirpated, he noted that "[e]stablishment of a viable reproducing population through transplants is a long-term process." Nevertheless, survival of many taxa now categorized as Endangered, Threatened, or of Special Concern may well depend on translocating individuals or populations into rivers offering optimum habitat conditions that can be reasonably protected from future degradation.

In addition to translocation of one or a few species to a localized stretch of river that offers their best chance for survival and a habitat for successful reproduction, reclamation of extensive sections degraded by pollution, silting, or damming offers another possibly viable alternative. For example, since 1991 the Tennessee Valley Authority has initiated a method of improving the quality of water releases by use of weirs and other aeration systems at 13 of 16 major dams. It

has been estimated that these efforts have aided in recovery of some 150 to 200 miles of aquatic habitat previously lost from intermittent drying of riverbeds. In addition, by the time these improvements are completed, it has been predicted that dissolved oxygen levels will increase considerably over 300 miles of river below TVA dams. Assuming these efforts are successful, the abundance and possible diversity of the fish fauna and other aquatic organisms may well increase, including the re-establishment of mussel assemblages.

However, as pointed out by Neves et al. (1997), "The piecemeal approach to conservation, focused on particular species and habitats, has not been effective." Recently new initiatives that are being considered by various state and federal agencies involve addressing habitat and biodiversity issues on a watershed or ecosystem level. Neves et al. (1997) add: "Thus an ecosystem approach to fish and wildlife conservation will enable natural resource agencies to conserve and restore the structure, function and natural assemblage of biota in ecosystems, while accommodating sustainable economic development." Whether a goal of such magnitude or complexity can be achieved in time to stave off extinction of so many mussel species whose present status is threatened or endangered remains to be seen. Nevertheless, this approach, in combination with translocation, local habitat improvement, careful monitoring and regulation of commercial exploitation of any kind, continued research to determine host fish for glochidia, and artificial propagation and maintenance of populations for eventual release will be the answer for survival of many species. For others, such as the White Wartyback and Ringpink, which are considered "functionally extinct," all such efforts at saving them may be to no avail. However, with the current impetus on all biological aspects of freshwater mussel research, including the reintroduction of large quantities of laboratory-reared juvenile mussels, critical assessment of aquatic habitats (natural and artificial), and a view toward restoring ecosystems and preserving their total biota, the majority of extant freshwater mussel species have a good chance to survive and flourish well into the twenty-first century.

ACCOUNTS
OF SPECIES

Family Margaritiferidae

Cumberlandia monodonta (Say, 1829)
Spectaclecase

RANGE MAP 1; PLATE 1

Synonymy:
Unio monodonta Say, 1829; Say, 1829:293; Say, 1830a:pl. 5
Unio soleniformis Lea, 1831; Lea, 1831:87, pl. 10, fig. 17
Margaritana soleniformis (Lea, 1831); Paetel, 1890:173
Unio monodontus Say, 1829; Say, 1834:no pagination
Alasmidonta monodonta (Say, 1829); Férussac, 1835:26
Margarita (Unio) monodontus (Say, 1829); Lea, 1836:40
Margaron (Unio) monodontus (Say, 1829); Lea, 1852c:39
Margaritana monodonta (Say, 1829); Conrad, 1853:262
Margaritania monodonta (Say, 1829); Ortmann, 1912a:233
Cumberlandia monodonta (Say, 1829); Ortmann, 1912b:13
Margaritifera (Cumberlandia) monodonta (Say, 1829); Haas, 1969a:14

Type Locality: Falls of the Ohio and Wabash River.

General Distribution: Cumberland and Tennessee River systems; Ohio and Mississippi River drainages from Minnesota and western Pennsylvania south to the Gulf of Mexico.

Tennessee Distribution: *Cumberlandia monodonta* has been collected from the Tennessee River and its headwater tributaries and in the Clinch River from Knox County to Hancock County. It has been found in the Powell River, Claiborne County. The Spectaclecase was reported from the Holston River, Knox County, upriver to Hawkins County, and in the Tennessee River and lower Little River in Knox County (Walker, 1911; Ortmann, 1918; Bates and Dennis, 1978). *Cumberlandia monodonta* has also been reported from the Sequatchie River, Marion County (Ortmann, 1925). It was locally common in the Nolichucky River (Greene, Hamblen, Cocke counties) as late as the mid-1970s (Ahlstedt, 1991b). Specimens are taken occasionally by commercial shellers in the Tennessee River (below Pickwick Landing Dam), Hardin County. The Spectaclecase has not been reported from either the Duck or Buffalo rivers, although this primitive species has been collected in the Cumberland River from Davidson County downriver to Stewart County (Wilson and Clark, 1914). Stansbery (1966) obtained the Spectaclecase from the Stones River. It appears to be absent from Mississippi River tributaries in West Tennessee.

Description: The shell is greatly elongated, arcuate, and subinflated, being thin in young individuals, becoming thick and solid in old specimens. Senile individuals may become quite large, attaining a length of 160–170 mm. Beak sculpture evident in young individuals consists of 3–4 strong, straight ridges running parallel with the growth lines. Both anterior and posterior ends of the shell are rounded with a shallow depression near the

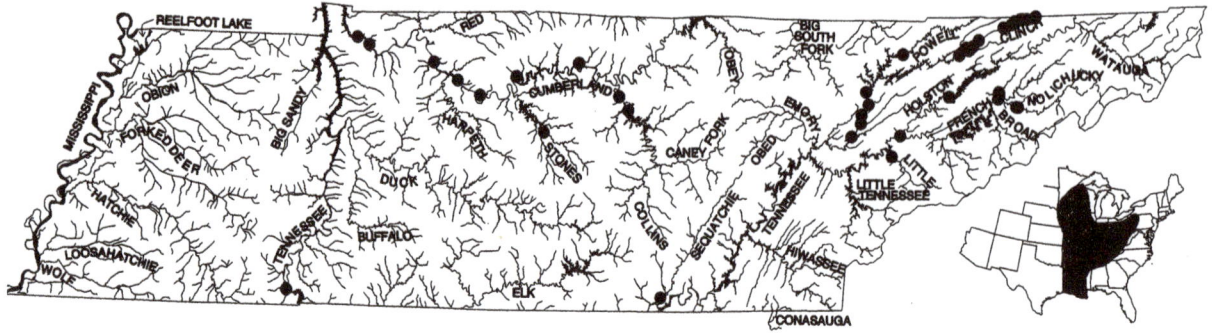

Range Map 1. *Cumberlandia monodonta* (Say, 1829), Spectaclecase.

center. The posterior ridge is low and broadly rounded. Rest lines are pronounced. The periostracum is rayless, greenish tan to brown in young shells, dull black in old shells. The pseudocardinal tooth in young individuals is small and tubercular. Two long, straight lateral teeth occur in the left valve, one in the right, but with age they merge into an elongated, slightly raised hinge line, becoming indistinct. Pseudocardinal teeth become greatly reduced with a single peglike tooth in the right valve fitting into a depression in the left. Muscle scars are large and impressed, the anterior adductor muscle scar is rough; the posterior is fairly smooth and elliptical. Nacre color is white, occasionally granular and pitted, mostly iridescent in young specimens but iridescent only posteriorly in old individuals.

Life History and Ecology: Howard (1915) and Gordon and Smith (1990) suggested that *Cumberlandia monodonta* produces two broods per season. Van der Schalie (1966) noted that the Spectaclecase may occasionally be hermaphroditic.

Cumberlandia monodonta has been collected from various types of stream bottoms, including gravel, sand, and mud, in medium-sized to large rivers. Stansbery (1966) reported finding the Spectaclecase in firm mud around roots of eel grass immediately adjacent to fast currents, and embedded in fine mud between large boulders adjacent to rapid current. It also may be found under large, flat rocks in swift current. Spectaclecase fish hosts have not been identified (Watters, 1994a). However, Knudsen and Hove (1997) have shown that some gravid females produce flat, white conglutinates with short branches, indicating that at least one species of fish may serve as a host for the glochidia.

Status: Threatened (Williams et al., 1993:10).

Plate 1. *Cumberlandia monodonta* (Say, 1829), Spectaclecase.

Family Unionidae

Actinonaias ligamentina (Lamarck, 1819)
Mucket

RANGE MAP 2; PLATE 2

Synonymy:

Unio crassus Say, 1817 non Retzius, 1778; Say, 1817:pl. 1, fig. 8

Mya crassa (Say, 1817); Eaton, 1826:217

Margarita (Unio) crassus (Say, 1817); Lea, 1836:24

Margaron (Unio) crassus (Say, 1817); Lea, 1852c:28

Unio ligamentina Lamarck, 1819; Lamarck, 1819:72

Unio ligamentinus Lamarck, 1819; Küster, 1852:23, pl. 3 fig. 3

Margaron (Unio) ligamentinus (Lamarck, 1819); Lea, 1852c:28

Lampsilis ligamentinus (Lamarck, 1819); Baker, 1898:108, pl. 16

Nephronaias ligamentina (Lamarck, 1819); Ortmann, 1912a:325

Actinonaias ligamentina (Lamarck, 1819); Ortmann, 1919:232, pl. 14, figs. 5, 6

Actinonaias ligamentina ligamentina (Lamarck, 1819); Stansbery, 1976a:47

Unio (Elliptio) fasciatus Rafinesque, 1820; Rafinesque, 1820:294

Unio (Elliptio) fasciatus var. *nigrofasciata* Rafinesque, 1820; Rafinesque, 1820:294

Unio (Elliptio) fasciatus var. *alternata* Rafinesque, 1820; Rafinesque, 1820:294

Unio (Elliptio) fasciatus var. *cuprea* Rafinesque, 1820; Rafinesque, 1820:294

Unio carinatus Barnes, 1823; Barnes, 1823:259, pl. 11, fig. 10

Mya carinata (Barnes, 1823); Eaton, 1826:220

Actinonaias carinata (Barnes, 1823); Ortmann and Walker, 1922:47

Lampsilis (Ortmanniana) carinata (Barnes, 1823); Frierson, 1927:79

Lampsilis (Ortmanniana) carinata carinata (Barnes, 1823); Murray and Leonard, 1962:108

Actinonaias carinata carinata (Barnes, 1823); Murray and Leonard, 1962:108–111

Actinonaias ligamentina carinata (Barnes, 1823); Oesch, 1984:146

Unio ellipticus Barnes, 1823—non Rafinesque, 1820; Barnes, 1823:259, pl. 19, figs. 19a–d (outline)

Mya elliptica (Barnes, 1823); Eaton, 1826:219

Unio (Obliquaria) calendis Rafinesque, 1831; Rafinesque, 1831:3

Lampsilis (Ortmanniana) carinata calendis (Rafinesque, 1831); Frierson, 1927:79

Unio ellipsarius Rafinesque, 1820; Say, 1834: Am. Conch 6: [misidentification]

Mya gravis Wood, 1856; Wood, 1856:199, pl. 1, fig. 6

Unio gravis (Wood, 1856); Paetel, 1890:154

Unio pinguis Lea, 1857; Lea, 1857c:84; Lea, 1858e:78, pl. 15, fig. 58

Margaron (Unio) pinguis (Lea, 1857); Lea, 1870:44

Lampsilis pinguis (Lea, 1857); Simpson, 1900a:540

Lampsilis (Ortmanniana) carinata pinguis (Lea, 1857); Frierson, 1927:80

Unio luteolus Lamarck, 1819; Sowerby, 1867:pl. 58, fig. 293a [misidentification]

Unio delodontus Lamarck, 1819; Sowerby, 1867:pl. 57, fig. 288 [misidentification]

Unio crassidens Lamarck, 1819; Sowerby, 1868:pl. 62, fig.
312 [misidentification]
Unio pictus Lea, 1834; Sowerby, 1868:pl. 62, fig. 313
[misidentification]
Unio venustus Lea, 1838; Sowerby, 1868:pl. 64, fig. 326
[misidentification]
Unio crassus Say, 1817; Sowerby, 1868:pl. 95, fig. 520
[misidentification]
Unio upsoni Marsh, 1887; Marsh, 1887:51
Lampsilis (Ortmanniana) carinata upsoni (Marsh, 1887);
Frierson, 1927:79
Lampsilis ligamentinus var. *gibbus* Simpson, 1900; Simpson,
1900a:540
Lampsilis ligamentinus gibba Simpson, 1900; Scammon,
1906:290–291
Nephronaias ligamentina gibba (Simpson, 1900); Utterback,
1916a:341
Lampsilis ligamentina var. *nigrescens* Simpson, 1914;
Simpson, 1914:82
Actinonaias carinata orbis Morrison, 1942; Morrison,
1942:361. Johnson; 1975:32, pl. 1, fig. 3
Lampsilis (Ortmanniana) carinata orbis (Morrison, 1942);
Haas, 1969a:461

Type Locality: Ohio River.

General Distribution: The Mucket ranges throughout
the Mississippi River Basin from western New York
to Minnesota south to Oklahoma and northern Louisiana, including the Tennessee and Cumberland River
drainages. It is also found in the St. Lawrence River
Basin in tributaries to Lake Michigan, Lake Erie, and
Lake Ontario.

Tennessee Distribution: The Mucket is widespread in
both the Cumberland and Tennessee River drainages.
It is found in the Powell, Clinch, Holston, French
Broad, Nolichucky, and Little Tennessee rivers of the
upper Tennessee River drainage. The Mucket inhabited the main channel of the Tennessee River, where
it is now rare, from Knoxville to Chattanooga and extended downstream from Savannah at least to the
mouth of the Duck River. It also occurred in the headwaters of the Elk and throughout the Duck and Buffalo River drainages. *Actinonaias ligamentina* has been
collected in the Big South Fork Cumberland River,
Obey, Caney Fork, Stones, Harpeth, and Red rivers
of the Cumberland River system. The Mucket has
been found in the main channel of the upper Cumberland River from the Tennessee/Kentucky border
downstream to Lake Barkley.

Description: Shells of the Mucket vary from oval or
oblong to rounded elliptical in outline; this is a large
species with adults reaching about 140 mm in length.
The anterior end is broadly rounded, and the posterior margin is bluntly rounded to somewhat pointed
in the middle of the posterior margin; the ventral margin is broadly rounded. The shell is stout, becoming
thicker with age but slightly compressed when young,
becoming somewhat more inflated in adult specimens.
The posterior ridge is low and rounded. Sexual dimorphism is rarely apparent, although some female shells
are slightly more inflated posteriorly. Beaks are broad
and only slightly inflated and low, barely elevated above
the hinge line; the sculpture consists of very faint, double-
looped, irregular ridges. The beak cavity is open and
fairly shallow.

The left valve has two short, striated, triangular
pseudocardinal teeth and two slightly curved short lateral teeth, somewhat distant from the pseudocardinal
teeth. The interdentum is often long and narrow. The
right valve has one erect, triangular striated pseudocardinal tooth, often with a small anterior lamellar
tooth, and one short curved striated lateral tooth. Adductor muscle scars are well impressed and smooth.

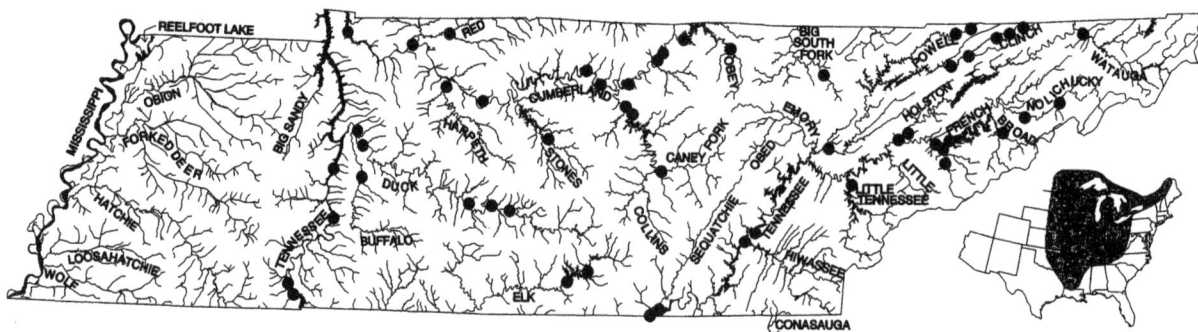

Range Map 2. *Actinonaias ligamentina* (Lamarck, 1819), Mucket.

Plate 2. *Actinonaias ligamentina* (Lamarck, 1819), Mucket.

The pallial line is well impressed. The shell surface is relatively smooth, but often marked with irregular raised concentric ridges. The periostracum is smooth and light yellowish brown to greenish, becoming a dull olive yellow to dark brown with age; the periostracum may be either variably covered with broad, dark green rays which may be interrupted, or rays may be absent. Nacre color is white, iridescent posteriorly.

Life History and Ecology: The Mucket is found usually at depths of three feet or less in sediments ranging from cobble and gravel in riffles with a strong current to quiet water in runs with coarse gravel to sand or mud bottoms. Surber (1912) reported the breeding season for the Mucket as from August to May. Watters (1994a) lists the following fish hosts for the glochidia of *Actinonaias ligamentina*: banded killifish *(Fundulus*

diaphanus), black crappie *(Pomoxis nigromaculatus)*, white crappie *(P. annularis)*, bluegill *(Lepomis macrochirus)*, green sunfish *(L. cyanellus)*, orangespotted sunfish *(L. humilis)*, largemouth bass *(Micropterus salmoides)*, smallmouth bass *(M. dolomieu)*, rockbass *(Ambloplites rupestris)*, sauger *(Stizostedion canadense)*, white bass *(Morone chrysops)*, and yellow perch *(Perca flavescens)*. Coker et al. (1921) also listed the American eel *(Anguilla rostrata)* and the tadpole madtom *(Noturus gyrinus)* as hosts for Mucket glochidia.

Status: Currently Stable (Williams et al. 1993:10).

Actinonaias pectorosa (Conrad, 1834)
Pheasantshell

RANGE MAP 3; PLATE 3

Synonymy:

Unio pectorosus Conrad, 1834; Conrad, 1834a:37, pl. 6, fig. 1 [May 1834]

Margarita (Unio) pectorosus (Conrad, 1834); Lea, 1836:23

Nephronajas pectorosa (Conrad, 1834); Ortmann, 1918:569

Actinonaias pectorosa (Conrad, 1834); Ortmann and Walker, 1922:48–49

Lampsilis (Lampsilis) pectorosa (Conrad, 1834); Frierson, 1927:69

Lampsilis (Lampsilis) pectorosa pectorosa (Conrad, 1834); Haas, 1969a:456

Unio perdix Lea, 1834; Lea, 1834:72, pl. 11, fig. 31 [August/ September 1834]

Margarita (Unio) perdix (Lea, 1834); Lea, 1836:23

Margaron (Unio) perdix (Lea, 1834); Lea, 1852c:26

Lampsilis perdix (Lea, 1834); Simpson, 1900a:542

Nephronajas perdix (Lea, 1834); Ortmann, 1912a:326

Unio biangularis Lea, 1840; Lea, 1840:288

Unio biangulatus Lea, 1842; Lea, 1842b:197, pl. 9, fig. 6

Margaron (Unio) biangulatus (Lea, 1842); Lea, 1852c:38 [Unjustified emendation of *U. biangularis* Lea, 1840]

Lampsilis biangulatus (Lea, 1842); Simpson, 1900a:533

Lampsilis (Lampsilis) pectorosa biangularis (Lea, 1842); Frierson, 1927:69

Type Locality: Elk River.

General Distribution: Tennessee and Cumberland River drainage systems.

Tennessee Distribution: This species was formerly known from the Tennessee River at and below Knoxville, and from the Clinch, Powell, North and South Forks of Holston, French Broad, Watauga, and lower Nolichucky rivers. It has also been recorded from the

Range Map 3. *Actinonaias pectorosa* (Conrad, 1834), Pheasantshell.

Little, Sequatchie, Elk, Duck, and Buffalo rivers; Cumberland River above Clarksville, Big South Fork Cumberland River, Obey, Caney Fork, Stones, and Red rivers (Starnes and Bogan, 1988). It still occurs in the Powell, Clinch, Elk, Duck, and Buffalo rivers (Ahlstedt, 1991a, b).

Description: The shell is large, elongate, elliptical, moderately solid to thick in old individuals, inflated with a well-developed posterior ridge. Mature individuals may attain a length of 140–150 mm. Beaks are rather full, extending only slightly beyond hinge line. The left valve has two equal, erect pseudocardinal teeth and two small, short lateral teeth widely separated from the pseudocardinals. The right valve has one high pseudocardinal tooth with a small tooth anterior and often a vestigial tooth posterior; the lateral tooth is short, straight, and thick. The interdentum is fairly wide and long. Muscle scars are large and impressed, and the beak cavity is moderately deep.

The surface has irregular growth lines concentrically ridged in front, is smooth across disc, but roughened on the posterior slope, usually appearing as parallel plications. Shells of males are similar to those of females; the latter becomes only slightly more rounded behind with a faint marsupial swelling. Shells of both sexes gape slightly anterior ventrally. The periostacum is a dirty yellowish green or tawny usually marked with faint, wide broken rays; senile individuals become a dark brown to black in color. The nacre is bluish white to creamy or silvery, with a wide prismatic border. Some specimens show a light salmon wash, primarily in the beak cavity area.

Life History and Ecology: The Pheasantshell occurs in sand-gravel bottoms in riffles with fast current, typically in water less than three feet in depth. Ortmann

Plate 3. *Actinonaias pectorosa* (Conrad, 1834), Pheasantshell.

(1921) reported gravid females in September with eggs and ripe glochidia in May, thus the species is bradytictic. Host fish unknown.

Status: Special Concern (Williams et al., 1993:10).

Alasmidonta atropurpurea (Rafinesque, 1831)
Cumberland Elktoe

RANGE MAP 4; PLATE 4

Synonymy:

Alasmodon atropurpureum Rafinesque, 1831; Rafinesque, 1831:5

Alasmodonta atropurpureum (Rafinesque, 1831); Lea, 1836:58

Alasmodon (Decurambis) atropurpureum Rafinesque, 1831; Simpson, 1900a:672

Alasmidonta (Decurambis) marginata Say, 1818; Ortmann and Walker, 1922:38–39 [in part]

Decurambis marginata atropurpureum (Rafinesque, 1831); Frierson, 1927:21

Alasmidonta (Decurambis) atropurpurea (Rafinesque, 1831); Clarke, 1981b:68

Type Locality: Cumberland River.

General Distribution: Cumberland River system, Kentucky and Tennessee.

Tennessee Distribution: White Oak Creek and Clear Fork River, Fentress, Morgan, and Scott counties; South Fork Cumberland River, Scott County; Collins River, Grundy County.

Description: The shell is subovate, rather uniformly thin but not fragile. Old individuals may reach a length of 100 mm. Beak sculpture, distinct only in juveniles, consists of 3–4 parallel, slightly double-looped low bars; in some specimens, additional indistinct bars may extend onto the disc. The anterior end is sharply rounded, the ventral margin nearly straight or slightly curved; the posterior margin is bluntly pointed, biangulate below and rounded above (Clarke, 1981b). The posterior ridge is distinct, somewhat of a gradual slope in

Plate 4. *Alasmidonta atropurpurea* (Rafinesque, 1831), Cumberland Elktoe.

juveniles but developing more of a sharp angle in mature individuals. Distinct but low parallel corrugations are present on the posterior slope of most individuals.

Pseudocardinal teeth are variable, but generally well developed; the single, elongated, thickened tooth in the

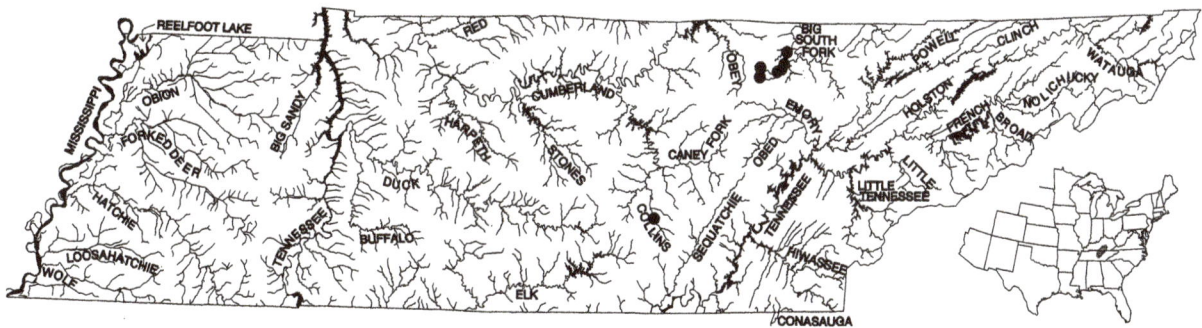

Range Map 4. *Alasmidonta atropurpurea* (Rafinesque, 1831), Cumberland Elktoe.

right valve is more pronounced than the usually two, less distinct pseudocardinal teeth in the left. Lateral teeth are hardly more than a thickening of the hinge line in each valve. Muscle scars are quite shallow; the anterior adductor muscle scar is the most deeply impressed. Periostracum color in juveniles is generally a dull yellowish tan marked with thin, often broken dark green rays (which are sometimes closely placed so as to appear as wide bands); with maturity the rays become indistinct and the shell appears as a uniform dark brown or black. The nacre is shiny, generally tinted with bluish or bluish white; some specimens exhibit a salmon or pink wash in the beak cavity area.

Life History and Ecology: Until around 1980, *A. atropurpurea* was considered a rare species and little was known about its life history. Knowledge of its breeding cycle, longevity, and specific habitat requirements is still limited. Ahlstedt (pers. comm., 1995) found it living in the cracks of bedrock ledges in the Clear Fork River. Gordon and Layzer (1993) found that of the 12 fish species tested as potential hosts for this mussel, only the northern hog sucker *(Hypentelium nigricans)* was suitable. With the discovery of locally abundant viable populations in several tributaries of the Cumberland River (Call and Parmalee, 1981), further studies of this naiad will provide more comprehensive life history data. Presently it reaches its greatest local abundance in stretches of rivers such as Clear Fork and White Oak Creek that have a slow current and contain an abundance of large cobbles and a sand and mud substrate. Typically it occurs at depths of one to two feet.

Status: Endangered (Williams et al., 1993:10).

Alasmidonta marginata Say, 1818
Elktoe
RANGE MAP 5; PLATE 5

Synonymy:
Alasmidonta marginata Say, 1818; Say, 1818b:459, 460
Margarita (Margaritana) marginata (Say, 1818); Lea, 1836:43
Margaron (Margaritana) marginata (Say, 1818); Lea, 1852c:42
Margaritana marginata (Say, 1818); Küster, 1862:297
Unio marginatus (Say, 1818); Sowerby, 1866:pl. 51, fig. 267
Alasmodonta marginata (Say, 1818); Baker, 1898:62
Alasmidonta (Decurambis) marginata Say, 1818; Ortmann, 1919:181
Margarita (Margaritana) truncata "Say" Lea, 1838; Lea, 1838b:135

Decurambis marginata (Say, 1818); Frierson, 1927:21
? *Mya rugulosa* Wood, 1828; Wood, 1828:182–183
Alasmodon (Decurambis) scriptum Rafinesque, 1831; Rafinesque, 1831:5
Alasmodon truncata ("Say" Lea, 1838); Gould, 1841:171
Alasmodonta truncata ("Say" Lea, 1838); Conrad, 1853:262
Margaritana marginata "var. *truncata*" B. H. Wright, 1898; B. H. Wright, 1898b:124
Alasmidonta truncata (B. H. Wright, 1898); Simpson, 1900a:671
Alasmidonta (Decurambis) marginata susquehannae Ortmann, 1919; Ortmann, 1919:187, pl. 12, fig. 4
Alasmidonta marginata var. *variabilis* F. C. Baker, 1928; F. C. Baker, 1928a:194, pl. 69, figs. 4–9

Type Locality: Scioto River, Chillicothe, Ohio.

General Distribution: Upper Interior drainage of the Ohio, Cumberland, and Tennessee River systems; St. Lawrence drainage from Lake Huron to the Ottawa River; Susquehanna River drainage (Burch, 1975a).

Tennessee Distribution: Small and medium-sized streams of the Tennessee and Cumberland River drainages, East and Middle Tennessee. Included in its present range are the unimpounded stretches of the Clinch and Powell rivers, the Nolichucky (Ahlstedt, 1991b), Hiwassee (Parmalee and Hughes, 1994), and the Duck and Red rivers. Individuals or small populations may become established in small tributary streams of these rivers. Formerly found in the Watauga, Elk, and Buffalo rivers and in preimpoundment main channels of the Tennessee and Cumberland rivers.

Description: The shell is elongated, somewhat rhomboid, inflated, thin when young, thick and solid when old. Mature individuals attain an average length of about 75 mm. The anterior end is sharply rounded; the ventral margin is straight and the posterior margin nearly so, and they meet in a blunt, squared point. The posterior ridge is high and sharply angled, producing a broadly truncated posterior end; there are numerous fine, radial ridges on the posterior slope extending upward toward the margin. Beaks are large, inflated, elevated, nearly centrally located on hinge line; sculpture consists of 3–4 heavy, rounded, usually doublelooped bars. The pseudocardinal teeth are thin, elongated, low; one in the right, sometimes the one in the left, has a partially divided pseudocardinal which appears as an additional interdental projection. Lateral teeth are lacking, but appear as a thickened hinge line.

Range Map 5. *Alasmidonta marginata* Say, 1818, Elktoe.

The beak cavity is moderately deep, and there is no interdentum. The periostracum is yellowish brown or greenish, usually marked with numerous greenish or blackish rays, plus many darker spots which appear in connection with the rays. The nacre is bluish white with a slight iridescence, occasionally with shades of pink.

Plate 5. *Alasmidonta marginata* Say, 1818, Elktoe.

Life History and Ecology: The Elktoe reaches its greatest abundance in small, shallow rivers with a moderately fast current such as the upper Clinch and Powell rivers in East Tennessee. A mixture of fine gravel and sand comprises the most suitable substrate for this mussel. Baker (1928a) records this species as being bradytictic in Wisconsin, the reproductive season extending from mid-July to mid-June. The white sucker *(Catostomus commersoni),* northern hog sucker *(Hypentelium nigricans),* shorthead redhorse *(Moxostoma macrolepidotum),* rockbass *(Ambloplites rupestris),* and warmouth *(Lepomis gulosus)* have been reported by Howard and Anson (1922) as hosts for the glochidia of this mussel.

Status: Special Concern (Williams et al., 1993:10).

Alasmidonta raveneliana (Lea, 1834)
Appalachian Elktoe

RANGE MAP 6; PLATE 6

Synonymy:
Margaritana raveneliana Lea, 1834; Lea, 1834:106, pl. 17, fig. 50
Alasmodonta raveneliana (Lea, 1834); Férussac, 1835:26
Margarita (Unio) raveneliana (Lea, 1834); Lea, 1836:44
Margaron (Unio) raveneliana (Lea, 1834); Lea, 1852c:42
Baphia raveneliana (Lea, 1834); H.& A. Adams, 1857:500
Strophitus ravenelianus (Lea, 1834); Conrad, 1853:263
Unio swananoensis Hanley, 1843; Hanley, 1843:211, pl. 23, fig. 39 [new name for *Unio ravenelianus* (Lea, 1834)]
Decurambis marginata atropurpureum (Rafinesque, 1831); Frierson, 1927:21 [in part]
Alasmidonta (Rugifera) raveneliana (Lea, 1834); Simpson, 1900a:671
Alasmidonta (Decurambis) marginata Say, 1818; Haas, 1969a:394 [in part]
Alasmidonta (Decurambis) raveneliana (Lea, 1834); Clarke, 1981b:72–75

Family Unionidae 57

Type Locality: French Broad and Swananoe [*sic*] rivers, North Carolina.

General Distribution: The Appalachian Elktoe is restricted to the tributaries of the Tennessee River in East Tennessee and western North Carolina.

Tennessee Distribution: This rare species has been reported only from the Holston and upper Nolichucky (Unicoi County) rivers, but historically may have occurred in the lower portions of the Nolichucky, French Broad, and Little Tennessee rivers (Clarke, 1981b).

Description: The Appalachian Elktoe is oblong, somewhat kidney-shaped in outline, moderately inflated, and thin-shelled but not fragile. The anterior margin is sharply rounded, and the posterior margin is broadly rounded, coming to a rounded point close to the posterior ventral margin. The ventral margin is nearly straight or slightly concave, and the dorsal margin is nearly straight posterior to the beaks. The posterior ridge is rounded and often double; the posterior slope is slightly concave, but not as acute as in the Elktoe *(Alasmidonta marginata)*. The beaks are moderately full, rounded, and situated on the anterior third of the shell and slightly above the hinge line. Beak sculpture consists of a few fairly heavy straight or slightly double looped bars which terminate at the posterior ridge. The Appalachian Elktoe reaches a maximum length of about 80 mm.

The left valve has a single small, compressed, pyramidal pseudocardinal tooth; the lateral tooth is reduced to a swelling or ridge and is not an articulating tooth. The left valve also has a moderate-sized interdental projection. The right valve has a single small compressed pyramidal pseudocardinal tooth and a single reduced ridge along the hinge line in place of

Plate 6. *Alasmidonta raveneliana* (Lea, 1834), Appalachian Elktoe.

the lateral tooth. Beak cavity is quite shallow. Adductor muscle scars are shallow, becoming somewhat deeper and more distinct in large mature individuals. Dorsal muscle scars are present and consist of one or two short grooves. The pallial line is absent in some individuals, complete and distinct in others. The periostracum varies from yellowish brown in younger specimens to dark brown or black in adults with faint, often interrupted

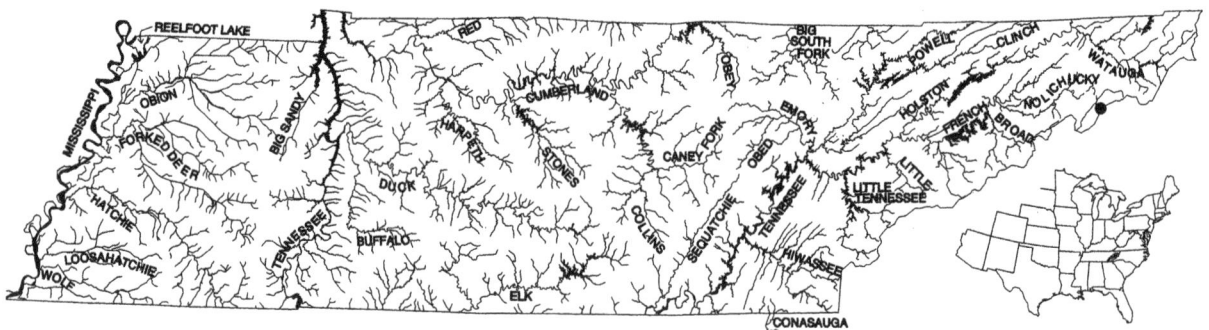

Range Map 6. *Alasmidonta raveneliana* (Lea, 1834), Appalachian Elktoe.

green rays. The surface is mostly smooth, but interrupted by concentric growth lines. There may be some fine plications on the posterior slope on juvenile specimens. Nacre color varies from a uniform bluish white to lavender, sometimes with a purplish tint, to salmon or pinkish in the center of the shell and beak cavity.

Life History and Ecology: *Alasmidonta raveneliana* may be locally common in some rivers, such as the Little Tennessee and Nolichucky in North Carolina, where it inhabits a sand and gravel substrate among cobbles and boulders and under flat rocks, usually in moderate current at depths of less than three feet. Ortmann (1921) reported that the breeding season ended in May, the species being bradytictic. Watters (1994a) lists the fish host for the glochidia of *Alasmidonta raveneliana* as the banded sculpin *(Cottus carolinae).*

Status: Endangered (Williams et al. 1993:10). The U.S. Fish and Wildlife Service has developed a recovery plan for this species (U.S. Fish and Wildlife Service, 1996).

Alasmidonta viridis (Rafinesque, 1820) Slippershell Mussel

RANGE MAP 7; PLATE 7

Synonymy:
Unio viridis Rafinesque, 1820; Rafinesque, 1820:293
Alasmidonta (Pressodonta) viridis (Rafinesque, 1820); Clarke, 1981b:17
Unio calceolus Lea, 1828; Lea, 1828:265, pl. 4, fig. 1
Unio calceola Lea, 1828; Deshayes, 1835:546
Margarita (Margaritana) calceola (Lea, 1828); Lea, 1836:45
Margaron (Margaritana) calceola (Lea, 1828); Lea, 1852c:43
Strophitus calceolus (Lea, 1828); Conrad, 1853:262
Baphia calceola (Lea, 1828); H. and A. Adams, 1857:500
Margaritana calceola (Lea, 1828); Küster, 1862:299, pl. 99, fig. 6
Alasmidonta calceola (Lea, 1828); Simpson, 1900a:668
Alasmidonta (Pressodonta) calceolus (Lea, 1828); Haas, 1969a:393
Alasmodonta truncata Conrad, 1834; Conrad, 1834a:73 [nomen nudum]
Margarita (Margaritana) deltoidea Lea, 1836; Lea, 1836:44 [nomen nudum]
Margaritana deltoidea Lea, 1838; Lea, 1838b:43, pl. 13, fig. 38
Unio deltoidea (Lea, 1838); Hanley, 1843:211, pl. 22, fig. 50
Margaron (Margaritana) deltoidea (Lea, 1838); Lea, 1852c:42
Strophitus deltoidea (Lea, 1838); Conrad 1853:263
Baphia deltoidea (Lea, 1838); H. and A. Adams, 1857:499
Alasmodonta deltoidea (Lea, 1838); Baker, 1898:63, pl. 6, fig. 2, pl. 7, fig. 4
Unio deltoideus (Lea, 1838); Sowerby 1868:pl. 76, fig. 395

Calceola angulata Swainson, 1840; Swainson, 1840:382
Margaritana minor Lea, 1845; Lea, 1845:166; Lea, 1848:82, pl. 8, fig. 26
Strophitus minor (Lea, 1845); Conrad, 1853:263
Margaron (Margaritana) minor (Lea, 1845); Lea, 1852c:42
Baphia minor (Lea, 1845); H. and A. Adams, 1857:499
Alasmidonta minor (Lea, 1845); Simpson, 1900a:668
Unio diversus Conrad, 1856; Conrad, 1856:172, outline figure
Comment: Ortmann (1925:345) remarked: *"("Unio diversus"* Conrad *(Alasmidonta diversa* Simpson, '14 p. 500), a spurious species from Shoals Creek, [Alabama] has never been recognized, and it is even doubtful, whether it is an *Alasmidonta.*)" Johnson and Baker (1973) noted that the type was lost. This taxon probably belongs here.
Alasmidonta diversa (Conrad, 1856); Simpson, 1900a:669
Alasmidonta (Sulcularia) diversa (Conrad, 1856); Frierson, 1927:19
Lasmigona (Sulcularia) diversa (Conrad, 1856); Haas, 1969a:399–400

Type Locality: Ohio River.

General Distribution: Upper Mississippi River drainage; Ohio, Cumberland, and Tennessee rivers; lower and middle sections of the St. Lawrence River systems: Lake Huron, Lake St. Clair, and Lake Erie drainages in Canada (Clarke, 1981a).

Tennessee Distribution: A species inhabiting headwater creeks and small streams in primarily East and, to a lesser extent, Middle Tennessee. Ortmann (1918) reported it from Big Creek, Hawkins County; Little Pigeon River and Boyds Creek, Sevier County; Little River and Pistol Creek, Blount County; Conasauga Creek, Monroe County—all East Tennessee records. Also previously reported from the Clinch, French Broad, Holston, Hiwassee, Sequatchie, Elk, Duck, and Buffalo rivers (Starnes and Bogan, 1988). Since 1960 *A. viridis* has been recorded for the Stones, Harpeth, and Red rivers, tributaries of the Cumberland River in Middle Tennessee.

Description: The shell is small, rhomboid, moderately solid, and slightly inflated. Maximum shell length of adult specimens seldom exceeds 55 mm. The anterior end is rounded; the posterior end is squared or obliquely truncated. The posterior ridge is high, rounded, and usually ends as a blunt point at the base of the shell. Beaks are moderately swollen, only slightly elevated; sculpture consists of 5–6 irregular, heavy loops, the first one or two are diagonal to the hinge line, and

Range Map 7. *Alasmidonta viridis* (Rafinesque, 1820), Slippershell Mussel.

the others are somewhat concentric. The surface has uneven growth lines, the rest periods appear as raised, dark-lined ridges. The pseudocardinal teeth in both valves are somewhat rudimentary, or they appear as elevated, triangular projections, usually doubled in the left valve. Lateral teeth are indistinct, being represented as a slight swelling of the hinge line. The beak cavity is relatively shallow; the interdentum is narrow or absent. The periostracum in young shells is a dull eggshell white, greenish or yellowish, with numerous wavy green rays; the colors are darker and the rays less distinct in old shells. The nacre is a dull white, and the posterior margin is slightly iridescent.

Life History and Ecology: A species of small creeks and shallow streams today, *A. viridis* once inhabited the shoals and riffles of large rivers such as the French Broad and Holston before impoundment. The Slippershell Mussel may typically be found living in a substrate composed of sand and fine gravel, although in stretches where there is a continuous current this small naiad will thrive in a mud and sand bottom among the roots of aquatic vegetation. Host fish for the glochidia include the banded sculpin *(Cottus carolinae)* and probably the mottled sculpin *(C. bairdi)* and johnny darter *(Etheostoma nigrum)* (Zale and Neves, 1982c; Watters, 1994a). Baker (1928a:186) states that the species in Wisconsin is "[p]robably bradytictic, with mature glochidia in the fall (September)." Individuals of this species have been observed spawning in January and February in the upper Little Tennessee River, North Carolina (Ahlstedt, pers. comm., 1994). Females lay on the substrate surface while spawning.

Status: Special Concern (Williams et al., 1993:10).

Plate 7. *Alasmidonta viridis* (Rafinesque, 1820), Slippershell Mussel.

Amblema plicata (Say, 1817)
Threeridge

RANGE MAP 8; PLATE 8

Synonymy:

Unio plicata Say, 1817; Say, 1817:[11–12]

Mya plicata (Say, 1817); Eaton, 1826:219

Margarita (Unio) plicatus (Say, 1817); Lea, 1836:12

Unio (Theliderma) plicata Say, 1817; Swainson, 1840:271, fig. 54e

Unio plicatus Say, 1817; Hanley, 1843:175, pl. 21, fig. 21

Margaron (Unio) plicatus (Say, 1817); Lea, 1852c:20

Plectomerus plicatus (Say, 1817); Conrad, 1853:261

Quadrula plicata (Say, 1817); Baker, 1898:pl. 25, fig. 271

Crenodonta plicata (Say, 1817); Ortmann, 1912a:246

Amblema plicata (Say, 1817); Utterback, 1915:115

Amblema (Amblema) plicata (Say, 1817); Haas, 1930:327

Unio peruvianus Lamarck, 1819; Lamarck, 1819:71

Quadrula peruviana (Lamarck, 1819); Walker, 1918a:168

Amblema peruviana (Lamarck, 1819); Utterback, 1915:115

Amblema rariplicata peruviana (Lamarck, 1819); Utterback, 1915:116

Crenodonta peruviana peruviana (Lamarck, 1819); Murray and Leonard, 1962:47

Unio rariplicata Lamarck, 1819; Lamarck, 1819:71

Quadrula rariplicata (Lamarck, 1819); Walker, 1918a:168

Amblema rariplicata (Lamarck, 1819); Baker, 1926:107

Amblema costata Rafinesque, 1820; Rafinesque, 1820:315, pl. 82, figs. 13, 14

Unio costatus (Rafinesque, 1820); Conrad, 1836:17, pl. 7

Plectomerus costatus (Rafinesque, 1820); Conrad, 1853:260

Amblema plicata costata Rafinesque, 1820; Utterback, 1915:121

Crenodonta peruviana costata (Rafinesque, 1820); Murray and Leonard, 1962:49

Unio undulatus Barnes, 1823; Hanley, 1843:175, pl. 20, fig. 26

Mya undulata (Barnes, 1823); Eaton, 1826:219

Unio undulata Barnes, 1823; Valenciennes, 1827:229, pl. 54, figs. 3, 3a, 3b

Quadrula undulata (Barnes, 1823); Baker, 1898:82, pl. 22, figs. 1,2; pl. 12, fig. 1

Crenodonta undulata (Barnes, 1823); Ortmann, 1912a:246

Unio gigantea Barnes, 1823; Lea, 1834:31, 35 [misidentification]

Unio heros Say, 1829; Küster, 1856:136, pl. 40, figs. 1, 2 [misidentification]

Unio perplicatus Conrad, 1841; Conrad, 1841:19; Conrad, 1850:276, pl. 38, fig. 2

Margaron (Unio) perplicatus (Conrad, 1841); Lea, 1852c:20

Plectomerus perplicatus (Conrad, 1841); Conrad, 1853:261

Quadrula perplicata (Conrad, 1841); Simpson, 1900a:767

Crenodonta perplicata (Conrad, 1841); Ortmann, 1912a:247

Amblema perplicata (Conrad, 1841); Utterback, 1915:118

Amblema plicata perplicata (Conrad, 1841); Frierson, 1927:61

Unio hippopoeus Lea, 1845; Lea, 1845:163; Lea, 1848:67, pl. 1, fig. 1

Margaron (Unio) hippopaeus (Lea, 1845); Lea, 1852c:21 [misspelling]

Quadrula plicata var. *hippopaea* (Lea, 1845); Simpson, 1900a:767 [misspelling]

Quadrula plicata hippopoea (Lea, 1845); Sterki, 1907:390

Quadrula undulata form *hippopoea* (Lea, 1845); Sterki, 1914:271

Amblema plicata hippopaea (Lea, 1845); Frierson, 1927:61 [misspelling]

Unio pearlensis Conrad, 1855; Conrad, 1855:256; Reeve, 1864:pl. 40, fig. 42

Unio perlensis (Conrad, 1855); Paetel, 1890:163 [misspelling]

Unio brazosensis Lea, 1868; Lea, 1868a:144; Lea, 1868c:309, pl. 48, fig. 122

Margaron (Unio) brazosensis (Lea, 1868); Lea, 1870:31

Unio lincecumii Lea, 1868; Lea, 1868a:144; Lea, 1868c:312, pl. 49, fig. 125

Margaron (Unio) lincecumii (Lea, 1868); Lea, 1870:31

Unio lincecurii Lea, 1868; Paetel, 1890:157 [misspelling]

Unio pauciplicatus Lea, 1872; Lea, 1872b:156; Lea, 1874:29, pl. 9, fig. 26

Unio quintardii Cragin, 1887; Cragin, 1887:6; Pilsbry, 1892:131, pl. 7, figs. 1–3

Quadrula perplicata var. *quintardii* (Cragin, 1887); Simpson, 1900a:768

Amblema perplicata quintardi (Cragin, 1887); Utterback, 1915:119 [misspelling]

Amblema perplicata quintardii (Cragin, 1887); Utterback, 1916c:20

Amblema plicata quintardi (Cragin, 1887); Frierson, 1927:61 [misspelling]

Unio pilsbryi Marsh, 1891; Marsh, 1891:1; Pilsbry, 1892:131, pl. 8, figs. 7, 8

Quadrula undulata var. *pilsbryi* (Marsh, 1891); Simpson, 1900a:769

Amblema plicata pilsbryi (Marsh, 1891); Frierson, 1927:61

Cokeria southalli Marshall, 1916; Marshall, 1916:133

Amblema plicata southalli (Marshall, 1916); Frierson, 1927:61

Type Locality: Lake Erie.

General Distribution: This species can be found in the Mississippi River system north to Manitoba, in the Great Lakes and St. Lawrence River system, Red River of the North, Saskatchewan River, and Lake Winnipeg, and in the Gulf drainages from the Pearl River in Mississippi and Louisiana west to eastern Texas.

Tennessee Distribution: *Amblema plicata* is a very widely distributed species in the state and has been reported from the Clinch, Emory, French Broad, Holston, Nolichucky, Powell, Little, Hiwassee, and Sequatchie rivers and the main channel of the Tennessee River in East Tennessee. The Threeridge is found in the Elk and Duck rivers and the main channel of the lower Tennessee in western Tennessee. In the Cumberland River system in Tennessee it is known from the Obey, Stones, Red, and Harpeth rivers as well as the main channel of the Cumberland River. *Amblema plicata* has been reported from the direct tributaries of the Mississippi River in western Tennessee, including the North Fork Obion, Hatchie, and Loosahatchie rivers, as well as Reelfoot Lake.

Description: The Threeridge varies from quadrate to subrhomboid in outline and is compressed to inflated and solid. The anterior end is narrowly to broadly rounded, with the ventral margin slightly curved to straight; the posterior portion of the dorsal margin may be extended into a low wing, the posterior margin being truncated. The posterior ridge is low and rounded; above it there is often a broad, shallow radial depression. Beaks vary from depressed to considerably elevated, inflated, and turned forward, with a well-marked lunule; sculpture consists of a few distinct, coarse, irregular ridges evident only in young specimens. The beak cavity is deep and fairly compressed with a wide interdentum. Maximum shell length seldom exceeds 170 mm.

The left valve has two large, roughened, elevated divergent triangular pseudocardinal teeth and two long and nearly straight lateral teeth. The right valve has a single, thick, serrated triangular pseudocardinal tooth, preceded by a small lamellar ridge or tooth and one well-developed lateral tooth. The anterior adductor muscle scar is shallow and heavily corrugated; the posterior adductor muscle scar is shallow. The pallial line is deeply impressed. The surface has uneven concentric growth lines and usually has several (1–8) weak to strongly developed rounded folds or ridges parallel with the posterior ridge. The periostracum is unrayed; color in young animals is green or yellowish green, becoming brownish or blackish with increasing age.

Plate 8. *Amblema plicata* (Say, 1817), Threeridge.

The posterior slope of the shell is with or without elongated pustules, plications, or ridges. The nacre color is white, becoming iridescent posteriorly and occasionally tinged with pinkish or purple.

This species is extremely variable in terms of size, inflation of the shell, inflation of the beaks, and devel-

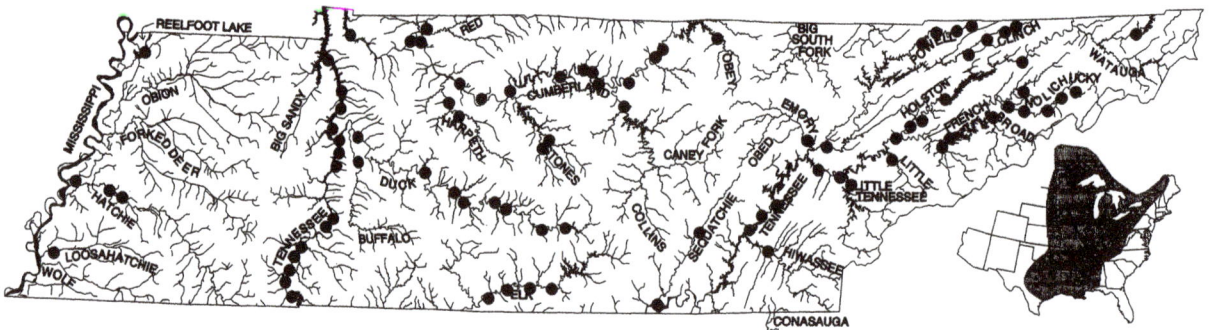

Range Map 8. *Amblema plicata* (Say, 1817), Threeridge.

opment of the plications, ranging from almost smooth to having seven or eight distinct plications running across the disk of the shell. Plications along the posterior ridge vary from strong and numerous to absent.

Life History and Ecology: *Amblema plicata* can be found in a variety of habitats, ranging from small streams to big rivers, and from locations such as lakes, rivers, and streams with little or no current to areas of very swift current. It is found in a variety of substrates including clay, mud, sand, sand mixed with gravel, and gravel. This animal is most common on bottoms composed of sand and gravel in one to three feet of water, but has been taken at depths of 30 feet. Breeding season of the Three-ridge is reported as June to early August (tachytictic) (Baker, 1928a; Frierson, 1904; Ortmann, 1912a, 1914). Watters (1994a) lists the following fish hosts for the glochidia of *Amblema plicata*: black crappie *(Pomoxis nigromaculatus)*, bluegill *(Lepomis macrochirus)*, flathead catfish *(Pylodictis olivaris)*, green sunfish *(Lepomis cyanellus)*, largemouth bass *(Micropterus salmoides)*, northern pike *(Esox lucius)*, pumpkinseed *(Lepomis gibbosus)*, rockbass *(Ambloplites rupestris)*, sauger *(Stizostedion canadense)*, shortnose gar *(Lepisosteus platostomus)*, white bass *(Morone chrysops)*, white crappie *(Pomoxis annularis)*, and the yellow perch *(Perca flavescens)*. In addition to this rather long list of fish host taxa, Weiss and Layzer (1995) add the following: goldeye *(Hiodon tergisus)*, emerald shiner *(Notropis atherinoides)*, spotfin shiner *(Cyprinella spiloptera)*, steelcolor shiner *(Cyprinella whipplei)*, streamline chub *(Erimystax dissimilis)*, black redhorse *(Moxostoma duquesnei)*, golden redhorse *(M. erythrurum)*, northern hog sucker *(Hypentelium nigricans)*, channel catfish *(Ictalurus punctatus)*, logperch *(Percina caprodes)*, and freshwater drum *(Aplodinotus grunniens)*.

Status: Currently Stable (Williams et al., 1993:10).

Anodonta suborbiculata Say, 1831
Flat Floater

RANGE MAP 9; PLATE 9
Synonymy:
Anodonta suborbiculata Say, 1831; Say, 1831b:Jan. 29; Say, 1831a:pl. 11
Margarita (Anodonta) suborbiculata (Say, 1831); Lea, 1836:52
Anodon suborbiculata (Say, 1831); Catlow and Reeve, 1845:68

Margaron (Anodonta) suborbiculata (Say, 1831); Lea, 1852c:51
Anodon suborbiculatus (Say, 1831); Sowerby, 1867:pl. 5, fig. 2
Anodonta (Nayadina) venusta De Gregorio, 1914; De Gregorio, 1914:65, pl. 12, fig. 2
Anodonta (Utterbackiana) suborbiculata Say, 1831; Frierson, 1927:17
Anodonta (Nayadina) suborbiculata Say, 1831; Haas, 1969a:360
Anodonta (Utterbackia) suborbiculata Say, 1831; Johnson 1980:113, pl. 9 fig. 5

Type Locality: Ponds near the Wabash River, Indiana.

General Distribution: This species can be found in the Mississippi River drainage from Nebraska east to Iowa, Illinois, and Ohio south to Louisiana, as well as in the Mississippi River, Wisconsin (Mathiak, 1979), and Escambia River system, Alabama (Johnson, 1969a).

Tennessee Distribution: The Flat Floater is found primarily in the larger lakes and rivers of West and Middle Tennessee. It is abundant in Reelfoot Lake and, in addition to the Tennessee River and its backwater bays (Hardin County to Stewart County), *A. suborbiculata* occurs in the Obion, Hatchie, and other Mississippi River tributaries and their drainages. It has now extended its range up the Tennessee River to the confluence of the Hiwassee River (Meigs County, East Tennessee), where it occurs in the impounded lower Hiwassee River upstream to about Mile 15.0. It is present in Fort Loudoun Lake (Knox County) and fairly common in Tellico Lake, Loudon County (Parmalee and Hughes, 1993).

Description: The shell is thin to moderately solid, nearly circular in outline, and rather compressed. The anterior and ventral margins are broadly curved; the posterior margin is nearly straight and slightly incurved; the dorsal margin is straight. The posterior ridge is low but distinct, with the dorsal section of the shell behind the beaks alate. Beaks are low and flattened; sculpture consists of 4–5 pairs of slight nodules, which appear as irregular or broken double-looped ridges. Old individuals may attain a length of 180–200 mm. Both valves are edentulous; the hinge line is only very slightly thickened. The beak cavity is wide but shallow. The periostracum is light yellow to dark brown; occasionally young shells have fine green rays; the surface is smooth and shiny, except for the darker posterior slope. The nacre is white, and most are iridescent, often tinged with salmon near the beaks.

Range Map 9. *Anodonta suborbiculata* Say, 1831, Flat Floater.

Life History and Ecology: The Flat Floater occurs most commonly in lakes, sloughs, and shallow backwater bay areas (usually not over three feet in depth) of the larger rivers, primarily those that have been impounded. Populations reach their greatest density in a substrate composed of mud. As silting continues to increase in many of the Tennessee River reservoirs, this *Anodonta* appears to be increasing in abundance and extending its range upstream in the Tennessee River and some of its larger tributaries, such as the Hiwassee and Buffalo rivers. Utterback (1915–1916a) reported the Flat Floater as bradytictic. Barnhart et al. (1995) were able to infect, under laboratory conditions, the following species of fish with glochidia of *Anodonta suborbiculata:* golden shiner *(Notemigonus crysoleucas),* warmouth *(Lepomis gulosus),* white crappie *(Pomoxis annularis),* and largemouth bass *(Micropterus salmoides).*

Status: Currently Stable (Williams et al., 1993:10).

Anodontoides ferussacianus (Lea, 1834) Cylindrical Papershell

RANGE MAP 10; PLATE 10

Synonymy:

Anodonta ferussaciana Lea, 1834; Lea, 1834:45, pl. 6, fig. 15
Margarita (Anodonta) ferussaciana (Lea, 1834); Lea, 1836:51
Anodon ferussaciana (Lea, 1834); De Kay, 1843:200, pl. 16, fig. 230
Margaron (Anodonta) ferussaciana (Lea, 1834); Lea, 1852c:50
Unio ferussacianus Lea, 1834; Lea, 1868b:255
Margaron (Unio) ferussacianus (Lea, 1834); Lea, 1870:46
Anodontoides ferussacianus (Lea, 1834); Baker, 1898:72, pl. 3, fig. 6, pl. 5, fig. 2
Anodontoides ferussacianus ferussacianus (Lea, 1834); Haas, 1969a:371
Anodonta buchanensis Lea, 1838; Lea, 1838b:47, pl. 14, fig. 43
Margarita (Anodonta) buchanensis Lea, 1836; Lea, 1836:54 [nomen nudum]
Margarita (Anodonta) buchanensis (Lea, 1838); Lea 1838a:32
Margaron (Anodonta) buchanensis (Lea, 1838); Lea, 1852c:51
Anodontoides ferussacianus var. *buchanensis* (Lea, 1838); Simpson, 1914:469
Anodontoides ferussacianus buchanensis (Lea, 1838); La Rocque, 1967:200

Plate 9. *Anodonta suborbiculata* Say, 1831, Flat Floater.

Anodonta argentea Lea, 1840; Lea, 1840:289; Lea, 1842b:223, pl. 19, fig. 41

Margaron (Anodonta) argentea (Lea, 1840); Lea, 1852c:50

Strophitus argenteus (Lea, 1840); Conrad, 1853:262

Anodonta ferruginea Lea, 1840; Lea, 1840:289; Lea, 1842b:225, pl. 19, fig. 43

Margaron (Anodonta) ferruginea (Lea, 1840); Lea, 1852c:50

Anodon plicatus Haldeman, 1842; Haldeman, 1842:201

Margaron (Anodonta) plicata (Haldeman, 1842); Lea, 1870:79

Anodonta denigrata Lea, 1852; Lea, 1852b:285, pl. 25, fig. 45

Margaron (Anodonta) denigrata (Lea, 1852); Lea, 1852c:50

Anodonta oblita Lea, 1852; Lea, 1852b:290, pl. 28, fig. 52

Margaron (Anodonta) oblita (Lea, 1852); Lea, 1852c:50

Anodonta subcylindracea Lea, 1838; Lea, 1838b:106, pl. 24, fig. 117

Margaron (Anodonta) subcylindracea (Lea, 1838); Lea, 1852c:51

Anodon subcylindracea (Lea, 1838); De Kay, 1843:200, pl. 16, fig. 229

Anodontoides ferussacianus var. *subcylindraceus* (Lea, 1838); Simpson, 1900a:660

Anodonta (Anodontoides) ferussacianus subcylindracea Lea, 1838; Frierson, 1927:17

Anodontoides ferussacianus subcylindracea (Lea, 1838); La Rocque, 1967:200

Anodontoides ferussacianus subcylindraceus (Lea, 1838); Haas, 1969a:371

Anodonta modesta Lea, 1857; Lea, 1857c:84; Lea, 1860e:364, pl. 63, fig. 189

Margaron (Anodonta) modesta (Lea, 1857); Lea, 1870:79

Anodon modestus (Lea, 1857); Sowerby, 1867:pl. 10, fig. 26

Anodontoides ferussacianus var. *modestus* (Lea, 1857); Simpson, 1900a:660

Anodontoides modestus (Lea, 1857); Simpson, 1914:470

Anodonta (Anodontoides) modesta Lea, 1857; Frierson, 1927:17

Anodontoides ferussacianus modestus (Lea, 1857); La Rocque, 1967:200

Anodontoides birgei F. C. Baker, 1923; F. C. Baker, 1923a:123–125

Anodonta (Anodontoides) birgei (F. C. Baker, 1923); Frierson, 1927:17

Anodontoides ferussacianus birgei F. C. Baker, 1923; Haas, 1969a:372

Type Locality: Ohio River, Cincinnati, Ohio.

General Distribution: Mississippi River system from Pennsylvania and Tennessee west to Minnesota and Colorado; St. Lawrence River system and the Great Lakes (Burch, 1975a). James Bay and Hudson Bay drainage from central Ontario to southeastern Saskatchewan (Clarke, 1981a).

Tennessee Distribution: Ortmann (1918:559–560) noted that "Lewis doubtfully reports this *[A. ferussacianus]* from the Tennessee below Knoxville, but this probably is a mistake. Originally, *A. oblita* Lea and *Anodonta denigrata* Lea, which are this species, have been described from Campbell County, Tenn.: this is surely in the Cumberland drainage, since Wilson and Clark report it ('14) from Clear Fork, at Jellico, Campbell Co., Tenn. . . . ."

Description: The shell is thin, elliptical, elongated, and moderately inflated; the anterior end is rounded; the posterior end is bluntly pointed, more inflated and rounded in females; the posterior ridge is rounded but usually prominent. The dorsal margin is straight; the ventral margin has a slight indentation or depression, presenting a somewhat "pinched" appearance at midpoint. Beaks are slightly to moderately swollen, flattened, and only slightly raised above hinge line; sculpture consists of 3–4 very fine, sharp, concentric ridges, bent up sharply behind. The first two ridges are usually oblique to the hinge line; the others form a distinct angle. Both valves are edentulous, but rudimentary pseudocardinal teeth are often indicated by irregular swellings on the hinge line beneath the beaks. The remaining hinge line is slightly swollen. The beak cavity is shallow. Mature individuals reach a length of about 80 mm. The periostracum is light green to a yellowish brown, lighter on the beaks, usually with numerous green rays, which are often very faint in old shells. Black concentric bands on the surface are indicative of rest periods. The nacre is bluish white, silvery, iridescent at both ends, and the beak cavity is occasionally cream or tinted with salmon.

Life History and Ecology: In Wisconsin, Mathiak (1979:26) found that the Cylindrical Papershell "favors silt areas in shallow water near shore," while Parmalee (1967) reported it as living in small, quiet streams in Illinois in a substrate of sand or fine gravel. Clear Fork, Campbell County, appears to be the only stream in Tennessee from which *A. ferussacianus* has been collected, and no habitat discussion for the species was given by Wilson and Clark (1914) who first reported it. In Canada, Clarke (1981a) states that the brooding period of this bradytictic species extends from August until May. Fuller (1974), in citing the records from Wilson and Ronald (1967) and Clarke and Berg (1959), list the sea lamprey *(Petromyzon marinus)* and mottled sculpin *(Cottus bairdi)* as host fish for this mussel. The bluntnose minnow *(Pimephales notatus),*

Range Map 10. *Anodontoides ferussacianus* (Lea, 1834), Cylindrical Papershell.

common shiner *(Luxilus cornutus)*, Iowa darter *(Etheostoma exile)*, and white sucker *(Catostomus commersoni)* have also been recorded as hosts for the glochidia of the Cylindrical Papershell (Fuller, 1978). Hove et al. (1995) identified the black crappie *(Pomoxis nigro-*

Plate 10. *Anodontoides ferussacianus* (Lea, 1834), Cylindrical Papershell.

maculatus) and spotfin shiner *(Cyprinella spiloptera)* as hosts for the glochidia of the Cylindrical Papershell. Watters (1995) added the largemouth bass *(Micropterus salmoides)* and bluegill *(Lepomis macrochirus)* as other host fish parasitized by this mussel.

Status: Currently Stable (Williams et al., 1993:10). No other specimens have been collected since the Wilson and Clark records (1914), and in all probability the species is now extirpated in Tennessee.

Arcidens confragosus (Say, 1829)
Rock Pocketbook

RANGE MAP 11; PLATE 11

Synonymy:

Alasmidonta confragosa Say, 1829; Say, 1829:339; Say, 1831:pl. 21

Unio confragosa (Say, 1829); Deshayes, 1835:552

Margarita (Margaritana) confragosa (Say, 1829); Lea, 1836:43

Unio confragosus (Say, 1829); Catlow and Reeve, 1845:57

Margaron (Margaritana) confragosa (Say, 1829); Lea, 1852c:42

Baphia confragosa (Say, 1829); H. and A. Adams, 1857:500

Margaritana confragosa (Say, 1829); Calkins, 1874:46

Arcidens confragosa (Say, 1829); Simpson, 1900a:662

Arcidens (Arcidens) confragosus (Say, 1829); Clarke, 1981b:89

Arcidens confragosa jacintoensis Strecker, 1931; Strecker, 1931:13

Type Locality: Bayou Teche, St. Mary Parish, Louisiana.

General Distribution: Mississippi River drainage, from southern Ohio west to eastern Kansas, north to Minnesota; south to Louisiana and southwest to eastern Texas and Oklahoma.

Tennessee Distribution: The Rock Pocketbook occurs in the impounded stretches of the Tennessee River, Hardin County to Stewart County (Kentucky Lake), and locally in the Cumberland River from Stewart County (Lake Barkley) upstream above Nashville to at least Smith County. In West Tennessee it occurs in Reelfoot Lake, Obion and Lake counties, and in the lower Hatchie River, Tipton County. *Arcidens confragosus* appears to thrive in large reservoirs that retain some current, and thus has expanded its range in the Cumberland and Tennessee River systems as a result of impoundment.

Description: The shell is thin to moderately solid, squarish to nearly rhomboid, and inflated. The beaks are high and full; the sculpture is strong and pronounced, consisting of irregular nodules or corrugations which form two loops. The base of the loops are swollen into knobs, which continue onto the umbones and out in two radiating rows, developing into several large, rounded ridges or folds on the posterior third of the shell. Mature individuals may attain a maximum length of about 150 mm.

The left valve has two elongated, compressed pseudocardinal teeth; the posterior one under the beak is considerably enlarged and curves upward, fitting into the hinge line in the right valve; sometimes there are two short, poorly developed lateral teeth, often only a finely serrated thickening of the hinge line. The right valve has a large, erect triangular pseudocardinal tooth; the lateral tooth is incomplete, or only a striated thickening of the hinge line. The beak cavity is shallow. The periostracum of juveniles varies from a dull brown to a more common dark green, typically becoming black with maturity. Some individuals exhibit subtle green rays on portions of the shell. The nacre is white and iridescent over much of the surface.

Plate 11. *Arcidens confragosus* (Say, 1829), Rock Pocketbook.

Life History and Ecology: The Rock Pocketbook typically inhabits medium-sized to large rivers, reaching its greatest abundance in stretches with reduced or slow current and a substrate of mud or mud and fine sand. Once established, this mussel will often flourish in a lake habitat devoid of any current; under ideal

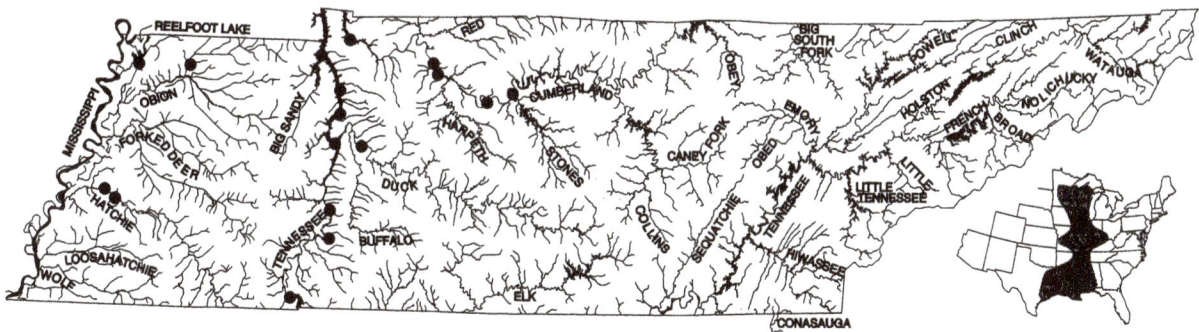

Range Map 11. *Arcidens confragosus* (Say, 1829), Rock Pocketbook.

conditions it may grow to over 170 mm in length and become greatly inflated. Baker (1928a) indicated that the species is bradytictic, breeding from September to June. Varied fish hosts for the glochidia, e.g., gizzard shad *(Dorosoma cepedianum),* rockbass *(Ambloplites rupestris),* white crappie *(Pomoxis annularis),* and freshwater drum *(Aplodinotus grunniens* [Fuller, 1978]), are a probable factor in its wide distribution and adaptability to varied habitats.

Status: Currently Stable (Williams et al, 1993:10).

Cyclonaias tuberculata (Rafinesque, 1820)
Purple Wartyback

RANGE MAP 12; PLATE 12

Synonymy:

Obliquaria (Rotundaria) tuberculata Rafinesque, 1820; Rafinesque, 1820:103
Unio tuberculosa (Rafinesque, 1820); Valenciennes, 1827:232 [misspelling]
Unio tuberculatus (Rafinesque, 1820); Conrad, 1836:43, pl. 22
Rotundaria tuberculata (Rafinesque, 1820); Agassiz, 1852:48
Quadrula (Rotundaria) tuberculata (Rafinesque, 1820); Simpson, 1900a:795
Cyclonaias tuberculata (Rafinesque, 1820); Ortmann and Walker, 1922:18
Quadrula (Cyclonaias) tuberculata (Rafinesque, 1820); Frierson, 1927:52
Cyclonaias tuberculata tuberculata (Rafinesque, 1820); Haas, 1969:606
Unio verrucosus Barnes, 1823; Barnes, 1823:123, pl. 5, fig. 6
Mya verrucosa (Barnes, 1823); Eaton, 1826:216
Margarita (Unio) verrucosus (Barnes, 1823); Lea, 1836:16
Margaron (Unio) verrucosus (Barnes, 1823); Lea, 1852c:22
Quadrula verrucosa (Barnes, 1823); F. C. Baker, 1898:85
Unio verrucosus purpureus Hildreth, 1828; Hildreth, 1828:281
Unio graniferus Lea, 1838; Lea, 1838b:69, pl. 19, fig. 60
Margarita (Unio) graniferus (Lea, 1838); Lea, 1838a:15
Margaron (Unio) graniferus (Lea, 1838); Lea, 1852c:22
Quadrula (Rotundaria) granifera (Lea, 1838); Simpson, 1900a:795
Rotundaria granifera (Lea, 1838); Grier, 1922:11–31
Cyclonaias granifera (Lea, 1838); Grier and Mueller, 1922:48
Cyclonaias tuberculata granifera (Lea, 1838); F. C. Baker, 1928a:107, pl. 51, figs. 3, 4
Quadrula (Cyclonaias) tuberculata granifera (Lea, 1838); Frierson, 1927:52
Quadrula (Rotundaria) granifera var. *pusilla* Simpson, 1900; Simpson, 1900a:795
Quadrula (Cyclonaias) tuberculata pusilla Simpson, 1900; Frierson, 1927:52
Cyclonaias tuberculata pusilla (Simpson, 1900); Haas, 1969a:606

Plethobasus cooperianus (Lea, 1834); Utterback, 1915:183, pl. 20, fig. 57 [misidentification]
Quadrula (Cyclonaias) tuberculata utterbackiana Frierson, 1927; Frierson, 1927:52
Cyclonaias tuberculata utterbackiana (Frierson, 1927); Haas, 1969a:607

Type Locality: Ohio River and its tributaries.

General Distribution: Upper Mississippi River drainage generally; Lake St. Clair drainage, and from Pennsylvania northwest to southern Michigan and northwestern Wisconsin (Mathiak, 1979), south to Iowa, Missouri, and Arkansas. In Canada, Lake Erie and the Sydenham River in southern Ontario (Clarke, 1981a). It occurs throughout the Tennessee and Cumberland river drainages.

Tennessee Distribution: The Purple Wartyback occurs in the main channels of both the Tennessee and Cumberland rivers in Middle and East Tennessee, as well as in most of their major tributaries. Although an inhabitant of the French Broad, Sequatchie, and Obey rivers prior to 1960, it is apparently now extirpated from these systems (Starnes and Bogan, 1988).

Description: The shell is compressed (among specimens in streams) to slightly inflated (among specimens in large rivers), solid, subquadrate to circular in outline. Mature individuals may attain a length of 130 mm. Beaks are depressed (in stream forms) to moderately swollen, rather prominent (in large river forms); sculpture consists of numerous fine, irregular, broken ridges, each made up of alternating zigzag bars and loops which continue down the valve until the nodules appear; there is often a shallow, narrow furrow or depression centrally on the umbone area. There is a winglike depression above the low dorsal ridge, more extensive in stream forms. Center and posterior surfaces are covered with rounded or elongated tubercles that parallel the growth lines; tubercles are more numerous in stream forms. The anterior end and ventral margins are broadly rounded; the posterior end is occasionally rounded, usually squarely or obliquely truncated.

The left valve has two narrow but heavy, divergent pseudocardinal teeth, deeply serrated between, sometimes almost meeting anteriorly; the two lateral teeth are short, heavy, and slightly curved. The right valve has a single, massive, ragged, slightly triangular-shaped

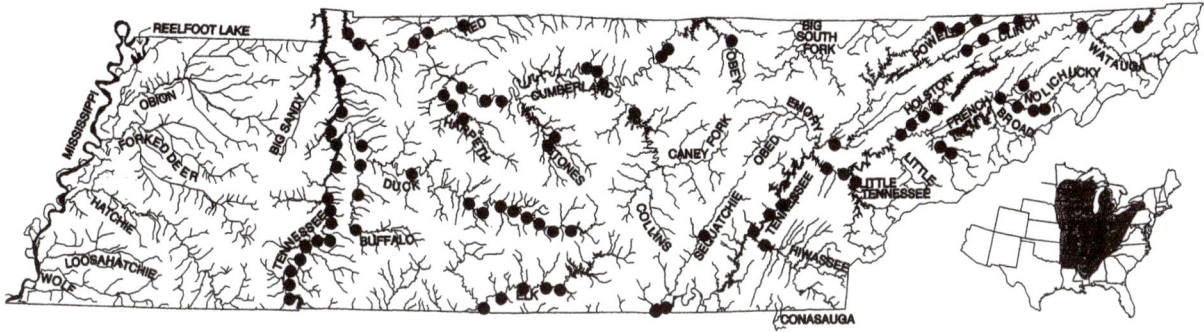

Range Map 12. *Cyclonaias tuberculata* (Rafinesque, 1820), Purple Wartyback.

pseudocardinal tooth, usually with a small tubercular tooth on either side. The beak cavity is compressed and deep; the interdentum is wide and flat. The periostracum is a dull yellowish brown to dark brown; young shells occasionally have traces of greenish rays. The nacre varies from a uniform deep purple, often with a coppery tinge along the margin, to light purple with the center (within the pallial line) nearly white; the posterior margin is iridescent.

Life History and Ecology: Two rather distinct forms of *Cyclonaias tuberculata* occur in Tennessee; the shell of one, which is found in the Duck and Harpeth rivers and other medium-sized to small streams, is characterized by its massive size, squarish outline, compressed valves, and numerous tubercles. The other, typical of the main channel stretches of the Cumberland and Tennessee rivers, is usually smaller, more rounded or oval, and inflated. This mussel typically inhabits a gravel/mud bottom, usually in areas of current; the former may be found in water less than two feet in depth, while the latter can occur at depths up to 20 feet. The species is tachytictic, and the reproductive period lasts from June to August. Initial studies by Hove et al. (1994) showed the channel catfish *(Ictalurus punctatus)* and yellow bullhead *(Ameiurus natalis)* as suitable hosts for glochidia of this mussel. Additional testing of the yellow bullhead confirmed it as a valid host species (Hove et al. 1994b). Subsequently, Hove (1997) has reconfirmed the channel catfish as a host species and has added two new host species for the glochidia of the Purple Wartyback: the flathead catfish *(Pylodictis olivaris)* and the black bullhead *(Ameiurus melas)*.

Status: Special Concern (Williams et al., 1993:11).

Plate 12. *Cyclonaias tuberculata* (Rafinesque, 1820), Purple Wartyback.

Cyprogenia stegaria (Rafinesque, 1820)
Fanshell

RANGE MAP 13; PLATE 13

Synonymy:

Obovaria stegaria Rafinesque, 1820; Rafinesque, 1820:312, pl. 82, figs. 4, 5

Unio stegarius (Rafinesque, 1820); Say, 1834:no pagination

Cyprogenia stegaria (Rafinesque, 1820); Ortmann, 1919:218, pl. 13, fig. 5

Unio verrucosus albus Hildreth, 1828; Hildreth, 1828:281

Unio irroratus Lea, 1828; Lea, 1828:269, pl. 5, fig. 5

Unio irrorata Lea, 1828; Deshayes, 1830:579

Margarita (Unio) irroratus (Lea, 1828); Lea, 1836:16

Theliderma irrorata (Lea, 1828); Swainson, 1840:271

Margaron (Unio) irroratus (Lea, 1828); Lea, 1852c:22

Unio stegarius var. *irroratus* (Lea, 1828); Paetel, 1890:168

Cyprogenia irrorata (Lea, 1828); Simpson, 1900a:610

Cyprogenia irrorata var. *pusilla* Simpson, 1900; Simpson, 1900a:610

Cyprogenia irrorata pusilla Simpson, 1900; Frierson, 1927:66

Type Locality: Ohio.

General Distribution: Ohio, Cumberland, and Tennessee River systems (Simpson, 1914).

Tennessee Distribution: Although generally considered a big river species, the Fanshell may be found inhabiting shallow, unimpounded upper stretches of the Clinch River as well as portions of the impounded Tennessee and Cumberland rivers. Starnes and Bogan (1988) record *Cyprogenia stegaria* as having also once occurred in the Holston, Powell, Little Tennessee, and Duck rivers. Shells of this species have been recovered from aboriginal middens along the French Broad River and the Little Pigeon River, both sites in Sevier County (Parmalee, 1988).

Description: The shell is inflated, thick, solid, somewhat rounded or subcircular. The ventral margin is long, broadly rounded; the posterior margin is bluntly angled or slightly truncated. The posterior ridge is well developed, producing a sharp angle behind the umbones, becoming rounded toward the ventral-posterior margin; sometimes the posterior ridge has a shallow sulcus between the ridge and dorsal margin (on the flattened posterior slope). The posterior two-thirds of the shell is covered with numerous rounded pustules and irregular knobs; those on the center of the valve often appear in rows, being located on the distinct, elevated concentric ridges indicative of growth periods. Beaks are elevated and full; sculpture consists of a few indistinct, feeble ridges. Maximum length of individuals is about 70 mm.

The left valve has two low, thick, divergent, roughened pseudocardinal teeth; the two lateral teeth are slightly curved, short, and heavy (the inner tooth is much broader than the outer) and roughened. The right valve has a low, triangular, deeply serrated pseudocardinal tooth; the lateral tooth is short, low, finely striated with an indication of a second inner, flattened tooth. The interdentum is wide and flat; the beak cavity is very shallow. The periostracum is a pale greenish yellow, covered with a pattern of darker green flecks or dots which may appear as rays; knobs and the ventral area are often lighter with less green.

Life History and Ecology: Viable populations of *Cyprogenia stegaria* in Tennessee are restricted primarily to the unimpounded stretches of the Clinch River in Claiborne and Hancock Counties (Ahlstedt, 1991a). Here they may be found, usually at depths less than three feet, on a substrate of coarse sand and gravel. Ortmann (1919:220) noted that "[i]n the Tennessee-drainage I found it fre-

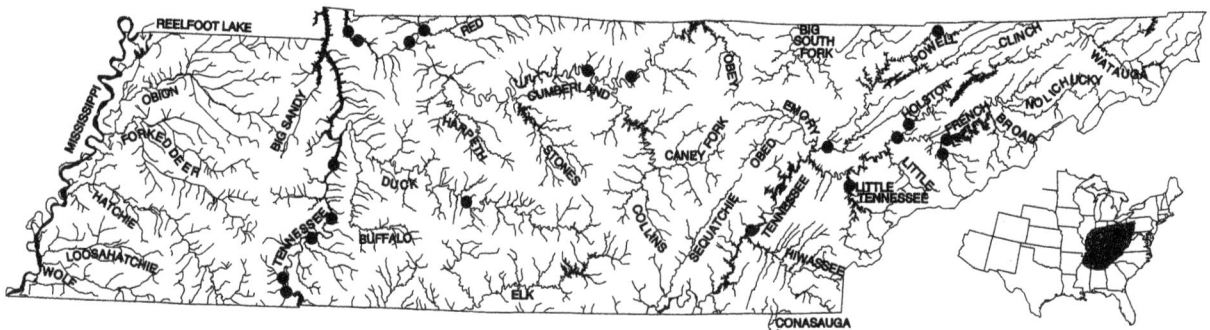

Range Map 13. *Cyprogenia stegaria* (Rafinesque, 1820), Fanshell.

Plate 13. *Cyprogenia stegaria* (Rafinesque, 1820), Fanshell.

quently in firmly packed gravel, in strongly flowing water, in rivers of medium size (Clinch, Holston)." A few relic individuals are still taken by shellers in sections of the Cumberland River (Smith County) and, even more rarely, in the Tennessee River (Hardin County). Shells of these individuals become extremely thickened and inflated. Reproduction by this species may be occurring in the Tennessee River below Pickwick Landing Dam. Ortmann (1919) recorded the Fanshell as bradytictic. Host fish for the glochidia unknown. Fuller (1974), however, recorded the goldfish (*Carassius auratus*) as a host fish for *C. irrorata* (=*C. stegaria*) based on the work of Chamberlain (1934), but Chamberlain's study dealt with the Western Fanshell, *Cyprogenia aberti* (Conrad, 1850).

Status: Endangered (Williams et al., 1993:11). U.S. Fish and Wildlife Service has developed a recovery plan for this species (U.S. Fish and Wildlife Service, 1991a).

Dromus dromas (Lea, 1834)
Dromedary Pearlymussel
RANGE MAP 14; PLATE 14

Synonymy:
Unio dromas Lea, 1834; Lea, 1834:70, pl. 10, fig. 29
Unio caperatus Lea, 1845; Lea, 1845:164; Lea, 1848:75, pl. 5, fig. 14
Margaron (Unio) caperatus (Lea, 1845); Lea, 1852c:22
Dromus caperatus (Lea, 1845); Simpson, 1900a:615
Dromus dromas caperatus (Lea, 1845); Ortmann, 1918:566
Unio abacoides Haldeman, 1846; Haldeman, 1846:75
Margarita (Unio) dromas (Lea, 1834); Lea, 1836a:16
Cyprogenia dromas (Lea, 1834); Agassiz, 1852:48
Margaron (Unio) dromas (Lea, 1834); Lea, 1852c:23
Dromus dromas (Lea, 1834); Simpson, 1900a:615
Dromus dromas (Lea, 1834); Hinkley, 1906:53
Conchodromus dromas (Lea, 1834); Haas, 1930:317
Comment: Baker (1964b) observed that *Dromus* Simpson, 1900 is not a junior homonym, is an available nomen and *Conchodromus* Haas, 1930 is an unnecessary substitute and is a junior synonym of *Dromus*.
Dromus dromas form *dromas* (Lea, 1834); Warren 1975:97

Type Locality: Harpeth River; Cumberland River, Nashville, Tennessee.

General Distribution: Tennessee and Cumberland River systems.

Tennessee Distribution: The Dromedary Pearlymussel was found in the Clinch River from about Anderson County upriver to Claiborne County and in the Holston River from Knox County to about as far as Grainger County. The Powell River contained only the form *D. d. caperatus* from its mouth upstream to Hancock County (Ortmann, 1918; Pilsbry and Rhoads, 1896). *Dromus dromas* has been identified from archaeological deposits along the Little Tennessee River, Monroe County, and from the Hiwassee River, Bradley and McMinn counties (Parmalee and Hughes, 1994). This mussel has been reported from the lower Elk River in Giles County, but not from the Duck or Buffalo rivers (Ortmann, 1925; Isom et al., 1973). It was described from the Harpeth River and reported from the main channel of the Cumberland River from Clay County downriver at least to Davidson County (Wilson and Clark, 1914).

Description: Shells of *Dromus dromas,* which may attain a length of 90–100 mm at maturity, are rounded to subtriangular or subelliptical in outline with rather full, high beaks set well forward. Valves are solid,

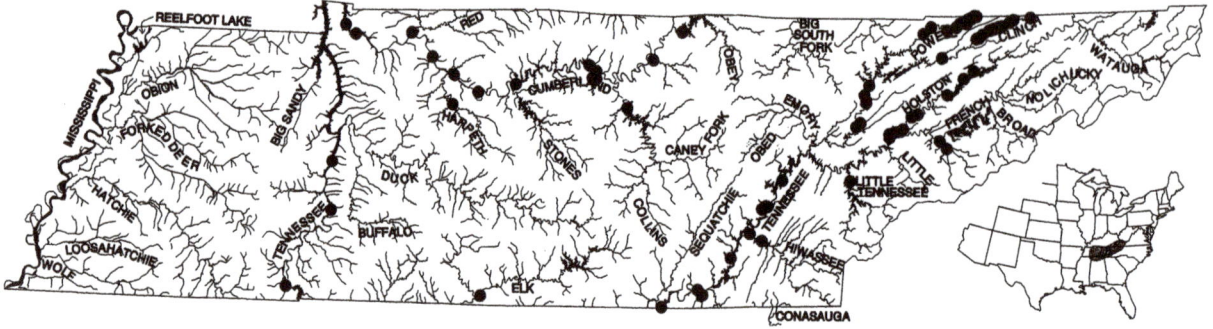

Range Map 14. *Dromus dromas* (Lea, 1834), Dromedary Pearlymussel.

somewhat inequilateral and inflated. Beak sculpture consists of fine ridges running parallel with the growth lines with the furrows interrupted at the posterior ridge. The surface is marked by irregular concentric growth lines with a decided strong concentric ridge or hump at the point of earlier growth of the shell. This ridge is a hump or irregular knob on the median line of the shell; sometimes there is a curved row of smaller knobs along the midline running from the umbo area to the ventral margin. Posterior dorsal and ventral margins are rounded, while the anterior margin is somewhat projected and rounded. The hinge ligament is rather short and slightly curved.

The left valve has two low, rough pseudocardinals, the posterior-dorsal tooth being the larger. There are two straight, relatively short lateral teeth, separated from the pseudocardinals by a broad, flat interdentum. The right valve has three pseudocardinal teeth, the middle tooth being much larger than the other two. This species has a single lateral tooth, but some specimens may develop a vestigial lower tooth. Beak cavities are deep and compressed. Muscle scars are small and impressed with the anterior adductor muscle scar typically having a characteristic rough surface. The pallial line is distinct from the margin anteriorly, becoming faint posteriorly. The periostracum is a shiny tawny or yellowish green in color, usually with two sets of broken green rays. There are several contiguous narrow rays of dots or broken lines, alternating with wide rays made of irregular blotches and flecks of green; alternating rays are particularly pronounced anteriorly, but the ray pattern is highly variable. The nacre is white, cream, pink, or salmon.

The above description is for typical mainstream *Dromus dromas*. The headwaters form is similar, but more compressed and somewhat less solid. The pos-

terior ridge is sharp above but fades out below. The "hump" of *D. dromas* is usually lacking in *D. d. caperatus,* but the shell still has concentric growth lines, those representing early growth being especially strong, often with several low knobs along the median line. Coloration of the epidermis tends to be fainter and less distinct in *D. d. caperatus* (Simpson, 1914); the nacre is more often pinkish to deep purple.

Plate 14. *Dromus dromas* (Lea, 1834), Dromedary Pearlymussel.

Life History and Ecology: The Dromedary Pearlymussel was recorded from the Clinch River, Anderson County, in a gravel and sand substrate in about three feet of water (Hickman, 1937). It has been collected in the upper Powell and Clinch rivers in similar habitat along shoals and riffles. The big river form *(Dromus dromas)* was apparently an inhabitant of shoals and riffles such as those once occurring at Muscle Shoals, Alabama. *Dromus d. caperatus* inhabited smaller rivers, usually in the headwater streams (Ortmann, 1925). The big river form is represented by relict populations (Parmalee et al., 1980) in the middle Cumberland River, Smith County, and the Tennessee River in Meigs County (Ahlstedt and McDonough, 1994). The small river form exists only in the upper Powell and Clinch rivers, Tennessee and Virginia. Based on samples of shell recovered from aboriginal sites, prehistorically *D. dromas* was one of the most abundant species in the Tennessee River.

This mussel is bradytictic with gravid females having been observed during September. As yet fish hosts for glochidia of the Dromedary Pearlymussel have not been identified, but Neves (1991), based on personal communication with B. Yeager, suggests the gilt darter *(Percina evides)* as one possibility.

Status: Endangered (Williams et al., 1993:11). U.S. Fish and Wildlife Service has developed a recovery plan for this species (U.S. Fish and Wildlife Service, 1983b) and has created a watershed implementation schedule for the recovery plan (U.S. Fish and Wildlife Service, 1989b).

Ellipsaria lineolata (Rafinesque, 1820)
Butterfly
RANGE MAP 15; PLATE 15

Synonymy:
Obliquaria (Plagiola) depressa Rafinesque, 1820; Rafinesque, 1820:302, pl. 31, figs. 5–7.
Comment: Conrad (1834a) as first revisor choose *lineolata* as the senior taxon and placed the other Rafinesque taxa in synonymy. Baker (1964a) pointed out *Ellipsaria* Rafinesque, 1820 is the correct generic placement of *Obliquaria lineolata* with the type species by absolute tautonymy *Obliquaria ellipsaria*.
Obliquaria (Ellipsaria) ellipsaria Rafinesque, 1820; Rafinesque, 1820:303
Obliquaria (Plagiola) lineolata Rafinesque, 1820; Rafinesque, 1820:303
Unio lineolata (Rafinesque, 1820); Say, 1834:no pagination

Unio lineolatus (Rafinesque, 1820); Conrad, 1834a:70
Plagiola lineolata (Rafinesque, 1820); Agassiz, 1852:48
Ellipsaria lineolata (Rafinesque, 1820); H. B. Baker, 1964a:141
Unio securis Lea, 1829; Lea, 1829:437, pl. 11, fig. 17
Margarita (Unio) securis (Lea, 1829); Lea, 1836:19
Margaron (Unio) securis (Lea, 1829); Lea, 1852c:24
Crenodonta securis (Lea, 1829); Herrmannsen, 1852:38
Plagiola securis (Lea, 1829); Smith, 1899:291, pl. 80
Plagiolopsis securis (Lea, 1829); Thiele, 1934:834

Type Locality: Falls of the Ohio River, Louisville, Kentucky.

General Distribution: Mississippi River drainage from western Pennsylvania west to Minnesota, south to eastern Iowa, Kansas, Arkansas, and Oklahoma; in the Southeast in the Tombigbee and Alabama River systems (Simpson, 1900a).

Tennessee Distribution: Widespread in the large and medium-sized rivers in the state including the Tennessee, Cumberland, Elk, and Stones; prior to 1960, *Ellipsaria lineolata* was known to have occurred in the Clinch and Duck rivers (Starnes and Bogan, 1988).

Description: The shell is solid, heavy, and subtriangular. Shells of adult females are generally smaller in relation to the length and height of those of males, as well as being more inflated and swollen posteriorly. Simpson (1914) noted that the females show a distinct gape in front and behind. Male shells are more compressed and flattened in the disc area. Adult males reach a maximum length of about 110 mm; females are about 70 mm. The anterior end is broadly rounded; the ventral is margin slightly curved; the posterior end is bluntly pointed (in males) to broadly rounded (in females). The posterior ridge is distinct and sharply angled along the anterior-dorsal half of the shell, becoming broadly rounded toward the posterior end.

The left valve has two massive, triangular, deeply serrated pseudocardinal teeth; the two lateral teeth are short, low, widely separated. The right valve has one heavy, deeply serrated pseudocardinal tooth, sometimes a small laminated pseudocardinal tooth is present dorsally and anteriorly; there is one low lateral tooth. Beak cavity is shallow; the interdentum is short and wide; muscle scars are deeply impressed. The periostracum is a dull straw yellow or greenish; most specimens have widely spaced fine dark green rays; the most pronounced rays are often broken and appear as rows

Range Map 15. *Ellipsaria lineolata* (Rafinesque, 1820), Butterfly.

of dots or small broken chevrons. Females are occasionally rayless or often less heavily patterned than the males. The surface is roughened by close, concentric ridges. The nacre is silvery white.

Life History and Ecology: The Butterfly reaches its greatest abundance in large rivers in stretches with pronounced current and a substrate of coarse sand and gravel. This species appears to have been successful in adapting to impoundment conditions in the Cumberland and Tennessee rivers where it is locally common and where it may be found at depths of up to 20 feet. Morrison (1942:361), in his study of the mollusks from aboriginal mounds in the Pickwick Basin in northern Alabama (Tennessee River), stated that "[t]his showy, deep-water species *[Plagiola (=Ellipsaria) lineolata]* was found only as single scattered individuals in two of the mounds. . . ." Parmalee et al. (1980) did not recover valves of this mussel in aboriginal deposits along the Cumberland River, Smith County. These data tend to point to the fact that the Butterfly has extended its range upstream in both the Tennessee and Cumberland rivers in recent historic times. Ortmann (1919) recorded bradytictic females being gravid from August to June or July. Fish hosts for the glochidia include the green sunfish *(Lepomis cyanellus),* sauger *(Stizostedion canadense),* and freshwater drum *(Aplodinotus grunniens)* (Fuller, 1978).

Status: Special Concern (Williams et al., 1993:11).

Elliptio arca (Conrad, 1834)
Alabama Spike
RANGE MAP 16; PLATE 16
Synonymy:
Unio arcus Conrad, 1834; Conrad, 1834b:340, pl. 50, fig. 8
Margarita (Unio) arcus (Conrad, 1834); Lea, 1836:38
Margaron (Unio) arcus (Conrad, 1834); Lea, 1852c:38
Unio (Elliptio) gibbosus var. *arcus* Conrad, 1834; Simpson, 1900a:704
Elliptio (Eurynia) dilatata arcus (Conrad, 1834); Frierson, 1927:34

Plate 15. *Ellipsaria lineolata* (Rafinesque, 1820), Butterfly.

Elliptio (Eurynia) nasutus arcus (Conrad, 1834); Haas, 1969a:245

Elliptio arcus (Conrad, 1834); Stansbery, 1976a:47

Unio lazarus Lea, 1852; Sowerby, 1868:pl. 68, fig. 348 [misidentification]

Unio luridus Lea, 1852; Lea, 1852a:251 [nomen nudum]

Unio luridus Lea, 1852; Lea, 1852b:273, pl. 20, fig. 29

Margaron (Unio) luridus (Lea, 1852); Lea, 1852c:30

Unio (Elliptio) luridus Lea, 1852; Simpson, 1900a:705

Unio rufus Lea, 1857; Lea 1857f:171; Lea, 1858:85, pl. 17, fig. 65

Margaron (Unio) rufus (Lea, 1857); Lea, 1870:61

Unio subgibbosus Lea, 1857; Lea, 1857f:169; Lea, 1858e:53, pl. 6, fig. 36

Margaron (Unio) subgibbosus (Lea, 1857); Lea, 1870:61

Unio (Elliptio) gibbosus var. *subgibbosus* Lea, 1857; Simpson, 1900a:704

Type Locality: Alabama River.

General Distribution: Mobile Bay basin west to the Pearl River in Mississippi and the Amite River in Louisiana.

Tennessee Distribution: Conasauga River, Bradley and Polk counties.

Description: The shell is fairly thin but solid, elliptical, and compressed (in juveniles) to moderately inflated. Adult individuals from the Conasauga River in Tennessee and northern Georgia reach a maximum length of about 75 mm. The anterior end is bluntly rounded, the dorsal and ventral margins are straight, and the posterior end is bluntly pointed or biangulated at or near the base of the shell. Beaks are fairly low, compressed, and not projected beyond the hinge line; sculpture consists of 3–4 strong longitudinal bars which are sometimes double-looped. The posterior ridge may be either elevated and fairly sharply angled, low and rounded, or entirely wanting.

Plate 16. *Elliptio arca* (Conrad, 1834) Alabama Spike

The left valve has two stumpy, triangular low pseudocardinal teeth, usually widely divergent and deeply serrated between. The two lateral teeth are long and straight; the inner tooth is thinner and more elevated than the dorsal tooth. The right valve has a single low, serrated triangular pseudocardinal tooth. The single lateral tooth is straight, thin, and elevated. Adductor muscle scars and the pallial line are deeply impressed

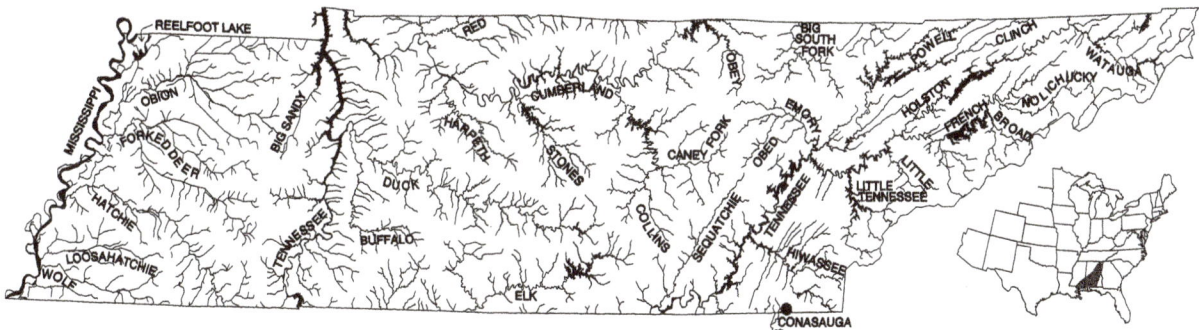

Range Map 16. *Elliptio arca* (Conrad, 1834) Alabama Spike

in mature specimens. There is no beak cavity; the interdentum is narrow or wanting. The surface is roughened by concentric growth and rest lines. The periostracum is rayless; color varies from a dull olive yellow in juveniles to dark brown or blackish in old adults. The nacre color is a dull bluish white, often with a salmon wash.

Life History and Ecology: The Alabama Spike has been found inhabiting shallow riffles as well as slow moving stretches of the Conasauga River in a substrate composed of fine gravel, sand, and silt, sometimes in areas mixed with mud. The breeding season and host fish for the glochidia of *Elliptio arca* are unknown, although they may be similar to those of the Spike *(Elliptio dilatata)* to which it appears closely related.

Status: Threatened (Williams et al., 1993:11). Since the mid-1980s populations of this once common mussel have shown drastic declines locally.

Elliptio arctata (Conrad, 1834)
Delicate Spike
RANGE MAP 17; PLATE 17
Synonymy:
Unio arctatus Conrad, 1834; Conrad, 1834b:340, pl. 1, fig. 9
Margarita (Unio) arctatus (Conrad, 1834); Lea, 1836:38
Margaron (Unio) arctatus (Conrad, 1834); Lea, 1852c:38
Elliptio (Elliptio) arctatus (Conrad, 1834); Frierson, 1927:28
Unio strigosus Lea, 1840; Lea, 1840:287; Lea, 1842b:198, pl. 9, fig. 9
Margaron (Unio) strigosus (Lea, 1840); Lea, 1852c:36
Unio tortivus Lea, 1840; Lea, 1840:287; Lea, 1842b:204, pl. 12, fig. 17
Margaron (Unio) tortivus (Lea, 1840); Lea, 1852c:33
Unio arctatus var. *tortivus* Lea, 1840; Simpson, 1900a:718
Elliptio (Elliptio) tortivus (Lea, 1840); Frierson, 1927:28
Unio purpurellus Lea, 1857; Lea, 1857f:171; Lea, 1859f:198, pl. 23, fig. 81
Margaron (Unio) purpurellus (Lea, 1857); Lea, 1870:53
Unio tetricus Lea, 1857; Lea, 1857f:170; Lea, 1859f:195, pl. 22, fig. 78
Margaron (Unio) tetricus (Lea, 1857); Lea, 1870:53
Unio viridans Lea, 1859; Lea, 1859d:170; Lea, 1860e:337, pl. 54, fig. 162
Margaron (Unio) viridans (Lea, 1859); Lea, 1870:52
Unio merceri Lea, 1862; Lea, 1862a:169; Lea, 1862c:209, pl. 31, fig. 278
Margaron (Unio) merceri (Lea, 1862); Lea, 1870:61

Type Locality: Black Warrior and Alabama rivers.

General Distribution: Alabama–Coosa River system; Escambia River system, east to the Apalachicola River system (but discontinuous); west to the Pearl River system in Mississippi. Johnson (1970) includes the upper Savannah River system, South Carolina; Catawba River and lower Cape Fear River system, North Carolina, but this range is doubtful.

Tennessee Distribution: Conasauga River, Bradley and Polk counties.

Description: The shell is small, mature specimens attaining a length of about 80 mm; the shell is subsolid, compressed to slightly inflated, elongated, and elliptical. The anterior end is narrowly rounded, the ventral margin is straight, and the posterior end is somewhat truncated; the posterior ridge is low and rounded. Beaks are low, not extending beyond the hinge line; beak sculpture consists of corrugated ridges which nearly follow the growth lines (Simpson, 1914). Pseudocardinal teeth, one in the right valve, two in the left, are triangular, low, and stumpy. There is a single lateral tooth in the right valve and two in the left which are elevated, thin, long, and straight to slightly curved. The beak cavity is very shallow to absent; the interdentum is narrow to wanting. The periostracum is fairly smooth, but some old individuals are roughened with irregular growth lines; color varies from a dull yellowish green in juveniles to dark brown or blackish in old adults. Rays are lacking in Tennessee specimens. The nacre is a dull bluish white, often with a pale salmon wash in the beak area.

Life History and Ecology: At normal water level, the Delicate Spike may be found at depths of less than three feet in a substrate composed of coarse sand and gravel. In the Conasauga River, this species becomes locally common in this type of substrate when in association with cobbles and a strong current. Hurd (1974) noted that individuals of this species may be found together and packed vertically under large rocks in swift water. The reproductive season and fish host for the glochidia of *Elliptio arctata* are unknown.

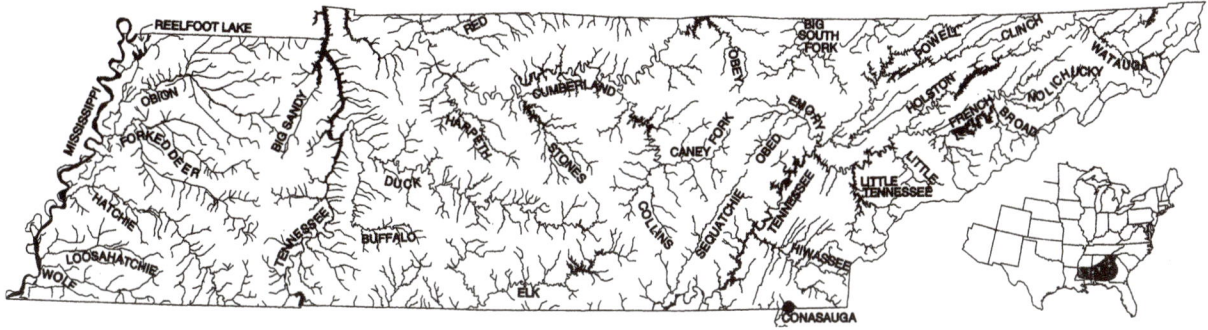

Range Map 17. *Elliptio arctata* (Conrad, 1834), Delicate Spike.

Plate 17. *Elliptio arctata* (Conrad, 1834), Delicate Spike.

Status: Special Concern (Williams et al., 1993:11). Since the study by Hurd (1974), the delicate spike appears to have been extirpated from stretches of the Conasauga River in Tennessee and northern Georgia.

Elliptio crassidens (Lamarck, 1819)
Elephantear

RANGE MAP 18; PLATE 18

Synonymy:

Unio crassidens Lamarck, 1819; Lamarck, 1819:71

Margarita (Unio) crassidens (Lamarck, 1819); Lea, 1836:19

Margaron (Unio) crassidens (Lamarck, 1819); Lea, 1852c:24

Unio (Elliptio) crassidens (Lamarck, 1819); Simpson, 1900a:706

Elliptio crassidens (Lamarck, 1819); Ortmann, 1912a:266

Elliptio (Elliptio) crassidens var. *crassidens* (Lamarck, 1819); Johnson, 1970:305–307

Unio (Elliptio) nigra Rafinesque, 1820; Rafinesque, 1820:291, pl. 80, figs. 1–4

Unio (Elliptio) nigra var. *fusca* Rafinesque, 1820; Rafinesque, 1820:291

Unio (Elliptio) nigra var. *maculata* Rafinesque, 1820; Rafinesque, 1820:291

Unio niger (Rafinesque, 1820); Say, 1834:no pagination

Elliptio nigra (Rafinesque, 1820); Ortmann, 1919:91

Elliptio (Elliptio) niger (Rafinesque, 1820); Frierson, 1927:25

Unio cuneatus Barnes, 1823; Barnes, 1823:263

Mya cuneata (Barnes, 1823); Eaton, 1826:220

Obliquaria venus Rafinesque, 1831; Rafinesque, 1831:3

Unio incrassatus Lea, 1840; Lea, 1840:286; Lea, 1842b:217, pl. 16, fig. 34

Margaron (Unio) incrassatus (Lea, 1840); Lea, 1852c:24

Unio (Elliptio) incrassatus Lea, 1840; Simpson, 1900a:707

Unio (Elliptio) crassidens var. *incrassatus* Lea, 1840; Simpson, 1914:608

Elliptio (Elliptio) niger incrassatus (Lea, 1840); Frierson, 1927:25

Elliptio crassidens var. *incrassatus* (Lea, 1840); Clench and Turner, 1956:171

Unio crassus Say, 1817; Sowerby, 1868:pl. 95, fig. 520 [misidentification]

Unio discus Lea, 1838; Sowerby, 1868:pl. 62, fig. 310 [misidentification]

Unio lehmanii S. H. Wright, 1897; S. H. Wright, 1897:138; Simpson, 1900b:80, pl. 4, fig. 9

Unio polymorphus B. H. Wright, 1899; B. H. Wright, 1899:42; Johnson, 1967b:8, pl. 6, fig. 2
Unio danielsii B. H. Wright, 1899; B. H. Wright, 1899:31; Johnson, 1967b:5, pl. 6, fig. 1
Elliptio (Elliptio) niger danielsi (B. H. Wright, 1899); Frierson, 1927:25 [misspelling]
Elliptio (Elliptio) crassidens var. *danielsii* (B. H. Wright, 1899); Haas, 1969a:215

Type Locality: Mississippi and other rivers and lakes.

General Distribution: Mississippi River drainage, from western Pennsylvania west to Wisconsin, south to Missouri (Meramec River, Oesch, 1984), the Alabama River system, and Georgia and northern Florida (Ochlockonee River system).

Tennessee Distribution: The Elephantear formerly occurred in most medium-sized and large rivers in Tennessee, but it is now limited to primarily the Cumberland and Tennessee rivers and their larger tributaries (e.g., the Elk, Duck, Big South Fork Cumberland, and Holston rivers) in East and Middle Tennessee. It is still present in the Loosahatchie River, West Tennessee. Starnes and Bogan (1988) note that, prior to 1960, *Elliptio crassidens* inhabited the Emory, French Broad, Hiwassee, Sequatchie, Obey, and Red rivers.

Description: The shell is thick, heavy, and solid; older valves are nearly rhomboid and elongated. Individuals may become quite large, even those inhabiting small rivers. Mature individuals may reach a length of 140–150 mm. The anterior end is rather broadly rounded, the ventral margin is straight or somewhat curved, the posterior end is obliquely truncated with the posterior ridge prominent and typically sharply angled; the dorsal margin and posterior ridge usually end in a point.

Plate 18. *Elliptio crassidens* (Lamarck, 1819), Elephantear.

Beaks are large, flattened, and slightly elevated; sculpture consists of 2–3 coarse loops parallel to the growth lines, but is usually evident only in very young shells. The surface is marked with prominent growth lines, often giving it a rough appearance in old valves.

The left valve has two heavy, roughened, triangular, divergent pseudocardinal teeth; the lateral teeth are

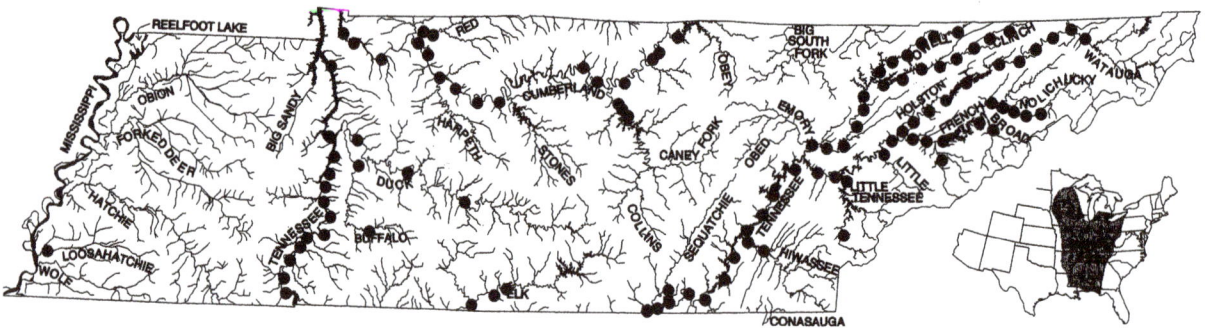

Range Map 18. *Elliptio crassidens* (Lamarck, 1819), Elephantear.

short, heavy, wide, and nearly straight. The right valve has a heavy, triangular, serrated pseudocardinal tooth; the lateral tooth is short and finely serrated. The interdentum is narrow, the beak cavity being shallow to nearly absent in old shells. The periostracum is thick and reddish brown to black; some young shells are marked with dark green rays. The nacre may be white or various shades of salmon or purple, the latter color being the most common in Tennessee specimens.

Life History and Ecology: Although generally considered a big river species, individuals and occasionally small populations of *Elliptio crassidens* may become established in relatively small streams such as the Tellico River and the upper Powell and Clinch rivers, and the local abundance of the Elephantear in some of these rivers testifies to its adaptability to impoundment conditions. However, many beds in these reservoirs are composed of only relict, nonreproducing individuals (Ahlstedt and McDonough, 1994). The Elephantear may occur at depths of 20 feet or more, usually in strong current, and in a substrate composed of sand and coarse gravel, often with a high percentage of mud. Fuller (1974) records the skipjack herring *(Alosa chrysochloris)* as the host fish for the glochidia of this mussel. According to Baker (1928a), the species is tachytictic, the reproductive period comprising the months of June and July.

Status: Currently Stable (Williams et al., 1993:11).

Elliptio dilatata (Rafinesque, 1820) Spike

RANGE MAP 19; PLATE 19

Synonymy:

Unio nasuta Lamarck, 1819 non Say, 1817; Lamarck, 1819:75

Unio nasutus Lamarck, 1819; Agassiz, 1852:50

Elliptio (Eurynia) nasutus nasutus (Lamarck, 1819); Haas, 1969a:244

Unio (Eurynia) dilatata Rafinesque, 1820; Rafinesque, 1820:297

Unio dilatatus Rafinesque, 1820; Say, 1834:no pagination

Elliptio dilatata Rafinesque, 1820; Utterback, 1915:201–203

Elliptio (Elliptio) dilatatus (Rafinesque, 1820); Ortmann and Walker, 1922:30

Elliptio (Eurynia) dilatatus (Rafinesque, 1820); Frierson, 1927:33

Unio mucronatus Barnes, 1823; Barnes, 1823:266, pl. 13, fig. 13

Mya mucronata (Barnes, 1823); Eaton, 1826:221

Unio gibbosus Barnes, 1823; Barnes, 1823:262, pl. 11, fig. 12

Mya gibbosa (Barnes, 1823); Eaton, 1826:220

Margarita (Unio) gibbosus (Barnes, 1823); Lea, 1836:38

Margaron (Unio) gibbosus (Barnes, 1823); Lea, 1852:c38

Unio (Elliptio) gibbosus Barnes, 1823; Simpson, 1900a:703

Elliptio gibbosus (Barnes, 1823); Ortmann, 1912a:271

Obliquaria violacea Rafinesque, 1831; Rafinesque, 1831:3

Unio fulvus Rafinesque, 1831; Rafinesque, 1831:3 *[Eurynia fulva]*

Elliptio (Eurynia) dilatatus var. *fulvus* (Rafinesque, 1831); Frierson, 1927:34

Elliptio (Eurynia) nasutus fulvus (Rafinesque, 1831); Haas, 1969a:244

Unio bicolor Rafinesque, 1831; Rafinesque, 1831:3

Margarita (Unio) arctior Lea, 1836; Lea, 1836:39 [nomen nudum]

Unio arctior Lea, 1838; Lea, 1838b:10, pl. 4, fig. 10

Margaron (Unio) arctior (Lea, 1838); Lea, 1852:c38

Unio stonensis Lea, 1840; Lea, 1840:286; Lea, 1842b:195, pl. 8 fig. 5

Margaron (Unio) stonensis (Lea, 1840); Lea, 1852c:27

Elliptio dilatata var. *subgibbosa* (Lea, 1857) Utterback, 1915:202–203 [misidentification]

Unio gibbosus var. *armathwaitensis* B. H. Wright, 1898; B. H. Wright, 1898b:123; Johnson, 1967b:5, pl. 3, fig. 1

Elliptio (Eurynia) dilatatus armathwaitensis (B. H. Wright, 1898); Frierson, 1927:34

Elliptio (Eurynia) nasutus var. *armathwaitensis* (B. H. Wright, 1898); Haas, 1969a:245

Unio (Elliptio) gibbosus var. *delicatus* Simpson, 1900; Simpson, 1900a:704

Elliptio (Eurynia) dilatatus delicatus (Simpson, 1900); Frierson, 1927:34

Elliptio (Elliptio) dilatatus delicatus (Simpson, 1900); F. C. Baker, 1928a:128, pl. 56, figs. 7, 8, pl. 57, figs. 1, 2

Elliptio (Eurynia) nasutus delicatus (Simpson, 1900); Haas, 1969a:245

Elliptio dilatatus var. *sterkii* Grier, 1918; Grier, 1918:9–10

Elliptio dilatatus sterkii Grier, 1918; Ortmann, 1919:101, pl. 8, fig. 3

Elliptio (Elliptio) dilatatus sterkii Grier, 1918; F. C. Baker, 1928a:130, pl. 56, figs. 3–6

Type Locality: *Unio dilatatus:* no type locality published, but generally considered to be Ohio River; *Unio gibbosus:* Wisconsin.

General Distribution: Entire Mississippi River drainage from the St. Lawrence River and its tributaries south to northern Louisiana and west to the tributaries of the Red River, Oklahoma. Howells et al. (1996) report a single locality on the San Marcos River in Texas. With reference to Canada, Clarke (1981a:268) states that the Spike is "[c]ommon in the Great Lakes and their tributaries from Lake Michigan to Lake Erie; uncommon in Lake Ontario and in the St. Lawrence River."

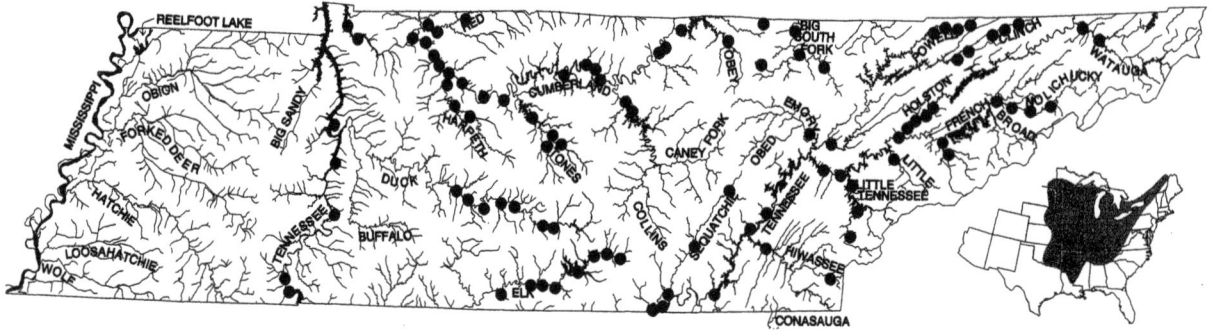

Range Map 19. *Elliptio dilatata* (Rafinesque, 1820), Spike.

Tennessee Distribution: The Spike occurs in most small streams and large rivers throughout East and Middle Tennessee, from the upper Clinch and Powell rivers in East Tennessee to the Stones, Elk, and Duck rivers in the central part of the state. Starnes and Bogan (1988) list *Elliptio dilatata* as having occurred in the Watauga, French Broad, Little, Sequatchie, Obey, Harpeth, and Red rivers prior to 1960; it now appears extirpated from these rivers.

Description: The shell is compressed to slightly inflated, attenuate, solid, and thick. Mature individuals inhabiting impounded stretches of large rivers become especially large and develop extremely thick shells, many attaining a length of 120 mm. Beaks are depressed and flattened; sculpture consists of 4–5 pronounced loops, running parallel with the growth lines. The anterior end is broadly rounded, the ventral margin is straight or slightly curved, and the posterior end is sharply pointed and often compressed behind the strongly developed, rounded posterior ridge. Concentric rest lines are often prominent.

The left valve has two triangular, divergent, compressed pseudocardinal teeth, usually roughened or finely serrated; the two lateral teeth are heavy, rough, widely separated, and the inner surface is usually rough. The right valve has a heavy, triangular pseudocardinal tooth; the lateral tooth is low, thick, and roughened. The interdentum is moderately wide; the beak cavity is usually very shallow or absent. Periostracum of young shells is light brown, greenish, or yellowish green, often faintly rayed; old valves are dark greenish brown to black. The nacre varies from white through salmon to deep purple.

Plate 19. *Elliptio dilatata* (Rafinesque, 1820), Spike.

Life History and Ecology: *Elliptio dilatata* is somewhat generalized relative to the size of rivers which it inhabits and depths at which it may occur. In contrast to populations in the Tennessee and Cumberland rivers (reservoirs) living at depths of up to 24 feet, those of small streams such as the Tellico River or the upper Clinch and Powell rivers may be found, often in great abundance, in less than two feet of water. A firm substrate composed of coarse sand and gravel with moderately strong current appears to provide the most suitable habitat. Baker (1928a) indicated the species was tachytictic, the reproductive season occurring from mid-May to August. The gizzard shad *(Dorosoma cepedianum)*, flathead catfish *(Pylodictis olivaris)*, white crappie *(Pomoxis annularis)*, black crappie *(P. nigromaculatus)*, and the yellow perch *(Perca flavescens)* have been listed by Fuller (1974) as host species for the glochidia. Under laboratory conditions Luo (1993) was able to infect the rainbow darter *(Etheostoma caeruleum)*, banded sculpin *(Cottus carolinae)*, and rockbass *(Ambloplites rupestris)* with glochidia of the Spike.

Status: Currently Stable (Williams et al., 1993:11).

Epioblasma arcaeformis (Lea, 1831)
Sugarspoon

RANGE MAP 20; PLATE 20

Synonymy:
Unio arcaeformis Lea, 1831; Lea, 1831:116, pl. 17, fig. 44
Margarita (Unio) arcaeformis (Lea, 1831); Lea, 1836:18
Margaron (Unio) arcaeformis (Lea, 1831); Lea, 1852c:23
Truncilla arcaeformis (Lea, 1831); Agassiz, 1852:44
Dysnomia (Truncillopsis) arcaeformis (Lea, 1831); Ortmann, 1925:359
Dysnomia (Penita) arcaeformis (Lea, 1831); Frierson, 1927:94
Dysnomia arcaeformis (Lea, 1831); Morrison, 1942:363
Epioblasma arcaeformis (Lea, 1831); Stansbery, 1973:22
Plagiola (Plagiola) arcaeformis (Lea, 1831); Johnson, 1978:257, pl. 11, figs. 1–4, pl. 8. fig. B
Unio nexus Say, 1831; Say, 1831c:527; Say, 1834:pl. 6, pl. 51

Type Locality: Tennessee River.

General Distribution: A Cumberlandian species formerly restricted to medium-sized and large rivers of the Tennessee and Cumberland River drainages. It occurred in the lower Clinch River, the Holston and French Broad rivers, and in the Tennessee River (Ortmann, 1918). This naiad was apparently uncommon in the upper Cumberland River and the Big South Fork Cumberland River in Kentucky. Its range extended down the Cumberland and Tennessee rivers to within close proximity to their confluence with the Ohio River in Kentucky (Casey, 1986).

Tennessee Distribution: *Epioblasma arcaeformis* occurred in the lower reaches of the Clinch, Holston, and French Broad rivers, and in the Tennessee River from Knoxville to northern Alabama (Ortmann, 1918). Based on archaeological records, it occurred downstream in the Tennessee River as far as TRM 119.0, Perry County. It was collected before impoundment from the Cumberland River near Nashville, and prehistorically it was found throughout the lower stretches of the Cumberland River, although there are apparently no specimens extant from any of the other tributaries of the Cumberland in Tennessee (Johnson, 1978).

Description: The shell is medium-sized, reaching a maximum length of about 70 mm. The outline of the shell is usually quadrate or rhomboidal, the valves being inequilateral, very solid, and greatly inflated, especially in females. The anterior end is evenly rounded; the posterior end is more broadly rounded. The posterior ridge is full, high and angulate, the slope being somewhat wavy and giving the appearance of being double or triple; it is truncate behind the ridge. The posterior ridge in females became a marsupial swelling at about the time the animal was half grown, the swelling slightly serrated along the margin with the appearance of faint tooth sculpturing at rest lines. The marsupial swelling flattens below but does not project

Range Map 20. *Epioblasma arcaeformis* (Lea, 1831), Sugarspoon.

below the ventral margin. The dorsal slope above the upper posterior ridge is usually a shallow furrow. Beaks are full and high, almost touching one another, possessing only a faint sculpture which consists of undulating ridges. The surface of the shell is marked with irregular growth lines; the texture of the periostracum is smooth or satinlike.

The left valve has two stumpy, rough, triangular pseudocardinal teeth of about equal size; the space between the pseudocardinal teeth is triangular, extending to the hinge. The pseudocardinal teeth are separated from two nearly straight, short, heavy, obliquely sculptured lateral teeth by a short and narrow interdentum. The right valve has one large pseudocardinal, sometimes with a smaller tooth before or behind it. The pit before the interdentum is deep. There is one well-developed lateral, sometimes with a vestigial tooth below, in the right valve. Beak cavities are shallow. Anterior muscle scars are small and deep; posterior scars in the male shell are shallow, deeply impressed in females. The marsupial area of the female shell is much thinner than anterior portions of the valve. Both male and female shells often have a shallow radial furrow above the posterior ridge. Males and females are similar in general outline, although females are less elevated and more inflated. The periostracum is tawny to yellowish green with fine green rays over most of the surface. The nacre color is white.

Life History and Ecology: The Sugarspoon was found in large to medium-sized rivers and inhabited shoal and riffle areas (Ortmann, 1918). The breeding season and fish host for the glochidia of this mussel are unknown.

Status: Endangered, Possibly Extinct (Williams et al., 1993:11). "The entire range of this species is now under a series of impoundments. It has not been collected in over half a century and hence is presumed extinct" (Stansbery, 1970:19).

Epioblasma biemarginata (Lea, 1857)
Angled Riffleshell

RANGE MAP 21; PLATE 21

Synonymy:

Unio biemarginatus Lea, 1857; Lea, 1857b:83; Lea, 1866:47, pl. 16, fig. 45

Margaron (Unio) biemarginatus (Lea, 1857); Lea, 1870:38

Truncilla (Pilea) biemarginata (Lea, 1857); Simpson, 1900a:524

Truncilla biemarginata (Lea, 1857); Hinkley, 1906:52

Dysnomia (Pilea) biemarginata (Lea, 1857); Ortmann, 1925:361

Dysnomia (Torulosa) biemarginata (Lea, 1857); Frierson, 1927:95

Dysnomia biemarginata (Lea, 1857); Morrison, 1942:364

Epioblasma biemarginata (Lea, 1857); Stansbery, 1976a:50

Plagiola (Torulosa) biemarginata (Lea, 1857); Johnson, 1978:268, pl. 13, figs. 1, 2

Type Locality: Tennessee River, Florence (Lauderdale County, Alabama).

General Distribution: The Angled Riffleshell formerly occurred in the Clinch, Holston, Elk, and Sequatchie rivers in Tennessee and in the Paint Rock River, Jackson County, and the Flint River, Madison County, Alabama. It also inhabited the main channel of the Tennessee River at Muscle Shoals, Colbert and Lauderdale counties, Alabama (Ortmann, 1925; Johnson, 1978).

Plate 20. *Epioblasma arcaeformis* (Lea, 1831), Sugarspoon.

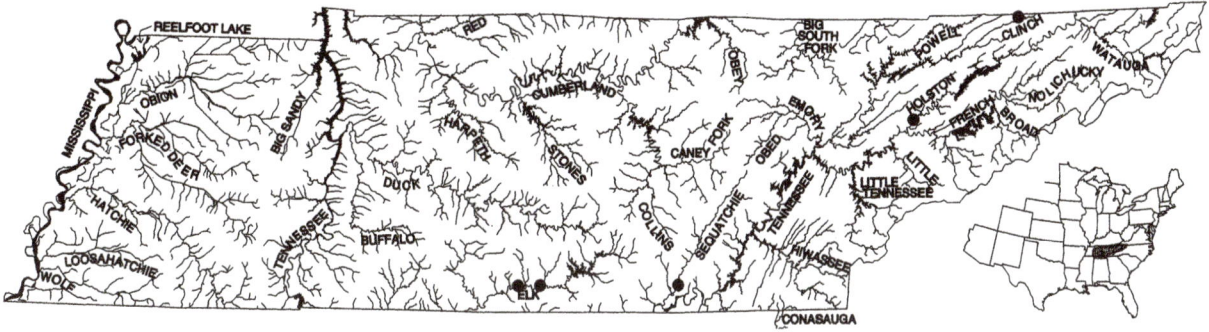

Range Map 21. *Epioblasma biemarginata* (Lea, 1857), Angled Riffleshell.

It has been collected in the Cumberland River drainage only from the Big South Fork Cumberland River, Pulaski County, Kentucky.

Tennessee Distribution: Only four collection locales for *Epioblasma biemarginata* have been recorded in Tennessee. These are the Clinch River, Hancock County; Holston River, Knox County; Sequatchie River, Sequatchie County; and the Elk River, Lincoln County (Ortmann, 1925; Johnson, 1978).

Description: The shell is small, seldom reaching more than 50 mm in length, triangular, rhomboid, or irregularly obovate in outline. Valves are solid and slightly inflated, inequilateral, and with moderately full and elevated beaks. The anterior end is evenly rounded; the posterior end of the shell of males is truncated while the females are evenly rounded. The posterior ridge of valves of males is biangulate and rounded with a moderately developed median ridge. The radial groove or depression between the two ridges is wide and shallow, fading out on the marsupial swelling in females. Growth lines are strong and irregular. The posterior edge of males and marsupial swelling in females are thinner than the rest of the shell.

The left valve has two triangular pseudocardinal teeth and two long, almost straight lateral teeth. The interdentum is narrow, and the beak cavity is shallow. The right valve has a single lateral tooth and three pseudocardinal teeth, the center one being a large triangular tooth with usually a smaller tooth before and behind it. The pallial line and anterior and posterior adductor muscle scars are deeply impressed.

Male shells are subtriangular with a decidedly biangulate posterior ridge ending in a biangulation below with a wide radial depression in front. Female shells are obovate, having a flattened, rounded marsupial swelling extending from the middle of the ventral margin. A shallow radial furrow is evident between the slight medial and posterior ridges, but fades out on the marsupial swelling, which is a darker green than the

Plate 21. *Epioblasma biemarginata* (Lea, 1857), Angled Riffleshell.

rest of the shell. The periostracum is yellowish green with numerous green rays of varying width over the entire surface. The nacre color is bluish white to creamy.

Life History and Ecology: *Epioblasma biemarginata* apparently was a riffle species inhabiting relatively shallow, fast-moving water in medium-sized to large rivers. Little else is known about the species, including its breeding season and host fish for the glochidia.

Status: Endangered, Possibly Extinct (Williams et al., 1993:11). Stansbery (1976a) considered the species to be extinct.

Epioblasma brevidens (Lea, 1831)
Cumberlandian Combshell

RANGE MAP 22; PLATE 22

Synonymy:

Obliquaria (Plagiola) interrupta Rafinesque, 1820 [of authors];
Unio interrupta (Rafinesque, 1820) [of authors]; Conrad, 1834a:69
Truncilla interrupta (Rafinesque, 1820) [of authors]; Agassiz, 1852:44
Plagiola (Plagiola) interrupta (Rafinesque, 1820) [of authors]; Johnson, 1978:252, pl. 20, figs. 5–7
Epioblasma interrupta (Rafinesque, 1820)[of authors]; Bogan and Parmalee, 1983:27
Comment: Bogan (1997) has determined the identification of O. (Plagiola) interrupta to be *Ptychobranchus fasciolaris.*
Unio brevidens Lea, 1831; Lea, 1831:75
Margarita (Unio) brevidens (Lea, 1831); Lea, 1836a:29
Margaron (Unio) brevidens (Lea, 1831); Lea, 1852d:32
Truncilla brevidens (Lea, 1831); Simpson, 1900a:517
Dysnomia (Truncillopsis) brevidens (Lea, 1831); Ortmann and Walker, 1922:66
Dysnomia (Penita) brevidens (Lea, 1831); Frierson, 1927:94
Dysnomia brevidens (Lea, 1831); Hickman, 1937:53
Epioblasma brevidens (Lea, 1831); Stansbery, 1973:22

Type Locality: Ohio; Cumberland River, Tennessee.

General Distribution: A Cumberlandian species once found throughout both the Cumberland and Tennessee River systems.

Tennessee Distribution: This medium-sized naiad is restricted to the Tennessee and Cumberland River drainages. *Epioblasma brevidens* inhabits the headwater streams of the Tennessee River, including the Powell River in Claiborne and Campbell counties, the Clinch River in Hancock and Anderson counties, the Holston River (formerly) from Knox County upstream to Hancock County, the Nolichucky River, a tributary of the French Broad River, and formerly in the Tennessee River at Knoxville, Knox County (Ortmann, 1918; Stansbery, 1973). *Epioblasma brevidens* has been collected in the Elk River in Lincoln County and the Duck River in Marshall and Maury counties (Ortmann, 1924a; Isom and Yokley, 1968a; Isom et al., 1973; Van der Schalie, 1973). It occurred in the Stones River, Rutherford and Davidson counties, and in the Caney Fork River, Putnam County, both tributaries of the Cumberland River (Wilson and Clark, 1914). *Epioblasma brevidens* also occurred in the mainstem of the Cumberland River at Nashville, Davidson County (Johnson, 1978), and downstream, based on archaeological records, as far as Stewart County.

Description: The shell is of medium size, quadrangular or rhomboid in outline, and may reach over 80 mm in length, although most mature specimens average only about 50 mm. The valves are very solid, somewhat inequilateral. In males the anterior end is evenly rounded with the posterior broadly rounded. The umbonal region is elongated, flattened, and located anteriorly; the hinge ligament prominent. Beak sculpture consists of feeble, double-looped bars. The posterior ridge is broadly curved and often appears faintly double in males. In females it becomes a sharply elevated marsupial swelling, serrated along the ventral margin, with remains of former (growth line) serrations along the marsupial swelling. The marsupial swelling separated from the rest of the shell by two distinct sulci, one anterior and one posterior to the marsupium, and it often projects below the ventral margin, producing a semicircular outline.

The left valve has two ragged, triangular pseudocardinal teeth, the anterior tooth being narrow, straight, and directed obliquely forward. There is a very short, narrow interdentum separating the pseudocardinal teeth from the two very short, heavy lateral teeth, which are strongly and obliquely sculptured. The right valve has one large pseudocardinal tooth, separated from the interdentum by a deep pit, and one well-developed lateral tooth with a vestigial tooth below. The beak cavity is shallow. The anterior adductor muscle scars are small and deeply impressed; the posterior adductor muscle scars are large and deep; the pallial line is distinct. The periostracum is smooth and satiny, yellowish, tawny, or tawny brown in color with narrow

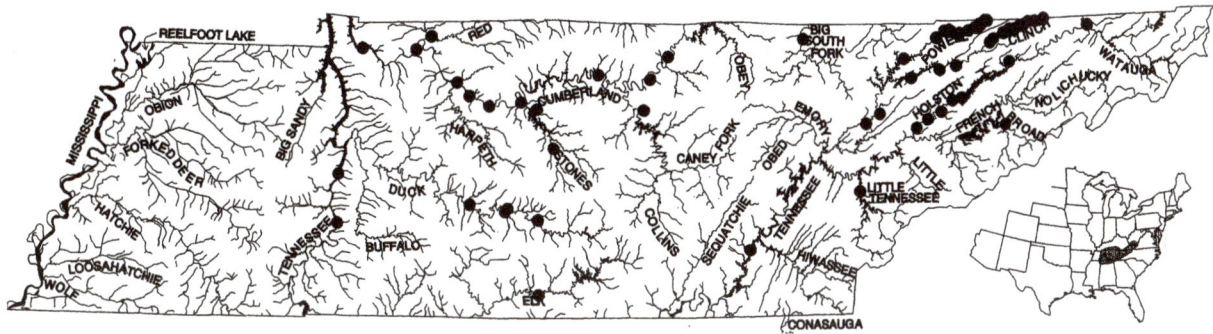

Range Map 22. *Epioblasma brevidens* (Lea, 1831), Cumberlandian Combshell.

green broken rays which sometimes appear as large dots, especially posteriorly. The nacre color is white.

Life History and Ecology: Ortmann (1918) reported this unionoid from medium-sized and large rivers. Hickman (1937) noted taking live specimens of *Epioblasma brevidens* in water about two feet deep from a

sand and gravel bottom in the Clinch River. Wilson and Clark (1914), collecting in the Cumberland River and its tributaries, observed *E. brevidens* in moderate-sized, clear streams with rocky bottoms; it appears to be absent in the smaller tributary streams. This mussel is bradytictic, retaining glochidia in the gills over winter with gravid individuals being reported in May and June (Ahlstedt, 1991a). Reported fish hosts include the greenside darter *(Etheostoma blenioides),* spotted darter *(E. maculatum),* redline darter *(E. rufilineatum),* Tennessee snubnose darter *(E. simoterum),* logperch *(Percina caprodes),* and banded sculpin *(Cottus carolinae)* (Yeager, 1987; Neves, 1991). Yeager and Saylor (1995) found the wounded darter *(Etheostoma vulneratum)* also to serve as a host for the glochidia of the Cumberlandian Combshell.

Status: Endangered (Williams et al., 1993:11).

Epioblasma capsaeformis (Lea, 1834)
Oyster Mussel

RANGE MAP 23; PLATE 23

Synonymy:
Unio capsaeformis Lea, 1834; Lea, 1834:31, pl. 2, fig. 4
Margarita (Unio) capsaeformis (Lea, 1834); Lea, 1836:24
Margaron (Unio) capsaeformis (Lea, 1834); Lea, 1852c:27
Truncilla (Pilea) capsaeformis (Lea, 1834); Simpson, 1900a:524
Truncilla capsaeformis (Lea, 1834); Ortmann, 1912a:359
Dysnomia capsaeformis (Lea, 1834); Wilson and Clark, 1914:46
Dysnomia (Pilea) capsaeformis (Lea, 1834); Ortmann, 1924a:38
Dysnomia (Capsaeformis) capsaeformis (Lea, 1834); Frierson, 1927:95
Epioblasma capsaeformis (Lea, 1834); Stansbery, 1973:22
Plagiola (Torulosa) capsaeformis (Lea, 1834); Johnson, 1978:269

Plate 22. *Epioblasma brevidens* (Lea, 1831), Cumberlandian Combshell.

Family Unionidae 85

Type Locality: Cumberland River, Tennessee.

General Distribution: Found throughout the Tennessee River system in Virginia, Tennessee, and northern Alabama, and in the Cumberland River system in Kentucky and Tennessee (Johnson, 1978).

Tennessee Distribution: Formerly known from the whole length of the Clinch River, French Broad, North and South Forks of the Holston River, Little River in Blount County, Little Pigeon River, the Nolichucky (where it may still occur) and Powell rivers, the headwaters of the Tennessee River, and from the main channel of the Tennessee River. It has also been recorded from the Little Tennessee River and from the Elk and Duck rivers. It has been reported from the Cumberland River, the Big South Fork Cumberland River, and the Caney Fork River, Obey, and Harpeth rivers (Ortmann, 1918; Johnson, 1978; Starnes and Bogan, 1988).

Description: The shell is elliptical or irregularly obovate in outline and of medium size: maximum length is about 70 mm. Valves are subsolid and somewhat inequilateral. Beaks are moderately full and elevated, located slightly anterior of the middle of the shell in males and within the anterior third of the shell in females; the sculpture is feeble, consisting of 2–4 faint parallel loops discernible only in very young juveniles. The surface of the shell is covered with irregular growth lines. The hinge ligament is short. The dorsal shell margin is straight, and the anterior end is regularly rounded; the posterior end of males is slightly protruding, females are more broadly rounded. The ventral margin in males is slightly curved; in females, it is straight and anterior to the sulcus, but behind the sulcus there is a pronounced, rounded marsupial swelling extending well below the base. Marsupial swelling is thin, slightly inflated, often offset from the rest of the shell by an anterior and posterior sulcus, sometimes toothed along the margin. Male shells are almost regularly elliptical, with a double posterior ridge, slightly biangulate behind. Female shells are obovate with a distinct thin, slightly inflated marsupial swelling, darker than the rest of the shell.

The left valve has two small, triangular, subcompressed pseudocardinal teeth, a narrow interdentum, with two short, slightly curved lateral teeth. The right valve has a single stout, triangular, subcompressed pseudocardinal tooth, sometimes with a small tooth on either side, a narrow interdentum, and a single short, slightly curved lateral tooth. The beak cavity is shallow; anterior adductor muscle scars are small, well-impressed; posterior adductor muscle scars are longer and shallow. The pallial line is distinct anteriorly, fading posteriorly. The periostracum subshiny, yellowish green, with fine green rays over entire shell. The marsupial swelling of the female is usually dark green, sometimes almost black. The nacre color is bluish white to creamy.

Life History and Ecology: The Oyster Mussel can be found living in shallow riffles in fast water less than three feet in depth in a gravel and sand substrate. S. A. Ahlstedt (pers. comm., 1992) noted that the Oyster Mussel females, when releasing glochidia, move up onto the surface with the ventral margin uppermost, the mantle margins being visible at some distance. Ortmann (1924a:53) observed that the "pad" of the mantle margin in females collected from the Duck River was "grayish to blackish, and never of the peculiar bluish or greenish-white so often seen in upper Tennessee specimens." The Oyster Mussel is a long-term brooder (bradytictic): gravid individuals have been observed

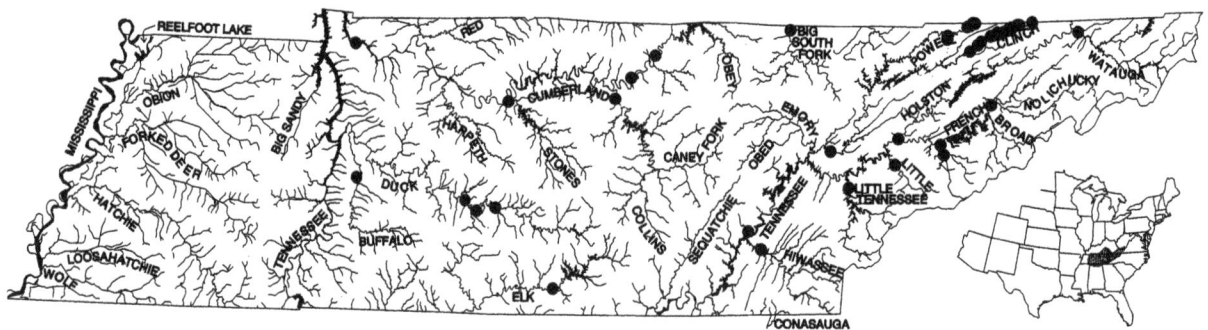

Range Map 23. *Epioblasma capsaeformis* (Lea, 1834), Oyster Mussel.

Plate 23. *Epioblasma capsaeformis* (Lea, 1834), Oyster Mussel.

from early spring into fall. Fish species identified as hosts for the glochidia include the spotted darter *(Etheostoma maculatum)*, redline darter *(E. rufilineatum)*, dusky darter *(Percina sciera)*, banded sculpin *(Cottus carolinae)* (Yeager, 1987; Neves, 1991), and wounded darter *(E. vulneratum)* (Yeager and Saylor, 1995).

Status: Endangered (Williams et al., 1993:11). Dennis (1987) reported a dramatic change in populations of *E. capsaeformis* in the upper Clinch River at Kyles Ford, Hancock County, Tennessee, showing a decline from being a dominant species in 1973–1976 to a very scarce species in 1986. She further documented this decline at Pendleton Island, Clinch River, in southwestern Virginia (Dennis, 1989). Recent (1993–1995) collections made from Kyles Ford downstream to about Sneedville (Hancock County), however, have shown extant viable local populations in this stretch of the river.

Epioblasma flexuosa (Rafinesque, 1820)
Leafshell

RANGE MAP 24; PLATE 24

Synonymy:
Obliquaria flexuosa Rafinesque, 1820; Rafinesque, 1820:306
Dysnomia flexuosa (Rafinesque, 1820); Agassiz, 1852:43
Dysnomia (Dysnomia) flexuosa (Rafinesque, 1820); Ortmann and Walker, 1922:70
Dysnomia flexuosa var. *flexuosa* (Rafinesque, 1820); Warren, 1975:53
Epioblasma flexuosa (Rafinesque, 1820); Stansbery, 1976a:50
Plagiola (Epioblasma) flexuosa (Rafinesque, 1820); Johnson, 1978:283, pl. 15, figs. 5–8, pl. 5
Unio foliatus Hildreth, 1828; Hildreth, 1828:284, fig. 16
Margarita (Unio) foliatus (Hildreth, 1828); Lea, 1836:13
Margaron (Unio) foliatus (Hildreth, 1828); Lea, 1852c:20
Truncilla foliata (Hildreth, 1828); Simpson, 1900a:521
Truncilla (Dysnomia) foliata (Hildreth, 1828); Simpson, 1914:18
Epioblasma biloba Rafinesque, 1831; Stansbery, 1976a:50
[in part]

Type Locality: Kentucky, Salt and Green rivers.

General Distribution: Tennessee River system, Tennessee and Alabama; Cumberland River system, Kentucky; Ohio River system from the lower Wabash River, Indiana, to the Ohio River, Jefferson County, Ohio (Johnson, 1978).

Tennessee Distribution: Known only from archaeological deposits along the lower Tennessee River upstream to the mouth of the Duck River (Perry and Decatur counties), and from the Cumberland River at the Penitentiary Branch (Jackson County) (Breitburg, 1983b) and Hogan (Stewart County) sites. Not reported from the Cumberland River by Neel and Allen (1964) or Wilson and Clark (1914). Recorded by Johnson (1978) from the Clinch and Powell rivers (Claiborne County) and the Tennessee and Holston rivers above Knoxville (Knox and Grainger counties).

Description: The outline of male shell is rhomboid or quadrate; the female shell is irregularly quadrate to triangular depending upon the development of the marsupial swelling. Maximum length attained about 75 mm. Valves are nearly equilateral, solid, and laterally compressed, and the central area of the disk is inflated. Beaks are somewhat elevated above hinge line, compressed, and elongated, situated about middle of the valve. Beak sculpture is slightly corrugated and faint. The posterior ridge is well developed and narrowly

Range Map 24. *Epioblasma flexuosa* (Rafinesque, 1820), Leafshell.

rounded, becoming flattened posteriorly. The posterior end of male is truncate; the female is extended and pointed. The anterior end of the male and female shells is evenly rounded. The male shell has two small emarginations at the ventral margin of a broad medial sulcus; the female has a smaller emargination. The dorsal margin in males is broadly curved but short, straight in females. The marsupial swelling is large, rounded, elongated, and tapered, projecting ventrally and slightly posteriorly from the middle of the valve. A central, broad, flat-bottomed sulcus is present in both male and juvenile female shells; as females mature, the sulcus becomes projected more posteriorly.

The left valve has two elevated, compressed, and serrated pseudocardinal teeth separated by a deep gape extending to the hinge; the interdentum is long and moderately wide. The two lateral teeth are short, straight, and finely serrated. The right valve has a small anterior and a large posterior triangular pseudocardinal tooth; the lateral tooth high, thin, and short, sometimes with a vestigial second ventral, lateral tooth. The interdentum is short, wide, and flat; the beak cavity is shallow. Anterior and posterior adductor muscle scars are well impressed; the pallial line is clear and distinct. The periostracum is dull yellow, yellowish-green to dark horn in color, usually with numerous but subtle dark green rays; growth lines are numerous and darker. The nacre color is white, iridescent posteriorly.

Life History and Ecology: Call (1900:511) remarked, "The shell should be sought in deep and muddy bottoms." Stansbery (1970:19) contradicts this, noting "it was apparently a species of shallow riffles in big rivers, a habitat which has been totally eliminated." Nothing is known about the breeding biology or host fish of this species.

Plate 24. *Epioblasma flexuosa* (Rafinesque, 1820), Leafshell.

Status: Endangered, Possibly Extinct (Williams et al., 1993:11). This mussel, thought to be uncommon throughout its former range, was last collected in 1900 (Stansbery, 1976a) and is now considered extinct.

Epioblasma florentina florentina (Lea, 1857)
Yellow Blossom

Epioblasma florentina walkeri (Wilson and Clark, 1914)
Tan Riffleshell

RANGE MAP 25; PLATE 25; PLATE 26

Synonymy:

Unio florentinus Lea, 1857; Lea, 1857b:83; Lea, 1862b:64, pl. 5, fig. 213

Margaron (Unio) florentinus (Lea, 1857); Lea, 1870:42

Truncilla florentina (Lea, 1857); Simpson, 1900a:524

Dysnomia (Pilea) florentina (Lea, 1857); Ortmann, 1925:362

Dysnomia (Capsaeformis) florentina (Lea, 1857); Frierson, 1927:95

Dysnomia florentina (Lea, 1857); Morrison, 1942:364

Dysnomia florentina florentina (Lea, 1857); Stansbery, 1971:15, 18b, figs. 15, 16

Epioblasma florentina (Lea, 1857); Stansbery, 1976a:51

Plagiola (Torulosa) florentina (Lea, 1857); Johnson, 1978:271, pl. 13, figs. 5–14

Truncilla walkeri Wilson and Clark, 1914; Wilson and Clark, 1914:46, pl. 1, fig. 1

Unio saccatus Küster, 1861; Küster, 1861:263, pl. 39, fig. 2

Unio sacculus Reeve, 1864; Reeve, 1864:pl. 15, species 67

Dysnomia (Pilea) florentina walkeri (Wilson and Clark, 1914); Ortmann, 1924a:50–52

Dysnomia florentina walkeri (Wilson and Clark, 1914); Frierson, 1927:95

Dysnomia (Capsaeformis) walkeri (Wilson and Clark, 1914); Haas, 1969a:488

Dysnomia walkeri (Wilson and Clark, 1914); Stansbery, 1971:15

Epioblasma walkeri (Wilson and Clark, 1914); Stansbery, 1973:22

Truncilla curtisii Frierson and Utterback in Utterback, 1916; Utterback, 1916a:453–455, pl. 6, fig. 14a–d, pl. 28, fig. 109a–d

Epioblasma florentina curtisi (Frierson and Utterback in Utterback, 1916); Nordstrom et al., 1977:18

Plagiola curtisi (Frierson and Utterback in Utterback, 1916); Oesch, 1984:235–237

Type Locality: Tennessee River, Florence, Alabama *(Epioblasma florentina florentina)*; East Fork Stones River, Rutherford County, Tennessee *(Epioblasma florentina walkeri)*.

General Distribution: Tennessee River drainage and the Cumberland River (Burch, 1975a:18). It was widespread throughout the Tennessee River drainage and has been collected from the upper Clinch River, Tazewell County, Virginia, Middle Fork Holston River in Smyth and Washington counties, Virginia and in the South Fork Holston River, Washington County, Virginia. Tributaries of the Tennessee River in Alabama once contained populations of *E. f. florentina*, including the Flint River and Hurricane Creek, a tributary of the Flint River in Madison County, and Bear Creek, Franklin County, Alabama. The only collections of *E. f. florentina* from the main channel of the Tennessee River were from the Muscle Shoals area and in archaeological deposits in Lauderdale and Colbert counties, Alabama (Ortmann, 1925; Morrison, 1942). *Epioblasma florentina* ranged throughout several of the tributaries of the Cumberland, but its total distribution is not well known. It was collected from Buck Creek, Pulaski County, and Beaver Creek, Russell County, and in the Cumberland River, Pulaski and Russell counties, Kentucky. Another subspecies of this mussel, *E. f. curtisi* (Utterback, 1916a, b) is known from southwestern Missouri and northwestern Arkansas. It may represent a disjunct population that was once continuous with the subspecies east of the Mississippi River.

Tennessee Distribution: The form *Epioblasma florentina florentina* inhabited the larger rivers and graded into the headwaters form, *Epioblasma florentina walkeri*. This small naiad was collected from the South Fork Holston River from Grainger County downstream to Knox County (Ortmann, 1918). H. D. Athearn (pers. comm., 1978) found specimens of *E. florentina* in Citico Creek and the Little Tennessee River in Monroe County. *Epioblasma florentina* occurred in the Elk River, Lincoln and Franklin counties, and in the Duck River, Marshall and Maury counties (Ortmann, 1924a; 1925; Isom et al., 1973). Van der Schalie (1973) obtained specimens of *E. f. walkeri* from the Buffalo River, Perry County. It once inhabited portions of the Cumberland River drainage, including the Obey River, Pickett County; Stones River, Davidson County; East Fork of the Stones River, Rutherford County; Harpeth River, Davidson County; and the lower Red River, Montgomery County (Stansbery, 1976e; Johnson, 1978). A small population of *Epioblasma florentina walkeri* was reported from the Hiwassee River, Polk County (Parmalee and Hughes, 1994).

Description: A relatively small naiad, it seldom exceeds 60 mm in length. The shell outline is irregularly elliptical or obovate; valves are inequilateral, subinflated,

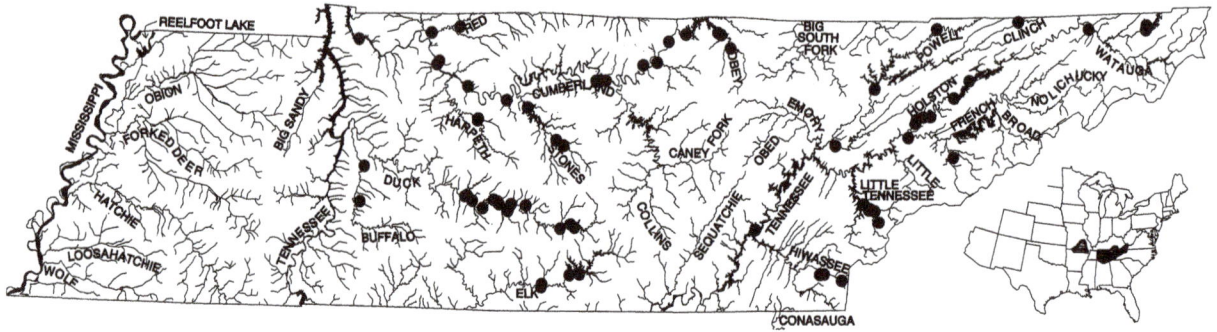

Range Map 25. *Epioblasma florentina* (Lea, 1857), Yellow Blossom, and Tan Riffleshell.

and rather solid. The anterior end is regularly rounded; the posterior end in males protrudes only slightly, while in females the posterior margin is more rounded. The hinge ligament is short. The posterior ridge of the male shell appears faintly doubled, ending in a slight biangulation posteriorly. This ridge is scarcely visible in female shells. There is a shallow radial depression in front of the posterior ridge. Female shells have a somewhat pronounced, rounded, thin marsupial swelling, usually not darker in color than the rest of the shell. This swelling is defined from the rest of the shell by anterior and posterior sulci; the ventral margin is often serrated. Beaks are quite full, elevated, and located slightly anterior of the middle in the male shell and in the anterior third of the female shell. The surface of the shell is usually broken by uneven growth lines. The beak cavity is shallow.

Pseudocardinal teeth are small, triangular to compressed, double in the left valve, single in the right, with a narrow interdentum separating them from the lateral teeth. Lateral teeth are short and curved; they are double in the left valve and single or sometimes double in the right valve. Anterior muscle scars are well impressed, and the posterior scars are shallow; the pallial line is distinct only anteriorly. The nacre color is a bluish white. The periostracum a dull brownish green or yellowish green in color with numerous feeble green rays more or less evenly distributed over the entire surface. The posterior end of the female shell is especially thin and iridescent.

Male shells are irregularly elliptical with a double posterior ridge and a wide, shallow, radial depression. Female shells are irregularly obovate with a thin, slightly inflated marsupial swelling. This swelling may be considerably expanded, extending below the ventral margin, and may be serrated along the margin (Simpson, 1914; Johnson, 1978). The big river form, *Epioblasma florentina,* is more swollen in contrast to the medium-sized form, *Epioblasma florentina walkeri.* Males of the two forms differ in the percentage of shell height to shell length, while females differ in the percentage of shell diameter to shell length (Ortmann, 1924a).

Plate 25. *Epioblasma florentina florentina* (Lea, 1857), Yellow Blossom.

Plate 26. *Epioblasma florentina walkeri* (Wilson and Clark, 1914), Tan Riffleshell.

Life History and Ecology: This ecologically variable species ranged from the headwaters of small rivers to Muscle Shoals in the main channel of the Tennessee River. Considering the naiads listed as associated with this species, it was probably a riffle and shoal species living in sand and gravel substrates. Extant populations of *Epioblasma florentina walkeri* in the Clinch (Virginia) and Hiwassee (Tennessee) rivers occur in a substrate of coarse sand, gravel, and some silt, in current and in less than three feet of water.

Life history information is lacking for this species. It can be inferred that life history data for other species of *Epioblasma* may apply to *Epioblasma florentina*; it is probably a long-term brooder, releasing glochidia in late spring and summer. Recent laboratory studies conducted by Watson and Neves (1996) have shown

that the sculpin (*Cottus* sp.), greenside darter *(Etheostoma blennioides),* fantail darter *(E. flabellare),* and redline darter *(E. rufilineatum)* serve as suitable hosts for the glochidia of the Tan Riffleshell.

Status: Both forms are listed as Endangered by Williams et al., (1993:11), but the big river form, *Epioblasma florentina florentina,* is probably extinct. The U.S. Fish and Wildlife Service has developed a recovery plan for the taxa *E. florentina curtisi* (U.S. Fish and Wildlife Service, 1986), *E. florentina florentina* (U.S. Fish and Wildlife Service, 1985a), and *E. florentina walkeri* (U.S. Fish and Wildlife Service, 1984a) and has created a watershed implementation schedule for the *E. florentina walkeri* recovery plan (U.S. Fish and Wildlife Service, 1989b).

Epioblasma haysiana (Lea, 1834)
Acornshell
RANGE MAP 26; PLATE 27

Synonymy:
Unio haysiana Lea, 1834; Lea, 1834:35, pl. 3, fig. 7
Margarita (Unio) haysianus (Lea, 1834); Lea, 1836:22
Margaron (Unio) haysianus (Lea, 1834); Lea, 1852c:26
Scalenaria haysiana (Lea, 1834); Agassiz, 1852:48
Truncilla (Scalenaria) haysiana (Lea, 1834); Simpson, 1900a:520
Truncilla haysiana (Lea, 1834); Hinkley, 1906:52
Dysnomia haysiana (Lea, 1834); Walker, 1910b:78
Truncilla (Scalenilla) haysiana (Lea, 1834); Ortmann, 1925:361
Dysnomia (Penita) haysiana (Lea, 1834); Frierson, 1927:94
Epioblasma haysiana (Lea, 1834); Stansbery, 1973:22
Plagiola (Pilea) haysiana (Lea, 1834); Johnson, 1978:280, pl. 15, figs. 1–4
Unio sowerbyanus Lea, 1834; Lea, 1834:68, pl. 10, fig. 28
Margarita (Unio) sowerbyanus (Lea, 1834); Lea, 1836:20
Margaron (Unio) sowerbyanus (Lea, 1834); Lea, 1852c:25

Type Locality: Cumberland River, Tennessee.

General Distribution: The Acornshell was a Cumberlandian species, restricted to the Tennessee River system, Virginia, Tennessee, and Alabama (above Muscle Shoals), and the Cumberland River system, Kentucky and Tennessee (Johnson, 1978).

Tennessee Distribution: This small species has been collected from the headwaters of the Tennessee River (Clinch, Powell, and Holston rivers) in northeastern

Tennessee, the Little Tennessee, and the main channel of the Tennessee River in East Tennessee (Ortmann, 1918). Based on archaeological specimens, prehistorically this mussel occurred downstream in the Tennessee River as far as Perry and Decatur counties. The Acornshell also inhabited the Elk River in Middle Tennessee. *Epioblasma haysiana* was present in the Caney Fork River (a tributary to the Cumberland River), Smith County, and in the main channel of the Cumberland River in Tennessee downriver to about Clarksville, Montgomery County (Wilson and Clark, 1914; Johnson, 1978).

Description: The shell is small, not over 40 mm in length, triangular, inflated, and solid. The anterior end is evenly rounded; the posterior end of the male shell is somewhat elongated and slightly more broadly rounded than in females. Beaks are full and high with only weak sculpturing. The anterior half of the shell is smooth, but sometimes is rendered subnodulous by deep lines resulting from rest periods. There is a radial furrow in the front of the posterior ridge; it is shallow in males while narrow and deeper in females. The posterior ridge of the male is faint, but in the female it is elevated into a marsupial swelling. Female shells show pronounced serrations along the margin of the marsupium.

The left valve has two chunky, triangular, divergent pseudocardinal teeth of about equal size and two short, straight lateral teeth. The right valve has a partly doubled pseudocardinal tooth and one short lateral tooth. The beak cavity is fairly shallow; the interdentum is short and wide; muscle scars are deeply impressed. Males attained a greater size than females, the shells being more subtriangular in shape; females are suborbicular in outline.

Marsupial swelling is raised and thinner posteriorly. Because of these pronounced differences in the shape

Plate 27. *Epioblasma haysiana* (Lea, 1834), Acornshell.

of the valves between sexes, Lea described the female of this species as *Unio haysianus* and the male as *U. sowerbyanus*. The periostracum is tawny to chestnut, shiny anteriorly, rough, and often of a lighter shade behind. Some individuals exhibit a few subtle thin green rays. The nacre color ranges from white to a deep purple.

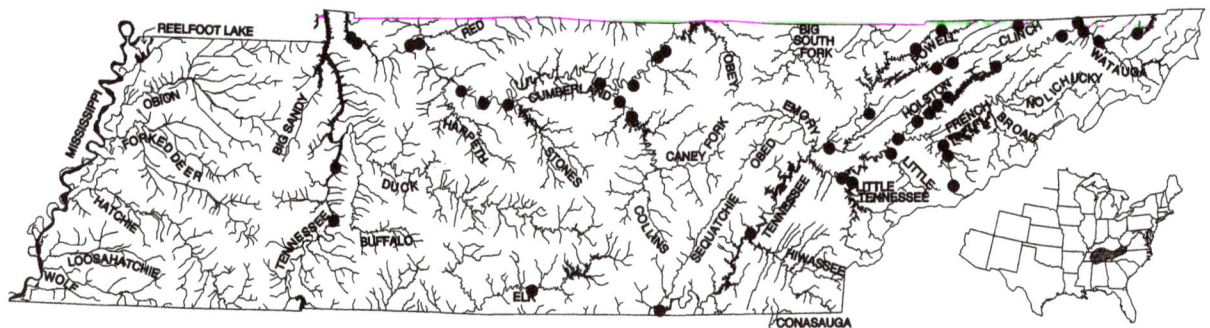

Range Map 26. *Epioblasma haysiana* (Lea, 1834), Acornshell.

Life History and Ecology: Based upon the lists provided by Ortmann (1913a, 1918), the Acornshell apparently lived in areas with a gravel and sand substrate in riffle and shoal portions of clear, cool, high-gradient streams and rivers. The breeding season and host fish for the glochidia of this species are unknown.

Status: Endangered, Possibly Extinct (Williams et al., 1993:11). The Acornshell was listed by Stansbery (1970; 1971) as endangered and, finally, as extinct (Stansbery, 1976a).

Epioblasma lenior (Lea, 1842)
Narrow Catspaw
RANGE MAP 27; PLATE 28

Synonymy:
Unio lenis Lea, 1840; Lea, 1840:286 (non Conrad, 1838—changed to *Unio lenior*)
Unio lenior Lea, 1842; Lea, 1842b:204, pl. 12, fig. 18
Margaron (Unio) lenior (Lea, 1842); Lea, 1852c:39
Truncilla lenior (Lea, 1842); Simpson, 1900a:518
Dysnomia (Truncillopsis) lenior (Lea, 1842); Ortmann, 1924a:49
Dysnomia (Penita) lenior (Lea, 1842); Frierson, 1927:94
Dysnomia lenior (Lea, 1842); Hickman, 1937:54, figs . 27a, b
Epioblasma lenior (Lea, 1842); Stansbery, 1973:22
Plagiola (Plagiola) lenior (Lea, 1842); Johnson, 1978:259, pl. 7, fig. c; pl. 11, figs. 5, 6

Type Locality: Stones River, Tennessee.

General Distribution: Tennessee River system, Virginia, Tennessee, and Alabama. Cumberland River system, restricted to the Stones River, Tennessee (Johnson, 1978).

Tennessee Distribution: This Cumberlandian species occurred in the headwaters of the Clinch River, North Fork and South Fork Holston River, the Holston River in northeast Tennessee, and in the Tennessee River at Knoxville (Ortmann, 1918; Stansbery, 1973). Ortmann noted that it occurred in the Paint Rock River and the Duck River; Hinkley and Marsh also collected *Epioblasma lenior* from the Duck River (Ortmann, 1924a; 1925).

Description: The shell is quite thin (subsolid) and small, seldom exceeding 35 mm in length. Outline of the male is elliptical, that of the female quadrate. Valves are subinflated to inflated with slightly inflated beaks; sculpture consists of irregular, double-looped bars. The anterior end of the shell is evenly rounded; the posterior end of the male shell is slightly more broadly rounded, while the posterior end is truncated in females. The ventral margin is slightly curved, while the dorsal margin is almost straight. The marsupial swelling is well developed, marked off behind by the sulcus with the posterior end of the shell radially sculptured and serrated.

The left valve has two subcompressed, roughened pseudocardinal teeth; there is no interdentum between the pseudocardinals and the two short laterals. The right valve has one pseudocardinal and one short lateral tooth. The beak cavity is very shallow. The anterior adductor scar is impressed; the posterior scar is barely apparent. The periostracum is a pale ashy green or yellow green with numerous narrow green rays on the posterior half, lightly sculptured by growth lines. The nacre color varies from flesh-colored to bluish white.

Life History and Ecology: Hickman (1937) collected *Epioblasma lenior* in the Clinch River from a sand and gravel bottom in two feet of water. The rivers in which it occurred were all clear, fast-flowing, and small to medium-sized, such as the Stones River in Middle Tennessee. The breeding season and host fish for the glochidia of this species are unknown.

Range Map 27. *Epioblasma lenior* (Lea, 1842), Narrow Catspaw.

Plate 28. *Epioblasma lenior* (Lea, 1842), Narrow Catspaw.

Status: Endangered, Possibly Extinct (Williams et al., 1993:11). What was apparently the last known population (Stones River) of *Epioblasma lenior* was destroyed with completion of the J. Percy Priest Dam (1967); this mussel is presumed extinct (Stansbery, 1971).

Epioblasma lewisii (Walker, 1910)
Forkshell

RANGE MAP 28; PLATE 29

Synonymy:

Truncilla lewisii Walker, 1910; Walker, 1910a:42, pl. 3, fig. 3, female; figs. 4, 5, male

Dysnomia (Dysnomia) lewisi (Walker, 1910); Frierson, 1927:93

Dysnomia flexuosa lewisi (Walker, 1910); Morrison, 1942:366

Dysnomia lewisi (Walker, 1910); Neel and Allen, 1964:450

Epioblasma lewisi (Walker, 1910); Stansbery, 1973:22

Plagiola (Epioblasma) flexuosa (Rafinesque, 1820); Johnson, 1978:283, pl. 15, figs. 5–8 [in part]

Type Locality: Holston River, Tennessee.

General Distribution: Tennessee and Cumberland River systems. The Tennessee River in northern Alabama (Ortmann, 1918) and the Cumberland River, Pulaski and Russell counties, Kentucky (Neel and Allen, 1964).

Tennessee Distribution: This small naiad once occurred in the Clinch, Powell, and Holston rivers in East Tennessee. *Epioblasma lewisii* was collected from the mouth of the Holston River and in the Tennessee River at Little River Shoals, Knox County (Ortmann, 1918). Two specimens were recovered at the aboriginal McCrosky site located at the confluence of the Little Pigeon and French Broad rivers, Sevier County. The species probably occurred in the Tennessee River downstream to the Chattanooga area, and although a few specimens have been recovered from aboriginal middens by the authors (one a left valve in a Middle–Late Woodland midden, 40DR305, Decatur County, TRM 143.0), apparently no live individuals were ever collected in historic times in this section of the river. There are no records of this naiad from the Cumberland drainage in Tennessee, although it was recovered from the main channel of the Cumberland River in Kentucky. Athearn obtained specimens from below the Center Hill Dam on the Caney Fork River before the dam was built (Johnson, 1978).

Description: Male shells are quadrate, slightly inflated, thick, and solid. Female shells are triangular but somewhat variable, depending on the extent of the marsupial swelling; female shells are generally thinner than those of males and proportionately wider and compressed with the posterior ridge being more oblique and extended. The anterior ridge is greatly extended and pronounced beyond the ventral margin as a triangular projection. This marsupial swelling or expansion is thinner than the surrounding shell and is dark green in color. The base is deeply emarginate between the marsupial expansion and the posterior ridge. Mature individuals attained a maximum length of about 50 mm. The shell surface is marked by uneven concentric sculpturing.

Beaks are laterally compressed, only slightly elevated, and located near the middle of the dorsal shell margin; the sculpture is faint and corrugated. The anterior margin of the male shell is evenly rounded, the dorsal

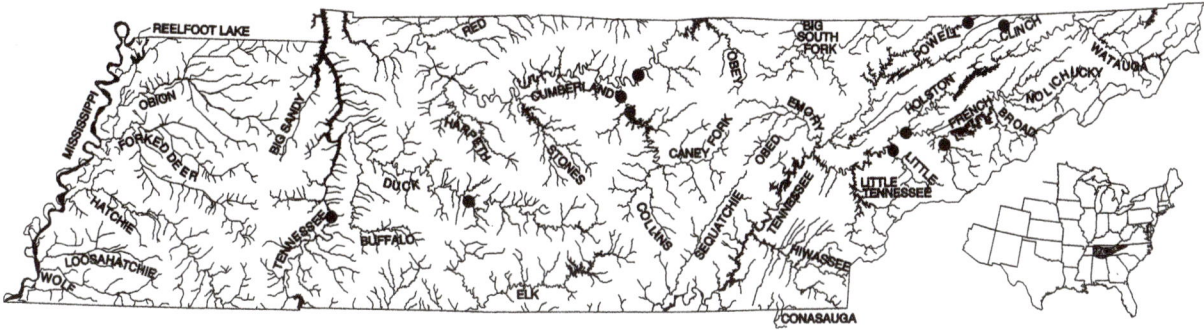

Range Map 28. *Epioblasma lewisii* (Walker, 1910), Forkshell.

margin is curved, and the posterior end is subtruncated, and the ventral margin has two slight emarginations. Male shells have a broad, flat groove extending from the beaks to the ventral margin, becoming wider and deeper as it approaches the ventral margin. The posterior ridge is distinct and rounded toward the umbos, becoming flattened as it approaches the posterior end where it terminates in a slight projection below the ventral margin. There is a strong anterior ridge in front of the medial groove which becomes more pronounced as it approaches the ventral margin. Shells of the two sexes are essentially alike until the animal is about one-third grown. Females then developed the rounded and elongated marsupial swelling.

The left valve has two ragged pseudocardinal teeth of about equal size; the anterior tooth is very narrow and directed obliquely forward, the posterior tooth triangular. The space between them is triangular and extends to the hinge line. The interdentum is rather long, narrow, and rounded and parallel with the hinge. Two nearly straight, granular lateral teeth bend obliquely downward from the hinge line. The right valve has a long triangular pseudocardinal tooth which is separated from the interdentum by a deep groove. There is one well-developed lateral tooth, often with a parallel vestigial tooth below. The beak cavity is shallow. The anterior and posterior muscle scars are well impressed; the pallial line is distinct. The periostracum is a light greenish yellow with faint, radiating darker green rays. The nacre color is white.

Life History and Ecology: The Ohio River species *Epioblasma flexuosa* inhabited muddy bottoms in deep water while *Epioblasma lewisii* was apparently a species of shallow riffles in big rivers such as the Tennessee and Cumberland (Stansbery, 1970). Johnson (1978) synony-mized *E. lewisii* under *E. flexuosa*, claiming that *E. lewisii* is only the small stream shell form ("ecophenotypic variant") of *E. flexuosa,* a big river species. Little else is known of this formerly uncommon naiad, including its breeding season and host fish for the glochidia.

Status: Endangered, Possibly Extinct (Williams et al., 1993:11). Stansbery (1970:19) commented that *Epioblasma lewisii* "has not been collected in over 20 years and hence is presumed extinct."

Plate 29. *Epioblasma lewisii* (Walker, 1910), Forkshell.

Family Unionidae 95

Epioblasma metastriata (Conrad, 1838)
Upland Combshell

RANGE MAP 29; PLATE 30

Synonymy:

Unio metastriatus Conrad, 1838; Conrad, 1838:No. 11 [back cover]; 1840:104, pl. 57, fig. 2

Truncilla metastriata (Conrad, 1838); Simpson, 1900a:519

Dysnomia (Penita) metastriata (Conrad, 1838); Frierson, 1927:93

Epioblasma metastriata (Conrad, 1838); Stansbery, 1976a:49

Unio compactus Lea, 1859; Lea, 1859b:154; Lea, 1859f:218, pl. 38, fig. 98

Margaron (Unio) compactus (Lea, 1859); Lea, 1870:36

Truncilla penita (Conrad, 1834); Simpson, 1900a:518 [in part]

Plagiola (Plagiola) penita (Conrad, 1834); Johnson, 1978:254 [in part]

Type Locality: Black Warrior River near Blount Springs, Alabama.

General Distribution: Historically found in the Alabama and Black Warrior River systems.

Tennessee Distribution: Conasauga River, Polk County.

Description: The shell outline is elliptical to subtriangular to subquadrate. Maximum length attained is about 40 mm. The valves are solid, inflated, and somewhat inequilateral, beaks are moderately full, situated anterior of middle of the valve and prominent; the hinge is short. The anterior end is evenly rounded. The posterior margin of the male shell is straight, that of the female rounded; ventral margin of the male is very broadly rounded, and the female is anteriorly rounded, interrupted by marsupial swelling that is preceded by a sulcus. The posterior slope is evenly rounded. Marsupial swelling is full, strongly sculptured, and not situated in the extreme posterior region of the shell.

Plate 30. *Epioblasma metastriata* (Conrad, 1838), Upland Combshell.

The left valve has two sculpted pseudocardinal teeth of about equal size; the interdentum is short and narrow. The two short, almost straight lateral teeth exhibit heavy serrations. The right valve has a small anterior tooth and an elongated large posterior pseudocardinal tooth, separated by a deep pit; there is one strong lateral tooth, often accompanied by a vestigial tooth ventrally. The male shell is subtriangular and striate

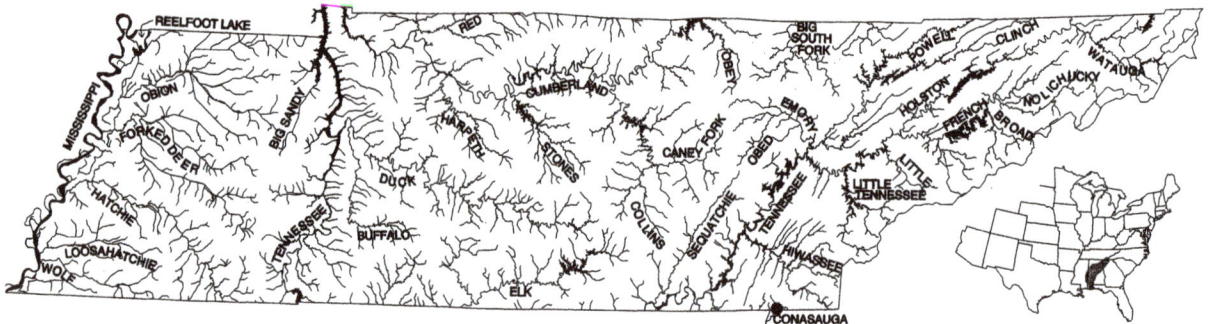

Range Map 29. *Epioblasma metastriata* (Conrad, 1838), Upland Combshell.

posteriorly, while the female shell is more quadrate, narrow in front. The beak cavity is very shallow. The anterior and posterior adductor muscle scars are well impressed; the pallial line is distinct throughout. The periostracum is smooth, shiny, yellowish to greenish yellow with numerous light, narrow green rays. The nacre is white to bluish white.

Life History and Ecology: The life history, including the fish host for the glochidia of this species, is unknown. It has been found living in sand and gravel substratum in riffle sections of small to medium-sized rivers.

Status: Endangered (Williams et al., 1993:11). Stansbery (1976) listed the species as Threatened in Alabama, while Athearn (1970) reported it as Endangered. Hurd (1974) shows one record from Tennessee but did not collect it during his survey. This species is probably extinct in the Tennessee portion of its range (Conasauga River). The U.S. Fish and Wildlife Service has developed a recovery plan for this species (U.S. Fish and Wildlife Service, 1994).

Epioblasma obliquata (Rafinesque, 1820) Catspaw

RANGE MAP 30; PLATE 31

Synonymy:
Obliquaria obliquata Rafinesque, 1820; Rafinesque, 1820b:309
Unio obliquatus (Rafinesque, 1820); Say, 1834:6
Plagiola (Pilea) obliquata (Rafinesque, 1820); Johnson, 1978:278, pl. 14, figs. 10–12
Epioblasma obliquata obliquata (Rafinesque, 1820); Watters, 1993:38
Unio sulcatus Lea, 1829; Lea, 1829:430, pl. 8, fig. 12
Margarita (Unio) sulcatus (Lea, 1829); Lea, 1836:22
Margaron (Unio) sulcatus (Lea, 1829); Lea, 1852c:26
Truncilla (Scalenaria) sulcata (Lea, 1829); Simpson, 1900a:520

Truncilla sulcata (Lea, 1829); Daniels, 1903:646
Dysnomia (Scalenaria) sulcata (Lea, 1829); Ortmann and Walker, 1922:68
Dysnomia (Pilea) sulcata (Lea, 1829); Frierson, 1927:96
Dysnomia sulcata (Lea, 1829); Morrison, 1942:364
Epioblasma sulcata sulcata (Lea, 1829); Stansbery, 1976a:50
Unio perplexus Say, 1829; Say, 1829:309; Say, 1830a:pl. 5 (non Lea, 1831)
Unio ridibundus Say, 1829; Say, 1829:308; Say, 1830a:pl. 5
Unio flagellatus Say, 1830; Say, 1830a:pl. 5
Unio gibbosus perobliquus Conrad, 1836; Conrad, 1836:51, pl. 27, fig. 2
Unio perobliquus Conrad, 1836; Conrad, 1837:(back cover)
Dysnomia sulcata perobliquus (Conrad, 1836); Stansbery, 1970:19
Epioblasma obliquata perobliquus (Conrad, 1836); Watters, 1993:39
Unio pectitis Conrad, 1853; Conrad, 1853:225 [nomen nudum]
Unio pectitis Conrad, 1854; Conrad, 1854:297. pl. 27, fig. 4
Dysnomia sulcata pectita (Conrad, 1854); Frierson, 1927:96
Truncilla (Scalenaria) sulcata var. *delicata* Simpson, 1900; Simpson, 1900a:520
Dysnomia sulcata delicata (Simpson, 1900); Walker, 1913:21
Unio stewardsonii stewardsonii (Lea, 1852); De Gregorio, 1914:45, pl. 6, fig. 3 [misidentification]
Unio propeculcatus De Gregorio, 1914; De Gregorio, 1914:60, pl. 10, fig. 2

Type Locality: Kentucky River.

General Distribution: The Catspaw was found throughout the eastern portion of the Interior Basin, including the Ohio and Cumberland River systems, up into the St. Lawrence River system, and in the Tennessee River, but only in the vicinity of Muscle Shoals, northern Alabama (Johnson, 1978). In the Cumberland River system it occurred in the Caney Fork and Harpeth rivers (Johnson, 1978).

Tennessee Distribution: The Catspaw was widely distributed throughout the eastern United States, but in

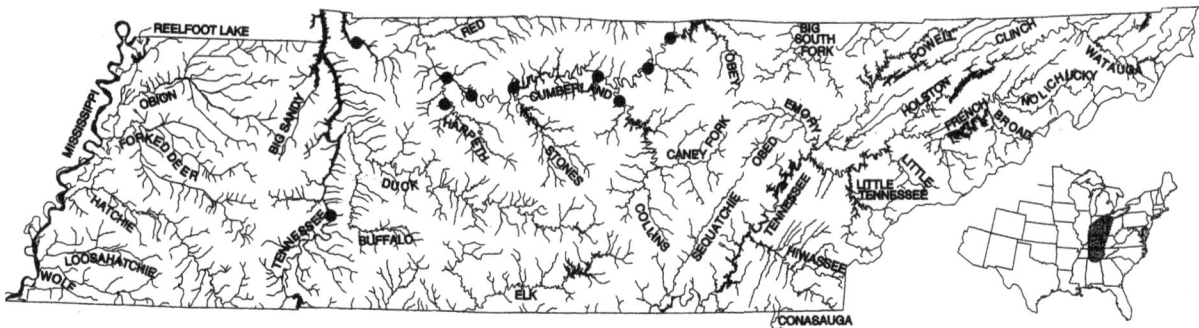

Range Map 30. *Epioblasma obliquata* (Rafinesque, 1820), Catspaw.

Tennessee it has been collected historically only from the Cumberland River system. However, on the basis of four specimens recovered from a Mississippian culture shell midden site (40DR305) located along the Tennessee River, Decatur County, prehistorically the Catspaw may have been present in localized populations throughout the lower stretches. Specimens have been obtained from the Harpeth and Caney Fork rivers, tributaries to the Cumberland River (Johnson, 1978), and from the main channel of the Cumberland River in Jackson, Smith, Davidson, and Cheatham counties (Wilson and Clark, 1914). Fresh specimens were taken by commercial shellers in 1976–1978 from the Cumberland River, Smith County (Isom et al., 1979; Parmalee et al., 1980).

Description: Shells of the Catspaw are medium-sized, subquadrate, trapezoid, or quadrate in outline. Relict males from the Cumberland River below Carthage, Smith County, Tennessee, reach a maximum length of about 70 mm; females are about 50 mm. Valves are very inequilateral, inflated, and solid. The anterior end of the male is squarely truncate, regularly rounded in females. The posterior end of the male is somewhat bluntly pointed above, sloping and truncated below; in the female it is truncated posteriorly. The ventral shell margin of males is broadly curved, almost straight in females. Posterior ridge of the male is doubled, rather low, and faint; in females it is somewhat obscured by a sharp sulcus before the marsupial swelling. Beaks are elevated, prominent, located at the extreme anterior end; sculpture consists of a few faint corrugations, sometimes broken or occasionally double-looped. The surface is interrupted with numerous distinct rest lines.

The left valve has two subtriangular, ragged, elevated pseudocardinal teeth and two short, straight lateral teeth. The right valve usually has one pseudocardinal tooth, but sometimes three: one large triangular tooth with a smaller tooth before and behind it. The lateral tooth is short, often with a vestigial tooth below it. Beak cavities are rather shallow; the interdentum is short and wide; muscle scars and the pallial line are deeply impressed. The periostracum is smooth and shiny, yellowish green, yellow, or brownish, usually with numerous fine, faint, wavy green rays. The shell often has a satinlike epidermis when young, but with age it becomes brownish and the rays become obscured. The nacre varies from light to deep purple or bluish white to white.

The shells of males attain a greater size than those of females; they are subtrapezoid, the dorsal and ventral margins are lightly curved, the anterior end squarely truncate, and the posterior end bluntly pointed. They have a widely separated pair of posterior ridges, often with a radial sulcus. Female shells are quadrate in outline and truncated posteriorly; dorsal and ventral margins are curved, but the anterior margin is subtruncate. The marsupial area, which occupies the space between the lower posterior and medial ridges, is inflated, rounded, and distinct from the shell by two sulcations, the posterior sulcus being the more acute, ending in an acute emargination. Marsupial swelling is strongly serrated along its margin and sculptured by former serrations at rest periods. The shell of the marsupium is thin and convex.

Life History and Ecology: All of the records for the Catspaw in Tennessee are from medium-sized and large

Plate 31. *Epioblasma obliquata* (Rafinesque, 1820), Catspaw.

rivers. Specimens obtained by the authors from shellers' cull piles, taken in about 15 to 20 feet of water in a swift current in the Cumberland River, Smith County, Tennessee, represent a relict population. Originally this naiad would have belonged to the riffle fauna of big rivers (e.g., the Tennessee at Muscle Shoals). The only recorded data on the life history of the Catspaw are those of Surber (1912), who found gravid specimens of *T. sulcatus* (=*E. obliquata*) during September and October (bradytictic). Host fish for the glochidia unknown.

Status: Endangered (Williams et al., 1993:11). The U.S. Fish and Wildlife Service has developed a recovery plan for this species (U.S. Fish and Wildlife Service, 1990a).

Epioblasma othcaloogensis (Lea, 1857)
Southern Acornshell

RANGE MAP 31; PLATE 32

Synonymy:
Unio othcaloogensis Lea, 1857; Lea, 1857a:32; Lea, 1858e:74, pl. 14, fig. 54
Margaron (Unio) othcaloogensis (Lea, 1857); Lea, 1870:38
Truncilla othcaloogensis (Lea, 1857); Simpson, 1900a:521
Dysnomia (Penita) othcaloogensis (Lea, 1857); Frierson, 1927:94
Dysnomia othcaloogensis (Lea, 1857); van der Schalie, 1938:16
Epioblasma othcaloogensis (Lea, 1857); Stansbery, 1976a:48
Unio modicellus Lea, 1859; Lea 1859d:171; Lea, 1860e:347, pl. 57, fig. 172
Margaron (Unio) modicellus (Lea, 1859); Lea, 1870:41
Truncilla modicella (Lea, 1859); Simpson, 1900a:518
Dysnomia (Penita) modicella (Lea, 1859); Frierson, 1927:94
Dysnomia modicella (Lea, 1859); Stansbery, 1971:15
Plagiola (Plagiola) penita (Conrad, 1834); Johnson, 1978:254 [in part]

Type Locality: Othcalooga Creek, Gordon County, Georgia.

General Distribution: *Epioblasma othcaloogensis* is known only from the upper Coosa River System in Alabama, Georgia, and Tennessee.

Tennessee Distribution: *Epioblasma othcaloogensis* is found only in southeastern Tennessee in the short stretch of the Conasauga River in Polk County.

Description: The Southern Acornshell is small: the maximum shell length seldom exceeds 25 mm and is quadrate, inflated, and solid. The anterior end is evenly rounded; the ventral margin is nearly straight; the posterior margin is almost squarely truncated above and rounded below; the dorsal margin is slightly arched. The posterior ridge is rounded. Male shells are more rounded posteriorly, while those of the female are marked by a slight marsupial swelling along the posterior-ventral margin. Beaks are somewhat inflated and rise well above the hinge line; sculpture has not been reported. The beak cavity is shallow.

The left valve has two elevated, ragged, slightly compressed pseudocardinal teeth and two short lateral teeth. The right valve has a single slightly compressed pseudocardinal tooth and one short lateral tooth. Adductor muscle scars are well impressed; the pallial line is impressed. The periostracum is nearly smooth with a shiny yellowish color and no rays; sculpture of the marsupial swelling is scarcely visible. The marsupial area of the female shell is thinner than the rest of the shell. The nacre color is white.

Life History and Ecology: In Tennessee the Southern Acornshell was found inhabiting a fine gravel substrate in stretches of river with moderate current and typically at a depth of less than three feet. Life history data are totally lacking other than a presumption that, based

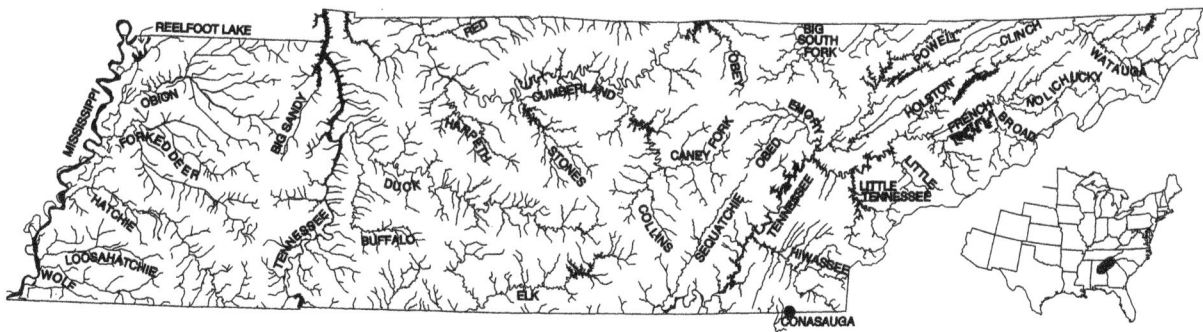

Range Map 31. *Epioblasma othcaloogensis* (Lea, 1857), Southern Acornshell.

Plate 32. *Epioblasma othcaloogensis* (Lea, 1857), Southern Acornshell.

on breeding habits of other members of the genus (Heard and Guckert, 1970), this naiad was a winter brooder. Host fish for *Epioblasma othcaloogensis* unknown.

Status: Endangered (Williams et al., 1993:11). The Southern Acornshell has not been seen in recent years in the Conasauga River in Tennessee and is presumed extirpated in the state. The U.S. Fish and Wildlife Service has developed a recovery plan for this species (U.S. Fish and Wildlife Service, 1994).

Epioblasma personata (Say, 1829)
Round Combshell

RANGE MAP 32; PLATE 33

Synonymy:
Unio personata Say, 1829; Say, 1829:309
Margarita (Unio) personatus (Say, 1829); Lea, 1836:33
Scalenaria personata (Say, 1829); Agassiz, 1852:43
Margaron (Unio) personatus (Say, 1829); Lea, 1852c:35
Truncilla (Pilea) personata (Say, 1829); Simpson, 1900a:522
Dysnomia (Pilea) personata (Say, 1829); Ortmann, 1925:361
Dysnomia personata (Say, 1829); Morrison, 1942:364
Epioblasma personata (Say, 1829); Stansbery, 1976a:43
Plagiola (Pilea) personata (Say, 1829); Johnson, 1978:276–278
Unio pileus Lea, 1831; Lea, 1831:119, pl. 18, fig. 47
Margarita (Unio) pileus (Lea, 1831); Lea, 1836:20
Margaron (Unio) pileus (Lea, 1831); Lea, 1852c:24
Unio capillaris Lea, 1834; Lea, 1834:29, pl. 2, fig. 2

Type Locality: Wabash River.

General Distribution: Ohio, Wabash, and Tennessee rivers. The Round Combshell is recorded historically from the Clinch River in the headwaters of the Tennessee River system. Morrison (1942) reported it from Pickwick Landing archaeological deposits along the Tennessee River at Muscle Shoals and at Florence, northwestern Alabama. *Epioblasma personata* was found in the White River, a tributary to the Wabash River, and in the lower Wabash River at New Harmony, Posey County, Indiana. Patch (1976) identified a single valve of this species from archaeological deposits along the Green River, Kentucky. In the Ohio River it was collected at Cincinnati.

Tennessee Distribution: *Epioblasma personata* is known historically from the Clinch River and from the Cumberland River near Priestly Shoals, Davidson County (Johnson, 1978). Three specimens were recovered in 1995 at the Hogan Site (40SW24), a Mississippian culture shell midden along the Cumberland River at Dover, Stewart County.

Description: Shells of the Round Combshell are solid and inflated. Maximum shell length is about 60 mm. Female shells are typically circular in outline, while males are somewhat triangular to quadrate. The anterior margin of the shell is evenly rounded; the posterior margin of the male shell is evenly rounded to slightly truncate, and female shells are slightly rounded above the median. Beaks are high and full and slightly turned over; the hinge ligament is short. The posterior ridge is faintly double in males with a narrow, shallow radial sulcus. The female shell has the area of the radial sulcus filled by the marsupial swelling, which is thin, slightly swollen, and radially sculptured, with the surface marked with denticulations along rest lines. The dorsal margin is rounded in males, broadly rounded in female shells.

The left valve has two triangular, upright pseudocardinal teeth, a short interdentum, and two short, almost straight lateral teeth. The right valve has one large, triangular pseudocardinal tooth, often with a smaller tooth anterior and/or posterior to the large tooth. The usual single, almost straight, bladelike lateral tooth may be supplemented by a smaller tooth ventral to it. The beak cavity is open and shallow. Adduc-

Range Map 32. *Epioblasma personata* (Say, 1829), Round Combshell.

tor muscle scars are impressed; the pallial line is distinct and impressed anteriorly, becoming less distinct posteriorly. The periostracum is silky or satinlike, the color a greenish yellow to brown, becoming a darker brownish green in old adult specimens, usually marked with faint wavy green rays. The surface of the shell is irregularly and concentrically sculptured. The nacre color varies from white to flesh-colored.

Life History and Ecology: *Epioblasma personata* was a species associated with riffle and shoal areas of medium-sized and big rivers, especially the latter, and probably inhabited stretches with moderately swift current. Little is known of its habitat requirements or life history, including host fish for the glochidia and breeding season.

Status: Endangered, Possibly Extinct (Williams et al., 1993:10). Stansbery (1976a) reported that there had not been a specimen of the Round Combshell collected in this century and that the species is presumed extinct.

Epioblasma propinqua (Lea, 1857)
Tennessee Riffleshell

RANGE MAP 33; PLATE 34

Synonymy:
Unio propinquus Lea, 1857; Lea, 1857b:83; Lea, 1862b:63, pl. 5, fig. 212
Margaron (Unio) propinquus (Lea, 1857); Lea, 1870:34
Truncilla (Pilea) propinqua (Lea, 1857); Simpson, 1900a:523
Truncilla propinqua (Lea, 1857); Simpson, 1914:27
Dysnomia (Pilea) torulosa propinqua (Lea, 1857); Ortmann, 1925:363
Dysnomia (Torulosa) propinqua (Lea, 1857); Frierson, 1927:95
Dysnomia propinqua (Lea, 1857); Morrison, 1942:365
Epioblasma propinqua (Lea, 1857); Stansbery, 1973:22
Plagiola (Torulosa) propinqua (Lea, 1857); Johnson, 1978:266, pl. 12, figs. 8, 9, pl. 6, fig. B

Type Locality: [Tennessee River] Florence [Lauderdale County] and Tuscumbia [Colbert County], Alabama.

Plate 33. *Epioblasma personata* (Say, 1829), Round Combshell.

General Distribution: Found historically in the lower Clinch and Holston rivers and in the Tennessee River downstream from Knoxville to Muscle Shoals, northwestern Alabama. Known from the Cumberland River at Nashville, Tennessee, the Wabash River at New Harmony, Indiana, and from the Ohio River at Cincinnati, Ohio (Johnson, 1978).

Tennessee Distribution: Formerly collected from the lower Clinch River and lower Holston River downstream in the Tennessee River from Knoxville to the Tennessee/Alabama border. Prehistorically it inhabited the lower Tennessee at least as far as Decatur and Perry counties. Known historically from the Cumberland River only at Nashville. Based on material from the Hogan site (40SW24), Stewart County, *E. propinqua* probably occurred throughout the mainstem of the Cumberland River. Archaeological specimens have been identified from sites along the Little Tennessee River (Bogan and Bogan, 1985; Bogan, 1987; Starnes and Bogan, 1988).

Description: The shells are very solid, inflated, and inequilateral, generally triangular or elliptical in shape with a maximum length of about 55 mm. Beaks are full and high, and the sculpture is weakly corrugated. The anterior ends are rounded: in the male the posterior end is broadly rounded to pointed, but it is more evenly rounded in females. Dorsal margins are almost straight; ventral margins are curved. The posterior ridge is low, narrow, and somewhat rounded; the medial ridge is full, and usually both ridges are slightly knobbed. Between the two ridges is an often broad furrow, which is broadest in male specimens. Marsupial swelling in females is rounded and expanded, with the medial sulcus becoming obscured posteriorly and often darker than the rest of the shell.

Plate 34. *Epioblasma propinqua* (Lea, 1857), Tennessee Rifleshell.

The left valve has two chunky, erect, serrated pseudocardinal teeth, the dorsal one largest. The lateral teeth are double, widely separated, strong, and curved. The right valve has a single large pseudocardinal tooth, usually with a smaller tooth dorsal and anterior to it.

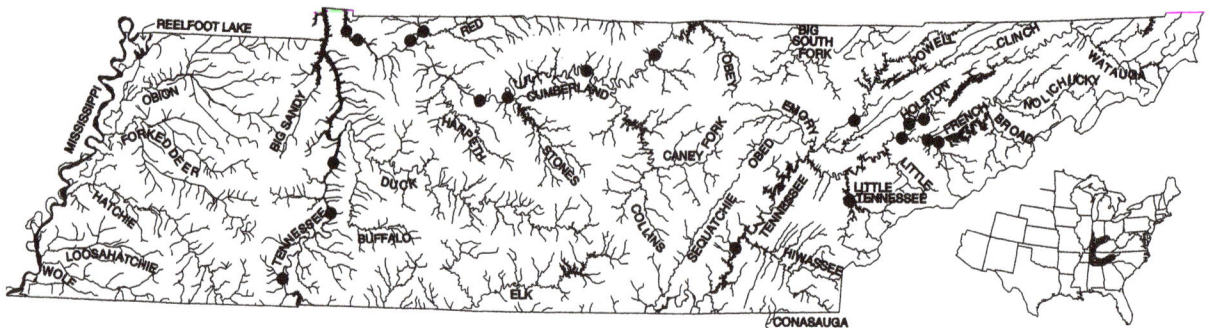

Range Map 33. *Epioblasma propinqua* (Lea, 1857), Tennessee Rifleshell.

The lateral tooth is strong and curved; some specimens have a second low ventral lateral. Male shells are more triangular with a deep sulcus; female shells are more elliptical with the well-rounded marsupial expansion obscuring the medial sulcus. The interdentum is short and fairly wide; the beak cavity is shallow. Anterior and posterior adductor muscle scars are small but well impressed. The pallial line is deeply impressed. The periostracum color varies from straw-colored to tawny to yellowish green, subshiny to silky, often with weak green rays. The nacre is a dull white, iridescent posteriorly.

Life History and Ecology: There is little information available on the biology and life history of this mussel, including the host fish for its glochidia. It is known only from large river habitats, living in riffles in strong current.

Status: Endangered, Possibly Extinct (Williams et al., 1993:11). In listing 11 species in the genus *Dysnomia (=Epioblasma)* presumed to be extinct, Stansbery (1971:8) noted that "[a]ll species in this genus are characteristic riffle or shoal species inhabiting those parts of streams which are shallow with sandy-gravel substrate and rapid currents. The eight species presumed extinct were, with few exceptions, recorded from riffles of our largest rivers. This type of habitat has become nearly eliminated in the conversion of our rivers into barge canals and impoundments."

Epioblasma stewardsonii (Lea, 1852)
Cumberland Leafshell

RANGE MAP 34; PLATE 35

Synonymy:
Unio stewardsonii Lea, 1852; Lea, 1852b:278, pl. 23, fig. 36
Margaron (Unio) stewardsoni (Lea, 1852); Lea, 1852c:20
Truncilla stewardsoni (Lea, 1852); Simpson, 1900a:521
Truncilla (Dysnomia) stewardsoni (Lea, 1852); Simpson, 1914:21–22
Dysnomia (Dysnomia) stewardsoni (Lea, 1852); Ortmann, 1925:364
Dysnomia stewardsoni (Lea, 1852); Morrison, 1942:365
Epioblasma stewardsoni (Lea, 1852); Stansbery, 1976a:50
Plagiola (Epioblasma) stewardsoni (Lea, 1852); Johnson, 1978:285, pl. 15, figs. 9, 10, pl. 8, fig. C

Type Locality: Chattanooga River (Tennessee River at Chattanooga?; Chattanooga Creek ?), Tennessee.

General Distribution: Tennessee River system (Simpson, 1900a). Hurd (1974) listed this species from the Coosa River, Alabama, and from the Etowah River, Georgia, but these are probably spurious records.

Tennessee Distribution: Historical and/or archaeological records of this mussel are known from the Clinch, Holston, Nolichucky, Little Pigeon, and Little Tennessee rivers; it also occurred in the Little River and the Tennessee River downstream from Knoxville to the Alabama border (Johnson, 1978; Starnes and Bogan, 1988; Parmalee, 1988). Johnson (1978) reported a specimen from the Cumberland River at Barton's Creek, Lebanon, Wilson County, and recent (1995) collections from the Hogan Site (40SW24), a Mississippian culture shell midden along the Cumberland River near Dover, Stewart County, contained numerous valves of this mussel. However, Wilson and Clark (1914) and Neel and Allen (1964) failed to report this species from the Cumberland River system.

Description: The shell is small, varying from subtriangular to irregularly rhomboidal in outline, with a maximum length of usually less than 50 mm. Valves are solid, somewhat inflated, and equilateral. The beak area is compressed; beaks are slightly elevated above the hinge line and located near the middle of the shell. The hinge ligament is short and prominent. The dorsal margin is broadly curved in males, but short and straight in females. The ventral margin of male shells is emarginate, and this shallow notching is even more pronounced in female valves. The posterior ridge is well developed, rounded, and slightly inflated. The anterior ridge is rounded with a broad, rather shallow, medial sulcus. The posterior end of the male shell is truncated; in females it is considerably extended into a marsupial swelling. The anterior end of the male shell is rounded; the female shell is regularly and more evenly rounded. The periostracum is greenish yellow to yellowish tan; some specimens have faint or subtle green rays.

The left valve has two sculptured pseudocardinal teeth of about equal size; the space between them is triangular and extends to the hinge. Two lateral teeth are short, thick, almost straight. The right valve has a small anterior and large posterior pseudocardinal tooth; the lateral tooth is short and well developed. The beak cavity is shallow; the interdentum curved and long and narrow. Adductor muscle scars are well impressed. The

Range Map 34. *Epioblasma stewardsonii* (Lea, 1852), Cumberland Leafshell.

pallial line is distinct in males but marked only anteriorly in females. The nacre is white, and some individuals have a tinge of light salmon or buff.

Male and female shells are similar in shape until they are about two-thirds grown when the female begins to develop a centrally located marsupial swelling along the ventral margin. This marsupial swelling is often faintly sculptured and thin, with the periostracum becoming a dark green with age. The radial sulcus is deeper in the male shell.

Life History and Ecology: Nothing is known of the life history of this species, including the host fish for its glochidia. A generalized habitat for the Cumberland Leafshell consisted of a sand and gravel substrate in riffle sections of the upper Tennessee River and its tributaries and shoal areas downstream.

Status: Endangered, Possibly Extinct (Williams et al., 1993:11). The Cumberland Leafshell apparently was never common, and since it has not been collected for over 70 years it is now presumed extinct (Stansbery, 1970, 1971, 1976a).

Epioblasma torulosa torulosa (Rafinesque, 1820)
Tubercled Blossom

Epioblasma torulosa biloba Rafinesque, 1831
Northern Riffleshell

Epioblasma torulosa gubernaculum (Reeve, 1865)
Green Blossom

RANGE MAP 35; PLATE 36
Synonymy:
Unio abnormis "Say"; Haas, 1969a:329 [Nomen nudum]
Amblema torulosa Rafinesque, 1820; Rafinesque, 1820:314, pl. 82, figs. 11, 12

Plate 35. *Epioblasma stewardsonii* (Lea, 1852), Cumberland Leafshell.

Unio torulosus (Rafinesque, 1820); Conrad, 1853:259

Truncilla torulosa (Rafinesque, 1820); Ortmann, 1918:589

Dysnomia (Pilea) torulosa (Rafinesque, 1820); Ortmann and Walker, 1922:69

Dysnomia torulosa (Rafinesque, 1820); Ortmann, 1926b:182

Dysnomia (Torulosa) torulosa (Rafinesque, 1820); Frierson, 1927:94

Dysnomia torulosa torulosa (Rafinesque, 1820); Stansbery, 1970:20

Epioblasma torulosa torulosa (Rafinesque, 1820); Stansbery, 1976a:51

Epioblasma torulosa (Rafinesque, 1820); Cummings and Berlocher, 1990:92

Plagiola (Torulosa) torulosa (Rafinesque, 1820); Johnson, 1978:261, pl. 3, pl. 11, figs. 7–11; pl. 12, figs. 1–5

Amblema torulosa var. *angulata* Rafinesque, 1820; Rafinesque, 1820:315

Amblema gibbosa Rafinesque, 1820; Rafinesque, 1820:315

Amblema gibbosa var. *olivacea* Rafinesque, 1820; Rafinesque, 1820:315

Amblema gibbosa var. *radiata* Rafinesque, 1820; Rafinesque, 1820:315

Amblema gibbosa var. *difformis* Rafinesque, 1820; Rafinesque, 1820:315

Obliquaria reflexa Rafinesque, 1820; Simpson, 1900a:611 [in part]

Plethobasus cyphyum (Rafinesque, 1820); Haas, 1969a:250 [in part]

Epioblasma biloba Rafinesque, 1831; Rafinesque, 1831:2

Comment: Bogan (1997) has established the identity of *E. biloba* as *E. torulosa rangiana*. *Epioblasma biloba* Rafinesque, 1831 has priority over *rangiana* Lea, 1838, but *Amblema gibbosa* Rafinesque, 1820 may be the earliest name for the subspecies in the Ohio River Basin.

Unio perplexus Lea, 1831(non Say, 1829); Lea, 1831:112, pl. 17, fig. 42

Truncilla (Pilea) perplexa (Lea, 1831); Simpson, 1900a:522

Truncilla perplexa (Lea, 1831); Price, 1900:79

Dysnomia perplexa (Lea, 1831); Goodrich and van der Schalie, 1944:314

Unio phillipsii Conrad, 1835; Conrad, 1835b:9, pl. 4, fig. 1

Margarita (Unio) phillipsii (Conrad, 1835); Lea, 1836:15

Unio philipsii Conrad, 1835; Hanley, 1842a:178 [misspelling]

Margaron (Unio) phillipsii (Conrad, 1835); Lea, 1852c:22

Obliquaria reflexa var. *phillipsi* (Conrad, 1835); Simpson, 1914:331 [misspelling]

Pleurobema (Plethobasus) cyphyum phillipsi (Conrad, 1835); Frierson, 1927:45 [misspelling]

Epioblasma phillipsi (Conrad, 1835); Watters, 1993:41 [misspelling]

Unio rangianus Lea, 1838; Lea, 1838b:95, pl. 18, fig. 56

Truncilla (Pilea) perplexa rangiana (Lea, 1838); Simpson, 1900a:523

Truncilla perplexa rangiana (Lea, 1838); Price, 1900:79

Truncilla rangiana (Lea, 1838); Daniels, 1903:646

Dysnomia rangiana (Lea, 1838); La Rocque and Oughton, 1937:152

Dysnomia (Torulosa) rangiana (Lea, 1838); Frierson, 1927:95

Dysnomia perplexa rangiana (Lea, 1838); Goodrich, 1932:113

Dysnomia (Pilea) torulosa rangiana (Lea, 1838); La Rocque, 1967:283

Dysnomia (Torulosa) torulosa rangiana (Lea, 1838); Haas, 1969a:486

Dysnomia torulosa rangiana (Lea, 1838); Stein, 1965:23

Epioblasma rangiana (Lea, 1838); Cummings and Berlocher, 1990:92

Unio cincinnatiensis Lea, 1840; Lea, 1840:285; Lea, 1842b:194, pl. 8, fig. 4

Margaron (Unio) cincinnatiensis (Lea, 1840); Lea, 1852c:22

Truncilla (Pilea) perplexa cincinnatiensis (Lea, 1840); Simpson, 1900a:253

Dysnomia perplexa cincinnatiensis (Lea, 1840); Sterki, 1907:388

Dysnomia (Torulosa) torulosa cincinnatiensis (Lea, 1840); Frierson, 1927:94

Dysnomia torulosa cincinnatiensis (Lea, 1840); Morrison, 1942:365

Dysnomia (Pilea) torulosa cincinnatiensis (Lea, 1840); La Rocque, 1967:283

Unio obliquus Lamarck, 1819; Potiez and Michaud, 1844:153, pl. 48, figs. 3, 4 [misidentification]

Unio gubernaculum Reeve, 1865; Reeve, 1865:pl. 28, species 146

Truncilla torulosa gubernaculum (Reeve, 1865); Ortmann, 1918:590

Dysnomia torulosa gubernaculum (Reeve, 1865); Ortmann, 1926b:182

Dysnomia (Torulosa) torulosa gubernaculum (Reeve, 1865); Frierson, 1927:95

Epioblasma torulosa gubernaculum (Reeve, 1865); Stansbery, 1973:22

Type Locality: Ohio and Kentucky rivers.

General Distribution: Formerly very widespread throughout the eastern United States and portions of southern Ontario. *Epioblasma torulosa* has been recorded historically from only three major river drainages in eastern North America: the Tennessee, Ohio, and St. Lawrence; it has not been found in the Cumberland River drainage in historic times. Rivers in the St. Lawrence drainage where *E. torulosa* has been collected flow into lakes Michigan, Huron, and Erie. These include the St. Marys River in Indiana; the Grand, Black, Clinton, Detroit, and Huron rivers, and the River Raisin in Michigan; the Syndenham River in Ontario (La Rocque and Oughton, 1937; Johnson, 1978).

The form *Epioblasma torulosa biloba* (= *rangiana*) (Northern Riffleshell) is widespread throughout the Ohio River drainage, having been collected from the Allegheny River drainage in Pennsylvania above Pittsburgh and from the West Fork River, a tributary of the

Monogahela River, Harrison County, West Virginia (Ortmann, 1913a). Records of this naiad in the Ohio River system include the Ohio River at Cincinnati, the Little Miami, Scioto, Muskingham, and Olentangy rivers, and Big Darby Creek, a tributary to the Scioto River; the Big Beaver River in Ohio and Pennsylvania (Stein, 1965, 1972; Johnson, 1978). It is also known from a Lake Erie drainage (Fish Creek), and the Kanawha River, West Virginia (pers. comm., G. T. Watters, 1995). This form has been recovered from the Wabash River and its tributaries: the Tippecanoe River and the Eel, Blue and White rivers (Johnson, 1978). Distribution of *E. t. biloba* in Kentucky is based on specimens collected from the Kentucky, Licking, and Green River drainages (Ortmann, 1926b; Isom, 1974; Johnson, 1978).

The distribution of *Epioblasma torulosa* in the Tennessee River drainage reflects the clinal variation of *E. torulosa* discussed by Ball (1922). *E. torulosa gubernaculum* (Green Blossom) occurred in headwater tributaries of the Tennessee River, including the South and North Fork Holston rivers in southwestern Virginia; the Clinch River, Scott County, Virginia; and the Powell River, Lee County, Virginia. The form *E. t. torulosa* (Tubercled Blossom) inhabited the Tennessee River from Knox County, Tennessee, to Muscle Shoals in northern Alabama, and downstream probably to the Ohio River. It also was found in the Elk and Paint Rock rivers, northern Alabama (Ortmann, 1918, 1925; van der Schalie, 1939; Morrison, 1942; Stansbery, 1964; Isom et al., 1973; Johnson, 1978). Morrison (1942) noted the occurrence of archaeological specimens from the Pickwick Basin Mounds which fit the form *D. t. cincinnatiensis*, raising a question as to its relation to *E. torulosa*.

Tennessee Distribution: The distribution of *E. t. torulosa* and *E. t. gubernaculum* provides an interesting

study in molluscan clinal variation. The compressed form or variety *E. t. gubernaculum* inhabited headwaters of the Tennessee River, while *E. t. torulosa* occurred in the Tennessee River below Knoxville. *Epioblasma t. gubernaculum* has been collected from the Clinch River in Claiborne, Union, and Anderson counties; the North Fork and South Fork Holston rivers, as well as the Holston River, Jefferson and Knox counties; the Nolichucky River, Green and Hamblen counties; the Powell River, Claiborne County; and the Tennessee River, Knox County (Boepple and Coker, 1912; Ortmann, 1918; Stansbery, 1973; Johnson, 1978). The form *E. t. gubernaculum* apparently graded into the downriver form *E. t. torulosa* in the Tennessee River at about Knoxville (Ortmann, 1920; 1925). *Epioblasma t. torulosa* was the form inhabiting the Tennessee River at Chattanooga, Hamilton County, and the Elk River, Lincoln County (Ortmann, 1925; Isom et al., 1973; Johnson, 1978). H. D. Athearn (pers. comm., 1981) obtained specimens of *E. t. torulosa* from the Duck River, Marshall County.

The authors have identified archaeological specimens of *E. t. torulosa* recovered at an archaeological site (Eva) located on a small tributary of the Tennessee River, Benton County. Although no recent specimens of *E. torulosa* have been reported from the Cumberland River drainage, archaeological specimens have been found in two aboriginal sites in Davidson County. The archaeological materials from Davidson County also contain two specimens which may be referred to as *E. t. cincinnatiensis* (Parmalee et al., 1980).

Description: The shell is of medium size with old adults attaining a maximum length of about 65 mm. The outline is irregularly ovate, elliptical or obovate, the valves being inequilateral, moderately inflated, and solid. The

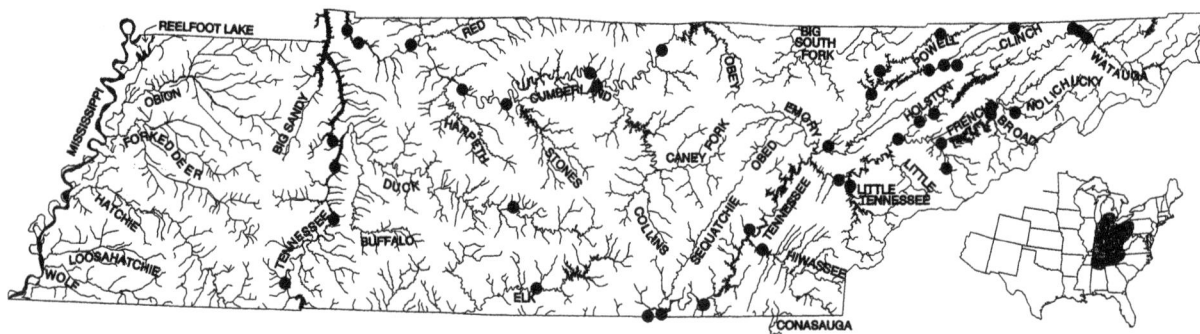

Range Map 35. *Epioblasma torulosa torulosa* (Rafinesque, 1820), Tubercled Blossom.

anterior end is rounded, and the posterior margin in female shells is broadly rounded. The hinge ligament is short. Beaks are full, somewhat turned forward, and located in the anterior third of the shell; sculpture is weak and corrugated. The posterior ridge in males is low, narrowly rounded, and separated from the medial ridge by a broad furrow which ends ventrally in an emargination between the two ridges. Both ridges vary from smooth to having elevated, torulus knobs. Female shells are generally larger than those of males and possess a large, flattened, rounded marsupial swelling which extends from the middle of the ventral margin to the upper part of the posterior end. Marsupial swelling is thin, usually dark green, and sometimes marked by small radial furrows which become obscured by pronounced irregular growth lines.

Plate 36. *Epioblasma torulosa torulosa* (Rafinesque, 1820), Tubercled Blossom.

The left valve has two triangular pseudocardinal teeth, a short but moderately wide interdentum, and two long, straight lateral teeth. The right valve has three triangular pseudocardinals, a large tooth with a smaller tooth before and behind, and one large lateral tooth, sometimes with a vestigial tooth below. The beak cavity is shallow; the pallial line and muscle scars are well impressed. The periostracum is smooth, shiny, and tawny, yellowish green, or straw-colored with numerous fine green rays. The nacre color varies from white to salmon.

Life History and Ecology: *Epioblasma torulosa* lived in riffle areas with swift currents in a substrate of coarse sand and gravel to a substrate of firmly packed fine gravel, typically in shallow water. However, it has been collected in water varying from only a few inches deep to about six feet (Ortmann, 1919; Hickman, 1937; Parmalee, 1967; Johnson, 1978). Dennis, in Neves (1991:299), stated that the life history of the subspecies *Epioblasma torulosa gubernaculum* is unknown, but that "it is probably a long-term brooder as are its congenerics. Glochidia have been observed for *Epioblasma torulosa* in September (Ortmann, 1919). Fish hosts have not been reported." However, Watters (1996b) recorded the banded sculpin *(Cottus carolinae)*, bluebreast darter *(Etheostoma camurum)*, banded darter *(E. zonale)*, and brown trout *(Salmo trutta)* as host fish for the glochidia of the subspecies *Epioblasma t. biloba (= rangiana)*, based on laboratory experiments.

Status: Both subspecies *Epioblasma torulosa torulosa* and *E. t. gubernaculum* are listed by Williams et al. (1993:11) as Endangered, Possibly Extinct. Dennis, in Neves (1991:299), commented that "[i]t is likely, based on current information, that the green blossom is already extinct." The big river nodulous subspecies *E. t. torulosa* is certainly extinct. However, the U.S. Fish and Wildlife Service has developed a recovery plan for the taxa *E. torulosa torulosa* (U.S. Fish and Wildlife Service, 1985a), *E. torulosa rangiana* (=biloba) (U.S. Fish and Wildlife Service, 1993), *E. torulosa gubernaculum* (U.S. Fish and Wildlife Service, 1983c) and has created a watershed implementation schedule for the *E. torulosa gubernaculum* recovery plan (U.S. Fish and Wildlife Service, 1989b).

Epioblasma triquetra (Rafinesque, 1820)
Snuffbox

RANGE MAP 36; PLATE 37

Synonymy:

Truncilla triqueter Rafinesque, 1820; Rafinesque, 1820:300, pl. 81, figs. 1–4

Unio triqueter (Rafinesque, 1820); Short and Eaton, 1831:79

Truncilla (Truncilla) triquetra Rafinesque, 1820; Simpson, 1900a:517

Truncilla triquetra Rafinesque, 1820; Scammon, 1906:283

Dysnomia triquetra (Rafinesque, 1820); Danglade, 1922:5

Dysnomia (Truncillopsis) triquetra (Rafinesque, 1820); Ortmann and Walker, 1922:65

Plagiola (Truncillopsis) triquetra (Rafinesque, 1820); Johnson, 1978:248, pl. 1, pl. 10, figs. 1–4

Plagiola triquetra (Rafinesque, 1820); Oesch, 1984:232

Unio triangularis Barnes, 1823; Barnes, 1823:272, pl. 13, figs. 17a, b

Mya triangularis (Barnes, 1823); Eaton, 1826:221

Margarita (Unio) triangularis (Barnes, 1823); Lea, 1836:18

Margaron (Unio) triangularis (Barnes, 1823); Lea, 1852c:23

Unio cuneatus Swainson, 1823; Swainson, 1823b:112

Unio formosus Lea, 1831; Lea, 1831:111, pl. 16, fig. 41

Unio triangularis var. *longisculus* De Gregorio, 1914; De Gregorio, 1914:40, pl. 4, fig. 5

Unio triangularis var. *pergibosus* De Gregorio, 1914; De Gregorio, 1914:40, pl. 4, fig. 4

Type Locality: Falls of the Ohio (River near Louisville, Jefferson County, Kentucky).

General Distribution: Tributaries to Lake Huron and Lake St. Clair, and lakes Michigan and Erie. Ohio River system from Pennsylvania to Indiana; Green River system, Kentucky; Tennessee and Cumberland River systems; Illinois River system, Illinois; Mississippi River system, Wisconsin and Iowa; Missouri River drainage, Kansas and Missouri; upper White River system, Missouri (Ortmann, 1919; Murray and Leonard, 1962; Johnson, 1978; Clarke, 1981a).

Tennessee Distribution: Known from throughout the Clinch, Powell, North and South Fork Holston, and lower Nolichucky rivers; Little River and the Tennessee River downstream from Knoxville; the Elk and Duck rivers; the Cumberland and Obey rivers (Johnson, 1978; Starnes and Bogan, 1988).

Description: The shells are solid, thick, much inflated, somewhat inequilateral, triangular. Beaks are located somewhat anterior to the middle of the shell, swollen, turned forward and inward over a distinct lunule, and elevated above the hinge line. Beak sculpture consists of 3–4 faint more or less double-looped bars. The anterior end of the shell is rounded, and the posterior end is truncated; the posterior ridge is sharply defined; the posterior slope is widely flattened. The ventral margin is slightly curved in males, almost straight in females. Shells of mature males on average are somewhat larger than females: maximum length of males is about 70 mm, of females about 45 mm.

The left valve has two elevated, strong, roughened, relatively thin pseudocardinal teeth and two short, strong, lateral teeth which are elevated and serrated. The right valve has two pseudocardinal teeth; the front tooth is thinner and less massive than the large elevated, serrated, triangular inner tooth. The single lateral tooth is short, erect, and heavy. The beak cavity is wide and deep. Anterior muscle scars are deeply impressed. The interdentum is narrow and short, or wanting. In female shells, the marsupial region is elevated into a narrow, rounded, radially sculptured ridge, projecting slightly below the base of the shell. Edges of the marsupial region are sometimes toothed and valves gape slightly. The periostracum is yellowish or yellowish green with broken dark green rays which often appear as squarish, triangular, or chevron-shaped spots. The surface is smooth except for occasional irregular growth lines. The nacre is white, often with a silvery luster and a gray-blue tinge in the beak cavity.

Life History and Ecology: These small mussels are found living in shallow riffles with swift current in a gravel and sand substrate. They are usually deeply buried (Baker, 1928a; Hickman, 1937). Ortmann (1919:329) observed that "only the truncated posterior slope is exposed, so that in the natural position the shell offers a very peculiar aspect; only the flat, broadly lanceolate posterior slope is exposed to view, differing entirely from the dark slit generally seen in other Naiades." The breeding season is bradytictic: females are gravid from September to May with glochidia present as early as mid-September, with discharge of glochidia taking place by late May (Ortmann, 1919). Yeager and Saylor (1995) reported the logperch *(Percina caprodes)* and banded sculpin *(Cottus carolinae)* as host fish for the glochidia of the Snuffbox.

Status: Threatened (Williams et al., 1993:11).

Range Map 36. *Epioblasma triquetra* (Rafinesque, 1820), Snuffbox.

Plate 37. *Epioblasma triquetra* (Rafinesque, 1820), Snuffbox.

Epioblasma turgidula (Lea, 1858)
Turgid Blossom

RANGE MAP 37; PLATE 38

Synonymy:

Unio turgidulus Lea, 1858; Lea, 1858a:40; Lea, 1862b:62, pl. 5, fig. 211

Truncilla turgidula (Lea, 1858); Hinkley, 1906:52

Dysnomia (Pilea) turgidula (Lea, 1858); Ortmann and Walker, 1922:69

Dysnomia (Capsaeformis) turgidula (Lea, 1858); Frierson, 1927:95

Dysnomia biemarginata turgidula (Lea, 1858); Stansbery, 1970:20

Dysnomia turgidula (Lea, 1858); Stansbery, 1964:27

Epioblasma turgidula (Lea, 1858); Stansbery, 1976a:51

Plagiola (Torulosa) turgidula (Lea, 1858); Johnson, 1978:274, pl. 14, figs. 1–6

Unio deviatus Reeve, 1864; Reeve, 1864:pl. 15, species 61

Margaron (Unio) deviatus (Reeve, 1864); Lea, 1870:42

Truncilla deviata (Reeve, 1864); Walker, 1910b:78, 81

Truncilla (Pilea) deviata (Reeve, 1864); Simpson, 1900a:524

Truncilla lefevrei Utterback, 1916; Utterback, 1916a:455, pl. 6, figs. 13a–d; pl. 28, figs. 108a–d

Dysnomia (Capsaeformis) lefevrei (Utterback, 1916); Frierson, 1927:95

Truncilla (Pilea) florentina (Lea, 1858); Simpson, 1900a:524 [in part]

Unio nux Lea, 1852; Küster, 1861:pt. 2, 218, pl. 73, fig. 2 [misidentification]

Type Locality: Cumberland River, Tennessee; [Tennessee River], Florence [Lauderdale County], Alabama.

General Distribution: This small unionoid had a disjunct distribution, occurring in the Cumberland and Tennessee rivers as well as in several rivers arising in the Ozark Mountain area including those of the Spring, Black, and White River systems in Arkansas and Missouri. *Epioblasma turgidula* has been recovered from tributaries of the Tennessee River in northern Alabama (Bear and Shoal creeks) as well as the main river. The Turgid Blossom formerly occurred in the Clinch, Emory, Holston, Elk, and Duck rivers in Tennessee.

Tennessee Distribution: The Tennessee distribution of this species follows the pattern characteristic of most naiads belonging to the genus. *Epioblasma turgidula* inhabited the Emory and Clinch rivers in Roane County and the Holston River in Knox and Hawkins counties (Ortmann, 1918). Johnson (1978) includes D. H. Stansbery's collection (Stansbery, 1976f) locality of the Turgid Blossom from the Elk River. Ortmann (1918) found this naiad in the Duck River in Bedford and Maury counties. One of the type localities for this species is listed as the Cumberland River (Lea, 1858a), but it is not known if the site was in Tennessee or Kentucky.

Description: The shell is rather small, elliptical, ovate, or obovate in shape. The Turgid Blossom reaches a maximum length of about 40 mm. Valves are inequilateral, rather solid, and only slightly inflated. The anterior end of the shell is rounded; the posterior end of male shells is pointed, while those of the female are broadly rounded. Male shells are elliptical with a distinct, raised double posterior ridge which ends in a biangulation near the base. There is a wide, shallow radial furrow and a faint medial ridge anterior to the posterior ridge. Female shells are somewhat obovate with the marsupial swelling obliterating the medial and posterior ridges, the shell being somewhat concave in the region of the posterior ridge. Beaks are moderately full and elevated, located at the anterior third of the shell. Beak sculpture is not reported.

The left valve has two rough compressed pseudo-cardinal teeth, lacking an interdentum that separates

Plate 38. *Epioblasma turgidula* (Lea, 1858), Turgid Blossom.

them from the two short, straight lateral teeth. The right valve has one small triangular pseudocardinal tooth parallel with the hinge line, sometimes with a second small tooth before it, and a single, short lateral tooth. Beak cavities are shallow. Anterior adductor muscle scars are well-impressed, but the posterior adductor muscle scars are only faintly outlined. The

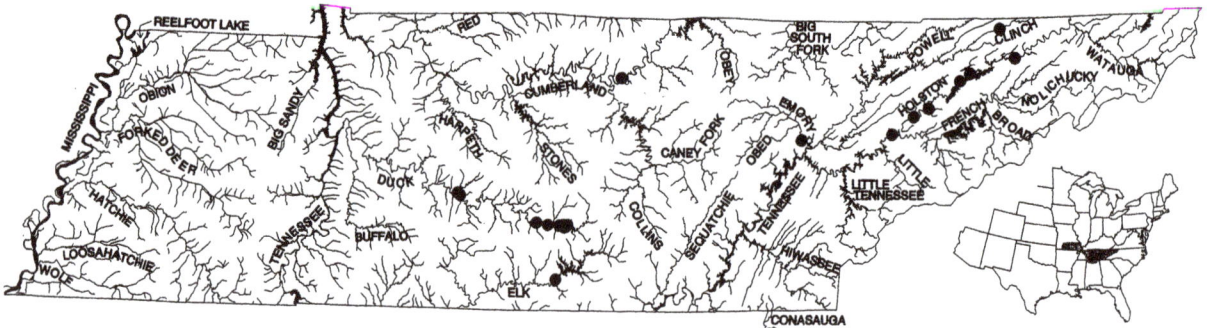

Range Map 37. *Epioblasma turgidula* (Lea, 1858), Turgid Blossom.

pallial line is distinct only in the anterior portion of the shell. The periostracum is rather shiny, having a yellowish green background color and covered with numerous fine green rays evenly distributed over the entire shell surface. The ground color of the marsupial swelling is sometimes slightly darker. The surface of the shell is marked with irregular growth lines. The nacre color is bluish white with the posterior end of female shells especially thin and iridescent.

Life History and Ecology: The localities listed by Johnson (1978) include mostly small to medium-sized rivers. This thin-shelled unionoid probably inhabited shoal areas. Life history data are totally lacking other than a presumption that, based on breeding habits of other members of the genus (Heard and Guckert, 1970), this naiad was a winter breeder. Host fish for the glochidia of *Epioblasma turgidula* unknown.

Status: Endangered, Possibly Extinct (Williams et al., 1993:11). The U.S. Fish and Wildlife Service has developed a recovery plan for this species (U.S. Fish and Wildlife Service, 1985a).

Fusconaia barnesiana (Lea, 1838)
Tennessee Pigtoe
RANGE MAP 38; PLATE 39
Synonymy:
Margarita (Unio) barnesianus Lea, 1836; Lea, 1836:20 [nomen nudum]
Unio barnesianus Lea, 1838; Lea, 1838b:31, pl. 10, fig. 26
Margaron (Unio) barnesianus (Lea, 1838); Lea, 1852c:24
Pleurobema barnesiana (Lea, 1838); Simpson, 1900a:751
Pleurobema barnesianum (Lea, 1838); Simpson, 1914:754
Fusconaia barnesiana (Lea, 1838); Ortmann, 1917:59
Quadrula (Pleuronaia) barnesiana (Lea, 1838); Frierson, 1927:58
Unio bigbyensis Lea, 1841; Lea, 1841a:30; Lea, 1842b:237, pl. 22, fig. 51
Margaron (Unio) bigbyensis (Lea, 1841); Lea, 1852c:24
Pleurobema bigbyensis (Lea, 1841); Simpson, 1900a:751
Pleurobema bigbyense (Lea, 1841); Simpson, 1914:756
Fusconaia barnesiana bigbyensis (Lea, 1841); Ortmann, 1917:59
Unio estabrookianus Lea, 1845; Lea, 1845:164; Lea, 1848:77, pl. 6, fig. 17
Margaron (Unio) estabrookianus (Lea, 1845); Lea, 1852c:24
Unio estabrokianus (Lea, 1845); Paetel, 1890:152 [misspelling]
Pleurobema estabrookiana (Lea, 1845); Simpson, 1900a:763

Fusconaia estabrookiana (Lea, 1845); Goodrich, 1913:93
Pleurobema estabrookianum (Lea, 1845); Simpson, 1914:803
Unio tumescens Lea, 1845; Lea, 1845:164; Lea, 1848:71, pl. 3, fig. 7
Margaron (Unio) tumescens (Lea, 1845); Lea, 1852c:25
Pleurobema tumescens (Lea, 1845); Simpson, 1900a:750
Fusconaia barnesiana tumescens (Lea, 1845); Ortmann, 1917:59
Quadrula (Pleuronaia) tumescens (Lea, 1845); Frierson, 1927:59
Unio meredithii Lea, 1858, Lea, 1858a:40; Lea, 1862b:65, pl. 6, fig. 214
Margaron (Unio) meredithii (Lea, 1858); Lea, 1870:35
Pleurobema meredithii (Lea, 1858); Simpson, 1914:791
Unio pudicus Lea, 1860; Lea, 1860a:92; Lea, 1860e:346, pl. 56, fig. 171
Margaron (Unio) pudicus (Lea, 1860); Lea, 1870:37
Pleurobema pudica (Lea, 1860); Simpson, 1900a:751
Pleurobema pudicum (Lea, 1860); Simpson, 1914:755
Unio ravenelianus Lea, 1834; Reeve, 1864:pl. XVI, fig. 70 [misidentification]
Unio lyonii Lea, 1865; Lea, 1865:89; Lea, 1868a:259, pl. 32, fig. 74
Margaron (Unio) lyonii (Lea, 1865); Lea, 1870:56
Pleurobema lyonii (Lea, 1865); Simpson, 1900a:751
Unio striatissimus Anthony, 1865; Anthony, 1865:156, pl. 12, fig. 1
Unio fassinans Lea, 1868; Lea, 1868a:143; Lea, 1868c:305, pl. 47, fig. 118
Margaron (Unio) fassinans (Lea, 1868); Lea, 1870:42
Pleurobema fassinans (Lea, 1868); Simpson, 1900a:762
Unio crudus Lea, 1871; Lea, 1871:190; Lea, 1874:14, pl. 4, fig. 10
Pleurobema cruda (Lea, 1871); Simpson, 1900a:751
Pleurobema crudum (Lea, 1871); Simpson, 1914:753
Unio radiosus Lea, 1871; Lea, 1871:192; Lea, 1874:13, pl. 3, fig. 9
Unio lenticularis Lea, 1872; Lea, 1872b:155; Lea, 1874:30, pl. 9, fig. 27
Pleurobema lenticulare (Lea, 1872); Simpson, 1914:790
Unio tellicoensis Lea, 1872; Lea, 1872b:155; Lea, 1874:31, pl. 10, fig. 28
Pleurobema fassinans var. *rhomboidea* Simpson, 1900; Simpson, 1900a:762
Pleurobema fassinans var. *rhomboideum* Simpson, 1900; Simpson, 1914:798

Type Locality: Cumberland River, Tennessee.

General Distribution: Cumberland and Tennessee River systems (Simpson, 1914).

Tennessee Distribution: The Tennessee Pigtoe occurs in many of the small to medium-sized rivers in East and Middle Tennessee including the upper Clinch and Powell, Little Pigeon, Nolichucky, Little, Elk, Duck,

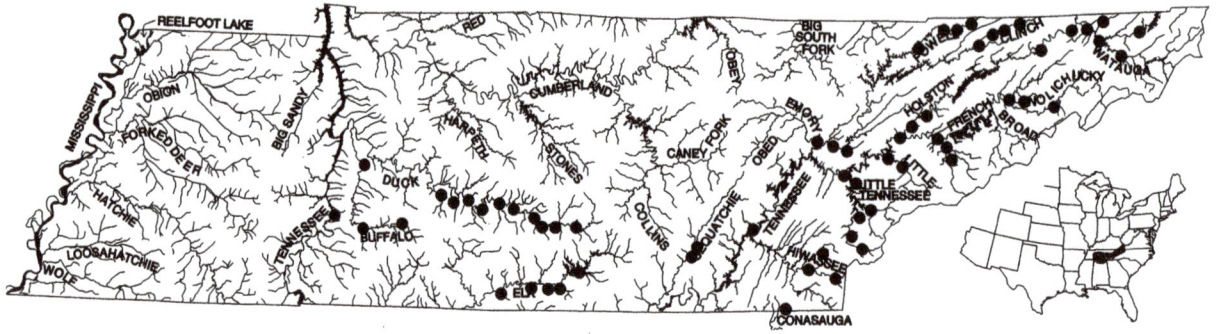

Range Map 38. *Fusconaia barnesiana* (Lea, 1838), Tennessee Pigtoe.

Buffalo, and Hiwassee rivers. Prior to 1960 it was found in the Emory, Watauga, French Broad, Holston, Sequatchie, and Tennessee rivers (Starnes and Bogan, 1988). Wilson and Clark (1914) did not find *Fusconaia barnesiana* in their survey of the Cumberland River, and it has not been collected since, so this mussel may never have occurred in that river.

Description: The shell of *Fusconaia barnesiana* is highly variable in outline, dimension, color, and pattern, factors that resulted in the recognition of three subspecies which may in reality reflect a headwaters-to-big-river cline and individual variations. Ortmann (1918:534) treated the problem in the following manner:

> Also here we have the phenomenon that flat and compressed forms *[Fusconaia b. bigbyensis]* are found in the headwaters, swollen forms *[F. b. tumescens]* in the larger rivers, with the intergrades between them ["typical" species *F. barnesiana*] in the rivers of medium size. The transition is complete and very gradual, so that it is extremely difficult to draw separating lines. The division into three local (or rather ecological) races, introduced here, is entirely arbitrary, but justified to a degree by the old division in "species", and the fact that it is just such an intergrade which has first received a name *(barnesianus)*. The difficulty of classifying these forms becomes greater yet, because these shells vary not only in obesity, but also in general outline, development of beaks, and color markings.

Shells of *Fusconaia barnesiana* are strong, occasionally very thick and heavy, oval, and somewhat truncated to triangular in outline, especially in *F. b. bigbyensis*. Mature specimens may reach a length of 90–95 mm. The posterior ridge is usually distinct but typically rounded; the anterior and ventral margins are broadly rounded, while the posterior margin is straight.

Beaks are only slightly inflated to rather high and full. The surface is usually evenly roughened with fine, uneven growth lines.

The left valve has two erect pseudocardinal teeth; the more dorsal one is often short and triangular, and the anterior tooth is elongated and bladelike. The two lateral teeth are moderately long and straight; in some

Plate 39. *Fusconaia barnesiana* (Lea, 1838), Tennessee Pigtoe.

individuals they are relatively short, slightly curved, and widely separated. The right valve has a large, erect, elongated pseudocardinal tooth, usually with a smaller low tooth on either side; the lateral tooth is long and straight, sometimes with a second low tooth present. Often the pseudocardinal teeth project ventrally at a 90° angle from the lateral teeth. The beak cavity is shallow to nearly wanting; the interdentum is usually wide. Muscle scars are deeply impressed. The periostracum of juveniles is a dull yellowish olive or brown with a satiny appearance, becoming dark brown to blackish with age. Some individuals are marked with a few to many dark green rays. The nacre color is white, and some specimens have a faint salmon wash.

Life History and Ecology: The big river form *Fusconaia barnesiana tumescens* apparently disappeared with the establishment of the Tennessee River reservoirs. The taxa *F. barnesiana* and *F. b. bigbyensis* still occur, often being one of the most abundant mussels present, in many of the small to medium-sized rivers in East and Middle Tennessee. Moderate current, water less than two feet deep, and a substrate composed of coarse sand, silt, and gravel appears to be the most favorable habitat for this species.

The Tennessee Pigtoe is probably tachytictic, spawning in late spring and being gravid into midsummer, based on data for other species of *Fusconaia (F. cor* and *F. cuneolus)* found living in the same rivers and under the same habitat conditions as *F. barnesiana* (Neves, 1991). Host fish for the glochidia unknown.

Status: Special Concern (Williams et al., 1993:11). Although extirpated over much of its former range—or former populations within that range are now greatly reduced or localized—the Tennessee Pigtoe may be found in considerable numbers in small streams and rivers such as the Little Pigeon and Hiwassee.

Fusconaia cor (Conrad, 1834)
Shiny Pigtoe

RANGE MAP 39; PLATE 40

Synonymy:

Unio cor Conrad, 1834; Conrad, 1834a:28, pl. 3, fig. 3
Margarita (Unio) cor (Conrad, 1834); Lea, 1836:21
Margaron (Unio) cor (Conrad, 1834); Lea, 1852c:26
Pleurobema cor (Conrad, 1834); Simpson, 1900a:754
Quadrula cor (Conrad, 1834); Frierson, 1927:59

Quadrula (Pleuronaia) cor (Conrad, 1834); Haas, 1969a:304
Fusconaia cor (Conrad, 1834); Ortmann, 1918:532
Unio edgarianus Lea, 1840; Lea, 1840:288; Lea, 1842b:214, pl. 15, fig. 20
Margaron (Unio) edgarianus (Lea, 1840); Lea, 1852c:25
Pleurobema edgarianus (Lea, 1840); Simpson, 1900a:747
Fusconaia edgariana (Lea, 1840); Goodrich, 1913:82
Unio obuncus Lea, 1871; Lea, 1871:192; Lea, 1874:9, pl. 2, fig. 5
Unio andersonensis Lea, 1872; Lea, 1872b:155; Lea, 1874:36, pl. 12, fig. 33
Fusconaia edgariana var. *analoga* Ortmann, 1918; Ortmann, 1918:533
Quadrula analogia (Ortmann, 1918); Frierson, 1927:60 [misspelling]

Type Locality: Holston River, Tennessee; Tennessee River, Florence, Alabama.

General Distribution: Tennessee River drainage at and above Muscle Shoals, Alabama (Bogan and Parmalee, 1983).

Tennessee Distribution: The Shiny Pigtoe once occurred in the Clinch River from Hancock County downstream to Roane County, and in the Powell River from Hancock County downstream to Campbell County; it is now restricted to the unimpounded stretches of these rivers in Claiborne and Hancock counties. Starnes and Bogan (1988) list it, prior to 1960, from the Tennessee and Holston rivers, and after that time in the Elk River.

Description: The shell of the Shiny Pigtoe is solid, heavy in old individuals, and typically subtriangular in outline. The anterior margin is broadly rounded, somewhat obliquely truncated above, and the ventral and posterior margins are nearly straight. This species reaches a maximum length of about 80 mm. The beaks are moderately high, full, and turned forward and possess a few broken subnodulous ridges apparent only in very young individuals. The posterior ridge is narrowly rounded or angular, sometimes double, ending in a point or biangulation at the base. Anterior to the posterior ridge there is a marked wide radial swelling, separated from it by a wide shallow sulcus; the diameter or width of the shell is thickest at this point. The surface is marked by uneven growth lines.

The left valve has two low, heavy triangular pseudocardinal teeth and two fairly long, stout lateral teeth. The right valve has one large, erect, rough triangular pseudocardinal tooth with one low broad tooth on either side; the single lateral tooth is heavy, sometimes

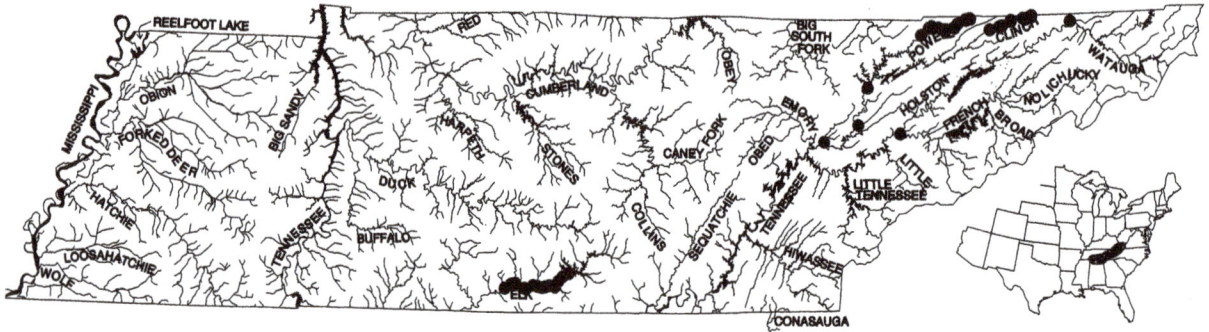

Range Map 39. *Fusconaia cor* (Conrad, 1834), Shiny Pigtoe.

with a partial low second tooth. The beak cavity is deep and compressed; the interdentum is wide; muscle scars are deeply impressed. The periostracum is greenish yellow or yellowish tan, often becoming dark brown to black in older specimens. Young shells generally have both wide and narrow broken dark green rays with a very shiny periostracum. Older specimens become dull, darker in color, and the rays become indistinct. The nacre color is a dull white, and the posterior third is iridescent.

Life History and Ecology: *Fusconaia cor* was collected in the Clinch River, Anderson County, in less than three feet of water on a sand and gravel bottom (Hickman, 1937). It typically is found living in shoal and riffle areas of clear streams with a moderate to fast current. The Shiny Pigtoe is tachytictic, spawning in late May to early June and is gravid from mid-June to mid-July (Kitchel, 1985). The following fish taxa were reported by Kitchel (1985) as being infested with glochidia in the upper North Fork Holston River, Virginia: telescope shiner *(Notropis telescopus),* warpaint shiner *(Luxilus coccogenis),* and Common Shiner *(L. cornutus).*

Status: Endangered (Williams et al., 1993:11). The Shiny Pigtoe remains as a relict population in many stretches of the Clinch and Powell rivers in Tennessee and Virginia, and in the Paint Rock River in northern Alabama (Ahlstedt, 1991b). A population in the lower Elk River has apparently been destroyed by quarry washings, although live specimens were found both above and below Fayetteville in 1980 (Ahlstedt, 1983). The only known population of any size was surviving in the Powell River (Stansbery, 1970), but it has now become extremely rare throughout its range, including Tennessee and Virginia (Neves, 1991). The U.S. Fish and Wildlife Service has developed a recovery plan for this species (U.S. Fish and Wildlife Service, 1983d) and has created a watershed implementation schedule for the recovery plan (U.S. Fish and Wildlife Service, 1989b).

Plate 40. *Fusconaia cor* (Conrad, 1834), Shiny Pigtoe.

Fusconaia cuneolus (Lea, 1840)
Finerayed Pigtoe

RANGE MAP 40; PLATE 41

Synonymy:

Unio cuneolus Lea, 1840; Lea, 1840:286; Lea, 1842b:193, pl. 7, fig. 3

Margaron (Unio) cuneolus (Lea, 1840); Lea, 1852c:24

Pleurobema cuneolus (Lea, 1840); Simpson, 1900a:748

Fusconaia cuneolus (Lea, 1840); Ortmann, 1918:530

Quadrula (Pleuronaia) cuneola (Lea, 1840); Frierson, 1927:58

Quadrula (Pleuronaia) cuneolus (Lea, 1840); Haas, 1969a:303

Unio appressus Lea, 1871; Lea, 1871:189; Lea, 1874:12, pl. 3, fig. 8

Pleurobema appressum (Lea, 1871); Simpson, 1900a:749 [in part]

Fusconaia appressa (Lea, 1871); Goodrich, 1913:93

Fusconaia cuneolus var. *appressa* (Lea, 1871); Ortmann, 1918:531

Unio tuscumbiensis Lea, 1871; Lea, 1871:191; Lea, 1874:11, pl. 46, fig. 116

Pleurobema tuscumbiense (Lea, 1871); Simpson, 1914:748

Unio flavidus Lea, 1872; Lea, 1872b:156; Lea, 1874:28, pl. 9, fig. 25

Type Locality: Holston River, Tennessee.

General Distribution: Tennessee River drainage, from the Clinch and Powell rivers in southwestern Virginia to Muscle Shoals (formerly), Alabama. Isom et al. (1973) reported it from the Paint Rock River, northern Alabama.

Tennessee Distribution: Starnes and Bogan (1988) list *Fusconaia cuneolus* from the Emory, Holston, and Tennessee rivers prior to 1960 and *F. c. appressa,* the inflated big river form, from the upper Tennessee, Clinch, Holston, and Nolichucky rivers. The inflated form was reported by Bogan and Parmalee (1983) as still present in the lower Clinch River, Anderson County, and lower Poplar Creek, a tributary to the Clinch, Roane County.

Neves (1991) records the Finerayed Pigtoe as still present in the North Fork Holston, Clinch, Powell, Sequatchie, Elk, and Little rivers in Tennessee.

Description: Shell characteristics of *Fusconaia cuneolus* are very similar to those of *F. cor:* they are subtriangular in outline, the anterior margin is broadly rounded and somewhat obliquely truncated above, and the ventral and posterior margins are nearly straight. Valves are solid with moderately full beaks. The posterior ridge is distinct and rather sharply angled behind the beaks, becoming slightly curved and more rounded toward the posterior base. Anterior to it is a wide, shallow sulcus. Large mature individuals may reach 80 mm in length. The surface has a satinlike appearance, but is roughened by low concentric rest lines.

The left valve has two fairly low, rough triangular pseudocardinal teeth; the two lateral teeth are fairly long, straight, or very slightly curved. The right valve has one large, rough pseudocardinal tooth with a small low tooth, or roughened swelling, on either side; the lateral tooth is partly doubled posteriorly. The interdentum is short and wide. Adductor muscle scars are deeply impressed; the pallial line is only distinct anteriorly. The beak cavity is moderately deep, being slightly more shallow than in *Fusconaia cor.* The periostracum is a dull yellowish green to olive green marked with often indistinct dark green rays of varying widths. The nacre color is white, iridescent posteriorly.

Life History and Ecology: Neves (1991:274) stated that this mussel "is a lotic, riffle-dwelling species that usually inhabits ford and shoal areas of rivers with moderate gradient." Hickman (1937) collected the Finerayed Pigtoe in about three feet of water from a sandy and rocky bottom in the Clinch River. Ortmann (1925:330)

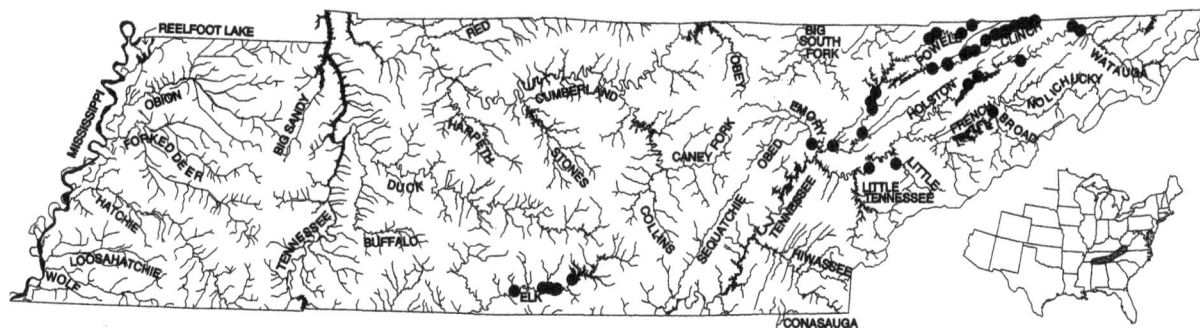

Range Map 40. *Fusconaia cuneolus* (Lea, 1840), Finerayed Pigtoe.

Plate 41. *Fusconaia cuneolus* (Lea, 1840), Finerayed Pigtoe.

noted that "[t]he fact the specimen from Flint Creek was found in a small creek, but in smooth sandy-muddy bottom, possibly indicates that not so much size of stream, as the character of the bottom and current determine the development of the big-river-type."

Fusconaia cuneolus is tachytictic, spawning in May and gravid until late July. The river chub *(Nocomis micropogon),* whitetail shiner *(Cyprinella galactura),* white shiner *(Luxilus albeolus),* telescope shiner *(Notropis telescopus),* Tennessee shiner *(N. leuciodus),* central stoneroller *(Campostoma anomalum),* fathead minnow *(Pimephales promelas),* and mottled sculpin *(Cottus bairdi)* have been shown, based on laboratory-induced infestations, to serve as hosts for the glochidia of the Finerayed Pigtoe (Bruenderman, 1989, Bruenderman and Neves, 1993).

Status: Endangered (Williams et al., 1993:11). The Finerayed Pigtoe has been extirpated throughout most of its former range, with the last remaining viable popu-

lations in Tennessee occurring in the Clinch (Hancock County) and Powell (Hancock and Claiborne counties) rivers. Survival of even these populations remains questionable. The U.S. Fish and Wildlife Service has developed a recovery plan for this species (U.S. Fish and Wildlife Service, 1984b) and has created a watershed implementation schedule for the recovery plan (U.S. Fish and Wildlife Service, 1989b).

Fusconaia ebena (Lea, 1831)
Ebonyshell

RANGE MAP 41; PLATE 42

Synonymy:

Unio obliqua Lamarck, 1819; Férussac, 1835:28 [in part]

Unio obliquus Lamarck, 1819; Conrad, 1837:77, pl. 43, fig. 2 [misidentification]

Obovaria pachostea Rafinesque, 1820; Rafinesque, 1820:312

Amblema antrosa Rafinesque, 1820; Rafinesque, 1820:322

Comment: Rafinesque created the new name for *Obovaria pachostea* Rafinesque, 1820. The species was renamed because he moved it to a new genus.

Quadrula (Fusconaia) antrosa (Rafinesque, 1820); Frierson, 1927:55

Fusconaia antrosa (Rafinesque, 1820); Haas, 1969a:314

Obovaria obovalis Rafinesque, 1820; Rafinesque, 1820:311

Unio obovalis (Rafinesque, 1820); Say, 1834:no pagination

Quadrula obovalis (Rafinesque, 1820); Vanatta, 1915:558

Unio ebenus Lea, 1831; Lea, 1831:84, pl. 9, fig. 14

Margarita (Unio) ebenus (Lea, 1831); Lea, 1836:34

Margaron (Unio) ebenus (Lea, 1831); Lea, 1852c:35

Quadrula (Fusconaia) ebenus (Lea, 1831); Simpson, 1900a:793

Quadrula ebena (Lea, 1831); Sterki, 1907:392

Fusconaja ebena (Lea, 1831); Ortmann, 1912a:245

Fusconaia ebena (Lea, 1831); F. C. Baker, 1920:382

Fusconaia ebenus (Lea, 1831); Ortmann and Walker, 1922:4

Unio mytiloides (Rafinesque, 1820); Swainson, 1840:270, figs. 52, 53 [misidentification]

Type Locality: Ohio River.

General Distribution: Mississippi River drainage from western New York and Pennsylvania, west and north to Minnesota and eastern South Dakota, south to east Texas and northern Louisiana. Alabama River system including the Tombigbee River.

Tennessee Distribution: The Ebonyshell occurs in the Hatchie River in West Tennessee and the main Tennessee and Cumberland rivers (reservoirs) in Middle Tennessee. Prior to 1960 it was found in the North Fork Obion River (Starnes and Bogan, 1988).

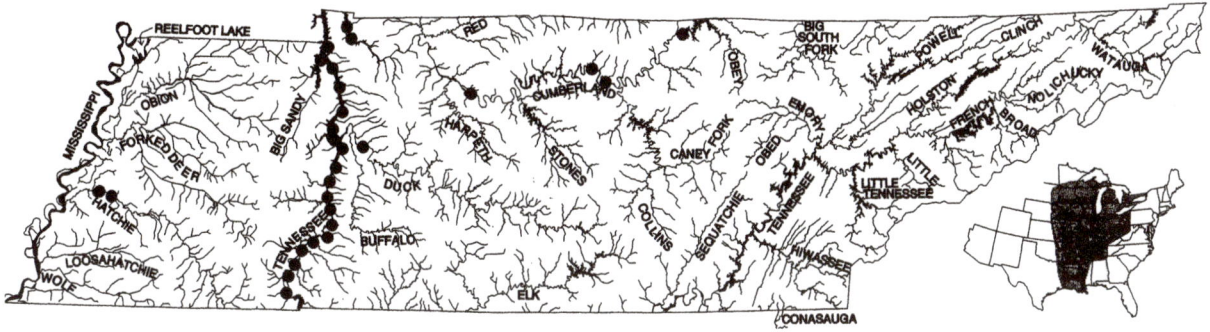

Range Map 41. *Fusconaia ebena* (Lea, 1831), Ebonyshell.

Description: Somewhat oblong to nearly circular in outline, the shells of *Fusconaia ebena* are heavy, solid, and inflated; valves are especially thick in the umbo region. Beaks are greatly elevated, full, and directed forward and inward; sculpture consists of a few weak, indistinct ridges which are noticeable only in very young shells. Rest periods are evident by the prominent concentric ridges, which give the surface a rough appearance. Adults attain a length of 100–110 mm.

The left valve has two massive, triangular pseudocardinal teeth that are usually joined anteriorly and appear as one, and the depression between is roughly serrated; the two lateral teeth are wide, nearly straight. The right valve has a large, ragged, triangular pseudocardinal tooth that appears to stem from a deep depression or pit; the lateral tooth is high and finely serrated. The pseudocardinal teeth are directed posteriorly and appear parallel with the lateral teeth. Muscle scars are deeply impressed; the anterior adductor muscle scar is oblong to circular, the base being flat and roughened or evenly serrated. The interdentum is wide; the beak cavity is deep. The periostracum is without rays and is reddish brown to black. Some very young juveniles are dull olive yellow on the beaks. The nacre is pearly white, and the posterior end is iridescent. In some individuals the nacre has a light pink cast.

Life History and Ecology: The Ebonyshell typically is a big river species, occurring in current at depths of 10 to 15 feet or more. A course sand and gravel substrate provides the most suitable habitat, although this mussel seems to thrive in rivers such as the Hatchie on a bottom composed of sand, silt, and mud.

Fusconaia ebena is tachytictic; breeding usually takes place in May with the females being gravid until early fall. Fish hosts for the glochidia include the skipjack herring *(Alosa chrysochloris)*, green sunfish *(Lepomis cyanellus)*, largemouth bass *(Micropterus salmoides)*, white crappie *(Pomoxis annularis)*, and black crappie *(P. nigromaculatus)* (Fuller, 1974).

Plate 42. *Fusconaia ebena* (Lea, 1831), Ebonyshell.

Status: Currently Stable (Williams et al., 1993:11). Viable populations of the Ebonyshell occur throughout major stretches of the Cumberland River and the middle and lower (Kentucky Lake) sections of the Tennessee River. Excessive harvesting by commercial shellers has had a detrimental effect on some local populations.

Fusconaia flava (Rafinesque, 1820)
Wabash Pigtoe

RANGE MAP 42; PLATE 43

Synonymy:

Quadrula (Fusconaia) obliqua (Lamarck, 1819); Simpson, 1900a:788 [in part]

Obliquaria flava Rafinesque, 1820; Rafinesque, 1820:305, pl. 81, figs. 13, 14

Unio flavus (Rafinesque, 1820); Conrad, 1837:74

Fusconaia flava (Rafinesque, 1820); Ortmann, 1919:14

Fusconaia flava flava (Rafinesque, 1820); Haas, 1969a:313

? *Obliquaria (Sintoxia) lateralis* Rafinesque, 1820; Rafinesque, 1820:310

Unio undatus Barnes, 1823; Barnes, 1823:121, pl. 4, fig. 4

Fusconaja undata (Barnes, 1823); Ortmann, 1912a:241

Quadrula (Fusconaia) undata (Barnes, 1823); Simpson, 1914:880

Fusconaia undata (Barnes, 1823); Ortmann and Walker, 1922:6

Fusconaia undata undata (Barnes, 1823); Haas, 1969a:311

Unio rubiginosus Lea, 1829; Lea, 1829:427, pl. 8, fig. 10

Margarita (Unio) rubiginosus (Lea, 1829); Lea, 1836:20

Margaron (Unio) rubiginosus (Lea, 1829); Lea, 1852c:24

Unio rubiginosa Lea, 1829; Deshayes, 1839:672

Unio flavus var. *rubiginosus* (Lea, 1829); Paetel, 1890:152

Quadrula rubiginosa (Lea, 1829); Baker, 1898:77, pl. 19, fig. 1, pl. 20, fig. 1

Quadrula (Fusconaia) rubiginosa (Lea, 1829); Simpson, 1900a:786

Fusconaja rubiginosus (Lea, 1829); Ortmann, 1912a:241

Unio trigonus Lea, 1831; Lea, 1831:110, pl. 16, fig. 40

Margarita (Unio) trigonus (Lea, 1831); Lea, 1836:18

Margaron (Unio) trigonus (Lea, 1831); Lea, 1852c:25

Quadrula trigona (Lea, 1831); Baker, 1898:76, pl. 15, fig. 5

Quadrula (Fusconaia) trigona (Lea, 1831); Simpson, 1900a:787

Fusconaia undata trigona (Lea, 1831); Grier and Mueller, 1926:16, 32

Quadrula (Fusconaia) undata trigona (Lea, 1831); Frierson, 1927:54

Unio triangularis Say, 1834; Küster, 1852:56, pl. 12, fig. 3 [misidentification]

Unio pilaris Lea, 1840; Reeve, 1865:pl. 27, fig. 138 [misidentification]

Fusconaia selecta Wheeler, 1914; Wheeler, 1914:76, pl. 4

Quadrula (Fusconaia) undata selecta (Wheeler, 1914); Frierson, 1927:55

Fusconaia undata selecta Wheeler, 1914; Haas, 1969a:312

Fusconaia flava var. *parvula* Grier, 1918; Grier, 1918:11

Fusconaia rubiginosa parvula Grier, 1918; Baker, 1922a:19

Fusconaia flava parvula Grier, 1918; Baker, 1924:133

Quadrula (Fusconaia) flava sampsoniana Frierson, 1927:55

Fusconaia flava sampsoniana (Frierson, 1927); Haas, 1969a:314

Fusconaia undata trigonoides Frierson in Utterback, 1915; Utterback, 1915:107, pl. 15, figs. 30a–d; pl. 4, figs. 9a–b

Quadrula (Fusconaia) undata trigonoides (Frierson in Utterback, 1915); Frierson, 1927:55

Fusconaia undata trigonoides (Frierson, 1915); Haas, 1969a:312

Fusconaia undata wagneri Baker, 1928; Baker, 1928a:64–66, pl. 40, figs. 1–3

Type Locality: Small tributaries of Kentucky, Salt, and Green rivers (Murray and Leonard, 1962).

General Distribution: The entire Mississippi River drainage from western New York to eastern Kansas, Nebraska and South Dakota, south to Texas and Louisiana. In Canada it occurs in the Lake Huron, Lake St. Clair, and Lake Erie drainage basins of Ontario, and in the Red River–Nelson River system of Manitoba (Clarke, 1981a). Tombigbee River, Alabama.

Tennessee Distribution: Starnes and Bogan (1988) list *Fusconaia flava* as occurring in the Cumberland, Stones, Harpeth, Hatchie, and lower Tennessee rivers. Formerly present at Union City, North Fork Obion River, Obion County (Ortmann, 1926a). Rare in Reelfoot Lake, Obion County.

Description: Shells of the Wabash Pigtoe vary in outline from quadrate to subtriangular, are moderately heavy to thick, and compressed (specimens from the Harpeth River) to considerably elevated (Cumberland River specimens). Mature individuals average about 70 mm in length. Beaks vary from compressed in small to medium-sized rivers to high, full, and considerably elevated in big rivers; sculpture consists of 3–5 nearly concentric bars that form an angle on the posterior ridge. The anterior end is rounded, the dorsal margin is squared, and the posterior end is obliquely truncated. The dorsal ridge is distinct but with a sloping angle; a wide, shallow depression is usually present anterior to the posterior ridge. The surface is often marked with growth lines that are occasionally raised and form slight ridges.

The left valve has two elevated, triangular, serrated, divergent pseudocardinal teeth; the two lateral teeth are thin, high, and nearly straight. The right valve has

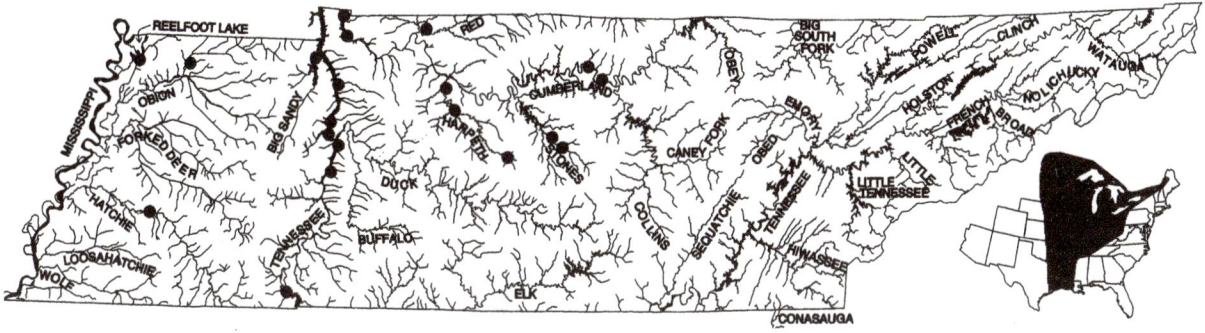

Range Map 42. *Fusconaia flava* (Rafinesque, 1820), Wabash Pigtoe.

a triangular, elevated, deeply serrated pseudocardinal tooth; the lateral tooth is high and thin, occasionally with a second smaller lateral tooth. The interdentum is fairly wide; the beak cavity is wide and moderately deep. The periostracum, typically without rays, is yellowish brown, becoming darker in old shells, and having a satinlike gloss. The nacre is whitish, often tinged with light pink or salmon; it is iridescent posteriorly. Ortmann (1926a) noted that specimens from the North Fork Obion River had a reddish nacre.

Life History and Ecology: *Fusconaia flava* is a species that may be found in medium-sized rivers like the Harpeth at depths of less than three feet, as well as in big rivers such as the Cumberland in water 12 to 15 feet deep. A stable substrate composed of coarse sand and gravel appear most suitable for this mussel. Manning (1989) collected live specimens in stretches of the Hatchie River with a firm clay and silt bottom. The Wabash Pigtoe is tachytictic, the breeding season lasting from May to August. Fuller (1974) lists the white crappie *(Pomoxis annularis),* black crappie *(P. nigromaculatus),* and bluegill *(Lepomis macrochirus)* as fish hosts for the glochidia.

Status: Currently Stable (Williams et al., 1993:11). Although adaptable to varying water depths, rates of flow and substrate composition, populations of *Fusconaia flava* in Tennessee appear local, and, but for a few exceptions, the number of individuals is low.

Fusconaia subrotunda (Lea, 1831)
Longsolid
RANGE MAP 43; PLATE 44
Synonymy:
? *Unio brevialis* Lamarck, 1819; Crouch, 1827:16, pl. 9, fig. 3 [misidentification]
Unio subrotundus Lea, 1831; Lea, 1831:117, pl. 18, fig. 454
Margarita (Unio) subrotundus (Lea, 1831); Lea, 1836:34
Margaron (Unio) subrotundus (Lea, 1831); Lea, 1852c:35

Plate 43. *Fusconaia flava* (Rafinesque, 1820), Wabash Pigtoe.

Quadrula (Fusconaia) subrotunda (Lea, 1831); Simpson,
  1900a:791
Quadrula subrotunda (Lea, 1831); Sterki, 1907:391
Fusconaja subrotunda (Lea, 1831); Ortmann, 1912a:294
Fusconaia subrotunda (Lea, 1831); Ortmann, 1919:7, pl. 1,
  fig. 2
Unio kirtlandianus Lea, 1834; Lea, 1834:98, pl. 14, fig. 41
Margarita (Unio) kirtlandianus (Lea, 1834); Lea, 1836:34
Unio kirklandianus Lea, 1834; Hanley, 1842a:203 [misspelling]
Margaron (Unio) kirtlandianus (Lea, 1834); Lea, 1852c:35
Quadrula (Fusconaia) kirtlandianus (Lea, 1834); Simpson,
  1900a:791
Quadrula kirtlandianus (Lea, 1834); Sterki, 1907:391
Fusconaja kirtlandiana (Lea, 1834); Ortmann, 1912a:245
Quadrula (Fusconaia) kirtlandiana (Lea, 1834); Simpson,
  1914:891
Fusconaia kirtlandiana (Lea, 1834); F. C. Baker, 1928a:136
Fusconaia subrotunda kirtlandiana (Lea, 1834); Ortmann,
  1919:11–14, pl. 1, figs. 3–5
Unio personatus Conrad, 1834 non Say, 1829; Conrad, 1834a:71
Unio politus Say, 1834; Say, 1834:no pagination
Unio lesueurianus Lea, 1840; Lea, 1840:286; Lea, 1842b:195,
  pl. 8, fig. 6
Margaron (Unio) lesueurianus (Lea, 1840); Lea, 1852c:35
Fusconaia pilaris lesueuriana (Lea, 1840); Ortmann, 1918:528
Unio pilaris Lea, 1840; Lea, 1840:285; Lea, 1842b:209, pl.
  14, fig. 24
Margaron (Unio) pilaris (Lea, 1840); Lea, 1852c:35
Quadrula (Fusconaia) pilaris (Lea, 1840); Simpson, 1900a:792
Fusconaia pilaris (Lea, 1840); Ortmann, 1918:527
Unio globatus Lea, 1871; Lea, 1871:191; Lea, 1874:5, pl. 1,
  fig. 1
Quadrula (Fusconaia) globata (Lea, 1871); Simpson, 1900a:793
Unio bursa-pastoris B. H. Wright, 1896; B. H. Wright,
  1896:133, pl. 3
Fusconaja bursapastoris (B. H. Wright, 1896); Ortmann,
  1913a:90–91
Quadrula (Fusconaia) bursa-pastoris (B. H. Wright, 1896);
  Simpson, 1900a:791
Fusconaia pilaris bursa-pastoris (B. H. Wright, 1896);
  Ortmann, 1918:529
Fusconaia bursapastoris (B. H. Wright, 1896); Haas, 1969a:316
Quadrula flexuosa Simpson, 1900; Simpson, 1900b:83, pl. 2,
  fig. 8
Quadrula (Fusconaia) kirtlandiana var. *minor* Simpson, 1900;
  Simpson, 1900a:791
Quadrula andrewsii Marsh, 1902; Marsh, 1902a:115
Quadrula andrewsae Marsh, 1902; Marsh 1902b:8, pl. 1,
  upper two figs. [unjustified emendation]
Quadrula (Fusconaia) andrewsii Marsh, 1902; Simpson,
  1914:895
Quadrula beauchampii Marsh, 1902; Marsh, 1902b:7, pl. 1,
  lower two figs.
Quadrula (Fusconaia) beauchampii Marsh, 1902; Simpson,
  1914:895
Fusconaja subrotunda var. *leucogona* Ortmann, 1913; Ortmann,
  1913b:89–90

Type Locality: *Fusconaia subrotunda*, Ohio; *F. pilaris*,
French Broad and Holston rivers, Tennessee; *F. p. lesueuri-
iana*, Caney Fork and Holston rivers, Tennessee; *F.
bursa-pastoris*, Powell River, Virginia.

General Distribution: Ohio, Cumberland, and Tennessee River systems (Simpson, 1914:893).

Tennessee Distribution: Considering the several subspecies or varieties of the Longsolid as one taxon, *Fusconaia subrotunda*, it is listed by Starnes and Bogan
(1988) as occurring in the Clinch, Powell, Little Tennessee (now extirpated), Elk, lower Tennessee, and
Cumberland rivers, and prior to 1960 it was known
from the French Broad, Holston, middle and upper
Tennessee, and Obey rivers. Occasional specimens may
still be found in the Holston River, Knox and Grainger
counties. It was formerly fairly common in the Tellico
River, Monroe County (Parmalee and Klippel, 1984),
and there is a small population of this mussel in a short
stretch of the Hiwassee River, Polk County (Parmalee
and Hughes, 1994).

Description: The big river form of *Fusconaia subrotunda* has a shell that is oval to broadly elliptical or
oblong in outline, solid, and inflated. Beaks are high, full,
and turned forward over the lunule; sculpture consists
of a few subnodular ridges or wrinkles. The anterior margin is broadly rounded, and the ventral and posterior
margins are slightly curved to nearly straight; lacking a
distinct posterior ridge, the dorsal slope is evenly curved.
The shell surface is generally sculptured with low, wide,
concentric ridges. Mature individuals may reach a length
of 100 mm. Shells of the medium-sized to small river
forms, *Unio pilaris* and *Unio lesueurianus*, are more con-
sistently oval in shape, compressed to only moderately
inflated, rayed (especially in juveniles), and olive yellow to light brown in color with a satinlike appearance. As Ortmann (1918:528) noted when discussing
mussels of the upper Tennessee River drainage, "The
typical form of *F. pilaris* passes into the following local races [*F. p. lesueuriana* and *F. p. bursa-pastoris*, the
compressed headwaters form] in an upstream direction, and it is hard to draw a line between them."

The left valve has two low, heavy triangular pseudocardinal teeth, widely separated with deep striations;
the two lateral teeth are moderately long and straight.
The right valve has a heavy, triangular serrated pseu-

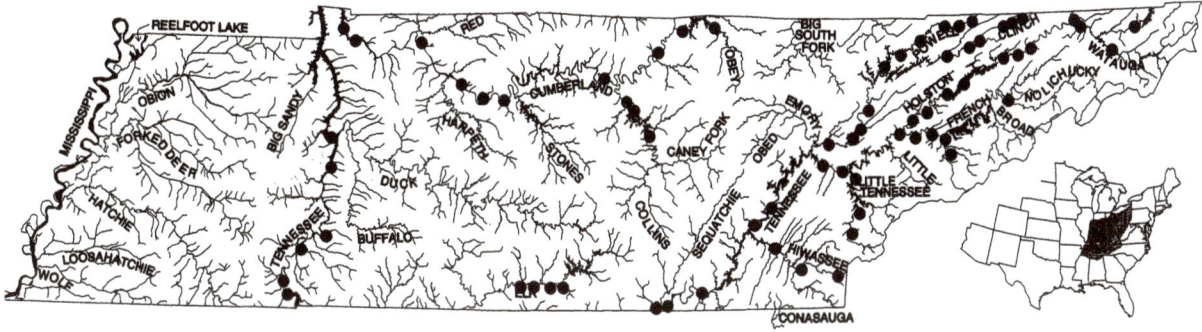

Range Map 43. *Fusconaia subrotunda* (Lea, 1831), Longsolid.

docardinal tooth usually with a low tooth or roughened area on either side. The lateral tooth is broad, serrated, and often doubled for most of its length. Muscle scars are deeply impressed, the interdentum is usually wide, and the beak cavity is deep and somewhat compressed. The periostracum is a dull straw yellow to greenish brown, becoming blackish in old individuals. Some may show subtle green rays, primarily on the umbos. The nacre is pearly white and iridescent posteriorly.

Life History and Ecology: The small to medium-sized river forms of *Fusconaia subrotunda* typically are found in current, usually in riffle areas, at a depth of less than two feet. The big river form, such as the one inhabiting the Tennessee and Cumberland River reservoirs, may live at depths of 12–18 feet, in current, and on a sand and gravel substrate. The Longsolid, like other species of *Fusconaia* for which the breeding season is known, is probably tachytictic, the females becoming gravid during the summer. Host fish for glochidia unknown.

Status: Special Concern (Williams et al., 1993:11). Commercial shellers working the Cumberland and lower Tennessee (Kentucky Lake) River reservoirs occasionally encounter individuals of the big river form of *Fusconaia subrotunda,* but it appears to be uncommon. In Kentucky, Cicerello et al. (1991:118) record it as "sporadic and rare in the Tennessee, upper Green, Licking, and Big Sandy rivers." Populations of medium-sized and small river forms, such as those found in the upper Powell and Clinch rivers, have remained stable and viable. Elsewhere in similar river habitats, such as the Hiwassee, it is very localized and uncommon.

Plate 44. *Fusconaia subrotunda* (Lea, 1831), Longsolid.

Hemistena lata (Rafinesque, 1820)
Cracking Pearlymussel

RANGE MAP 44; PLATE 45

Synonymy:

Anodonta (Lastena) lata Rafinesque, 1820; Rafinesque
  1820:317, pl. 82, figs. 17, 18
Unio latus (Rafinesque, 1820); Conrad, 1834a:70
Anodonta latus Rafinesque, 1820; Férussac, 1835:25
Leptodea lata (Rafinesque, 1820); Conrad, 1853:262
Anodon lata (Rafinesque, 1820); Sowerby, 1867:pl. 19, fig. 76
Lastena lata (Rafinesque, 1820); Simpson, 1900a:654
Hemistena lata (Rafinesque, 1820); Frierson, 1927:66
Unio dehiscens Say, 1829; Say, 1829:308; Say, 1830a:pl.24
Margarita (Unio) dehiscens (Say, 1829); Lea, 1836:35
Hemilastena dehiscens (Say, 1829); Agassiz, 1852:50
Margaron (Margaritana) dehiscens (Say, 1829); Lea, 1852c:43
Baphia dehiscens (Say, 1829); H. and A. Adams, 1858:499
Margaritana dehiscens (Say, 1829); Lewis, 1870:219
Anodonta dehiscens (Say, 1829); Paetel, 1890:178
Unio oriens Lea, 1831; Lea, 1831:68, 73, pl. 6, fig. 5
Odatelia radiata Rafinesque, 1832; Rafinesque, 1832:154
Unio hildrethi Delessert, 1841; Delessert, 1841:pl. 19, figs. 4a, b
Unio dehiscens orienopsis De Gregorio, 1914; De Gregorio,
  1914:39, pl. 7, figs. 2a, b

Type Locality: Kentucky River.

General Distribution: Ohio, Cumberland, and Tennessee River systems (Simpson, 1914).

Tennessee Distribution: The Cracking Pearlymussel has been recovered in the Cumberland River from Clay County, Tennessee, upstream to Burnside, Pulaski County, Kentucky, and from the Big South Fork Cumberland River (Wilson and Clark, 1914; Ortmann, 1924a; Neel and Allen, 1964). This deeply burrowing naiad was collected from the Clinch River from Anderson County, Tennessee, upstream to Russell County, Virginia (Goodrich, 1913; Ortmann, 1918; Stansbery,

1973; Bates and Dennis, 1978). Lewis (1870) and Call (1885) recorded *H. lata* from the Holston River in East Tennessee. *Hemistena* inhabited the Tennessee River from below Knoxville downstream to Muscle Shoals, Alabama (Ortmann, 1918, 1925; Morrison, 1942; Stansbery, 1964). The lower Elk River in Tennessee and Alabama formerly supported a population of *H. lata* (Ortmann, 1925; Isom et al., 1973). The Cracking Pearlymussel has been reported from the Duck and Buffalo rivers in Middle Tennessee (Marsh, 1885; Ortmann, 1924a, 1925; Isom and Yokley, 1968a; van der Schalie, 1973). A specimen of *H. lata* was collected at Diamond Island (Tennessee River Mile 196.0) below Pickwick Landing Dam, Hardin County, in November 1980.

Description: Valves of *Hemistena lata* are elongated, elliptical to subrhomboid, and slightly inflated. Mature specimens may reach a length of 90 mm, the shell being thin but fairly strong. The anterior end is evenly rounded; the posterior end is truncated or bluntly pointed. Beaks are flattened and sculptured with a few strong ridges. Valves do not meet but gape along the anterior and posterior margins. The posterior ridge is low and rounded, and in some specimens sculptured with a few strong ridges. The surface of the shell may be marked by uneven growth lines. Pseudocardinal teeth are represented by a single raised knob or ridge, while the lateral teeth appear as a thickened hinge line. Muscle scars are confluent but well defined. The beak cavity is very shallow or absent. The periostracum color varies from a dull yellow to brownish green to brown, usually with scattered, broken dark green rays. The nacre color is a pale bluish white, with the beak cavity being a dark purple.

Life History and Ecology: The Cracking Pearlymussel is found to be most numerous in medium-sized rivers,

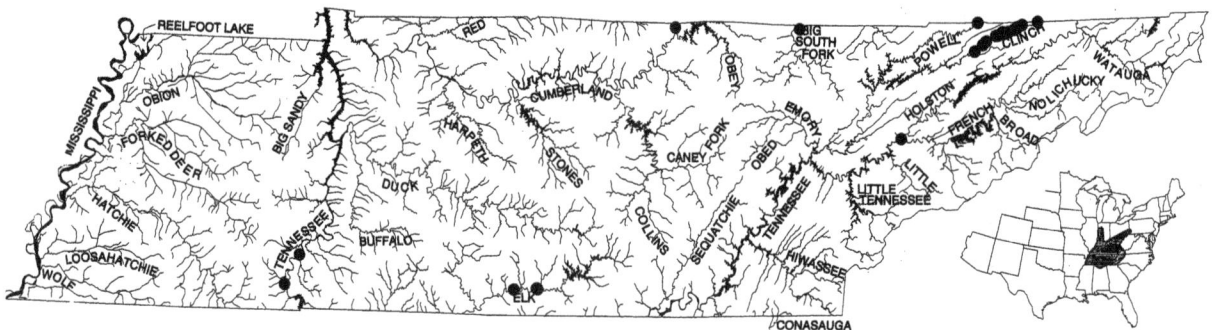

Range Map 44. *Hemistena lata* (Rafinesque, 1820), Cracking Pearlymussel.

Plate 45. *Hemistena lata* (Rafinesque, 1820), Cracking Pearlymussel.

such as the unimpounded stretches of the Clinch River in upper East Tennessee and southwestern Virginia. Usually occurring in less than two feet of water in moderate current, *Hemistena lata* spends most of its life deeply buried in a substrate composed of mud, sand, and fine gravel. In spite of this trait, muskrats prove effective in finding them.

Hemistena lata is tachytictic, glochidia having been observed during mid-May (Ortmann, 1915). Little else is known about this species, including host fish for the glochidia.

Status: Endangered (Williams et al., 1993:11). This mussel has been extirpated from most of its former range, although local and apparently viable populations survive in upper Clinch River (Hancock County) in East Tennessee. A population in the Elk River persisted until 1981 (Ahlstedt, 1983; Barr et al., 1993–1994). The U.S. Fish and Wildlife Service has developed a recovery plan for this species (U.S. Fish and Wildlife Service, 1990b).

Lampsilis abrupta (Say, 1831)
Pink Mucket
RANGE MAP 45; PLATE 46

Synonymy:
Unio orbiculatus Hildreth, 1828; Hildreth, 1828:284, fig. 15
Margarita (Unio) orbiculatus (Hildreth, 1828); Lea, 1836:25
Margaron (Unio) orbiculatus (Hildreth, 1828); Lea, 1852c:28
Lampsilis orbiculatus (Hildreth, 1828); Simpson, 1900a:540
Lampsilis orbiculata (Hildreth, 1828); Daniels, 1903:647
Lampsilis orbiculata (Hildreth, 1828); Stansbery, 1971:15
Unio abruptus Say, 1831; Say, 1831a:pl. 17
Lampsilis (Ortmanniana) abrupta (Say, 1831); Haas, 1969a:461
Toxolasma cyclips Rafinesque, 1831; Rafinesque, 1831:2
Unio cyclips (Rafinesque, 1831); Férussac, 1835:28
Unio crassus Say, 1817; Conrad, 1836:34, pl. 16 [in part]
Obovaria retusa (Lamarck, 1819); Frierson, 1927:89 [in part]

Type Locality: Muskingum River, Ohio.

General Distribution: Ortmann (1925:359) recorded this mussel as occurring in the Mississippi, Ohio, Cumberland, and Tennessee rivers, and in the Tennessee River "up to the lower Clinch, where it is very rare."

Tennessee Distribution: Typically a big river species, records of the Pink Mucket in the state primarily are from the Tennessee and Cumberland rivers. Except for an occasional relict individual (Ahlstedt and McDonough, 1994), it has about disappeared from the upper and middle stretches of the Tennessee River; probably the most stable population occurs below Pickwick Landing Dam, Hardin County. Populations of *Lampsilis abrupta* in the Cumberland River also tend to be localized, one of the larger occurring in the Carthage–Rome stretch, Smith County. Occasionally individuals become established in small to medium-sized tributaries of large rivers such as the Tennessee; Ahlstedt (1991a, b) cites several such localities that include, in addition to the Holston and French Broad rivers (that form the Tennessee River at Knoxville), the upper Clinch where it is rare.

Description: The shells of the Pink Mucket are somewhat inflated, especially so in large, mature females, and subquadrate or orbicular in outline. Valves become thick and heavy in mature individuals, and males (generally averaging larger than females) may reach a length of 110–120 mm. The anterior margins are evenly rounded, while the dorsal and ventral margins are slightly curved. The posterior margin of the female shell is slightly rounded to straight, while that of the male rounded

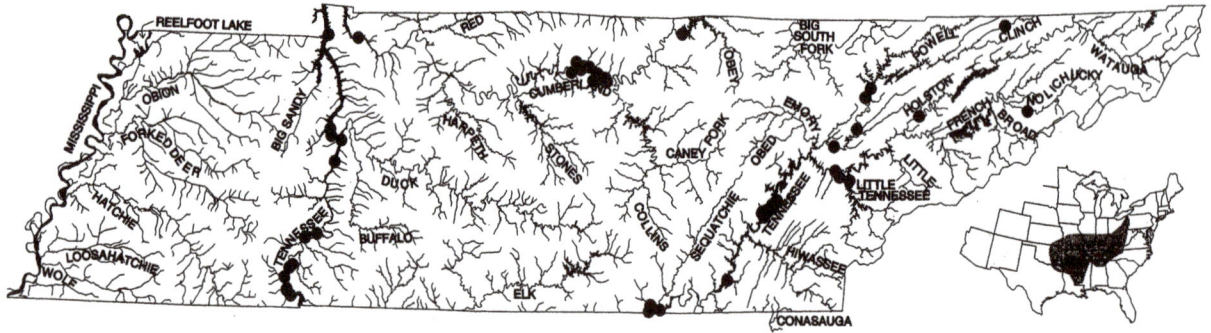

Range Map 45. *Lampsilis abrupta* (Say, 1831), Pink Mucket.

or bluntly pointed. A posterior ridge, well defined in males, is distinct along the dorsal margin. The two valves slightly gape along the anterior margin. Beaks are located in the anterior third of the shell, and in young individuals beaks are marked by faint, scarcely looped ridges. The surface is marked by uneven concentric rest lines.

The left valve has two large triangular pseudocardinal teeth separated from two strong, slightly curved lateral teeth by a short, broad interdentum. The right valve has one large triangular pseudocardinal tooth; sometimes there are smaller teeth before and behind the larger tooth. There is one large, slightly curved lateral tooth in the right valve. Anterior muscle scars are deep and rough, posterior muscle scars are well defined but shallow. The beak cavity is broad and deep. The periostracum color varies from a light yellow (in juveniles) or yellowish brown to dark brown and is occasionally marked with broken fine to fairly wide dark green rays. The nacre color varies from white to pink to salmon, with the posterior margin iridescent.

The posterior margin of the male shell is bluntly pointed; the female shell is squared off posteriorly and is inflated posteriorly to accommodate the marsupium (Hildreth, 1828; Simpson, 1914). Simpson (1900a:540) noted that "Some specimens can hardly be separated from *L. higginsii*"; in the case of the males, close similarities also exist between the shells of *L. abrupta* and *Actinonaias ligamentina*.

Life History and Ecology: Hickman (1937) collected specimens of the Pink Mucket in the Clinch and Holston rivers from locales with a rocky bottom and swift current, in less than three feet of water. Ortmann (1919) noted that he had collected *L. abrupta* from riffles with a strong current in large rivers (e.g., Ohio River). The

Pink Mucket is bradytictic; it becomes gravid in August, and females contain glochidia in September which are discharged the following June (Ortmann, 1912a, 1919). Fuller (1974) recorded the host fish for *L. abrupta* as the sauger *(Stizostedion canadense)* and freshwater drum *(Aplodinotus grunniens)*. However, Surber (1913)

Plate 46. *Lampsilis abrupta* (Say, 1831), Pink Mucket.

listed the sauger as the host fish for *L. higginsi,* and the records Fuller (1974) used are for *higginsi,* which he probably considered a synonym of *L. abrupta.*

Status: Endangered (Williams et al., 1993:11). This mussel has been found living in tailwaters of several dams, and there is a localized relict population in the Cumberland River, Smith County, but all individuals examined appear to be old adults. Recently collected specimens, those taken by commercial shellers, are mature and appear to be relict individuals of former (perhaps preimpoundment) populations. The U.S. Fish and Wildlife Service has developed a recovery plan for this species (U.S. Fish and Wildlife Service, 1985b) and has created a watershed implementation schedule for the recovery plan (U.S. Fish and Wildlife Service, 1989b).

Lampsilis altilis (Conrad, 1834)
Finelined Pocketbook
RANGE MAP 46; PLATE 47

Synonymy:
Unio altilis Conrad, 1834; Conrad, 1834a:43, 68, pl. 2, fig. 1
Margarita (Unio) altilis (Conrad, 1834); Lea, 1836:24
Margaron (Unio) altilis (Conrad, 1834); Lea, 1852c:27
Lampsilis altilis (Conrad, 1834); Simpson, 1900a:529
Lampsilis (Lampsilis) altilis (Conrad, 1834); Simpson, 1900a:529
Unio clarkianus Lea, 1852; Lea, 1852a:251; Lea, 1852b:273, pl. 21, fig. 30
Margaron (Unio) clarkianus (Lea, 1852); Lea, 1852c:27
Lampsilis clarkianus (Lea, 1852); Simpson, 1900a:532
Unio spillmanii Lea, 1861; Lea, 1861a:39; Lea, 1862b:98, pl. 15, fig. 246
Margaron (Unio) spillmanii (Lea, 1861); Lea, 1870:42
Lampsilis (Lampsilis) spillmani (Lea, 1861); Frierson, 1927:69 [misspelling]
Unio gerhardtii Lea, 1862; Lea, 1862a:168; Lea, 1862c:208, pl. 31, fig. 277
Margaron (Unio) gerhardtii (Lea, 1862); Lea, 1870:35
Lampsilis (Lampsilis) gerhardtii (Lea, 1862); Simpson, 1900a:532
Unio doliaris Lea, 1865; Lea, 1865:88; Lea, 1868b:260, pl. 32, fig. 75
Margaron (Unio) doliaris (Lea, 1865); Lea, 1870:42
Lampsilis (Lampsilis) doliaris (Lea, 1865); Simpson, 1900a:533

Type Locality: Alabama River, Claiborne, Alabama.

General Distribution: Alabama River drainage (Simpson, 1914).

Tennessee Distribution: Conasauga River, Polk and Bradley counties.

Description: The shell is fairly thin, subelliptical to ovate in outline; specimens from the Conasauga River are moderately inflated; those from small tributaries, such as Coahulla Creek, are more compressed. Mature individuals from the Conasauga River, Bradley County, Tennessee, attain a maximum length of about 85 mm. The anterior end is broadly rounded in creek forms, somewhat more narrowly so in Conasauga River specimens; the ventral margin is slightly curved; the posterior end is bluntly rounded. Some individuals develop a very low, rounded posterior ridge. Beaks are moderately swollen, only slightly projecting beyond the hinge line.

The left valve has two moderately thin, finely serrated pseudocardinal teeth, and the anterior one is generally more elevated; there are two thin lateral teeth which are widely spread and elevated. The right valve has two pseudocardinal teeth, the largest fairly heavy and triangular; the second tooth anterior to it is low, elongated, and thin. The single lateral tooth is straight, thin, and high. The interdentum is narrow, the beak cavity is moderately deep, and the anterior muscle scars

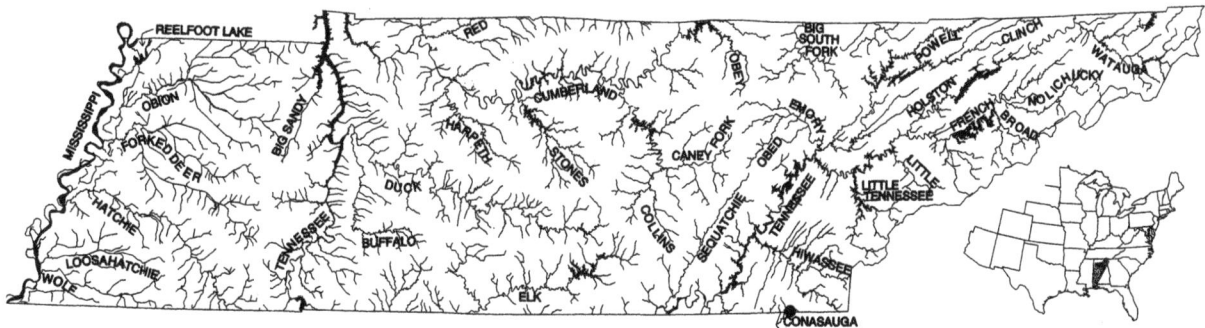

Range Map 46. *Lampsilis altilis* (Conrad, 1834), Finelined Pocketbook.

Plate 47. *Lampsilis altilis* (Conrad, 1834), Finelined Pocketbook.

are fairly deep. The periostracum is a dull straw yellow, and the majority of individuals have a few to many usually fine and often obscure greenish rays. The nacre is a dull bluish white, occasionally with a tint of pale pink or salmon.

Life History and Ecology: In the Conasauga River and its tributary, Coahulla Creek, the Fine-lined Pocketbook is found usually in a substrate composed of sand and mud mixed with some gravel, in moderate current, and at depths of about three feet or less. The reproductive season is unknown. Haag et al. (1997) have shown through laboratory fish host identification experiments that the redeye bass *(Micopterus coosae),* spotted bass *(M. punctulatus),* and largemouth bass *(M. salmoides)* may serve as hosts for the glochidia of *Lampsilis altilis.*

Status: Threatened (Williams et al., 1993:10). The U.S. Fish and Wildlife Service has developed a recovery plan for this species (U.S. Fish and Wildlife Service, 1994).

Lampsilis cardium Rafinesque, 1820
Plain Pocketbook

RANGE MAP 47; PLATE 48

Synonymy:

Lampsilis cardium Rafinesque, 1820; Rafinesque, 1820:298, pl. 80, figs. 16–19

Lampsilis cardia Rafinesque, 1831; Utterback, 1916c:15

Lampsilis (Lampsilis) cardium Rafinesque, 1820; Frierson, 1927:67

Unio cardium (Rafinesque, 1820); Conrad, 1834a:68

Unio ventricosus Barnes, 1823; Barnes, 1823:267, pl. 13, fig. 14

Mya ventricosus (Barnes, 1823); Eaton, 1826:221

Margarita (Unio) ventricosus (Barnes, 1823); Lea, 1836:23

Margaron (Unio) ventricosus (Barnes, 1823); Lea, 1852c:26

Lampsilis ventricosus (Barnes, 1823); F. C. Baker, 1898:94, pl. 12, figs. 3–5

Lampsilis ventricosa (Barnes, 1823); Stimpson, 1851:14

Unio cardium var. *ventricosus* (Barnes, 1823); Paetel, 1890:147

Lampsilis (Lampsilis) ventricosus (Barnes, 1823); Simpson, 1900a:526

Lampsilis ovata ventricosa (Barnes, 1823); Ortmann, 1913a:311

Unio occidens Lea, 1829; Lea, 1829:435; Lea, 1834:49, pl. 10, fig. 16

Margarita (Unio) occidens (Lea, 1829); Lea, 1836:23

Margaron (Unio) occidens (Lea, 1829); Lea, 1852c:26

Unio cardium var. *occidens* Lea, 1829; Paetel, 1890:147

Lampsilis ventricosa occidens (Lea, 1829); F. C. Baker, 1928a:286, pl. 92

Unio subovatus Lea, 1831; Lea, 1831:118, pl. 18, fig. 46

Margarita (Unio) subovatus (Lea, 1831); Lea, 1836:19

Margaron (Unio) subovatus (Lea, 1831); Lea, 1852c:24

Unio fasciolus (Rafinesque, 1820); Férussac, 1835:26 [in part]

Unio ovata Say, 1817; Deshayes, 1839:669 [misidentification]

Unio ovatus Say, 1817; Küster, 1852:55, pl. 12, fig. 1 [misidentification]

Unio latissimus Rafinesque, 1820; Sowerby, 1868:pl.66, fig. 337 [misidentification]

Unio dolabraeformis Lea, 1838; Sowerby, 1867:298 [misidentification]

Unio lenis Conrad, 1838; Conrad 1838:backcover No. 11; Conrad, 1840:106, pl. 58, fig. 2

Unio canadensis Lea, 1857; Lea, 1857d:85; Lea, 1860d:268, pl. 44, fig. 148

Margaron (Unio) canadensis (Lea, 1857); Lea, 1870:37

Lampsilis ventricosa canadensis (Lea, 1857); Ortmann, 1919:307, pl. 19, figs. 4, 5

Lampsilis ovata canadensis (Lea, 1857); La Rocque 1967:222

Lampsilis ventricosa cohongoronta Ortmann, 1912; Ortmann, 1912c:53

Lampsilis ventricosa var. *lurida* Simpson, 1914; Simpson, 1914:41

Lampsilis ventricosa lurida Simpson, 1914; F. C. Baker, 1928a:289, pl. 93, fig. 5, pl. 94, figs. 5, 6

Lampsilis ventricosa var. *perglobosa* F. C. Baker, 1928; F. C. Baker, 1928a:285, pl. 93, figs. 1–4

Lampsilis ventricosa var. *winnebagoensis* F. C. Baker, 1928; F. C. Baker, 1928a:291, pl. 94, figs. 1–4

Type Locality: Wisconsin River; Mississippi River, Prairie du Chien, Wisconsin.

General Distribution: The entire upper Mississippi River drainage from northern Arkansas and Tennessee north to Minnesota and Wisconsin, and from New York west to eastern Kansas (Murray and Leonard, 1962). The Winnipeg, Red, and Nelson River systems of central Canada; Great Lakes–St. Lawrence system throughout except most of Lake Superior (Clarke, 1981a).

Tennessee Distribution: Starnes and Bogan (1988) list *Lampsilis cardium* from most medium-sized and large rivers in East and Middle Tennessee including, among others, the Cumberland, Tennessee, French Broad, Holston, Clinch, Powell, Elk, Duck, Buffalo, Harpeth, and Stones.

Description: The Plain Pocketbook is one of the largest mussels occurring in Tennessee with old individuals reaching a length of 140–150 mm. Although thin when young, shells of mature individuals become solid and often very heavy; they are somewhat ovate or elliptical, inflated, swollen, and especially high in old females. The anterior end is sharply rounded, but the posterior end is bluntly pointed at the union of the dorsal and ventral margins; the posterior slope is flattened or slightly convex. The anterior area in front of umbones is often extended and somewhat alate. Beaks are swollen and elevated; sculpture consists of 4–5 coarse bars, of which the second and third may appear slightly double-looped, while the others are indistinct. Lines of growth are usually distinct, sometimes elevated into low ridges at rest periods.

The left valve has two heavy pseudocardinal teeth, which are somewhat compressed, elevated, and roughened in young individuals and squarish or stumpy in old specimens. The two lateral teeth are high, straight, fairly long, and striated. The right valve has a heavy, erect, high, and somewhat triangular pseudocardinal tooth, usually with a thin, elongated, moderately elevated tooth in front and occasionally a small peglike tooth behind. The lateral tooth is high, straight, and characteristically squared-off or abruptly truncated at the posterior end. The interdentum is very narrow, and the beak cavity is broad and deep. The periostracum is smooth, shiny, yellowish green, and the umbones often have a lighter tan. There are usually numerous dark green rays of varying numbers and widths distributed over most of the surface. The nacre is a dull white; some specimens have a light gun blue or salmon wash.

There has been some uncertainty as to the exact taxonomic relationship of *Lampsilis ventricosa* (=*L. cardium*) within the *Lampsilis ovata/cardium/satura/excavata* (=*Lampsilis ornata* (Conrad, 1835)) complex. In her study of the *Lampsilis ovata* complex in the Ohio River drainage system, Putnam (1971) concluded that *L. ovata* and *L. ventricosa* (=*L. cardium*) should be considered two distinct but closely related species. In an attempt to analyze the clinal possibilities as they relate to three nominal species of *Lampsilis* (*L. ovata, L. ventricosa, L. excavata* [=*ornata*]), Cvancara (1963:222) concluded:

> It has been a common experience to label northern forms *L. ventricosa*; that is, those forms found in the upper Mississippi River drainage (in latitudes encompassing New York, northern Ohio, Michigan, Wisconsin, and northern Illinois). On the other hand, specimens from Kentucky and Tennessee would be considered "southern" and with their more centrally placed beaks and their high and well-developed posterior ridges, these forms are identified as *L. ovata*. But in such intermediate areas as southern Ohio and Illinois intergrades are common and they have usually been named *L. ovata ventricosa*.

Range Map 47. *Lampsilis cardium* Rafinesque, 1820, Plain Pocketbook.

Based on the centrally placed, high, swollen beaks and sharp-angled, high posterior ridge, shells of *Lampsilis ovata* can usually be distinguished from those of *L. cardium*. As is the case with some specimens in rivers of the Midwest, certain individuals from Tennessee lakes (reservoirs) and rivers also appear intermediate between the two taxa. We are treating *L. cardium* and *L. ovata* as separate species.

Life History and Ecology: The Plain Pocketbook may be found at a depth of less than two feet in free-flowing rivers such as the Harpeth, and in 12 to 20 feet of water in the Tennessee and Cumberland River reservoirs. It usually occurs in sections of rivers with moderate to strong current in a substrate of coarse gravel and sand; like *Lampsilis ovata,* the Plain Pocketbook also seems to thrive on a stable substrate composed of a high percentage of mud and silt.

Plate 48. *Lampsilis cardium* Rafinesque, 1820, Plain Pocketbook.

Lampsilis cardium is bradytictic, a long-term breeder retaining glochidia from early August to early July (Baker, 1928a). Fish hosts for the glochidia (both *L. cardium* and *L. ovata*) listed by Fuller (1974) include white crappie *(Pomoxis annularis),* bluegill *(Lepomis macrochirus),* smallmouth bass *(Micropterus dolomieu),* largemouth bass *(M. salmoides),* yellow perch *(Perca flavescens),* and sauger *(Stizostedion canadense).* Waller et al. (1985), using artificial infestations, confirmed the largemouth bass as a host fish and added the walleye *(Stizostedion vitreum)* as a host fish for the glochidia of this species. Watters (1996a) added the green sunfish *(Lepomis cyanellus)* and banded killifish *(Fundulus diaphanus)* to the list of known host fish for glochidia of the Plain Pocketbook. Watters (1997) has identified larval tiger salamanders and six exotic fish species which act as surrogate hosts for the glochidia of *L. cardium* including the following: guppies, Siamese fighting fish, flame gourami, Panchax killifish, painted sword and lavender gourami.

Status: Special Concern (Williams et al., 1993:12). Pocketbooks appear rather tolerant of poor or polluted water conditions unsuitable for many mussel species; they seem to be maintaining viable populations throughout most of their range in Tennessee.

Lampsilis fasciola Rafinesque, 1820
Wavyrayed Lampmussel

RANGE MAP 48; PLATE 49

Synonymy:
Lampsilis fasciola Rafinesque, 1820; Rafinesque, 1820:299
Unio fasciolus (Rafinesque, 1820); Say, 1834:no pagination
Lampsilis (Lampsilis) fasciola Rafinesque, 1820; Frierson, 1927:69
Ligumia fasciola Rafinesque, 1820; Goodrich, 1932:109
Unio multiradiatus Lea, 1829; Lea, 1829:434, pl. 9, fig. 15
Margarita (Unio) multiradiatus (Lea, 1829); Lea, 1836:24
Margaron (Unio) multiradiatus (Lea, 1829); Lea, 1852c:31
Lampsilis (Lampsilis) multiradiatus (Lea, 1829); Simpson, 1900a:532
Lampsilis multiradiata (Lea, 1829); Sterki, 1907:388
Lampsilis (Lampsilis) multiradiata (Lea, 1829); Simpson, 1914:55
Unio perradiatus Lea, 1858; Lea, 1858a:40; Lea, 1862b:66, pl. 6, fig. 215
Margaron (Unio) perradiatus (Lea, 1858); Lea, 1870:37
Unio altilis Conrad, 1834; Reeve, 1865:pl. 23, fig. 109 [misidentification]
Unio perovalis Conrad, 1834; Sowerby, 1866:pl. 38, fig. 209 [misidentification]

Type Locality: Kentucky River.

General Distribution: "Great Lakes drainage in the tributaries of lake Michigan, Lake Huron, Lake St. Clair and Lake Erie, and Ohio–Mississippi drainage south to the Tennessee River system" (Clarke, 1981a:338).

Tennessee Distribution: The Wavyrayed Lampmussel occurs in a large number of the small creeks and medium-sized rivers throughout East and Middle Tennessee, as well as very locally in the main Cumberland and Tennessee River reservoirs. Starnes and Bogan (1988) list it from the Watauga, Emory, French Broad, Sequatchie, Buffalo, Obey, Harpeth, Red, and Roaring rivers prior to 1960; apparently it is now extirpated from these rivers. Small, localized populations still existed in the Harpeth River in 1973.

Description: The shell is elliptical or subovate, fairly thin to solid and heavy; it is generally inflated, especially so as a marsupial swelling in some females. Mature individuals may reach a length of 90–100 mm. Shells of some individuals in certain populations, such as those found in the Obed River (Cumberland County) and the Harpeth River (Cheatham County), may become extremely thick and reach a length of nearly 110 mm. The anterior end is broadly rounded; the ventral margin is straight to slightly curved; the posterior end is rounded (in some females) to bluntly pointed (in males). The posterior ridge is broadly rounded; the posterior-dorsal margin is occasionally compressed, appearing alate. Beaks are full, depressed, and only slightly elevated above the hinge line; sculpture consists of several indistinct, fine, wavy ridges. The surface is shiny, usually with numerous raised rest lines.

The left valve has two triangular, short, thick, rather

Plate 49. *Lampsilis fasciola* Rafinesque, 1820, Wavyrayed Lampmussel.

widely separated pseudocardinal teeth; the two lateral teeth are short, thick, nearly straight, and widely separated. The right valve has a large, coarsely serrated, heavy, erect pseudocardinal tooth, sometimes with a suggestion of a smaller tooth on either side as slightly roughened, raised areas; the lateral tooth is wide, short,

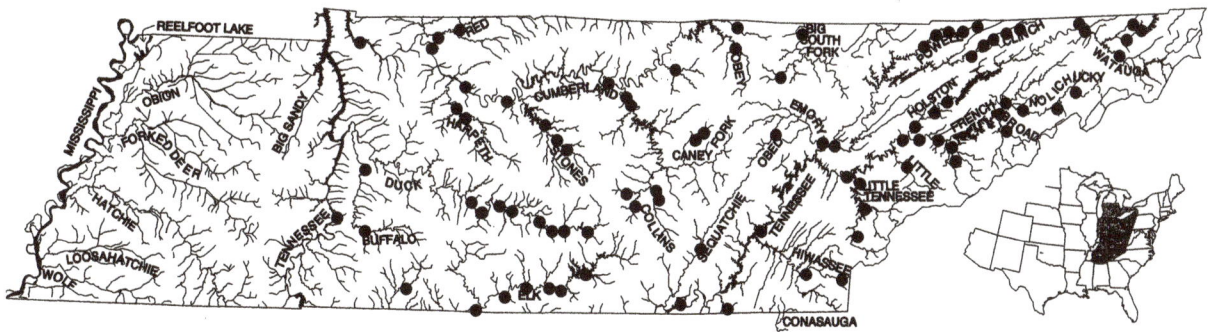

Range Map 48. *Lampsilis fasciola* Rafinesque, 1820, Wavyrayed Lampmussel.

and elevated. The interdentum is narrow or absent; the beak cavity is wide and moderately deep. The periostracum is light yellow or yellowish green, the beaks often tinged with reddish brown; the surface is densely patterned with green rays of varying widths, characteristically wavy in appearance, and often interrupted at the lines of growth. The nacre is white or bluish-white as a result of iridescence, especially posteriorly.

Life History and Ecology: *Lampsilis fasciola* is a species typical of small to medium-sized rivers, usually occurring at depths of three feet or less. Individuals encountered at depths of 10 to 18 feet in the Cumberland and Tennessee rivers (reservoirs) may be relicts from preimpoundment periods or simply of fortuitous occurrence. This mussel appears tolerant of habitat conditions unfavorable to many species, and, under favorable circumstances, including moderate current and a stable substrate composed of mud, sand, and gravel, it may become quite abundant locally. In spite of the extensive range and local abundance of the Wavyrayed Lampmussel, its reproductive period remains unknown, although there is some evidence to suggest it is bradytictic. In their detailed study of fish hosts of four species of Lampsiline mussels, Zale and Neves (1982a) found that glochidia of *Lampsilis fasciola* parasitized only smallmouth bass *(Micropterus dolomieu)*. Brian Watson, in experimenting with potential fish hosts for the glochidia of this mussel, was able to get glochidia to parasitize and transform on largemouth bass *(Micropterus salmoides)* (S. A. Ahlstedt, pers. comm., 1996)

Status: Currently Stable (Williams et al, 1993:12).

Lampsilis ornata (Conrad, 1835)
Southern Pocketbook

RANGE MAP 49; PLATE 50

Synonymy:
Unio ovatus var. *ornatus* Conrad, 1835; Conrad, 1835b:4
Unio ornatus Conrad, 1835; Sowerby, 1866:pl. 31. fig. 162
Lampsilis ornata (Conrad, 1835); Frierson, 1927:68
Lampsilis (Lampsilis) ornata (Conrad, 1835); Frierson, 1927:68; Haas, 1969a:455
Unio excavatus Lea, 1857; Lea, 1857a:32; Lea, 1858e:71, pl. 13, fig. 52
Margaron (Unio) excavatus (Lea, 1857); Lea, 1870:37
Lampsilis (Lampsilis) excavatus (Lea, 1857); Simpson, 1900a:528
Lampsilis (Lampsilis) excavata (Lea, 1857); Simpson, 1914:41

Type Locality: Othcalooga Creek, Georgia.

General Distribution: Tombigbee and Alabama River drainage (Simpson, 1914). Clench and Turner (1956:199) state that "[f]rom the Escambia river system of Alabama and western Florida this form ranges west to the Pearl River in Mississippi."

Tennessee Distribution: Conasauga River, Polk and Bradley counties.

Description: "Shell inflated, subsolid, the male irregularly ovate or rhomboid, the female obovate, with a high, decided posterior ridge" (Simpson, 1914:41). Beaks are high, full, and extend well beyond the hinge line and turn forward. Mature individuals from the Conasauga River, Murray County, Georgia, reach a length between 90 and 100 mm. The anterior end is broadly rounded, the ventral margin is evenly curved, and the posterior end is bluntly pointed. The female shell is somewhat inflated posteriorly, and the posterior end is more rounded and less sharp than in the male shell.

The left valve has two erect, usually roughened pseudocardinal teeth appearing as a compressed triangle, and the anterior tooth is the larger. The two lateral teeth are short, straight, and fairly high. The right valve has a large, erect, elongated pseudocardinal tooth, usually with a smaller, low, linear tooth anterior to it; the lateral tooth is short, high, and thin. The interdentum is narrow; the beak cavity is deep; anterior mussel scars are deeply impressed. The periostracum is a greenish yellow or light tan, and most individuals have one or a few widely separated narrow green rays. The disc section is usually smooth, the concentric rest lines producing a rough surface toward the shell margins. The nacre is white or bluish white.

Life History and Ecology: *Lampsilis ornata* occurs under the same habitat conditions in the Conasauga River as does the Fine-lined Pocketbook *(L. altilis)*: moderate current, water less than three feet deep, and a substrate of mixed sand, mud, and gravel. Host fish for the glochidia of this mussel and its breeding season are unknown. It may be bradytictic, as is the case in other species of *Lampsilis* where the reproductive period is known.

Status: Special Concern (Williams et al., 1993:12).

Range Map 49. *Lampsilis ornata* (Conrad, 1835), Southern Pocketbook.

Plate 50. *Lampsilis ornata* (Conrad, 1835), Southern Pocketbook.

Lampsilis ovata (Say, 1817)
Pocketbook

RANGE MAP 50; PLATE 51

Synonymy:
Unio ovatus Say, 1817; Say, 1817:pl. 2, fig. 7
Unio ovata Say, 1817; Lamarck, 1819:75
Lampsilis ovata (Say, 1817); Rafinesque, 1820:298
Mya ovata (Say, 1817); Eaton, 1826:218
Margarita (Unio) ovatus (Say, 1817); Lea, 1836:19
Aeglia ovata (Say, 1817); Swainson, 1840:266, fig. 49
Margaron (Unio) ovatus (Say, 1817); Lea, 1852c:24
Lampsilis (Lampsilis) ovatus (Say, 1817); Simpson, 1900a:530
Lampsilis (Lampsilis) ovata (Say, 1817); Simpson, 1914:48
Lampsilis ovata (Say, 1817); Frierson, 1927:67

Type Locality: Ohio River and its tributary streams.

General Distribution: Johnson (1970:388) gives the range of *Lampsilis ovata* as "Interior Basin: Mississippi and Ohio drainages. St. Lawrence drainage from Lake Superior to the Ottawa River and Lake Champlain. Hudson Bay drainage. Northern Atlantic slope: restricted to the Potomac River system, Maryland (introduced)." This extensive range, however, includes various closely related forms, subspecies, and/or species such as *L. ventricosa (=L. cardium)* and *L. satura*; the taxonomy of the *L. ovata* complex is far from clear. Goodrich and van der Schalie (1944:315) noted that "[a]s one progresses into the headwaters the sharp posterior ridge of the true *ovata* is seen to round off and we pass gradually to the more common form of the species in Indiana, known as *L. ovata ventricosa* (Barnes)." It has been suggested that *L. ovata* and *L. ventricosa (=L. cardium)* are ecophenotypes (Cvancara, 1963).

Tennessee Distribution: The Pocketbook occurs throughout the major river systems of East and Middle Tennes-

Range Map 50. *Lampsilis ovata* (Say, 1817), Pocketbook.

see, inhabiting shallow, free-flowing small to medium-sized rivers as well as the Cumberland and Tennessee River reservoirs. Some specimens from a few Middle Tennessee rivers, such as the Stones and Harpeth, appear similar in shape and pattern to *L. cardium* of the upper Midwest, rather than typical *L. ovata*.

Description: Although extremely thin and fragile as juveniles, shells of old adults become solid and heavy, somewhat ovate or elliptical, inflated, swollen, and especially high in old females. Mature individuals may reach a length of 160–170 mm. The anterior end is sharply rounded; the posterior end is bluntly pointed at the union of the dorsal and ventral margins; the posterior slope is flattened or slightly convex. The anterior area in front of umbones is often extended and somewhat alate. Beaks are swollen and elevated; sculpture consists of 4–5 coarse bars, of which the second and third may appear slightly double-looped, the others indistinct. Rest lines are usually distinct, sometimes elevated into low ridges.

The left valve has two heavy pseudocardinal teeth which are somewhat compressed, elevated, and roughened in young individuals and squarish or stumpy in old specimens; the two lateral teeth are high, straight, fairly short, and striated. The right valve has a heavy, erect, high, somewhat triangular pseudocardinal tooth, usually with a thin, elongated, moderately elevated tooth in front, and occasionally a small peglike tooth behind. The lateral tooth is high, straight, and characteristically squared-off or abruptly truncated at the posterior end. The interdentum is very narrow; the beak cavity is broad and deep. The periostracum is smooth in juveniles, but becomes somewhat roughened with age by concentric rest lines; the color of juveniles is a pale yellow, in adults a yellowish green to dark

olive. A few individuals exhibit narrow, often faint dark green rays on the umbo and disc. The nacre is pearly white, iridescent posteriorly.

Life History and Ecology: *Lampsilis ovata* is quite generalized in habitat preference; in many instances, it

Plate 51. *Lampsilis ovata* (Say, 1817), Pocketbook.

adapts well both to impoundment situations, such as the Cumberland and Tennessee River reservoirs, as well as free-flowing, shallow rivers characterized by the upper Clinch and Powell. The Pocketbook may be found in big rivers (reservoirs) at depths of 15 to 20 feet and in small streams in less than two feet of water. Although usually found in moderate to strong current, this mussel can adapt—or at least survive for a time—in standing water. The most suitable substrate consists of a mixture of gravel and coarse sand mixed with some silt or mud.

This species is probably bradytictic. Baker (1928a) noted that in the case of *Lampsilis ventricosa* (=*L. cardium),* females retained glochidia from early August to the following July. Fuller (1974), citing Coker et al. (1921) and others, records smallmouth bass *(Micropterus dolomieu),* largemouth bass *(M. salmoides),* bluegill *(Lepomis macrochirus),* white crappie *(Pomoxis annularis),* yellow perch *(Perca flavescens),* and sauger *(Stizostedion canadense)* as host fish for glochidia of the Pocketbook.

Status: Currently Stable (Williams et al., 1993:10). The Pocketbook appears to be common throughout its range in Tennessee.

Lampsilis siliquoidea (Barnes, 1823)
Fatmucket

RANGE MAP 51; PLATE 52

Synonymy:
Unio luteolus Lamarck, 1819 [of authors]; De Kay, 1843:190, pl. 20 fig. 241
Margarita (Unio) luteolus (Lamarck, 1819); [of authors] Lea, 1836:25
Margaron (Unio) luteolus (Lamarck, 1819) [of authors]; Lea, 1852c:28
Lampsilis luteola (Lamarck, 1819) [of authors]; Baker, 1898:103, pls. 11, 37, fig. 12
Lampsilis radiata luteola (Lamarck, 1819) [of authors]; Starrett, 1971:336
Comment: The name *Lampsilis radiata luteola* (Lamarck, 1819) has been used in place of *Lampsilis siliquoidea* (Barnes, 1823) and has resulted in some confusion (see Turgeon et al., n.d.).
Lampsilis (Ligumia) fasciata (Rafinesque, 1820) [of authors]; Frierson, 1927:71, 72
Unio siliquoideus Barnes, 1823; Barnes, 1823:269–270, pl. 13, fig. 150 (outline)
Mya siliquoidea (Barnes, 1823); Eaton, 1826:221
Unio siliquoides Barnes, 1823; Küster, 1852:30 pl. 5, fig. 2 [misspelling]
Lampsilis siliquoidea (Barnes, 1823); Stimpson, 1851:14

Lampsilis radiata siliquoidea (Barnes, 1823); Clarke and Berg, 1959:59
Ligumia siliquoidea (Barnes, 1823); Haas, 1969a:433
Unio inflatus Barnes, 1823; Barnes, 1823:266;
Mya inflata (Barnes, 1823); Eaton, 1826:221
Unio multiradiatus Lea, 1829; Sowerby, 1868:pl. 61, fig. 306 [misidentification]
Unio hydianus Lea, 1838; Küster, 1861:201, pl. 67, fig. 1 [misidentification]
Unio childreni Hanley, 1843 non Gray, 1834; Hanley, 1843:193, pl. 23, fig. 57
Unio rosaceus De Kay, 1843; De Kay, 1843:192, pl. 39, figs 355, 356, pl. 40, fig. 357
Lampsilis luteolus var. *rosaceus* (De Kay, 1843); Simpson, 1900a:535
Lampsilis luteola var. *rosacea* (De Kay, 1843); Simpson, 1914:62
Lampsilis luteola rosacea (De Kay, 1843); Walker, 1913:21
Lampsilis siliquoidea rosacea (De Kay, 1843); Baker, 1928a:277
Unio affinis Lea, 1852; Sowerby, 1868:pl. 69, fig. 307 [misidentification]
Unio distans Anthony, 1865; Anthony, 1865:156, pl. 13, fig. 2
Unio superiorensis Marsh, 1897; Marsh, 1897:103, pls. 1, 2, 5
Lampsilis superiorensis (Marsh, 1897); Simpson, 1900a:535
Lampsilis (Ligumia) fasciata superiorensis (Marsh, 1897); Frierson, 1927:72
Lampsilis siliquoidea pepinensis F. C. Baker, 1927; Baker, 1927:223
Lampsilis siliquoidea chadwicki F. C. Baker, 1928; Baker, 1928a:279, pl. 91, figs. 5–8

Type Locality: "Inhabits the Wisconsan" [*sic*] Wisconsin River.

General Distribution: *Lampsilis siliquoidea* occurs throughout the Mississippi River Basin except for the Tennessee and Cumberland River basins. The Fatmucket's range extends from western New York to Minnesota and south to Arkansas. It is found as far west as eastern Colorado and Montana and is widespread throughout the interior of Canada, including the western Hudson Bay drainages (Clarke, 1981a).

Tennessee Distribution: The Fatmucket's range in Tennessee is restricted to Reelfoot Lake (where it may be a relic or an extirpated species) and one tributary of the Mississippi River, the Wolf River.

Description: The shell is oblong and elongate and varies from moderately compressed to quite inflated with a maximum length about 130 mm. The anterior end is evenly rounded; the posterior margin is straight dorsally, angled where it meets the dorsal margin, and

Range Map 51. *Lampsilis siliquoidea* (Barnes, 1823), Fatmucket.

rounded below. Dorsal and ventral margins are straight and nearly parallel. The posterior ridge is rounded and not prominent. Growth lines are noticeable but not prominent. The male shell is less inflated than the female shell; the female shell has a more evenly rounded posterior margin, while the male shell has a bluntly pointed posterior end. The female shell is typically more inflated with a distinct posterior marsupial swelling, and the posterior bluntly rounded margin occurs more dorsally than in males. Beaks are broad and slightly raised above the hinge line; beak sculpture consists of fine, double-looped ridges with the posterior loops sometimes turned up. The beak cavity is rather shallow, often with four or five dorsal muscle scars.

The left valve has two short, erect, compressed pseudocardinal teeth and two short, slightly curved, low lateral teeth. The right valve has one large, erect pseudocardinal tooth, often with a small lamellar tooth anterior; there is one long, thin, slightly curved lateral tooth. Adductor muscle scars are large and well impressed. The interdentum is absent. The pallial line is lightly impressed. The periostracum is smooth and shining, usually yellowish to greenish yellow, often becoming brownish in old shells. Shells range from rayless to normally having bright green rays over the entire surface, varying in width from broad to narrow. The nacre color varies from white, to bluish white, to pink and is iridescent posteriorly.

Life History and Ecology: Baker (1928a) reported the Fatmucket as bradytictic, gravid from August to July. The Fatmucket can be found on a variety of substrates but usually prefers quiet or slow-moving water with a mud bottom, typically avoiding riffles. Watters (1994a) lists the following fish hosts for the glochidia of *Lamp-*

silis siliquoidea: black crappie *(Pomoxis nigromaculatus),* bluegill *(Lepomis macrochirus),* orangespotted sunfish *(L. humilis),* pumpkinseed *(L. gibbosus),* common shiner *(Luxilus cornutus),* largemouth bass *(Micropterus salmoides),* smallmouth bass *(M. dolomieu),* rockbass *(Ambloplites rupestris),* sauger *(Stizostedion*

Plate 52. *Lampsilis siliquoidea* (Barnes, 1823), Fatmucket.

canadense), walleye *(Stizostedion vitreum)*, white bass *(Morone chrysops)*, white crappie *(Pomoxis annularis)*, white sucker *(Catostomus commersoni)*, and the yellow perch *(Perca flavescens)*.

Status: Currently Stable (Williams et al., 1993:12). Based on the study of the Fatmucket in the Wolf River in Tennessee and Mississippi (Kesler and Manning, 1996), the mean age of live-collected specimens was 20 years or more. The population appears viable but is dominated by older individuals.

Lampsilis straminea claibornensis (Lea, 1838) Southern Fatmucket

RANGE MAP 52; PLATE 53

Synonymy:
Unio claibornensis Lea, 1838; Lea, 1838b:105, pl. 24, fig. 115
Margarita (Unio) claibornensis (Lea, 1838); Lea, 1838a:19
Margaron (Unio) claibornensis (Lea, 1838); Lea, 1852c:28
Lampsilis claibornensis (Lea, 1838); Simpson, 1900a:537
Unio obtusus Lea, 1840 Lea, 1840:287; 1842b:201, pl. 11, fig. 13, 1843
Margaron (Unio) obtusus (Lea, 1840); Lea, 1852c:39
Unio pallescens Lea, 1845; Lea, 1845:164; Lea 1848:79, pl. 7, fig. 20
Margaron (Unio) pallescens (Lea 1845); Lea, 1852c:27
Unio contrarius Conrad, 1849; Conrad, 1849:153
Lampsilis contrarius (Conrad, 1849); Simpson, 1900a:537
Unio perpastus Lea, 1861; Lea, 1861c:60; Lea, 1862b:69, pl. 7, fig. 219
Margaron (Unio) perpastus (Lea, 1861); Lea, 1870:43
Lampsilis perpastus (Lea, 1861); Simpson, 1900a:532
Lampsilis perpasta (Lea, 1861); Simpson, 1914:71
Lampsilis fasciata-claibornensis (Lea, 1838); Frierson, 1927:72
Lampsilis siliquoidea straminea (Conrad, 1834); Haas, 1969a:434 [in part]
Lampsilis straminea claibornensis (Lea, 1838); Hartfield, 1988

Type Locality: Alabama River, near Claiborne (Monroe County, Alabama).

General Distribution: Gulf Coast drainages from the Suwannee River west to the Pearl River and Lake Pontchartrain tributaries, Louisiana.

Tennessee Distribution: *Lampsilis straminea claibornensis* is known only from southeastern Tennessee in the Conasauga River, Polk County (Hurd, 1974).

Description: The Southern Fatmucket is solid, elliptical, and greatly inflated, a little thicker anteriorly. The anterior end is broadly rounded, the ventral margin is nearly straight, and the posterior ridge is rounded and sometimes double. The male shell is somewhat bluntly pointed in the middle of the posterior margin; the female shell is only slightly inflated posterior-ventrally, but with the posterior margin well rounded. Beaks are somewhat broad, moderately elevated, and inflated; sculpture has not been reported. The beak cavity is open and moderately deep. The maximum shell length is approximately 90 mm.

The left valve has two short, widely separated, thickened pseudocardinal teeth and two straight, heavy lateral teeth. The right valve has one large, triangular, thickened pseudocardinal tooth and one (anterior) thin pseudocardinal tooth, with a single erect, heavy lateral tooth. Adductor muscle scars are deep and smooth. The pallial line is impressed. The periostracum is concentrically sculptured but smooth, shiny on the disk, and somewhat roughened on the posterior slope. Color varies from a light to a dark yellowish brown, being usually rayless or with a few rays restricted to the posterior slope. Juvenile shells occasionally have faint green rays. The nacre color is whitish to light pink, iridescent posteriorly.

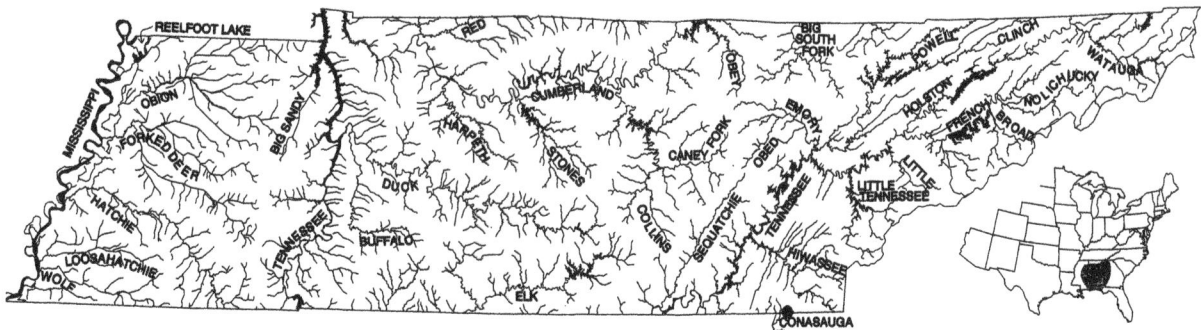

Range Map 52. *Lampsilis straminea claibornensis* (Lea, 1838), Southern Fatmucket.

Plate 53. *Lampsilis straminea claibornensis* (Lea, 1838), Southern Fatmucket.

Life History and Ecology: Clench and Turner (1956:204) reported the ecology of this species as "sandy areas where the bottom is fairly firm." Host fish for the glochidia of *Lampsilis straminea claibornensis* unknown.

Status: Currently Stable (Williams et al., 1993:12). The Southern Fatmucket is probably extirpated in the stretch of Conasauga River flowing through Tennessee.

Lampsilis teres (Rafinesque, 1820)
Yellow Sandshell

RANGE MAP 53; PLATE 54

Synonymy:
Elliptio teres Rafinesque, 1820; Rafinesque, 1820:321
Unio teres (Rafinesque, 1820); Say, 1834:no pagination
Lampsilis (Ligumia) teres (Rafinesque, 1820); Frierson, 1927:70
Ligumia teres teres (Rafinesque, 1820); Haas, 1969a:431

Lampsilis teres teres (Rafinesque, 1820); Oesch, 1984:207–209
Unio anodontoides Lea, 1834; Lea, 1834:81, pl. 8, fig. 11
Margarita (Unio) anodontoides (Lea, 1834); Lea, 1836:35
Margaron (Unio) anodontoides (Lea, 1834); Lea, 1852c:36
Lampsilis teres anodontoides (Lea, 1834); Oesch, 1984:209–12
Lampsilis anodontoides (Lea, 1834); Baker, 1898:100, pl. 10, fig. 1
Lampsilis anodontoides anodontoides (Lea, 1834); Murray and Leonard, 1962:143
Lampsilis anodontoides form *anodontoides* (Lea, 1834); Valentine and Stansbery, 1971:30
Unio oriens Lea, 1831; Sowerby, 1868:pl. 63, fig. 314 [mis-identification]
Unio floridensis Lea, 1852; Lea, 1852b:274, pl. 21, fig. 31
Margaron (Unio) floridensis (Lea, 1852); Lea, 1852c:39
Lampsilis (Eurynia) anodontoides var. *floridensis* (Lea, 1852); Simpson, 1900a:544
Lampsilis (Ligumia) teres floridensis (Lea, 1852); Frierson, 1927:70
Lampsilis anodontoides floridensis (Lea, 1852); Clench and Turner, 1956:201, pl. 3, fig. 1
Ligumia teres floridensis (Lea, 1852); Haas, 1969a:432
Lampsilis fallaciosus Smith, 1899; Smith, 1899:291, pl. 79
Lampsilis (Eurynia) fallaciosa Smith, 1899; Simpson, 1900a:544
Lampsilis (Ligumia) teres fallaciosa Smith, 1899; Frierson, 1927:70
Lampsilis anodontoides fallaciosa Smith, 1899; Grier and Mueller, 1926:18
Lampsilis anodontoides form *fallaciosa* Smith, 1899; Valentine and Stansbery, 1971:30

Type Locality: Mississippi, Alabama, and Ohio rivers.

General Distribution: Mississippi River drainage; north to eastern South Dakota, south to northern Mexico, and all of the Gulf drainages from the Withlacoochee River, Florida, to the Rio Grande (Baker, 1928a). This includes both forms or subspecies, *L. t. teres* and *L. t. fallaciosa*.

Tennessee Distribution: Ortmann (1926a:87) reported *Lampsilis anodontoides [=L. teres] fallaciosa* as "very abundant" in the Obion River, Union City, Obion County, and noted the record of it by Pilsbry and Rhoads (1896) from the Wolf River, at Raleigh (Memphis), Shelby County. Along with *Quadrula pustulosa*, the Yellow Sandshell is the most common and widespread unionoid in the Hatchie River, Lauderdale, and Tipton counties (Manning, 1989). It is also known from the Elk, Stones, Duck, Caney Fork, Cumberland, Red, and Tennessee (Kentucky Lake) rivers (Starnes and Bogan, 1988).

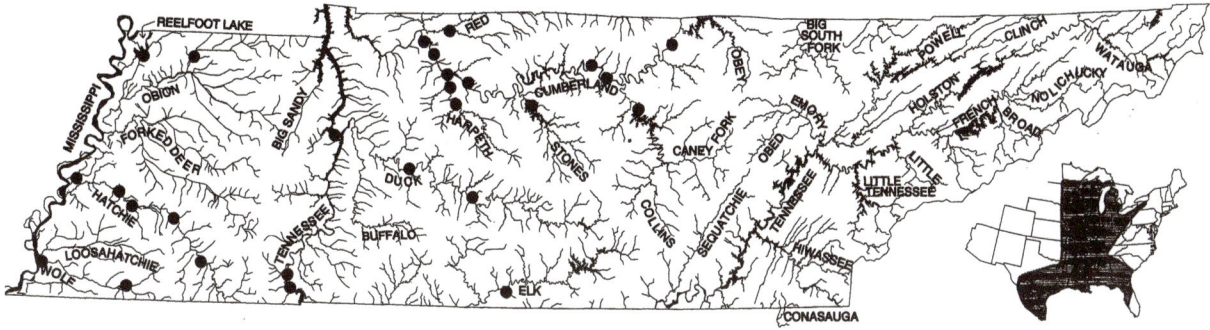

Range Map 53. *Lampsilis teres* (Rafinesque, 1820), Yellow Sandshell.

Description: Shells of mature specimens of the Yellow Sandshell are thick, moderately to greatly inflated and elongated, and reach a length of about 130–140 mm. The anterior end is rounded, and the posterior end is pointed, or somewhat truncated and inflated in mature females. The dorsal and ventral margins are straight, nearly parallel, and the posterior ridge is rounded and low. Beaks are full, not much elevated above the hinge line; sculpture consists of a few, indistinct ridges.

The left valve has two elevated, compressed, elongated, serrated pseudocardinal teeth; the two lateral teeth are long, nearly straight, and finely striated. The right valve has two pseudocardinal teeth; the front one is low and elongated; the posterior pseudocardinal is triangular, erect, heavy, and somewhat compressed. The lateral tooth in the right valve is long, solid, and roughened. There is no interdentum; the beak cavity is fairly shallow. The periostracum is a pale to bright yellowish, rarely faintly rayed when young, and smooth and shiny; the beak area is frequently washed with a reddish or brownish shade, this being especially pronounced in Hatchie River specimens. The nacre is silvery white, the beak cavity area sometimes tinged with cream or a pinkish color; it is iridescent posteriorly.

Shells of the Slough Sandshell, *Lampsilis teres fallaciosa* Smith, 1899, presently a "form" or subspecies once considered a species distinct from *L. anodontoides* (=*L. teres*), are similar to the Yellow Sandshell except somewhat more cylindrical and elongated. Mature specimens often possess a slight constriction or indentation about midpoint along the ventral margin, giving it a "pinched" appearance. Beak sculpture consists of 8–10 distinct ridges which are looped and drawn together in the middle; these ridges continue anteriorly,

but are often open and wavy posteriorly and are more distinct than those in the Yellow Sandshell. The periostracum is light yellow, more greenish yellow in young shells; the surface is smooth and shiny with numerous dark green rays covering the greater part. Pseudocardinal and lateral teeth and nacre color of the form *fallaciosa* are similar to *Lampsilis teres*.

Plate 54. *Lampsilis teres* (Rafinesque, 1820), Yellow Sandshell.

Life History and Ecology: As noted by Ortmann (1926a:93) *"fallaciosa* seems to be the form of quiet water and sandy-muddy bottom, while *anodontoides* is found in stronger current and gravel."* Specimens of *L. teres* taken by commercial shellers in the Cumberland (Smith County) and Tennessee (Hardin County) rivers, usually at 12–15 feet in depth, are typical Yellow Sandshells. Those collected from a mud, silt, and sand substrate in less than six feet of water in the Hatchie River are characteristic of the Slough Sandshell, although differences are not clear-cut.

Baker (1928a) records the breeding season of the Yellow Sandshell as bradytictic, retaining glochidia in the gills over winter. Based on data from Coker et al. (1921), Surber (1913), C. B. Wilson (1916), and others, Fuller (1974) lists the following fish hosts for the glochidia of *L. teres:* shovelnose sturgeon *(Scaphirhynchus platorynchus)*, longnose gar *(Lepisosteus osseus)*, shortnose gar *(L. platostomus)*, largemouth bass *(Micropterus salmoides)*, green sunfish *(Lepomis cyanellus)*, warmouth *(L. gulosus)*, orangespotted sunfish *(L. humilis)*, white crappie *(Pomoxis annularis)*, and black crappie *(P. nigromaculatus)*. Watters (1994a), citing data from Wilson (1916) and others, lists the alligator gar *(Lepisosteus [=Atractosteus] spatula)* as another host fish for Yellow Sandshell glochidia.

Status: Currently Stable (Williams et al., 1993:12). In Tennessee *Lampsilis teres* populations appear stable in the Hatchie River and locally in the Cumberland and Tennessee (Kentucky Lake) rivers.

Lampsilis virescens (Lea, 1858)
Alabama Lampmussel

RANGE MAP 54; PLATE 55

Synonymy:
Unio virescens Lea, 1858; Lea, 1858a:40; Lea, 1860e:341, pl. 55, fig. 166
Margaron (Unio) virescens (Lea, 1858); Lea, 1870:42
Lampsilis virescens (Lea, 1858); Simpson, 1900a:544
Ligumia virescens (Lea, 1858); Haas, 1969a:432

Type Locality: Tennessee River, Tuscumbia, Alabama.

General Distribution: Restricted to the Tennessee River drainage, northern Alabama, and East Tennessee. Specimens of *Lampsilis virescens* have been collected in the

Paint Rock River, Jackson County; Bear Creek, Colbert County; Little Bear Creek, a tributary to Bear Creek, Franklin County; and Spring Creek, Lauderdale County, all tributaries of the Tennessee River in northern Alabama (Call, 1885; Ortmann, 1925; Isom and Yokley, 1968b; Isom et al., 1973).

Tennessee Distribution: Formerly the Alabama Lampmussel occurred in the Emory River, Roane and Morgan counties, and in Coal Creek, a tributary of the Clinch River, Anderson County (Ortmann, 1918). At the time of this writing, it is found in the headwaters of the Paint Rock River on the Tennessee and Alabama border.

Description: Valves of the Alabama Lampmussel are elliptical or obovate in outline, subinflated, and have a low, rounded posterior ridge, and are relatively thin. Mature specimens rarely exceed 65–70 mm in length. The shell surface varies from smooth to slightly roughened from concentric rest lines. The anterior end of the shell is rounded and the dorsal margin is slightly curved, while the ventral margin is straight, curving up posteriorly. The posterior end of the shell in males is bluntly pointed; in females, it is slightly more inflated and rounded.

The left valve has two compressed, raised, bladelike pseudocardinal teeth separated from two slightly curved, delicate lateral teeth by a narrow, curved interdentum. The right valve has two bladelike pseudocardinal teeth; the anterior tooth is small and very compressed, while the posterior tooth is considerably larger and thicker. The lateral tooth is slightly curved. Beak cavities are broad and rather deep. Anterior muscle scars are well impressed, but the posterior muscle scars are shallow. The pallial line is faint. The periostracum is typically shiny, greenish to straw-colored and sometimes with thin green rays, especially on the posterior slope. The nacre color is bluish white and iridescent on the posterior two-thirds of the shell.

Life History and Ecology: Except for the fact that *Lampsilis virescens* inhabited small to medium-sized rivers in East Tennessee and that it apparently lived in sand and gravel substrates in shoal areas, little else is known about this taxon. Other species in the genus *Lampsilis* are bradytictic, thus suggesting this species may also be a winter breeder (Heard and Guckert, 1970). Host fish for the glochidia unknown.

Range Map 54. *Lampsilis virescens* (Lea, 1858), Alabama Lampmussel.

Status: Endangered (Williams et al., 1993:12). The Alabama Lampmussel is now probably extirpated throughout its former range in Tennessee. Ahlstedt (1991b) stated that it is now found only in the Paint Rock River and its largest tributary (Hurricane Creek) in north Alabama. The U.S. Fish and Wildlife Service has developed a recovery plan for this species (U.S. Fish and Wildlife Service, 1985c) and has created a watershed implementation schedule for the recovery plan (U.S. Fish and Wildlife Service, 1989b).

Plate 55. *Lampsilis virescens* (Lea, 1858), Alabama Lampmussel.

Lasmigona complanata (Barnes, 1823)
White Heelsplitter

RANGE MAP 55; PLATE 56

Synonymy:

Alasmodonta complanata Barnes, 1823; Barnes, 1823:278, pl. 13, fig. 21

Mya complanata (Barnes, 1823); Eaton, 1826:222

Symphynota complanata (Barnes, 1823); Lea, 1829:448

Margarita (Margaritana) complanata (Barnes, 1823); Lea, 1836:43

Alasmodon complanatus (Barnes, 1823); Sowerby, 1842:61

Unio complanata (Barnes, 1823); Hanley, 1843:210

Complanaria complanata (Barnes, 1823); Conrad, 1853:261

Baphia complanata (Barnes, 1823); H. and A. Adams, 1857:500

Margaritana complanata (Barnes, 1823); Calkins, 1874:46

Symphynota (Pterosygna) complanata (Barnes, 1823); Simpson, 1900a:665

Lasmigona complanata (Barnes, 1823); Utterback, 1916c:15

Lasmigona (Pterosyna) complanata (Barnes, 1823); Haas, 1969a:398

Lasmigona (Lasmigona) complanata complanata (Barnes, 1823); Clarke, 1985:25–36, figs. 7–9.

Unio katherinae Lea, 1838; Lea, 1838a:35 [nomen nudum]

Unio katherinae Lea, 1838; Lea, 1838:143

Symphynota (Pterosygna) complanata var. *katherinae* (Lea, 1838); Simpson, 1900:666

? *Unio gigas* Swainson, 1824; Swainson, 1824:15–17

Complanaria gigas Swainson, 1840; Sowerby, 1839:fig. 141

Megadomus gigas (Swainson, 1840); Swainson, 1840:265, 378

Lasmigona (Lasmigona) complanata alabamensis Clarke, 1985; Clarke, 1985:36–40, fig. 10

Type Locality: Fox River, Wisconsin.

General Distribution: Entire Mississippi River drainage from Lake Winnipeg–Nelson River system to western Ontario, Middle Great Lakes–St. Lawrence River system and tributaries of Lake Michigan, Lake St. Clair, and Lake Erie (Clarke, 1985); Pennsylvania west to Minnesota and Iowa south to Oklahoma and Louisiana, and in the Alabama River drainage (Burch, 1975a).

Tennessee Distribution: *Lasmigona complanata* is known to inhabit most medium-sized and large rivers in the state, including the Tennessee, Holston, Cumberland, Elk, Duck, Harpeth, Hatchie, and Stones. Prior to 1960 it was known to have occurred in the Buffalo, Caney Fork, North Fork Obion, and Red rivers (Starnes and Bogan, 1988).

Description: The shell is large, compressed, thin when young, thick when old, nearly rhomboid or irregularly elliptical in outline; including the dorsal wing, it is nearly as high as long. Mature individuals attain a length of 160–175 mm; old individuals under ideal habitat conditions may surpass 200 mm in length. The anterior end is rounded; the posterior end is squared or obliquely truncated; the dorsal margin is straight, ascending posteriorly to form a distinct wing which is often marked with several irregular ridges that extend toward the posterior margin. Beaks are depressed and flattened; sculpture consists of 4–5 distinct, heavy bars, the first two being simple, the others strongly double-looped. Numerous coarse lines on the shell surface are indicative of rest periods.

The left valve usually has two irregular, low, serrated pseudocardinal teeth which stem from the hinge line; lateral teeth are absent in both valves or represented as

Plate 56. *Lasmigona complanata* (Barnes, 1823), White Heelsplitter.

ridges or thickenings of the hinge line or interdentum. The right valve has a single (often appearing doubled) low, flattened, fan-shaped pseudocardinal tooth. The interdentum is narrow, appearing as part of the hinge line. The beak cavity is compressed and fairly shallow.

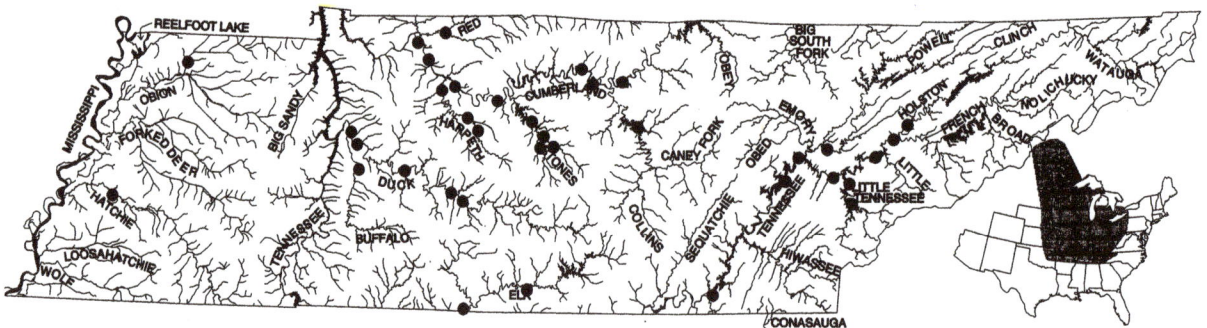

Range Map 55. *Lasmigona complanata* (Barnes, 1823), White Heelsplitter.

The periostracum is yellowish or green in young shells, often faintly rayed, and dark brown to black in old shells. The nacre is white, and most of the surface is iridescent; the anterior section is occasionally cream-colored or pinkish.

Life History and Ecology: The White Heelsplitter may be found inhabiting a variety of habitats, from medium-sized rivers like the Harpeth in Middle Tennessee to permanent sloughs, backwater bays, lakes, and reservoirs. It prefers quiet water, usually not over three feet in depth, although commercial shellers working the Tennessee and Cumberland River reservoirs do encounter this shell at depths of 15 to 20 feet. *Lasmigona complanata* seems to thrive well on a mud and fine sand substrate. Baker (1928a) recorded the reproductive period as bradytictic, lasting from September to April or May. Fish hosts for the glochidia of this species include the carp *(Cyprinus carpio)*, green sunfish *(Lepomis cyanellus)*, orangespotted sunfish *(L. humilis)*, banded killifish *(Fundulus diaphanus)*, largemouth bass *(Micropterus salmoides)*, and the white crappie *(Pomoxis annularis)* (Fuller, 1978; Watters, 1994a). Weiss and Layzer (1995) reported the longnose gar *(Lepisosteus osseus)*, gizzard shad *(Dorosoma cepedianum)*, river redhorse *(Moxostoma carinatum)*, and sauger *(Stizostedion canadense)* as also serving as hosts for the glochidia of this mussel.

Status: Currently Stable (Williams et al., 1993:13).

Lasmigona costata (Rafinesque, 1820) Flutedshell

RANGE MAP 56; PLATE 57

Synonymy:
Alasmidonta costata Rafinesque, 1820; Rafinesque, 1820:318, pl. 82, figs. 15, 16
Alasmodonta costata Rafinesque, 1820; Say, 1834:no pagination
Complanaria costata (Rafinesque, 1820); Conrad, 1853:261
Symphynota costata (Rafinesque, 1820); Simpson, 1900a:665
Lasmigona costata (Rafinesque, 1820); Ortmann and Walker, 1922:36
Lasmigona (Lasmigona) costata (Rafinesque, 1820); Frierson, 1927:19
Alasmodonta rugosa Barnes, 1823; Barnes, 1823:278, pl. 13, fig. 21
Mya rugosa (Barnes, 1823); Eaton, 1826:222
Margarita (Margaritana) rugosa (Barnes, 1823); Lea, 1836:44
Unio rugosa (Barnes, 1823); Hanley, 1843:211

Unio rugosus (Barnes, 1823); Küster, 1861:200, pl. 66, figs. 1–3
Alasmodon rugosa (Barnes, 1823); De Kay, 1843:196, pl. 14, fig. 226
Complanaria rugosa (Barnes, 1823); Stimpson, 1851:41
Margaron (Margaritana) rugosa (Barnes, 1823); Lea, 1852c:42
Baphia rugosa (Barnes, 1823); H. and A. Adams, 1857:500
Margaritana rugosa (Barnes, 1823); Calkins, 1874:46
Amblasmodon hians Rafinesque, 1831; Rafinesque, 1831:5
Alasmodonta hians (Rafinesque, 1831); Férussac, 1835:25
? *Alasmodon rugosum* Rafinesque, 1831; Rafinesque, 1831:5
Lasmigona costata var. *ereganensis* Grier, 1918; Grier, 1918:10
Lasmigona (Lasmigona) costata ereganensis Grier, 1918; Frierson, 1927:19
Lasmigona costata pepinensis F. C. Baker, 1928; F. C. Baker, 1928a:144
Lasmigona costata nuda F. C. Baker, 1928; F. C. Baker, 1928a:145

Type Locality: Kentucky River.

General Distribution: Entire Mississippi River drainage from western New York and Pennsylvania west to western Iowa and eastern Kansas, Oklahoma, and Texas; Wisconsin and southern Minnesota south to Louisiana, Alabama, and Mississippi (Tombigbee River system); Hudson Bay drainage in the Red and Winnipeg River systems, and in the Great Lakes–St. Lawrence system from southern Lake Huron and its tributaries to the Ottawa River and Lake Champlain (Clarke, 1985).

Tennessee Distribution: The Flutedshell is most common in the small to medium-sized rivers in East and Middle Tennessee. It occurs from the unimpounded stretches of upper Powell and Clinch rivers and other tributaries of the upper Tennessee River system, south and west to the Elk, Duck, Stones, Big South Fork Cumberland, and Harpeth rivers. Formerly *Lasmigona costata* inhabited the main channels of the Tennessee and Cumberland rivers, as well as medium-sized rivers such as the Obey, Caney Fork, Red, Roaring, Emory, Watauga, French Broad, and Holston (Starnes and Bogan, 1988).

Description: The shell is nearly rhomboid, elongated, solid, and somewhat compressed; the anterior end is rather sharply rounded, and the posterior end is obliquely truncated. The posterior ridge is usually well developed, with the dorsal/posterior end possessing numerous heavy, rounded flutings or ridges which are usually directed upward toward the margin. Mature individuals may attain a length of 190–200 mm. Beaks

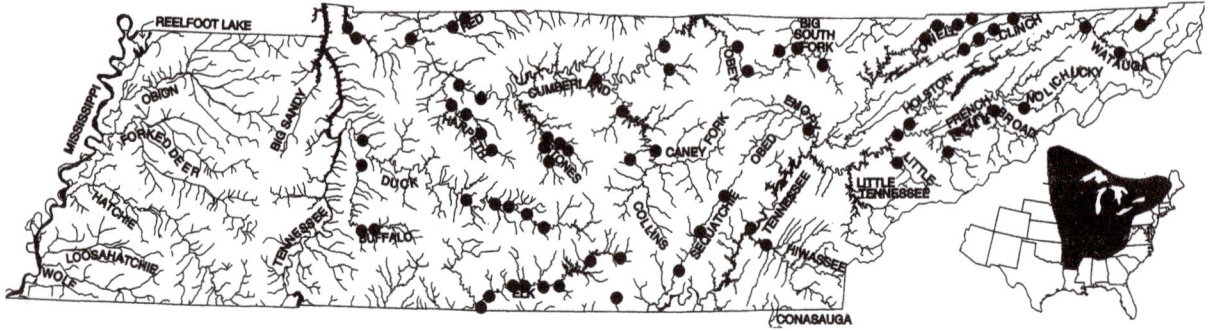

Range Map 56. *Lasmigona costata* (Rafinesque, 1820), Flutedshell.

are depressed and flattened; sculpture consists of 3–4 strongly developed, heavy bars, parallel with the hinge line. The first is curved, while the others are more or less double-looped. Surface sculpture consists of coarse lines of growth; rest periods are indicated by heavy, prominent concentric ridges edged with black.

The left valve has a single (two teeth fused into one) heavy, curved, pyramidal, elevated pseudocardinal tooth. The right valve has a heavy, low, somewhat elongated pseudocardinal tooth. Lateral teeth in both valves are represented by thickenings of the hinge line. The interdentum is narrow or absent; the beak cavity is very shallow. The periostracum is yellowish with numerous green rays in young shells; mature shells are a darker yellow, horn-colored or brownish, becoming black and rayless in old individuals. The nacre is white, with considerable variation in the amount of cream or salmon present; the area primarily between the pallial line and margin iridescent.

Life History and Ecology: Medium-sized rivers, such as the upper Powell and Clinch in East Tennessee and the Harpeth and Stones in Middle Tennessee, that possess a moderately strong current and a substrate composed of a coarse sand and gravel mixture provide the most suitable habitat for this species. In such rivers it is found typically at depths of three feet or less. Under ideal conditions growth will result in extremely large and thick-shelled individuals. The Flutedshell is bradytictic, the reproductive period beginning in August with the glochidia carried until May (Baker, 1928a). The carp *(Cyprinus carpio)* was reported by Fuller (1978) as the host for the glochidia of *Lasmigona costata*. Under laboratory conditions Luo (1993) was able to infect the rainbow darter *(Etheostoma caeruleum)*, fantail

darter *(E. flabellare)*, striped darter *(E. virgatum)*, green sunfish *(Lepomis cyanellus)*, longear sunfish *(L. megalotis)*, rockbass *(Ambloplites rupestris)*, smallmouth bass *(Micropterus dolomieu)*, banded sculpin *(Cottus carolinae)*, central stoneroller *(Campostoma anoma-*

Plate 57. *Lasmigona costata* (Rafinesque, 1820), Flutedshell.

lum), brown bullhead *(Ameiurus nebulosus),* and northern studfish *(Fundulus catenatus)* with glochidia of the Flutedshell. Since then the gizzard shad *(Dorosoma cepedianum)* and river redhorse *(Moxostoma carinatum)* have been shown to also serve as hosts for the glochidia of this species (Weiss and Layzer, 1995). Hove et al. (1994) have added six additional host species: bowfin *(Amia calva),* northern pike *(Esox lucius),* bluegill *(Lepomis macrochirus),* largemouth bass *(Micropterus salmoides),* yellow perch *(Perca flavescens),* and walleye *(Stizostedion vitreum).*

Status: Currently Stable (Williams et al., 1993:13).

Lasmigona holstonia (Lea, 1838)
Tennessee Heelsplitter

RANGE MAP 57; PLATE 58

Synonymy:

?*Alasmodon (Sulcularia) badium* Rafinesque, 1831; Rafinesque 1831:5

Lasmigona (Sulcularia) badia (Rafinesque, 1831); Ortmann, 1918:557

Alasmidonta (Sulcularia) badia (Rafinesque, 1831); Frierson, 1927:19

Margarita (Margaritana) holstonia Lea, 1836; Lea, 1836:46 [nomen nudum]

Margaritana holstonia Lea, 1838; Lea, 1838b:42, pl. 13, fig. 37

Unio holstonianus (Lea, 1838); Hanley, 1843:213 [unjustified emendation]

Margaron (Margaritana) holstonia (Lea, 1838); Lea, 1852c:44

Strophitus holstonia (Lea, 1838); Conrad, 1853:263

Baphia holstonia (Lea, 1838); H. and A. Adams, 1857:499

Margaritana holstoniana (Lea, 1838); Küster, 1862:302, pl. c, fig. 4 [unjustified emendation]

Alasmidonta holstonia (Lea, 1838); Simpson, 1900a:670

Symphynota (Alasminota) holstonia (Lea, 1838); Ortmann, 1914:43

Lasmigona (Alasminota) holstonia (Lea, 1838); Ortmann, 1914:42

Margaritana etowaensis Conrad, 1849; Conrad, 1849:154.

Margaritana etowahensis Lea, 1858 non Conrad, 1849; Lea, 1858b:138; Lea, 1859f:227, pl. 31, fig. 110

Margaritana (Alasmodonta) etowahensis (Conrad, 1849); Clessin, 1875:270, pl. 81, figs. 1, 2

Margaritana georgiana Lea, 1859; Lea, 1859e:280 [New name for *Margaritana etowahensis* Lea, 1858]

Margaron (Margaritana) georgiana (Lea, 1859); Lea, 1870:68

Alasmidonta (Alasmidonta) georgiana (Lea, 1859); Simpson, 1900a:670

Alasmodon impressa Anthony, 1865; Anthony, 1865:157, pl. 76, fig. 4

Type Locality: Holston River.

General Distribution: Upper Tennessee River drainage; headwaters of the Coosa River (Simpson, 1914).

Tennessee Distribution: Formerly in numerous small rivers and creeks in East Tennessee. Ortmann (1918) reported *Lasmigona holstonia* from the upper Holston River, Hawkins County, and the Hiwassee River, Polk County (specimens from adjacent sloughs), and from numerous small streams in Campbell, Knox, Cocke, Sevier, Rhea, and Monroe counties. It also occurs in the Conasauga River, Polk and Bradley counties, and in Hickory Creek, Coffee County.

Description: The shell is somewhat elongated, rhomboid, and moderately inflated; some very young juveniles are more oblong and compressed. Beaks are full but not high, projecting only slightly above the hinge line; sculpture consists of 4–5 strong, double-looped ridges; the last loop is low and almost straight. The anterior margin is broadly rounded, while the ventral margin is straight; the posterior end is broadly pointed to squared; the posterior ridge is pronounced but broadly rounded, in some specimens appearing doubled. The

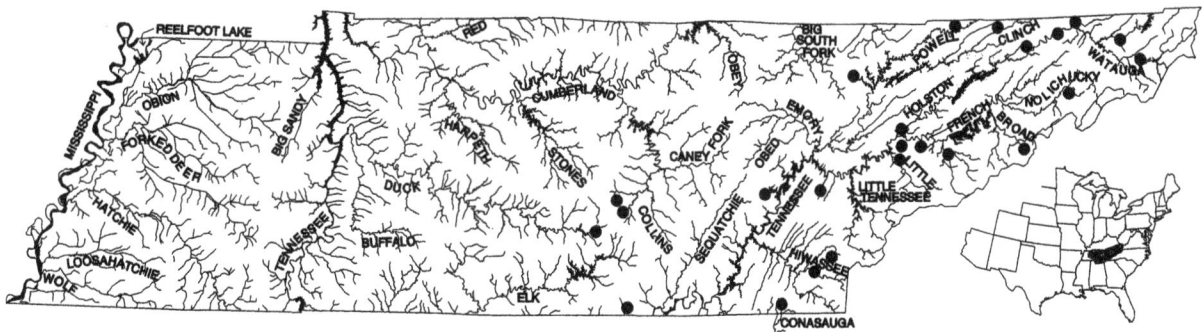

Range Map 57. *Lasmigona holstonia* (Lea, 1838), Tennessee Heelsplitter.

Plate 58. *Lasmigona holstonia* (Lea, 1838), Tennessee Heelsplitter.

shell is thin but not fragile. Mature specimens seldom exceed 75 mm in length. The surface is roughened with irregular, recessed, darkened rest lines. The right valve has a single compressed, but moderately heavy, pseudocardinal tooth; the left valve has two low, compressed pseudocardinal teeth, angled anteriorly and, in both valves, nearly parallel with the hinge line. Lateral teeth appear as a thickening of the hinge line in each valve. The periostracum is an almost uniform dull greenish brown or yellowish brown, and most shells become a dark brown or black with age. The nacre is bluish white, often with a pale salmon wash in the beak cavity area.

Life History and Ecology: *Lasmigona holstonia* is a species most often found inhabiting small shallow streams and headwater creeks with some current, and it may become locally abundant in stretches of substrate com-

posed of sand and mud. The reproductive period is unknown, but it is probably bradytictic. Host fish for the glochidia unknown.

Status: Special Concern (Williams et al., 1993:13).

Lasmigona subviridis (Conrad, 1835)
Green Floater
RANGE MAP 58; PLATE 59
Synonymy:
Unio subviridis Conrad, 1835; Conrad, 1835a:4 (appendix), pl. 9, fig. 1
Lasmigona (Platynaias) subviridis (Conrad, 1835); Ortmann, 1919:121
Unio viridis Rafinesque, 1820 [of authors]; Conrad, 1836:35, pl. 17, fig. 1
Symphynota (Symphynota) viridis (Rafinesque, 1820) [of authors]; Simpson, 1900a:663
Margarita (Unio) tappanianus Lea,1836; Lea, 1836:39 [nomen nudum]
Unio tappanianus Lea, 1838; Lea, 1838b:62, pl. 17, fig. 55 [replacement name for *Unio viridis* Rafinesque *sensu* Conrad, 1836]
Unio tappianus Lea, 1838; Catlow and Reeve, 1845:64 [misspelling]
Margaron (Unio) tappanianus (Lea, 1838); Lea, 1852c:39
Unio hyalinus Lea, 1845; Lea, 1845:164; Lea, 1848:69, pl. 2, fig. 4
Margaron (Unio) hyalinus (Lea, 1845); Lea, 1852c:39
Margaritana quadrata Lea, 1861; Lea, 1861a:41; Lea, 1862c:210, pl. 32, fig. 279
Margaron (Margaritana) quadrata (Lea, 1861); Lea, 1870:68
Margaritana (Alasmidonta) quadrata (Lea, 1861); Clessin, 1876:273, pl. 83, figs. 5, 6
Symphynota (Symphynota) quadrata (Lea, 1861); Simpson, 1900a:664
Lasmigona (Platynaias) quadrata (Lea, 1861); Frierson, 1927:20
Unio pertenuis Lea, 1863; Lea, 1863:193; Lea, 1866:8, pl. 2, fig. 4
Margaron (Unio) pertenuis (Lea, 1863); Lea, 1870:62

Type Locality: Schuylkill River, Juniata River, creeks in Lancaster County, Pennsylvania.

General Distribution: New and Greenbriar rivers of the upper Kanawha River drainage, Virginia and West Virginia. Upper Savannah River system of South Carolina north to the Hudson River system, and westward through the Mohawk River and the Erie Canal to the Genesee River of New York (Johnson, 1980; Clarke, 1985).

Tennessee Distribution: Watauga River, Johnson County.

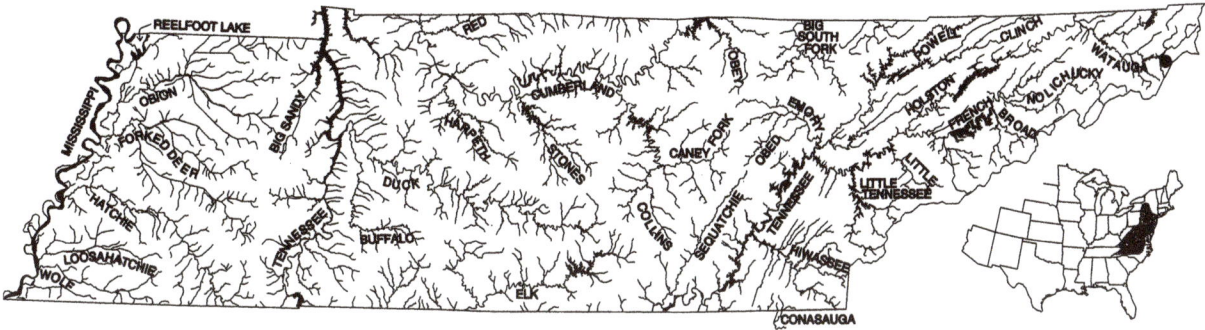

Range Map 58. *Lasmigona subviridis* (Conrad, 1835), Green Floater.

Description: The shell is thin and slightly inflated; it is subovate, narrower in front, higher behind, and the upper margin forms a blunt angle with the posterior margin (Ortmann, 1919). The posterior ridge is low, rounded, and appears more as a slight swelling than as a ridge. Beaks are low and not extended beyond the hinge line; sculpture consists of 4–5 nodulous bars, the first two concentric, the others deeply double-looped (Johnson, 1970). Mature individuals reach a length of about 60 mm. The lateral teeth, one in the right valve, two in the left, are long, straight, and thin. The left valve has two lamellate pseudocardinal teeth, and the right valve has one; pseudocardinals are directed forward of the beak and nearly parallel with the hinge line. The periostracum is a dull yellow or tan to brownish green, with variable concentrations of dark green rays. The nacre is a dull bluish white, often with mottled shades or tints of salmon in the general beak cavity area.

Life History and Ecology: Ortmann (1919:124) noted that *Lasmigona subviridis* is "adverse to very strong current, and prefers more quiet parts, pools or eddies with gravelly and sandy bottoms, and it also goes into canals, where it seems to flourish." In Tennessee it is restricted to perhaps no more than one or two miles of free-flowing stretches of the Watauga River above where it enters Watauga Lake. Pockets of sand and gravel among boulders provide a habitat for this mussel although it appears to be uncommon and localized. Ortmann (1919) stated that this species is normally hermaphroditic, and that it is bradytictic, with the reproductive season extending from August to May. Species of fish serving as hosts for the glochidia unknown.

Status: Threatened (Williams et al., 1993:13).

Plate 59. *Lasmigona subviridis* (Conrad, 1835), Green Floater.

Lemiox rimosus (Rafinesque, 1831)
Birdwing Pearlymussel

RANGE MAP 59; PLATE 60

Synonymy:
Unio (Lemiox) rimosus Rafinesque, 1831; Rafinesque, 1831:3
Lemiox rimosus (Rafinesque, 1831); Frierson, 1914:7
Unio coelatus Conrad, 1834; Conrad, 1834a:338, pl. 1, fig. 2
Unio caelatus Conrad, 1834; Conrad, 1834b:29, pl. 3, fig. 4 [corrected spelling]
Margarita (Unio) caelatus (Conrad, 1834); Lea, 1836:12
Margaron (Unio) caelatus (Conrad, 1834); Lea, 1852c:20
Micromya caelata (Conrad, 1834); Simpson, 1900a:525
Conradilla caelata (Conrad, 1834); Ortmann, 1921:90
Lemiox caelata (Conrad, 1834); Burch, 1973:99, fig. 93

Type Locality: Elk and Flint rivers, Tennessee.

General Distribution: Tennessee River system, including most of the major tributaries, downstream to Muscle Shoals, Alabama.

Tennessee Distribution: *Lemiox rimosus* is known from the Powell River, Hancock and Claiborne counties, and from the Clinch River, only in Hancock and Anderson counties. The Birdwing Pearlymussel has been collected in the North Fork Holston River, Hawkins County, and formerly from the Holston River in Hawkins, Grainger, and Knox counties, and the Nolichucky River (Stansbery, 1972, 1979). Formerly it inhabited the Tennessee River in the Knoxville area, Knox County (Ortmann, 1918; Bates and Dennis, 1978), and occurred in the Elk River, Giles and Lincoln counties, as recently as 1980 (Ahlstedt, 1983). Local viable populations are present in the Duck River in Maury and Marshall counties. A relic specimen was collected in the Sequatchie River in the 1980s by R. Biggins and C. Saylor (S. A. Ahlstedt, pers. comm., 1996). It was recovered in an archaeological site downstream in the Tennessee River at TRM 143, Decatur County.

Description: *Lemiox rimosus* is a small species, seldom exceeding 50 mm in length. Shells are subtriangular to subovate in outline, very thick and solid, and only slightly inflated. Beaks are high and turned forward, the sculpture consisting of three or four distinct double-looped bars, and the first one or two bars are sub-concentric. A posterior ridge is well developed, being somewhat rounded but distinct. The surface of the shell is marked by strong, irregular rest lines, and the posterior half or two-thirds is roughened by a strong, corrugated (rimose), subradial sculpture.

The left valve has two low, rugged pseudocardinal teeth separated from the two short, heavy, and slightly curved lateral teeth by a broad interdentum. The right valve has one to three pseudocardinal teeth, and the main tooth is ragged and triangular. Most specimens have a single lateral tooth, although some develop a vestigial tooth below. Muscle scars are small but deeply impressed, and the pallial line is impressed and distinct throughout. The beak cavity is very shallow or wanting. Shells are thickest anteriorly. Valves of the male possess a broad, shallow radial depression in front of the posterior ridge, while those of females are usually ovate, smaller than the male, and sometimes inflated with a weakly developed marsupial swelling along the posterior ventral margin. The periostracum is dull green or yellowish green and feebly rayed, becoming darker (nearly black) with age. The nacre is white and iridescent posteriorly.

Life History and Ecology: The Birdwing Pearlymussel occurs in riffle areas of small to medium-sized rivers, embedded in a sand and gravel substrate in moderate to fast currents. It formerly inhabited similar shoal areas of large rivers such as those once present in the Tennessee River at Knoxville, Tennessee, and at Muscle Shoals, Alabama.

Gravid specimens of *Lemiox rimosus* have been collected in mid-September, suggesting that this is a bradytictic species (Ortmann, 1916). Research studies by Tennessee Valley Authority biologists (Tennessee Valley Authority, 1986), resulting in laboratory-induced infestations of glochidia of *L. rimosus*, suggest that the greenside darter (*Etheostoma blennioides*) and the banded darter *(E. zonale)* are host fish for this mussel.

Status: Endangered (Williams et al., 1993:13). *Lemiox rimosus* is restricted to several small populations in the upper Powell and Clinch rivers, Tennessee and Virginia, and in the Duck River in Middle Tennessee (primarily Maury and Marshall counties). The U.S. Fish and Wildlife Service has developed a recovery plan for this species (U.S. Fish and Wildlife Service, 1983a) and has created a watershed implementation schedule for the recovery plan (U.S. Fish and Wildlife Service, 1989b).

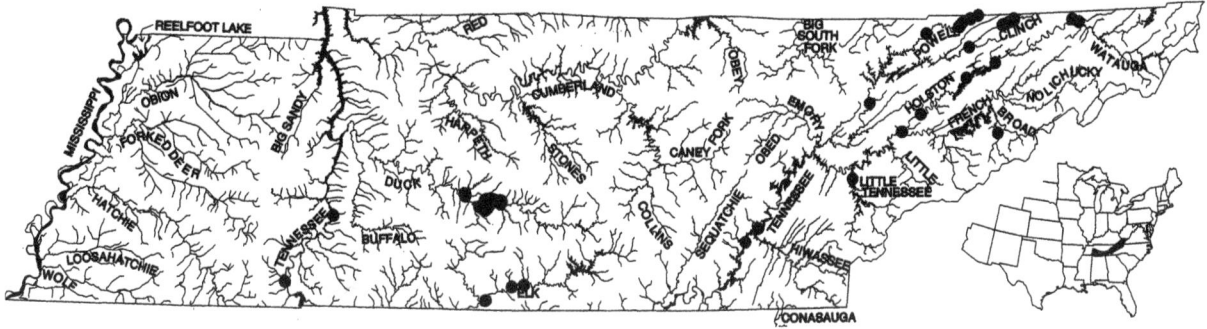

Range Map 59. *Lemiox rimosus* (Rafinesque, 1831), Birdwing Pearlymussel.

Plate 60. *Lemiox rimosus* (Rafinesque, 1831), Birdwing Pearlymussel.

Leptodea fragilis (Rafinesque, 1820)
Fragile Papershell

RANGE MAP 60; PLATE 61

Synonymy:

Unio (Leptodea) fragilis Rafinesque, 1820; Rafinesque, 1820:295

Unio fragilis Rafinesque, 1820; Swainson, 1823a:pl. 171

Symphynota fragilis (Rafinesque, 1820); Férussac, 1835:25

Metaptera fragilis (Rafinesque, 1820); Lapham, 1852:369

Unio fragilis var. *fragilis* Rafinesque, 1820; Jay, 1850:59

Paraptera fragilis (Rafinesque, 1820); Ortmann, 1918:572

Leptodea fragilis (Rafinesque, 1820); Ortmann and Walker, 1922:53

Lampsilis (Leptodea) fragilis (Rafinesque, 1820); Frierson, 1927:82

Unio gracilis Barnes, 1823; Barnes, 1823:274

Mya gracilis (Barnes, 1823); Eaton, 1826:222

Symphynota gracilis (Barnes, 1823); Lea, 1829:452

Margarita (Unio) gracilis (Barnes, 1823); Lea, 1836:11

Metaptera gracilis (Barnes, 1823); Stimpson, 1851:14

Margaron (Unio) gracilis (Barnes, 1823); Lea, 1852c:19

Unio fragilis var. *gracilis* Barnes, 1823; Paetel, 1890:153

Lampsilis gracilis (Barnes, 1823); Baker, 1898:99, pl. 19, fig. 1

Lampsilis (Proptera) gracilis (Barnes, 1823); Simpson, 1900a:573

Paraptera gracilis (Barnes, 1823); Ortmann, 1911:334

Proptera gracilis (Barnes, 1823); Haas, 1969a:416

Mya plana (Barnes, 1823); Eaton, 1826:221 [misidentification]

Unio leptodon var. *planus* Barnes, 1823; Jay, 1850:61 [misidentification]

Lasmonos fragilis Rafinesque, 1831; Rafinesque, 1831:5

Unio (Niaa) atrata Swainson, 1841; Swainson, 1841:pl. 171 non Sowerby, 1839

Unio atratus Swainson, 1841; Hanley, 1842a:199

Lampsilis simpsoni Ferriss, 1900; Ferriss, 1900:38

Lampsilis (Leptodea) fragilis simpsoni Ferriss, 1900; Frierson, 1927:82

Paraptera gracilis lacustris F. C. Baker, 1922; F. C. Baker, 1922b:131

Leptodea fragilis lacustris (F. C. Baker, 1922); F. C. Baker, 1924:132

Type Locality: Wisconsin River and "the lakes." Ohio River.

General Distribution: All of the Mississippi River drainage; Gulf of Mexico drainage from Alabama to Texas (Burch, 1975a). The entire Great Lakes–St. Lawrence system in Canada (Clarke, 1981a).

Tennessee Distribution: Statewide throughout the Cumberland, Tennessee, and Mississippi River drainages. Apparently absent now, the Fragile Papershell had been reported from Reelfoot Lake and the Wolf, Nolichucky, and Sequatchie rivers prior to 1960 (Starnes and Bogan, 1988).

Description: The shell is large: mature individuals attain a length of 150–160 mm; the shell is thin, brittle, compressed, somewhat elliptical, and oblong. The anterior end is rounded, the ventral margin is slightly curved, and the posterior end is obliquely truncated. The posterior ridge is indistinct; the dorsal area behind the umbones is flattened and extended into a wing; the alate condition reaches maximum development in mature shells, becoming greatly reduced or obliterated in old individuals. Beaks are flattened, only slightly elevated above the dorsal margin; sculpture consists of 3–4 feeble bars, the first being concentric, the others double-looped, but all usually very faint. The surface is marked with concentric growth lines, often dark at the rest periods.

Pseudocardinal teeth, two in the left valve and one in the right valve, are low, small, thin, and compressed and parallel with the hinge line. The left valve has two long, thin, compressed lateral teeth, and the inner one is often weakly developed; the right valve has one long, thin, elevated lateral tooth. There is no interdentum;

Plate 61. *Leptodea fragilis* (Rafinesque, 1820), Fragile Papershell.

the beak cavity is shallow. The periostracum is smooth, yellow, or yellowish green, often with numerous light green, indistinct rays, but rayless yellow shells are not uncommon. The nacre is silvery white, usually tinged with pink dorsally and posteriorly; much of the inner surface is iridescent.

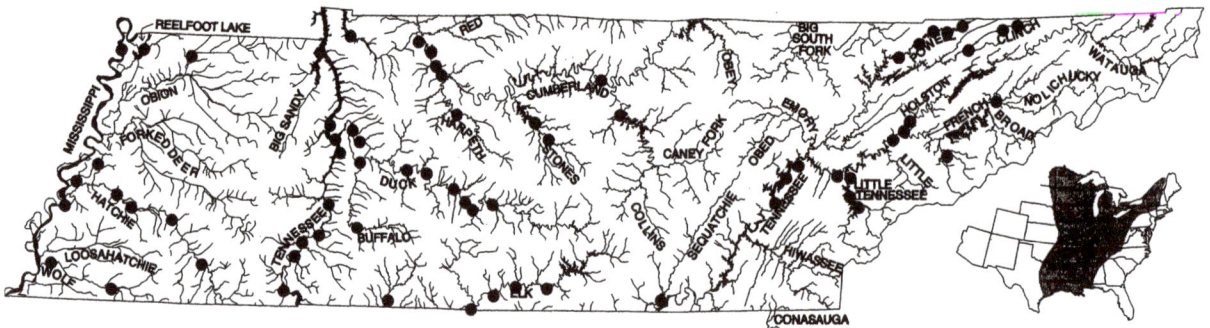

Range Map 60. *Leptodea fragilis* (Rafinesque, 1820), Fragile Papershell.

Life History and Ecology: *Leptodea fragilis* is a species tolerant of a variety of aquatic habitats. Although found in small streams in strong current with a coarse gravel and sand substrate, it reaches its maximum growth potential in rivers or river-lakes possessing slow current and a firm substrate composed of sand and mud. The Fragile Papershell may occur at depths of 15 to 20 feet, although it reaches its greatest population density at normal water levels of three feet or less in areas such as shallow embayments of Tennessee River reservoirs. Baker (1928a) lists this Papershell as being bradytictic, the reproductive period lasting from the end of August to about mid-July. The freshwater drum *(Aplodinotus grunniens)* has been recorded as the host fish for the glochidia of this mussel (Fuller, 1974).

Status: Currently Stable (Williams et al., 1993:13).

Leptodea leptodon (Rafinesque, 1820) Scaleshell

RANGE MAP 61; PLATE 62

Synonymy:

Unio (Leptodea) leptodon Rafinesque, 1820; Rafinesque, 1820:295, pl. 80, figs. 5–7
Unio leptodon Rafinesque, 1820; Say, 1834:no pagination
Symphynota leptodon (Rafinesque, 1820); Férussac, 1835:25
Alasmodonta leptodon (Rafinesque, 1820); Lapham, 1852:370
Leptodea leptodon (Rafinesque, 1820); Conrad, 1853:262
Lampsilis (Proptera) leptodon (Rafinesque, 1820; Simpson 1900:575
Proptera leptodon (Rafinesque, 1820); Sterki, 1907:393
Lampsilis leptodon (Rafinesque, 1820); Vanatta 1915:551
Lasmonos leptodon (Rafinesque, 1820); Utterback, 1916a:388
Paraptera leptodon (Rafinesque, 1820); Ortmann, 1918:571
Lampsilis (Leptodea) leptodon (Rafinesque, 1820); Frierson, 1927:82
Anodon purpurascens Swainson, 1823; Swainson, 1823a:pl. 160

Symphynota tenuissima Lea, 1829; Lea, 1829:453, pl. 11, fig. 21
Margarita (Unio) tenuissimus (Lea, 1829); Lea, 1836:38
Unio tenuissimus (Lea, 1829); Hanley, 1843:206, pl. 20, fig. 42
Margaron (Unio) tenuissimus (Lea, 1829); Lea, 1852c:38
Unio velum Say, 1829; Say, 1829:293
Leptodea velum (Say, 1829); Haas, 1969a:419
Lampsilis blatchleyi Daniels, 1902; Daniels, 1902:13, pl. 2
Lampsilis (Proptera) blatchleyi Daniels, 1902; Simpson, 1914:190
Leptodea blatchleyi (Daniels, 1902); Goodrich and van der Schalie 1944:316

Type Locality: Lower Ohio River.

General Distribution: Upper Mississippi River drainage, south to the Tennessee; Buffalo, New York; southern Michigan; Souris River, Manitoba (Simpson, 1914). Clarke (1981a), however, does not include this species in the naiad fauna of Canada.

Tennessee Distribution: Ortmann (1918:571) listed records of this mussel from the Tennessee River "below Knoxville," Clinch River (Union and Anderson counties), and the Holston River (Grainger County). The Scaleshell also was reported from the Duck River, Maury County (Ortmann, 1924a). Shoup et al. (1941) recorded this mussel from the East Fork Obey River, Putnam County; evidence of its former occurrence in the upper Tennessee and Cumberland rivers within the state is marginal. *Leptodea leptodon* was considered a rare species by these malacologists; although it may have been overlooked in the past four or five decades, the Scaleshell probably is no longer part of the mussel assemblage of Tennessee.

Description: The shell is elongate, ovate, and rhomboidal with the ventral margin nearly straight; it is thin and compressed. The anterior end is rounded, and the

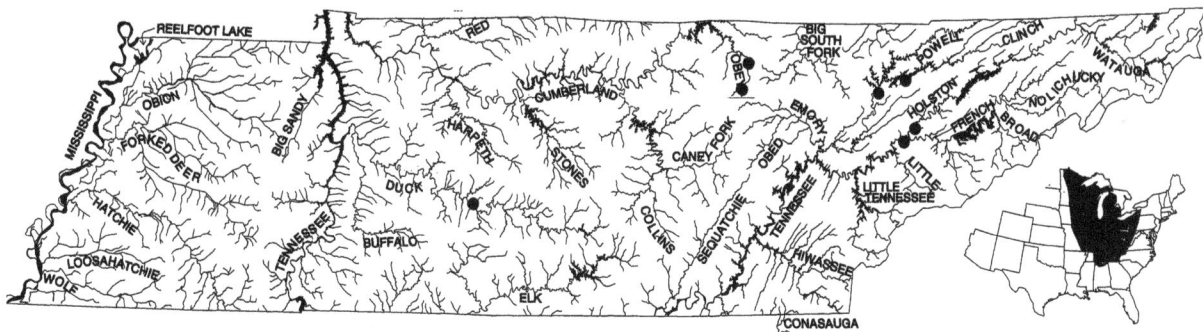

Range Map 61. *Leptodea leptodon* (Rafinesque, 1820), Scaleshell.

posterior end is bluntly pointed; the posterior ridge is distinct but low and rounded. Beaks are small, compressed, placed considerably forward, and about even with the hinge line. A pseudocardinal tooth in each valve is reduced to a very small, tubercular swelling; there are two low, incomplete, indistinct lateral teeth in the left valve, one slightly stronger lateral tooth in the right valve. Lateral teeth are long, and the anterior parts appear as a swelling of the hinge line. There is no interdentum; the beak cavity is very shallow or absent. The shell is small, although old individuals may reach 120 mm in length. The periostracum is yellowish or olive green, often with numerous, wide, faint green rays; the surface is roughened with growth lines. The nacre is purplish or salmon, especially the upper half; the rest is a bluish iridescence.

Life History and Ecology: In Missouri rivers, Oesch (1984:173) noted that "[t]his is a typical riffle shell, found only in clear, unpolluted water with a good current." He also made reference to the fact that this mus-

sel tends to bury itself down into the substrate several inches. Baker (1928a) indicated that this mussel is probably bradytictic. Host fish for the glochidia unknown.

Status: Endangered (Williams et al., 1993:13). Apparently highly susceptible to pollution and siltation, *Leptodea leptodon* probably has been extirpated in Tennessee rivers for at least several decades.

Lexingtonia dolabelloides (Lea, 1840)
Slabside Pearlymussel
RANGE MAP 62; PLATE 63
Synonymy:
Unio dolabelloides Lea, 1840; Lea, 1840:288; Lea, 1842b:215, pl. 15, fig. 31
Margaron (Unio) dolabelloides (Lea, 1840); Lea, 1852c:35
Pleurobema dolabelloides (Lea, 1840); Simpson, 1900a:750
Lexingtonia dolabelloides (Lea, 1840); Ortmann, 1920:294
Quadrula dolabelloides (Lea, 1840); Frierson, 1927:59
Quadrula (Pleuronaia) dolabelloides (Lea, 1840); Haas, 1969a:304
Pleurobema (Lexingtonia) dolabelloides (Lea, 1840); Burch, 1973:50, fig. 34
Unio thorntonii Lea, 1857; Lea, 1857b:83; Lea, 1866:38, pl. 14, fig. 36
Margaron (Unio) thorntonii (Lea, 1857); Lea, 1870:56
Unio mooresianus Lea, 1857; Lea, 1857b:83; Lea, 1866:39, pl. 14, fig. 37
Margaron (Unio) mooresianus (Lea, 1857); Lea, 1870:39
Unio recurvatus Lea, 1871; Lea, 1871:192; Lea, 1874:10, pl. 2, fig. 6
Unio circumactus Lea, 1871; Lea, 1871:192; Lea, 1874:15, pl. 4, fig. 11
Unio subglobatus Lea, 1871; Lea, 1871:191; Lea, 1874:7, pl. 1, fig. 3
Pleurobema subglobata (Lea, 1971); Simpson, 1900a:751
Pleurobema appressum (Lea); Simpson, 1900a:49 [of authors]
Pleurobema conradi Vanatta, 1915; Vanatta, 1915:559
Lexingtonia conradi (Vanatta, 1915); Ortmann and Walker, 1922:19

Type Locality: Holston River, Tennessee.

General Distribution: Tennessee River system, from Lee and Tazewell counties, southwestern Virginia to Mussel Shoals (formerly), Alabama.

Tennessee Distribution: The small river (tributary) form, *Lexingtonia d. conradi*, occurs in the headwaters of the Powell River, Claiborne and Hancock counties, and in the Clinch River from Hancock County

Plate 62. *Leptodea leptodon* (Rafinesque, 1820), Scaleshell.

downstream to Clinton, Anderson County. This form occurred throughout the North and South Forks of the Holston River, but was not found in the Holston River proper. *Lexingtonia dolabelloides* once inhabited the lower Clinch (intergrading with the headwaters form in Anderson County), the French Broad at Boyds Creek, Sevier County, and in the Tennessee River below Knoxville downstream as far as Hamilton County (Ortmann, 1918; 1925). The form *L. d. conradi* occurs in the headwaters of the Elk River in south-central Tennessee (Isom et al., 1973). The Duck River provides a good illustration of the clinal variation in *Lexingtonia,* with the form *L. d. conradi* being found in Coffee County and *L. d. dolabelloides* occurring in Maury County. The Buffalo River population also exhibited this clinal variation in the Slabside Pearlymussel (Ortmann, 1924a; Isom and Yokley, 1968a; van der Schalie, 1973).

Description: Although generally subtriangular in outline, this mussel exhibits considerable variability in shell shape. The majority of individuals possess a wide, flat disc, extending from the beak to the ventral margin. Valves are moderately inflated and very solid; large, mature specimens may reach a length of 85 mm. Beaks are prominent with the umbonal area arched forward and located near the anterior end. The anterior end of the shell is obliquely truncate above and rounded to the base; the posterior slope is truncated. The ventral margin of the shell is curved, and the dorsal slope is strongly curved. The posterior ridge is narrowly rounded but distinct, although not as elevated as the radial swelling in front of it. The surface of the shell is often irregularly and concentrically sculptured as a result of pronounced growth rings. Beak cavities are shallow; muscle scars are deep, and the pallial line is well impressed anteriorly. Sculpture consists of 6–8 fine, rather

Plate 63. *Lexingtonia dolabelloides* (Lea, 1840), Slabside Pearlymussel.

crowded, irregular, and wavy bands, which are distinct anteriorly, becoming indistinct in the middle.

The left valve possesses two pseudocardinal teeth; the upper tooth is triangular, and the lower tooth is bladelike, separated from two short, curved lateral teeth by a broad interdentum. The right valve has a triangular pseudocardinal tooth, occasionally with a

Range Map 62. *Lexingtonia dolabelloides* (Lea, 1840), Slabside Pearlymussel.

smaller tooth before and behind it. There is typically a single large lateral tooth, although occasionally a vestigial tooth below is present. The periostracum is greenish yellow (in juveniles) to tawny or brownish with a few broken green rays or blotches in some specimens, especially young individuals. The nacre color is white or, more rarely, straw-colored.

Life History and Ecology: *Lexingtonia dolabelloides* once occurred in shoal areas of the Tennessee River as well as in small to medium-sized streams and rivers such as the Clinch, Powell, Duck, and Hiwassee (the form *L. d. conradi*). Ortmann (1918) arbitrarily separated the big river species, *L. dolabelloides,* from the headwater form, *L. d. conradi,* on the basis of the shell diameter and length; specimens with a diameter which is 50% of the length or greater are *dolabelloides* and those less than 50% are *conradi.* A moderately strong current and a substrate composed of sand, fine gravel, and cobbles appear to provide the most suitable habitat for this species. It is probably tachytictic; Ortmann (1921) recorded finding unripe glochidia early in July. Six species of minnows have been found naturally infested with glochidia: popeye shiner *(Notropis ariommus),* rosyface shiner *(N. rubellus),* saffron shiner *(N. rubricroceus),* silver shiner *(N. photogenis),* telescope shiner *(N. telescopus),* and Tennessee shiner *(N. leuciodus)* (Kitchel, 1985; Kitchel in Neves, 1991).

Status: Threatened (Williams et al., 1993:13). *Lexingtonia dolabelloides* has been proposed as Endangered for inclusion in the federal list of *Endangered and Threatened Wildlife and Plants.* The Slabside Pearlymussel is restricted to thinly scattered or isolated populations in primarily the Clinch, Powell, Elk, Duck, and Hiwassee rivers in Tennessee, the North Fork and Middle Fork Holston rivers in Virginia, and the Paint Rock River in Alabama.

Ligumia recta (Lamarck, 1819)
Black Sandshell

RANGE MAP 63; PLATE 64

Synonymy:
Unio recta Lamarck, 1819; Lamarck, 1819:74
Unio (Ligumia) recta Lamarck, 1819; Swainson, 1840:267, 274, fig. 55
Unio rectus Lamarck, 1819; Conrad, 1836:33, pl. 15
Margarita (Unio) rectus (Lamarck, 1819); Lea, 1836:34

Margaron (Unio) rectus (Lamarck, 1819); Lea, 1852c:35
Lampsilis rectus (Lamarck, 1819); Smith, 1899:290, fig. 78
Lampsilis (Ligumia) recta (Lamarck, 1819); Haas, 1930:328
Eurynia (Eurynia) recta (Lamarck, 1819); Ortmann, 1912a:344, fig. 24
Eurynia recta (Lamarck, 1819); Baker, 1920:383
Ligumia recta (Lamarck, 1819); Ortmann and Walker, 1922:59
Ligumia (Ligumia) recta (Lamarck, 1819); Baker, 1928a:255, pl. 87, figs. 4, 5
Ligumia recta recta (Lamarck, 1819); Haas, 1969a:430
Unio (Eurynia) latissima Rafinesque, 1819; Rafinesque, 1819:426 [nomen nudum]
Unio (Eurynia) latissima Rafinesque, 1820; Rafinesque, 1820:297, pl. 80, figs. 14, 15
Eurynia (Eurynia) recta latissima (Rafinesque, 1820); Ortmann, 1919:276, pl. 16, fig. 12, 13
Ligumia recta latissima (Rafinesque, 1820); Ortmann and Walker, 1922:59
Ligumia (Ligumia) recta latissima (Rafinesque, 1820); Baker, 1928a:257, pl. 87, figs. 1–3; pl. 86, figs. 13, 14
Lampsilis (Ligumia) recta latissima (Rafinesque, 1820); Frierson, 1927:70
Unio praelongus Barnes, 1823; Barnes, 1823:261, pl. 13, fig. 11
Mya praelonga (Barnes, 1823); Eaton, 1826:220
Eurynea praelonga (Barnes, 1823); Stimpson, 1851:13
Unio sageri Conrad, 1836; Conrad, 1836:53, pl. 29, fig. 1
Lampsilis recta var. *sageri* (Conrad, 1836); Simpson, 1914:96
Unio arquatus Conrad, 1854; Conrad, 1854:297, pl. 26, fig. 8
Unio leprosus Miles, 1861; Miles, 1861:240

Type Locality: Lake Erie.

General Distribution: The Black Sandshell is widely distributed throughout the Mississippi River Basin from Minnesota to western New York and Pennsylvania southwest to Oklahoma and east to the Alabama River Basin, the Red River of the North, and the St. Lawrence River Basin.

Tennessee Distribution: *Ligumia recta* has been reported from throughout the Tennessee River system, including the Powell, Clinch, Holston, Nolichucky, French Broad, Little Tennessee, Hiwassee, and Duck rivers as well as the mainstem of the Tennessee River in East and Middle Tennessee. In the Cumberland River system, the Black Sandshell occurred in the Big South Fork Cumberland, Obey, Caney Fork, Stones, and Harpeth rivers as well as throughout the main channel of the Cumberland River.

Description: Shells of the Black Sandshell are large, solid, and elongate, elliptical in outline, and vary from compressed to moderately inflated. Maximum shell

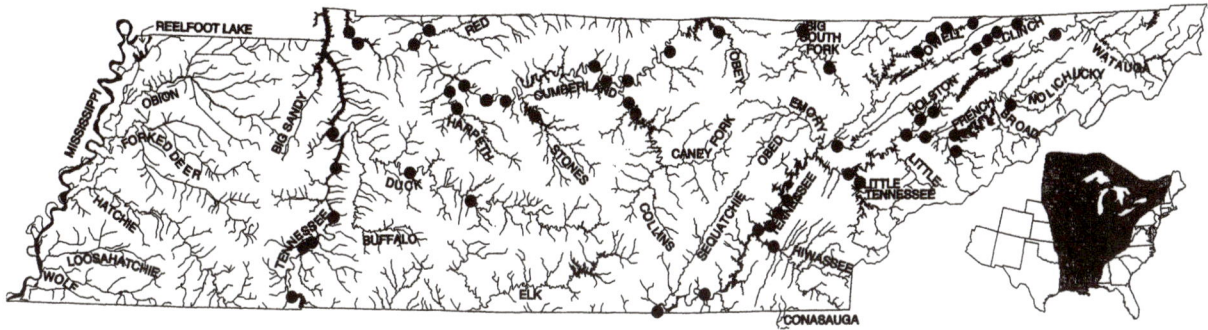

Range Map 63. *Ligumia recta* (Lamarck, 1819), Black Sandshell.

length is about 160 mm. The anterior end is evenly rounded; the dorsal and ventral margins are straight and parallel; the posterior end is pointed. The posterior ridge is rounded and most distinct close to the umbos. The male shell is drawn out posteriorly, ending in a blunt point near the middle of the posterior margin. The female shell has a long, rounded marsupial swelling and ends in a blunt point about two-thirds of the way up the posterior margin. The posterior slope of the shell is flattened. Beaks are low and only slightly elevated above the hinge line; sculpture consists of 3–5 indistinct, double-looped bars. The beak cavity is shallow.

The left valve has two diverging, compressed, serrated triangular pseudocardinal teeth and two long, straight lateral teeth. The right valve has a single heavy, erect, roughened, triangular pseudocardinal tooth, usually with a small, low, flattened tooth anteriorly; there is a single lateral tooth that is long and straight. The interdentum is very narrow or absent. Anterior adductor muscle scars are deep; posterior adductor muscle scars are shallow. The pallial line is impressed anteriorly. The periostracum is dark green to brown, becoming black with age. Young individuals have numerous faint green rays that become obliterated with maturity. The surface is marked by raised growth ridges. The nacre color is typically white with a pink wash near the beak cavity; in some individuals the nacre is completely pink or purple. There is some iridescence posteriorly.

Life History and Ecology: The Black Sandshell is typically found in medium-sized to large rivers in locations with strong current and substrates of coarse sand and gravel with cobbles in water depths from several inches to six feet or more. Ortmann (1919) reported gravid females from the middle of August to the last of July,

indicating that this is a bradytictic species. Watters (1994a) lists the following fish hosts for the glochidia of *Ligumia recta:* banded killifish *(Fundulus diaphanus),* bluegill *(Lepomis macrochirus),* green sunfish *(L. cyanellus),* orangespotted sunfish *(L. humilis),* largemouth bass *(Micropterus salmoides),* sauger *(Stizostedion canadense),* and white crappie *(Pomoxis annularis).* In addition, Hove et al. (1994) confirmed the

Plate 64. *Ligumia recta* (Lamarck, 1819), Black Sandshell.

largemouth bass as a host fish and added a new host fish, the walleye *(Stizostedion vitreum)*. Hove et al. (1994a) confirmed the early records of Surber (1913), Wilson (1916), and others that the bluegill *(Lepomis macrochirus)* could serve as a host fish.

Status: Special Concern (Williams et al., 1993:13).

Ligumia subrostrata (Say, 1831)
Pondmussel

RANGE MAP 64; PLATE 65

Synonymy:

Unio subrostratus Say, 1831; Say, 1831b:no pagination
Lampsilis subrostratus (Say, 1831); Simpson, 1900a:546
Lampsilis (Ligumia) subrostrata (Say, 1831); Frierson, 1927:77
Eurynia (Eurynia) subrostrata (Say, 1831); Ortmann, 1912a:344
Eurynia subrostrata (Say, 1831); Utterback, 1916c:15
Ligumia subrostrata (Say, 1831); Grier and Mueller, 1922:100
Unio nashvillianus Lea, 1834; Lea, 1834:100, pl. 14, fig. 43
Margarita (Unio) nashvillianus (Lea, 1834); Lea, 1836:26
Margaron (Unio) nashvillianus (Lea, 1834); Lea, 1852c:29
Margaron (Unio) nashvilliensis (Lea, 1834); Lea, 1870:45 [emendation]
Unio mississippiensis Conrad, 1850; Conrad, 1850:277, pl. 38 fig. 11
Margaron (Unio) mississippiensis (Conrad, 1850); Lea 1852:29
Unio rutersvillensis Lea, 1859; Lea, 1859c:155; Lea, 1860e:355, pl. 60, fig. 181
Margaron (Unio) rutersvillensis (Lea, 1859); Lea, 1870:43
Unio topekaensis Lea, 1868; Lea, 1868a:144; Lea, 1868c:313, pl. 49, fig. 126
Margaron (Unio) topekaensis (Lea, 1868); Lea, 1870:43
Unio cocoduensis Reeve, 1865; Reeve, 1865:pl. 24, fig. 117
Lampsilis subrostrata var. *furva* Simpson, 1914; Simpson, 1914:100

Type Locality: Wabash River.

General Distribution: The Pondmussel is found throughout the Mississippi River Basin from western Ohio and Michigan west to South Dakota, south to Oklahoma, Texas, Louisiana, and Mississippi. It also occurs in the lower Cumberland River system of Tennessee and Kentucky.

Tennessee Distribution: *Ligumia subrostrata* is known from lower Kentucky Lake, Reelfoot Lake and the Cumberland, Harpeth, and Hatchie rivers (Najarian, 1955).

Description: Shells of the Pondmussel are thin but stout, elongate to somewhat elliptical in outline, and moderately inflated. The anterior end is evenly rounded; the dorsal margin is slightly curved; the ventral margin is straight to slightly curved. The posterior ridge is moderately developed. Adult male shells are slightly larger than those of females with the posterior end sharply pointed. The female shell has a posterior margin that is bluntly truncated, with a very large, rounded marsupial swelling. Beaks are low, only slightly raised above the hinge line; sculpture consists of 6–10 fine, distinct lines drawn up in the middle. The beak cavity is shallow to moderately deep. Maximum shell length is usually less than 95 mm.

The left valve has two thin, compressed, erect pseudocardinal teeth which lie almost parallel with the hinge line and two slightly curved, long, thin lateral teeth. The right valve has a single thin, compressed, erect pseudocardinal tooth, with a second low thin tooth located anteriorly, and one long, thin, slightly curved lateral tooth. The interdentum is very narrow or absent. Anterior adductor muscle scars are deep, while the posterior adductor muscle scars are shallow. The pallial line is lightly impressed anteriorly. The periostracum is a dull greenish yellow, changing to black with age;

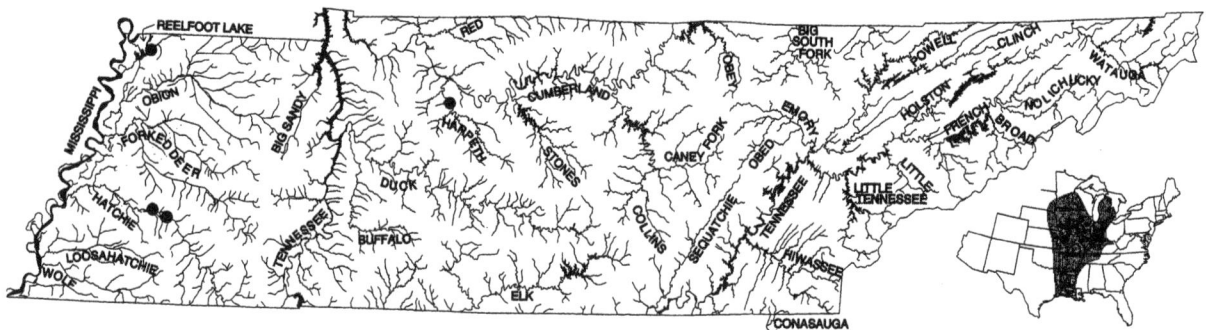

Range Map 64. *Ligumia subrostrata* (Say, 1831), Pondmussel.

Plate 65. *Ligumia subrostrata* (Say, 1831), Pondmussel.

dark green rays cover the shell but mostly are found on the posterior portion, becoming obscured with age. The nacre color is bluish white, iridescent posteriorly.

Life History and Ecology: The Pondmussel lives in small streams, in shallow portions of lakes and ponds, in sloughs, and in quiet water areas of larger rivers. It occurs in substrates of mud or sand, typically in a less than two feet of water. This species seems to adapt to newly created ponds and channels. Utterback (1916c) reported glochidia in June, suggesting that the Pondmussel is bradytictic. Watters (1994a) lists the following fish hosts for the glochidia of *Ligumia subrostrata:* largemouth bass *(Micropterus salmoides),* bluegill *(Lepomis macrochirus),* green sunfish *(L. cyanellus),* orangespotted sunfish *(L. humilis),* and warmouth *(L. gulosus).*

Status: Currently Stable (Williams et al., 1993:13).

Medionidus acutissimus (Lea, 1831)
Alabama Moccasinshell
RANGE MAP 65; PLATE 66
Synonymy:
Unio acutissimus Lea, 1831; Lea, 1831:89, pl. 10 fig. 18
Margarita (Unio) acutissimus (Lea, 1831); Lea, 1836:14
Margaron (Unio) acutissimus (Lea, 1831); Lea, 1852c:21
Medionidus acutissimus (Lea, 1831); Simpson, 1900a:590
Unio rubellinus Lea, 1857; Lea, 1857a:32; Lea, 1858e:70, pl. 13, fig. 51
Margaron (Unio) rubellinus (Lea, 1857); Lea, 1870:32
Unio rubellianus Lea, 1857; Sowerby, 1868:pl. 90, fig. 490 [misspelling]
Unio semiplicatus Troschel, 1841; Küster, 1862:279, pl. 94, fig. 4 [misidentification]

Type Locality: Alabama River.

General Distribution: The Alabama Moccasinshell is restricted to the Mobile Bay Basin in Alabama, Georgia, Mississippi, and Tennessee (Johnson, 1977; Williams et al., 1993).

Tennessee Distribution: *Medionidus acutissimus* is found only in southeastern Tennessee in the short stretch of the Conasauga River, Polk County.

Description: The Alabama Moccasinshell is small, rhomboid, and subinflated. Maximum shell length is about 55 mm. The anterior end is regularly rounded, and the ventral margin is slightly curved to incurved. The posterior ridge is high, sharp, and often double. Shells of males are arcuate in outline with the posterior end coming to a sharp point at the posterior-ventral margin. Female shells are swollen in the middle of the ventral margin, and the point at the posterior end is not as ventral as in the male. Beaks are slightly inflated and little elevated above the hinge line; sculpture is not reported. The beak cavity is open and shallow or entirely wanting.

The left valve has two short, triangular pseudocardinal teeth and two slightly curved lateral teeth. The right valve has one short, stubby pseudocardinal and one curved lateral tooth. The posterior slope of the shell is covered with parallel corrugated folds or ridges. The anterior adductor muscle scars are deep; the posterior adductor muscle scars are shallow. The pallial line is impressed anteriorly. The periostracum is fairly dull, yellowish to greenish to tan in color, and covered

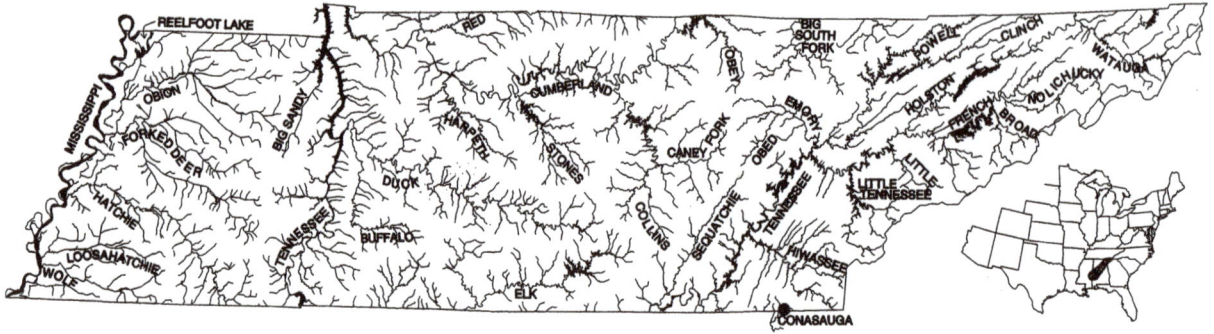

Range Map 65. *Medionidus acutissimus* (Lea, 1831), Alabama Moccasinshell.

with faint, broken, and fine green wavy rays. The nacre color is bluish green, salmon, flesh-colored, whitish, reddish, or purplish.

Life History and Ecology: Johnson (1977) noted that this species is typical of small streams. In the Conasauga River in Tennessee it inhabits gravel and cobble shoals in shallow water, typically in stretches with moderately strong current. Haag and Warren (1997) identified the blackspotted topminnow *(Fundulus oli-* *vaceous)*, Tuskaloosa darter *(Etheostoma douglasi)*, redfin darter *(Etheostoma whipplei)*, blackbanded darter *(Percina nigrofasciata)*, and logperch *(Percina* sp., cf. *P. caprodes)* as hosts for the glochidia. The breeding season for the Alabama Moccasinshell has not been documented, but it is assumed to be bradytictic, like *Medionidus conradicus*.

Status: Threatened (Williams et al., 1993:13). The U.S. Fish and Wildlife Service has developed a recovery plan for this species (U.S. Fish and Wildlife Service, 1994).

Plate 66. *Medionidus acutissimus* (Lea, 1831), Alabama Moccasinshell.

Medionidus conradicus (Lea, 1834)
Cumberland Moccasinshell

RANGE MAP 66; PLATE 67

Synonymy:

? *Unio plateolus* Rafinesque, 1831; Rafinesque, 1831:3
Medionidus plateolus (Rafinesque, 1831); Ortmann, 1918:575
Unio conradicus Lea, 1834; Lea, 1834:63, pl. 9, fig. 23
Margarita (Unio) conradicus (Lea, 1834); Lea, 1836:13
Margaron (Unio) conradicus (Lea, 1834); Lea, 1852c:21
Unio conradius Lea, 1834; Conrad, 1838:87, pl. 47, fig. 3 [misspelling]
Margaron (Unio) conradianus (Lea, 1834); Lea, 1870:32 [misspelling]
Unio conradianus Lea, 1834; B. H. Wright, 1888b:[2] [misspelling]
Medionidus conradicus (Lea, 1834); Simpson, 1900a:589

Type Locality: No locality given in the original description. Johnson (1977:165) lists it as "no locality [Caney Fork of the Cumberland River, Tennessee]," apparently supplying this locality as the type locality.

General Distribution: The Cumberland Moccasinshell is a Cumberlandian species endemic to the Tennessee and Cumberland River drainages.

Tennessee Distribution: This distinctive small shell was commonly found in the Powell, Clinch, Holston, Emory, Watauga, Little Pigeon, Little Tennessee, Tellico, Duck, and Little rivers, Conasauga Creek (Hiwassee River Basin), the main Tennessee River, and in numerous small streams, especially in upper East Tennessee. In the Cumberland River Basin it was found in the Obey, Collins, Roaring, West Fork Stones, and Stones rivers (Pilsbry and Rhoads, 1896; Wilson and Clark, 1914; Parmalee and Klippel, 1984; Parmalee, 1988; Starnes and Bogan, 1988). The Cumberland Moccasinshell has been reported from archaeological deposits along the Tellico River (Parmalee and Klippel, 1984), Little Pigeon River (Parmalee, 1988), Little Tennessee River, and the Elk and the Duck rivers (Robison, 1986; Bogan, 1990). Hurd (1974) identified specimens of *Medionidus* from the Coosa River drainage as *M. conradicus*, but Johnson (1977) felt these were errors in identification.

Description: The shell is usually elongate and elliptical in outline, becoming arcuate in adults; it is relatively thin, but becomes thicker with age. The Cumberland Moccasinshell is a small species, seldom exceeding 60 mm in length. Shells are rounded anteriorly with a posterior ridge ending in a rounded point at the posterior end. The ventral margin is straight, becoming incurved in adult specimens. Valves are subinflated and solid. The male shell is generally more arcuate and broader in the posterior area. Female shells are generally somewhat more inflated along the middle of the ventral margin; this area may be faintly radially grooved. Beaks are only slightly inflated and elevated, being marked by fine, irregular corrugated ridges often tending to double loops. The posterior slope is marked by wrinkles or corrugations which often extend onto the anterior portion of the shell.

Plate 67. *Medionidus conradicus* (Lea, 1834), Cumberland Moccasinshell.

The left valve has two short, stumpy pseudocardinal teeth and two slightly curved lateral teeth. The right valve has a single short, stumpy pseudocardinal tooth and a single lateral tooth. The beak cavity is very shallow, often lacking. Anterior adductor muscle scars are deep, and the posterior scars are only slightly impressed. The pallial line is impressed anteriorly where the shell is thicker. The periostracum is slightly shiny, tawny to

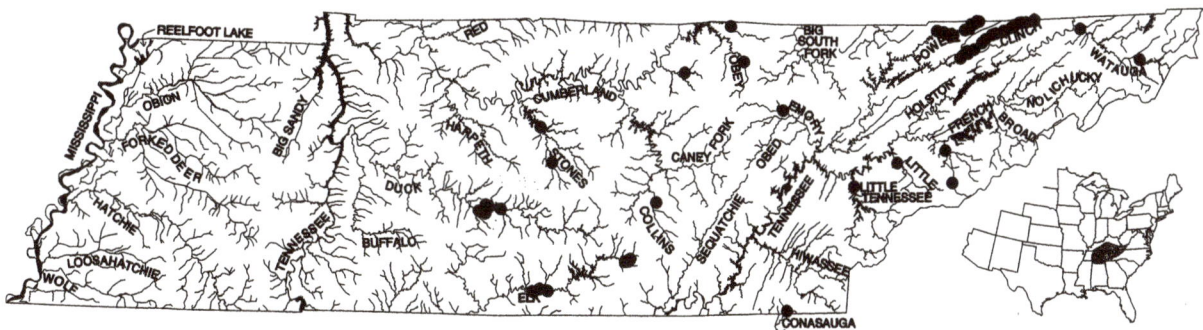

Range Map 66. *Medionidus conradicus* (Lea, 1834), Cumberland Moccasinshell.

yellowish green in color, and covered with weak, broken dark green rays which break into arrowhead-shaped markings. The nacre color is bluish to dirty white and may be iridescent posteriorly.

Life History and Ecology: The Cumberland Moccasinshell inhabits a substrate composed of sand and gravel, often living in cracks in the bedrock or under flat rocks. Usually it occurs at depths of less than three feet in moderate to strong current. Wilson and Clark (1914) and Ortmann (1918, 1924a, 1925) noted that this mussel is typically a small stream inhabitant. Zale and Neves (1982b) identified the fantail darter *(Etheostoma flabellare)* and the redline darter *(E. rufilineatum)* as hosts for the glochidia; Stern and Felder (1978) list the warmouth *(Lepomis gulosus)*. Luo (1993) was able to infect the rainbow darter *(Etheostoma caeruleum)* and striped darter *(E. virgatum)* with glochidia of the Cumberland Moccasinshell under laboratory conditions. The brooding period for *M. conradicus* begins with gravid females occurring from early September on with glochidia present in mid-September; they are not discharged until mid- to late May (Ortmann, 1921). The glochidia appear in river drift from January to May and June, but are absent in July and August; they reappear in river drift in September through early November (Zale and Neves, 1982a, b).

Status: Special Concern (Williams et al., 1993:13).

Medionidus parvulus (Lea, 1860)
Coosa Moccasinshell
RANGE MAP 67; PLATE 68

Synonymy:
Unio parvulus Lea, 1860; Lea, 1860c:307; Lea, 1866:45, pl. 16, fig. 43
Margaron (Unio) parvulus (Lea, 1860); Lea, 1870:32
Medionidus parvulus (Lea, 1860); Simpson, 1900a:590

Type Locality: Coosa River, Alabama; Chattanooga [Chattooga River], Georgia.

General Distribution: Coosa River Drainage of Georgia, Tennessee, and Alabama (Burch, 1975a; Williams et al., 1993).

Tennessee Distribution: *Medionidus parvulus* is known only from the Conasauga River, Polk County.

Description: The Coosa Moccasinshell is thin shelled, small, somewhat rhomboid, and slightly compressed. Maximum length is about 40 mm. The anterior end is evenly rounded, the posterior end is bluntly rounded to pointed, the ventral margin is slightly incurved to straight, and the dorsal margin is gently curved. The posterior ridge is rounded. The posterior slope is corrugated with fine ridges or plications, while the disk of the shell is smooth. Male shells are slightly arcuate while the female shell is not, and it is somewhat more inflated ventrally. Beaks are moderately inflated; sculpture is not reported.

The left valve has two small, compressed pseudocardinal teeth and two long, thin, nearly straight lateral teeth. The right valve has a single small, erect, but somewhat compressed pseudocardinal tooth and a single long, nearly straight lateral tooth. The beak cavity is very shallow. Adductor muscle scars are deeply to moderately impressed. The pallial line is lightly impressed. The periostracum is greenish to yellowish in color, marked with numerous faint green undulating or zigzag lines that sometimes form rays. The nacre color is bluish green and iridescent posteriorly.

Simpson (1914) noted that *M. parvulus* resembled *M. conradicus* and considered it a doubtful species. Ortmann (1923) separated *M. parvulus* and *M. acutissimus* on the shape of the posterior margin and ridge and strength of plications on the posterior ridge. He observed that the anatomy of *M. parvulus* was practically identical to that of *Medionidus conradicus*. However, van der Schalie (1938) commented that he saw intergrades between the two species in large collections of shells. Johnson (1977) included *M. parvulus* in the synonymy of *M. acutissimus*.

Life History and Ecology: In stretches of the Conasauga River in Tennessee where this species is reported to occur, normal water depth is less than three feet with moderate to strong current and a substrate consisting of a small cobble, gravel, and sand mixture. Host fish for *Medionidus parvulus* unknown. The breeding season for the Coosa Moccasinshell has not been documented, but it is assumed to be bradytictic like *Medionidus conradicus*.

Status: Endangered (Williams et al., 1993:13). The U.S. Fish and Wildlife Service has developed a recovery plan for this species (U.S. Fish and Wildlife Service, 1994).

Range Map 67. *Medionidus parvulus* (Lea, 1860), Coosa Moccasinshell.

Plate 68. *Medionidus parvulus* (Lea, 1860), Coosa Moccasinshell.

Megalonaias nervosa (Rafinesque, 1820)
Washboard

RANGE MAP 68; PLATE 69

Synonymy:

Unio (Elliptio) nervosa Rafinesque, 1820; Rafinesque, 1820:296, pl. 80, figs. 8–10

Megalonaias nervosa (Rafinesque, 1820); Morrison, 1969:24

Unio crassus var. *giganteus* Barnes, 1823; Barnes, 1823:119

Megalonaias gigantea (Barnes, 1823); Ortmann and Walker, 1922:7–8

Amblema gigantea (Barnes, 1823); Strecker, 1931:34–35

Amblema (Megalonaias) gigantea (Barnes, 1823); Frierson, 1927:62

Megalonaias gigantea gigantea (Barnes, 1823); Haas, 1969a:284–285

Unio heros Say, 1829; Say, 1829:291

Quadrula heros (Say, 1829); Simpson, 1900a:770

Crenodonta heros (Say, 1829); Ortmann, 1912a:248

Megalonaias heros (Say, 1829); Utterback, 1915:125–129, pl. 7, fig. 16; pl. 17, figs. 48a–f

Magnonaias heros (Say, 1829); Utterback, 1915:47

Unio undulatus Barnes, 1823; Say, 1831a:pl. 16 [misidentification]

Unio multiplicatus Lea, 1831; Lea, 1831:70, pl. 4, fig. 2

Margarita (Unio) multiplicatus (Lea, 1831); Lea, 1836:12

Margaron (Unio) multiplicatus (Lea, 1831); Lea, 1852c:20

Unio heros var. *multiplicatus* Lea, 1831; Paetel, 1890:155

Unio eightsii Lea, 1860; Lea, 1860c:306; Lea, 1860e:367, pl. 64, fig. 192

Margaron (Unio) eightsii (Lea, 1860); Lea, 1870:29

Unio eighti Lea, 1860; Paetel, 1890:151 [misspelling]

Unio atrocostatus Lea, 1848; Reeve, 1864:pl. 4, fig. 13 [changed in errata to *Unio heros*]

Unio triumphans B. H. Wright, 1898; Wright, 1898a:101

Quadrula triumphans (B. H. Wright, 1898); Simpson, 1900a:83, pl. 3, fig. 3

Amblema (Megalonaias) triumphans (B. H. Wright, 1898); Frierson, 1927:62

Megalonaias triumphans (B. H. Wright, 1898); Haas, 1969a:285

Type Locality: "rapides de l'Ohio."

General Distribution: The Washboard is widespread throughout the Mississippi River drainage. It inhabits the Gulf drainages from the Ochlockonee River system west to the Rio Grande and apparently extends into northeastern Mexico. Clarke (1973) discounted records of the Washboard in any of the rivers in Canada. M. Mulvey (pers. comm., 1996) has examined populations of *Megalonaias* identified as *nervosa* and *boykiniana* and has suggested that they appear to be the same species.

Tennessee Distribution: The Washboard occurs in the Hatchie and Obion rivers and Reelfoot Lake in West Tennessee, and it is found throughout the lower Tennessee River, extending upstream above Chattanooga to Watts Bar Dam (Ahlstedt and McDonough, 1994). *Megalonaias nervosa* also occurs throughout most of the middle and lower stretches of the Duck and Elk rivers. In the Cumberland River it inhabits the main channel; it was once numerous in many stretches of the Red, Harpeth, Stones, and lower Caney Fork rivers.

Description: The shell of the Washboard is rhomboid to elongate quadrate in outline, thick, solid, heavy, and subinflated to inflated. Shell length may reach 250 mm. The anterior end is evenly rounded, and the posterior end and dorsal margin are nearly straight, with the dorsal area appearing alate in younger specimens; the ventral margin is gently curved. The posterior ridge is rounded. The shell surface is often covered with oblique folds and nodules with the disk of younger shells usually densely covered with chevron-shaped, nodulous plications that may become obscured with growth. Beaks are narrow, depressed, not raised much above the hinge line; sculpture consists of heavy, double-looped ridges which become nodulous on the posterior slope. The beak cavity is slightly compressed and moderately deep.

The left valve has two heavy, roughened broadly triangular pseudocardinal teeth and two straight, heavy lateral teeth. The right valve has one thick, heavily striated triangular pseudocardinal tooth, often with a small tubercular tooth on either side, and one high, heavy, and straight lateral tooth. Lateral teeth may become slightly curved in older shells. The interdentum is usually narrow. Anterior adductor muscle scars are large and deeply impressed, and the bottom of the scar is flat and heavily sculptured; posterior adductor muscle scars are impressed and shallow. The pallial line is distinct but not deeply impressed. The posterior slope of the shell is rounded, appearing to merge with the folds on the surface, is covered with several folds or ridges and tubercles, and becomes indistinct with age; the surface is marked mainly with growth lines. The periostracum is a yellowish brown or dark green in juvenile shells and rayless, becoming dark reddish brown to black, in adult individuals. The nacre color is whitish, occasionally tinted with salmon and often blotched with brown spots, becoming iridescent posteriorly.

Life History and Ecology: The Washboard is typically a large river species, living in the main channel and in some of the overbank areas of reservoirs. However, in some instances it may also become established in medium-sized and even small rivers. It is found in areas with a slow current with muddy to coarse gravel substrates, often in water up to 50 feet in depth. This animal is bradytictic, bearing embryos in early fall and glochidia in late winter, but is barren from April to

Range Map 68. *Megalonaias nervosa* (Rafinesque, 1820), Washboard.

Plate 69. *Megalonaias nervosa* (Rafinesque, 1820), Washboard.

August (Utterback, 1915). Watters (1994a) lists the following fish hosts for the glochidia of *Megalonaias nervosa*: American eel *(Anguilla rostrata),* black bullhead *(Ameiurus melas),* black crappie *(Pomoxis nigromaculatus),* white crappie *(P. annularis),* bluegill *(Lepomis macrochirus),* green sunfish *(L. cyanellus),* brown bullhead *(Ameiurus nebulosus),* channel catfish *(Ictalurus punctatus),* flathead catfish *(Pylodictis olivaris),* freshwater drum *(Aplodinotus grunniens),* gizzard shad *(Dorosoma cepedianum),* tadpole madtom *(Noturus gyrinus),* and the white bass *(Morone chrysops).* Weiss and Layzer (1995) added the longnose gar *(Lepisosteus osseus),* warmouth *(Lepomis gulosus),* and spotted bass *(Micropterus punctulatus)* as fish hosts for the Washboard.

Status: Currently Stable (Williams et al., 1993:13).

Obliquaria reflexa Rafinesque, 1820
Threehorn Wartyback

RANGE MAP 69; PLATE 70

Synonymy:

Obliquaria (Quadrula) reflexa Rafinesque, 1820; Rafinesque, 1820:306

Unio reflexus (Rafinesque, 1820); Say, 1834:no pagination

Obliquaria reflexa Rafinesque, 1820; F. C. Baker, 1898:89, pl. 14, fig. 5, pl. 20, fig. 2

Unio cornutus Barnes, 1823; Barnes, 1823:122, pl. 4, figs. 5, 5a–c

Mya cornuta (Barnes, 1823); Eaton, 1826:216

Margarita (Unio) cornutus (Barnes, 1823); Lea, 1836:15

Theliderma cornuta (Barnes, 1823); Swainson, 1840:269

Margaron (Unio) cornutus (Barnes, 1823); Lea, 1852c:22

Unio torulosus (Rafinesque, 1820); Short and Eaton, 1831:75 [misidentification]

Obliquaria reflexa var. *conradi* Frierson, 1927; Frierson, 1927:65

Type Locality: Kentucky River and Rapids of Letart, Ohio.

General Distribution: Throughout most of the Mississippi River drainage from western Pennsylvania north into Michigan and Minnesota, southwest to eastern Kansas, Oklahoma, and Texas (Burch, 1975a). In Canada, Clarke (1981a) records it from Lake Erie and its tributaries. In the Southeast, *Obliquaria reflexa* occurs in the Coosa–Alabama River and Tombigbee River systems.

Tennessee Distribution: The Threehorn Wartyback now occurs throughout the Cumberland and Tennessee rivers and in several of their major tributaries (for example, the Duck, Elk, and Stones rivers). Manning (1989) records it as a relict in the Hatchie River and has not encountered it in any other West Tennessee streams. Starnes and Bogan (1988) note that, prior to 1960, the Threehorn Wartyback was found inhabiting the Clinch, Holston, Obey, and Caney Fork rivers. Like several other naiad species, such as *Megalonaias nervosa* (Washboard) and *Ellipsaria lineolata* (Butterfly), *Obliquaria reflexa* has expanded its range upstream in the Cumberland and Tennessee rivers (reservoirs) in historic times, apparently adapting well to a river/lake habitat. For example, since the Little Tennessee River was dammed by completion of Tellico Dam in 1979, the Threehorn Wartyback quickly invaded the newly formed Tellico Lake (Loudon County) and, as of this writing, is abundant throughout the lake (Parmalee and Hughes, 1993).

Range Map 69. *Obliquaria reflexa* Rafinesque, 1820, Threehorn Wartyback.

Description: The shell is solid, thick, inflated, and generally oval in outline; the anterior-ventral margin is broadly rounded; the posterior end is usually bluntly pointed and somewhat truncated. Mature individuals rarely exceed 80 mm in length. The posterior end is well developed and in most specimens rather sharp-angled, although in some individuals it is broadly rounded. The posterior ridge and slope are typically covered with low, rounded tubercles or a series of parallel raised plications. Occasionally scattered, low tubercles appear on other surface areas. A row of 3–5 large, rounded, elevated, elongated knobs are present and extend from the beaks to the center of the ventral margin; the knobs of one valve alternate in position with those of the other valve. Often a broad, shallow sulcus or depression is present between the row of knobs and the posterior ridge and between the largest raised knobs. Beaks are elevated, prominent, and curved inward; sculpture consists of 4–5 heavy, parallel ridges, low in front and curved upwards behind. Some individuals in populations adapted to the lake or large river embayment habitat often exhibit a reduction in the size and number of prominent tubercles, and the shell becomes more rounded.

The left valve has two strong, heavy, somewhat elevated, ragged pseudocardinal teeth, sometimes joined anteriorly; the two lateral teeth are straight or very slightly curved, short, wide, and serrated. The right valve has a massive, deeply serrated pseudocardinal tooth, having the appearance of rising from a pit, often with the area in front slightly elevated and roughened, producing a low, curved tooth. The lateral tooth is wide, short, striated, sometimes with a second low, incomplete, lateral tooth evident. The interdentum is flat, usually fairly wide; the beak cavity is moderately deep. The color of the periostracum is variable; in the majority of individuals, it is a dull straw yellow or has shades of light brownish yellow or green. Some speci-

Plate 70. *Obliquaria reflexa* Rafinesque, 1820, Threehorn Wartyback.

mens lack rays, but the majority are usually patterned with distinct, fine green rays; occasionally they appear as wide green bands. In some old individuals, the periostracum may become blackish. The nacre is a pearly white, iridescent posteriorly. Although not observed in Tennessee specimens, occasional individuals from the Tombigbee River, Alabama, possess a nacre color of varying shades of purple.

Life History and Ecology: The Threehorn Wartyback is a species typical of the larger rivers in the state, such as the Tennessee and Cumberland, where there is moderately strong current and a stable substrate composed of gravel, sand, and mud. Although found at depths up to 18 to 20 feet, this mussel seems to do well at a depth of no more than four to six feet, and it can become quite numerous locally in shallow, sand- and mud-bottom river embayments almost or entirely devoid of current (e.g., Tellico Lake; see Parmalee and Hughes, 1993). Possibly because of this adaptability, *Obliquaria reflexa* has greatly expanded its range upstream in the Tennessee River in historic times (it is absent in the Pickwick Basin aboriginal mounds below Muscle Shoals, Alabama; see Morrison, 1942) so that it now occurs in all reservoirs. The Threehorn Wartyback is considered tachytictic, and the reproductive period occurs during June, July, and August. Host fish for the glochidia unknown. However, certain factors involving glochidial development discussed by Utterback (1915–1916a:321) seem to suggest that "its metamorphosis may take place without parasitism."

Status: Currently Stable (Williams et al., 1993:13).

Obovaria jacksoniana (Frierson, 1912)
Southern Hickorynut

RANGE MAP 70; PLATE 71

Synonymy:
Unio castaneus Lea, 1831 (December) non Rafinesque, 1831 (November); Lea, 1831:91, pl. 11, fig. 21
Margarita (Unio) castaneus (Lea, 1831); Lea, 1836:22
Margaron (Unio) castaneus (Lea, 1831); Lea, 1852c:26
Obovaria (Pseudoon) castaneus (Lea, 1831); Simpson, 1900a:602
Unio (Obovaria) jacksonianus Frierson, 1912; Frierson, 1912:23, pl. 3, figs. 1–3.
Obovaria (Pseudoon) jacksoniana (Frierson, 1912); Simpson, 1914:301

Type Locality: Pearl River, Mississippi, and Yalabusha River, Mississippi.

General Distribution: The Southern Hickorynut is known from Alabama west to eastern Texas, and in the Mississippi embayment as far north as southeastern Missouri.

Tennessee Distribution: The only known location for the Southern Hickorynut in the state is the Hatchie River in West Tennessee (Manning, 1989).

Description: Shells of this small mussel usually measure less than 45 mm in length. The valves are solid, moderately inflated, and generally circular in outline. The anterior end is evenly rounded, and the ventral margin is rounded; the margin from the beak to the posterior point is almost straight. The posterior end is sharply pointed in males and bluntly truncated in females. The posterior ridge is not prominent. Beaks are broad and low; the cavity is open and shallow.

The left valve has two stout, rough, triangular pseudocardinal teeth and two long, sculptured, slightly curved lateral teeth. The right valve has one erect, sculptured pseudocardinal with a small, compressed emarginate tooth anterior to it; there is one roughened, slightly curved lateral tooth. The interdentum is short and moderately wide. Adductor muscle scars are small and impressed, and the pallial line is impressed anteriorly. The periostracum is light to dark brown with fine green rays covering the posterior portion of the shell in some individuals. The nacre color is white, iridescent posteriorly.

Life History and Ecology: Oesch (1984) reported the habitat of *Obovaria jacksoniana* in Missouri as water with slow to moderate current with a medium-sized gravel substrate. The few specimens collected by Manning (1989) in the Hatchie River were found in stretches of slow current at depths of about two feet in a stable fine gravel and silt substrate. Host fish for *Obovaria jacksoniana* unknown.

Status: Special Concern (Williams et al., 1993:13). This small mussel appears to be uncommon to rare throughout its historic range. With less than six specimens known from the Hatchie River, it certainly may be considered rare in Tennessee.

Range Map 70. *Obovaria jacksoniana* (Frierson, 1912), Southern Hickorynut.

Plate 71. *Obovaria jacksoniana* (Frierson, 1912), Southern Hickorynut.

Obovaria olivaria (Rafinesque, 1820)
Hickorynut

RANGE MAP 71; PLATE 72

Synonymy:

Amblema olivaria Rafinesque, 1820; Rafinesque, 1820:314

Unio olivarius (Rafinesque, 1820); Conrad, 1834a:70

Obovaria olivaria (Rafinesque, 1820); Vanatta, 1915:553; Ortmann and Walker, 1922:46

Obovaria (Pseudoon) olivaria (Rafinesque, 1820); Ortmann, 1919:229

Unio brevialis Lamarck, 1819; Sowerby and Sowerby, 1823:fig. [misidentification]

Unio ellipsis Lea, 1828; Lea, 1828:268, pl. 4, fig. 4

Margarita (Unio) ellipsis (Lea, 1828); Lea, 1836:22

Margaron (Unio) ellipsis (Lea, 1828); Lea, 1852c:26

Lampsilis ellipsis (Lea, 1828); Kelly, 1899:401

Obovaria (Pseudoon) ellipsis (Lea, 1828); Simpson, 1900a:602

Obovaria ellipsis (Lea, 1828); Scammon, 1906:305

Diplasma vitrea Rafinesque, 1831; Potiez and Michaud, 1844:150, pl. 58, figs. 1, 2 [misidentification]

Unio taitianus Lea, 1834; Sowerby, 1868:pl. 67, fig. 338 [misidentification]

Unio pealei Lea, 1871; Lea, 1871:191; Lea, 1874:26, pl. 8 fig. 23

Unio pearlii Lea, 1874; Scammon, 1906:305 [misspelling and error]

Type Locality: "le Kentucky."

General Distribution: *Obovaria olivaria* occurs throughout the Mississippi River drainage from western Pennsylvania and western New York west to Kansas, north to Minnesota, and south to Missouri, Arkansas, and Louisiana, and in the St. Lawrence River Basin from Lake Ontario to Quebec.

Tennessee Distribution: The Hickorynut has been found in the main channel of the Mississippi River in northwest Tennessee and the main channels of the lower Tennessee and Cumberland rivers.

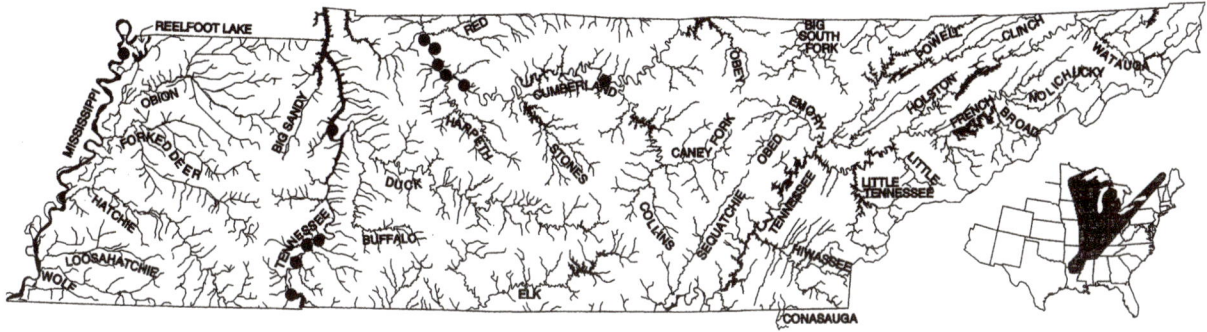

Range Map 71. *Obovaria olivaria* (Rafinesque, 1820), Hickorynut.

Description: The shell is thick, solid, and inflated, ovate or elliptical in outline, and reaches a maximum length of about 100 mm. The anterior end and ventral margin are broadly rounded; the posterior end is roundly pointed in male shells, but broadly rounded in female shells. Males tend to reach a somewhat greater size than females. The posterior ridge, when present, is rounded and only barely noticeable. Beaks are inflated, directed forward, and elevated above the hinge line; sculpture consists of 4–5 delicate bars with a sinus in the middle, presenting a double-looped shape. The beak cavity is relatively shallow with pronounced dorsal muscle scars.

The left valve has two heavy, roughened, triangular pseudocardinal teeth, often with a small, thin tooth anterior to these teeth, and two relatively long, raised, and striated lateral teeth. The right valve has one heavy, triangular, elevated, roughened pseudocardinal, with an erect, thin tooth present anteriorly, and a single straight, lateral tooth. The interdentum is short and broad. Anterior adductor muscle scars are very deep and roughened; posterior adductor muscle scars and pallial line are well impressed. The surface is marked by heavy growth lines raised as concentric ridges, often being dark in color. The periostracum ranges from olive green to yellowish brown, becoming very dark brown in old individuals; young shells have distinct fine green rays. The nacre color is bright white with iridescence posteriorly, sometimes with a pink or cream wash in the center of the valves.

Life History and Ecology: The Hickorynut is typically found on sand or gravel substrates in deep water, depths usually exceeding six to eight feet, with good current (Baker, 1928a). Scammon (1906:306) reported that the Hickorynut "is a lover of water of moderate depth and of sandy river beds." In large rivers it is often found in the large mussel beds in gravel bars in midriver. This species is bradytictic, with eggs and developing glochidia

Plate 72. *Obovaria olivaria* (Rafinesque, 1820), Hickorynut.

Family Unionidae 165

from August to June (Baker, 1928a). Watters (1994a) lists the shovelnose sturgeon *(Scaphirhynchus platorynchus)* as the fish host for glochidia of *Obovaria olivaria.*

Status: Currently Stable (Williams et al., 1993:13). The Hickorynut is presently extremely rare in the Tennessee River and possibly may be extirpated there.

Obovaria retusa (Lamarck, 1819)
Ringpink

RANGE MAP 72; PLATE 73

Synonymy:

Unio retusa Lamarck, 1819; Lamarck, 1819:72
Unio retusus Lamarck, 1819; Conrad, 1836:19, pl. 8
Obovaria torsa Rafinesque, 1820; Rafinesque, 1820:311, pl. 82, figs. 1–3
Unio torsus (Rafinesque, 1820); Potiez and Michaud, 1844:149, pl. 57, figs. 1, 2
Margarita (Unio) retusus (Lamarck, 1819); Lea, 1836:34
Margaron (Unio) retusus (Lamarck, 1819); Lea, 1852c:35
Obovaria (Obovaria) retusa (Lamarck, 1819); Simpson, 1900a:599
Obovaria retusa (Lamarck, 1819); Marsh, 1885:6

Type Locality: Lamarck (1819) erroneously gave the type locality as Nova Scotia. Johnson (1969) corrected this and designated the type locality as the Ohio River at Cincinnati, Ohio.

General Distribution: The Ringpink was found throughout the Ohio, Tennessee, and Cumberland River systems, including many of the major tributaries.

Tennessee Distribution: The Ringpink has been reported from the lower Holston River, Knox County, the Tennessee River at Knoxville, and the Clinch River at Clinton, Anderson County (Ortmann, 1918; Hickman,

1937). The authors obtained one fresh specimen (1973) of an extremely old, stunted individual from the lower Holston River, Knox County. Van der Schalie (1939) collected the Ringpink from the Tennessee River in Hardin and Humphreys counties and the authors more recently (1980–1990) found fresh adult specimens of the Ringpink in shellers' cull piles taken below Pickwick Landing Dam, Hardin County. Live specimens were reported from that stretch of river by TVA divers in 1987. Marsh (1885) and Ortmann (1924a) recorded *O. retusa* from the Duck River, Maury County. *Obovaria retusa* formerly inhabited the Cumberland River from Jackson County downriver to Stewart County (Wilson and Clark, 1914). Recovery of three specimens from shellers' cull piles below Hartsville, Trousdale County, indicates an extant population may still survive in the middle Cumberland River (Parmalee and Klippel, 1982).

Description: The shells of *Obovaria retusa* are inflated, solid, and thick and ovate to quadrate in outline. The ventral and posterior margins are evenly rounded. The posterior ridge in male shells is low and rounded. The female shell has a pronounced groove behind the posterior ridge, making it more distinct; the marsupial area is slightly more inflated. Beaks are swollen and high and turned forward over a lunule; sculpture consists of a few weak double-looped ridges. The shell of this species may become quite large, reaching a maximum length of about 95 mm in old relic individuals. Mature female shells are only slightly smaller than those of males.

The left valve has two heavy, triangular, sculptured pseudocardinal teeth separated from two short heavy, curved, lateral teeth by a short, wide interdentum. The right valve has one large triangular, sculptured pseudo-

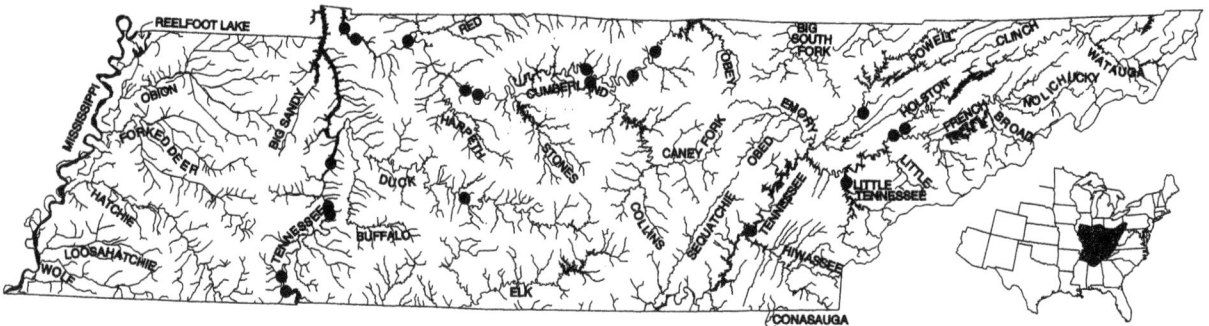

Range Map 72. *Obovaria retusa* (Lamarck, 1819), Ringpink.

Plate 73. *Obovaria retusa* (Lamarck, 1819), Ringpink.

cardinal tooth, often with a smaller tooth before and behind. It is separated from a single heavy, slightly curved lateral tooth by a short interdentum. The beak cavity is deep and compressed. A series of prominent dorsal scars are located under the pseudocardinal teeth. Muscle scars are small but deep; the pallial line is well impressed. The periostracum is shiny, lacking rays, and yellowish green to brown in color, becoming dark brown to black in old individuals. The surface of the shell is marked by low, irregular, concentric growth lines. The nacre color within the pallial line is a light pinkish salmon to deep purple, while the nacre outside the pallial line is white with a slight iridescence posteriorly.

Life History and Ecology: The Ringpink is typically a large river species inhabiting gravel bars (Neel and Allen, 1964). Hickman (1937) collected it in the Tennessee River from a gravel and sandy bottom in about two feet of water. *Obovaria retusa* is bradytictic (Surber, 1912).

Ortmann (1909a; 1912a) observed gravid females with eggs in late August and gravid females with glochidia in September. Host fish for the Ringpink unknown.

Status: Endangered (Williams et al., 1993:13). The U.S. Fish and Wildlife Service has developed a recovery plan for this species (U.S. Fish and Wildlife Service, 1991b).

Obovaria subrotunda (Rafinesque, 1820)
Round Hickorynut

RANGE MAP 73; PLATE 74

Synonymy:

Obliquaria (Rotundaria) subrotunda Rafinesque, 1820; Rafinesque, 1820:308, pl. 81, figs. 21–23

Obovaria subrotunda (Rafinesque, 1820); Vanatta, 1915:552; Ortmann and Walker, 1922:45

Obovaria (Obovaria) subrotunda (Rafinesque, 1820); Ortmann, 1918:567

Unio (Aximedia) levigata Rafinesque, 1820; Rafinesque, 1820:296, pl. 80, fig. 11

Unio laevigatus Rafinesque, 1820; Say, 1834: no pagination [misspelling]

Obovaria levigata (Rafinesque, 1820); Vanatta, 1915:552

Obovaria (Obovaria) levigata (Rafinesque, 1820); Haas, 1969a:422

Obovaria (Obovaria) subrotunda levigata (Rafinesque, 1820); Ortmann, 1918:568

Obovaria subrotunda levigata (Rafinesque, 1820); La Rocque, 1967:242

Obovaria striata Rafinesque, 1820; Rafinesque, 1820:311 Vanatta, 1915:552

Mya rotunda Wood, 1856; Wood, 1856:199, pl. 1 supp. fig. 1 [misspelling]

Unio circulus Lea, 1829; Lea, 1829:433, pl. 9, fig 14

Margarita (Unio) circulus (Lea, 1829); Lea, 1836:33

Margaron (Unio) circulus (Lea, 1829); Lea, 1852c:34

Unio subrotundus var. *circulus* Lea, 1829; Paetel, 1890:168

Obovaria circulus (Lea, 1829); Simpson, 1900a:600

Obovaria circula (Lea, 1829); Baker, 1920:383

Unio lens Lea, 1831; Lea, 1831:80, pl. 8, fig. 10

Margarita (Unio) lens (Lea, 1831); Lea, 1836:33

Margaron (Unio) lens (Lea, 1831); Lea, 1852c:34

Obovaria lens (Lea, 1831); Simpson, 1900a:600

Obovaria subrotunda lens (Lea, 1831); Goodrich and van der Schalie, 1944:318

Unio leibii Lea, 1862; Lea, 1862a:168; Lea, 1866:44, pl. 15, fig. 42

Margaron (Unio) leibii (Lea, 1862); Lea, 1870:36

Obovaria leibii (Lea, 1862); Simpson, 1900a:601

Obovaria circulus leibii (Lea, 1862); Sterki, 1907:390

Obovaria subrotunda leibii (Lea, 1862); Goodrich, 1932:103

Unio depygis Conrad, 1866; Conrad, 1866a:107, pl. 10, fig. 1

Obovaria lens var. *depygis* (Conrad, 1866); Simpson, 1900a:601

Obovaria subrotunda depygis (Conrad, 1866); Frierson, 1927:90

Obovaria lens var. *parva* Simpson, 1914; Simpson, 1914:294

Obovaria lens var. *elongata* Simpson, 1914; Simpson, 1914:294–295

Obovaria subrotunda globula Morrison, 1942; Morrison, 1942:360; Johnson, (1975:29, pl. 1, fig. 7).

Quadrula subrotunda globula (Morrison, 1942); Johnson, 1980:96

Type Locality: "l'Ohio."

General Distribution: *Obovaria subrotunda* is found throughout the Tennessee and Cumberland River systems and in the Ohio River system from western Pennsylvania and peninsular Michigan west to eastern Illinois. Clarke (1981a) lists the Round Hickorynut from Lake Erie and Lake St. Clair and their drainages in Ontario, Canada.

Tennessee Distribution: The Round Hickorynut was found throughout the Tennessee and Cumberland River systems in Tennessee. In the Tennessee River system, it was found in the Powell, Clinch, Holston, Pigeon, Little Tennessee, Sequatchie, Elk, Buffalo, and Duck rivers as well as the main channel of the Tennessee River. However, it has disappeared, or nearly so, from most of these rivers. In the Cumberland River system it occurs in the Obey, Stones, Harpeth, and Red rivers and in the mainstem of the Cumberland River.

Description: The shell is elliptical to circular in outline, solid, becoming inflated and heavy with age. Maximum shell length rarely exceeds 60 mm. All shell margins are rounded, although the posterior margin in females may be blunt. The posterior ridge is lacking. Beaks are centrally placed, full, high, curved inward, and elevated well above the hinge line; sculpture con-

Plate 74. *Obovaria subrotunda* (Rafinesque, 1820), Round Hickorynut.

sists of 4–5 weak bars, which are slightly sinuous centrally and angled posteriorly. The beak cavity is moderately deep with dorsal muscle scars partially in the cavity and partially up on the edge of the hingeplate.

The left valve has two short, thick, roughened, triangular pseudocardinal teeth and two slightly curved, short, strong lateral teeth. The right valve has one large,

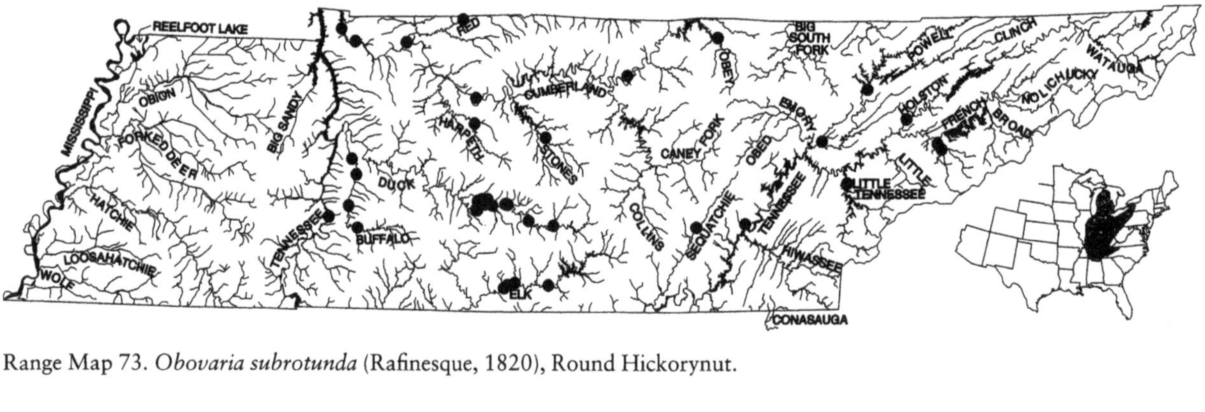

Range Map 73. *Obovaria subrotunda* (Rafinesque, 1820), Round Hickorynut.

massive, serrated triangular pseudocardinal tooth, usually with a small, low, compressed tubercular tooth on either side. There is one short, curved, thick, roughened lateral tooth, often with a secondary inner low, incomplete lateral tooth. Adductor muscle scars are distinct and rather deeply impressed. The interdentum is narrow or absent. The pallial line is impressed. The periostracum is greenish olive to dark brown, sometimes becoming blackish in old individuals. The posterior dorsal surface is marked by a distinct, lighter, yellowish streak or band; some young shells exhibit faint greenish rays but most are rayless. The nacre color is silvery white and iridescent posteriorly; some specimens have pink to purple shades inside the pallial line.

Life History and Ecology: Ortmann (1919) reported this species to be bradytictic with eggs present in September and glochidia in June. The Round Hickorynut is found in medium-sized to large rivers with sand and gravel substrates with moderate flow, usually at depths of less than six feet. Host fish for the Round Hickorynut unknown.

Status: Special Concern (Williams et al., 1993:13).

Pegias fabula (Lea, 1838)
Littlewing Pearlymussel
RANGE MAP 74; PLATE 75

Synonymy:
Margarita (Margaritana) fabula Lea, 1836; Lea, 1836:46 [nomen nudum]
Margaritana fabula Lea, 1838; Lea, 1838b:44, pl. 13, fig. 39
Unio fabula (Lea, 1838); Hanley, 1843:213, pl. 22, fig. 45
Margaron (Margaritana) fabula (Lea, 1838); Lea, 1852c:44
Micromya fabula (Lea, 1838); Agassiz, 1852:47
Strophitus fabula (Lea, 1838); Conrad, 1853:263
Baphia fabula (Lea, 1838); H. and A. Adams, 1858:499
Pegias fabula (Lea, 1838); Simpson, 1900a:660
Alasmidonta fabula (Lea, 1838); Ortmann, 1913a:311
Alasmidonta (Pegias) fabula (Lea, 1838); Ortmann, 1914:65
Margaritana curreyiana Lea, 1840; Lea, 1840:288; Lea, 1842b:223, pl. 18, fig. 40
Micromya curreyiana (Lea, 1840); Agassiz, 1852:47
Margaron (Margaritana) curreyiana (Lea, 1840); Lea, 1852c:42
Strophitus curreyiana (Lea, 1840); Conrad, 1853:263
Baphia curreyiana (Lea, 1840); H. and A. Adams, 1858:499
Unio curreyiana (Lea, 1840); Sowerby, 1868:pl. 63, fig. 319
Unio curreyianus (Lea, 1840); Hanley, 1856:386, pl. 24, fig. 10
Margaritana curreyiana Lea, 1840; Paetel, 1890:173
Unio propecaelatus De Gregorio, 1914; De Gregorio, 1914:30, pl. 8, figs. 1a–d

Type Locality: Stones River, Tennessee (Lea, 1842). Cumberland River, Tennessee (Simpson, 1914).

General Distribution: Tennessee and Cumberland River systems, formerly widespread from southern Kentucky (Rockcastle County) and southwestern Virginia (Lee, Russell, Smyth, Tazewell, and Washington counties), through the tributary stream system of the Tennessee and Cumberland rivers in East Tennessee and southwest to Lauderdale County, Alabama (Ortmann, 1925).

Tennessee Distribution: *Pegias fabula* has been collected from the Collins River in Warren County, in Cane Creek, a tributary to the Caney Fork River (Ahlstedt, 1986), and in Buck Creek and the Stones River, all in the Cumberland River drainage. Stansbery (1973; 1976a) reported the Littlewing Pearlymussel from the upper unimpounded reaches of the Clinch River. It has been found in the South Fork Holston River, Sullivan County, and in the Elk River, Franklin County (Ortmann, 1918; 1925).

Description: The Littlewing Pearlymussel is a small species: adults rarely exceed 35 mm in length. Valves are thickened anteriorly with a sharp posterior ridge, in front of which is a wide radial depression that ends in a basal sinus. There is another ridge above the posterior ridge, making the shell decidedly biangulate and truncate behind. Valves of the female possess a wider posterior slope behind the ridge and a more truncated posterior end than do those of the male. Beak sculpturing consists of heavy, subconcentric ridges, these being most prominent and persistent on the posterior ridges.

The left valve has an irregular triangular pseudocardinal tooth under the beak, sometimes with a vestige of another in front of it. Lateral teeth appear as short, faint, irregular ridges. The right valve has a single triangular pseudocardinal tooth in front of the beak. Beak cavities are rather deep and compressed; anterior muscle scars are deeply impressed. The periostracum is usually eroded away in mature individuals; a few dark brownish or olive green rays are apparent along the base of the shell in young specimens. The nacre is whitish on the anterior ventral third, flesh-colored or salmon in the beak cavities. The shells exhibit sexual dimorphism, one of the characters which separates *Pegias* from species belonging to the genus *Alasmidonta* (Simpson, 1900a, 1914; Stansbery, 1976b).

Range Map 74. *Pegias fabula* (Lea, 1838), Littlewing Pearlymussel.

Life History and Ecology: This very small unionoid inhabits cool, clear, high-gradient streams. It is usually found lying on top of or partially imbedded in sand and fine gravel between cobbles in only 6 to 10 inches of water, often just at the head of riffles. Ortmann (1914) collected a gravid female of this species in mid-September, suggesting that *Pegias fabula* is bradytictic, a winter breeder. Ahlstedt (in Neves, 1991:271) states that "[f]ish hosts are unknown for this species; however, based on field observations, the banded sculpin, *Cottus carolinae,* and redline darter, *Etheostoma rufilineatum,* may be likely hosts. These fish were observed under large flat rocks and were present on gravel shoals where live specimens of *Pegias fabula* were found." Layzer and Anderson (1992) identified the host fish for *Pegias fabula* as the greenside darter *(Etheostoma blennioides)* and emerald darter *(E. baileyi).*

Status: Endangered (Williams et al., 1993:13). The Littlewing Pearlymussel is reported to be a rare headwaters form. Stansbery (1976b) noted that many of the isolated populations of *Pegias fabula* have been extirpated by acid mine drainage, domestic pollution, and impoundment of rivers which it inhabited. A small population in the upper Caney Fork River drainage (Cane Creek, Sweetgum, Van Buren County: Ahlstedt, 1986) appears to be one of the last in Tennessee. The U.S. Fish and Wildlife Service has developed a recovery plan for this species (U.S. Fish and Wildlife Service, 1989a).

Plate 75. *Pegias fabula* (Lea, 1838), Littlewing Pearlymussel.

Plectomerus dombeyanus
(Valenciennes, 1827)
Bankclimber

RANGE MAP 75; PLATE 76

Synonymy:

Unio crassidens var. a Lamarck, 1819; Lamarck, 1819:71

Plectomerus crassidens var. a (Lamarck, 1819); Conrad, 1853:261

Unio dombeyana Valenciennes, 1827; Valenciennes, 1827:227, pl. 53, figs. 1, 1a, 1b

Unio dombeyanus Valenciennes, 1827; Barnes, 1828:360

Quadrula heros var. *dombeyana* (Valenciennes, 1827); Vanatta, 1910:102

Amblema (Plectomerus) dombeyana (Valenciennes, 1827); Frierson, 1927:62

Amblema dombeyana (Valenciennes, 1827); Strecker, 1931:35

Quadrula heros dombeyana (Valenciennes, 1827); Vanatta, 1910:102

Plectomerus dombeyanus (Valenciennes, 1827); Burch, 1973:12, fig. 19

Unio interruptus Say, 1831; Say, 1831c:525

Unio trapezoides Lea, 1831; Lea, 1831:69, pl. 3, fig. 1

Margarita (Unio) trapezoides (Lea, 1831); Lea, 1836:12

Margaron (Unio) trapezoides (Lea, 1831); Lea, 1852c:21

Quadrula (Crenodonta) trapezoides (Lea, 1831); Simpson, 1900a:772

Crenodonta trapezoides (Lea, 1831); Ortmann, 1912a:248, figs. 5, 5a

Plectomerus trapezoides (Lea, 1831); Ortmann and Walker, 1922:9

Quadrula trapezoides var. *pentagonoides* Frierson, 1902; Frierson, 1902:39

Quadrula (Crenodonta) trapezoides var. *pentagonoides* Frierson, 1902; Simpson, 1914:831

Type Locality: Lake St. Joseph, Louisiana.

General Distribution: Gulf drainage streams, from the Alabama River west to eastern Texas; northward in the Mississippi River system to northwest Tennessee (Simpson, 1914). A viable population recently has been discovered in Kentucky Lake (Tennessee River), Trigg County, Kentucky (Pharris et al., 1982).

Tennessee Distribution: The Bankclimber is common in the Hatchie River, southwestern Tennessee (Manning, 1989). Starnes and Bogan (1988) record it for the North Fork Obion River and Reelfoot Lake, northwestern Tennessee, prior to 1960. It has now extended its range upstream in the Tennessee River (Kentucky Lake) to the confluence of the Duck River (Humphreys County).

Description: The shell is heavy, solid, rhomboid, large, and moderately inflated. Mature individuals from the Hatchie River, Lauderdale County, Tennessee, vary between 130 and 140 mm in length. The anterior end is broadly rounded, the ventral margin is straight, and the posterior end is straight to slightly curved. A pronounced posterior ridge ends at the base of the shell in a point. The surface of the shell is rather variable; some individuals (especially juveniles) often exhibit a widely scattered series of low pustules or tubercles with a series of short, parallel plications on the disc area that radiate downward and forward from the beaks. Adults often develop one or two large, rounded plications nearly parallel and ventral to the posterior ridge. Low, rounded plications typically appear on the posterior slope.

The left valve has two erect triangular or elongated serrated pseudocardinal teeth; the two lateral teeth are long, narrow, and slightly curved. The right valve has one large, triangular, serrated pseudocardinal tooth, sometimes with a small, low, thin, elongated tooth anteriorly. The lateral tooth is long, thin, erect, and slightly curved. The interdentum is short, and pronounced; the beak cavity is shallow. The periostracum

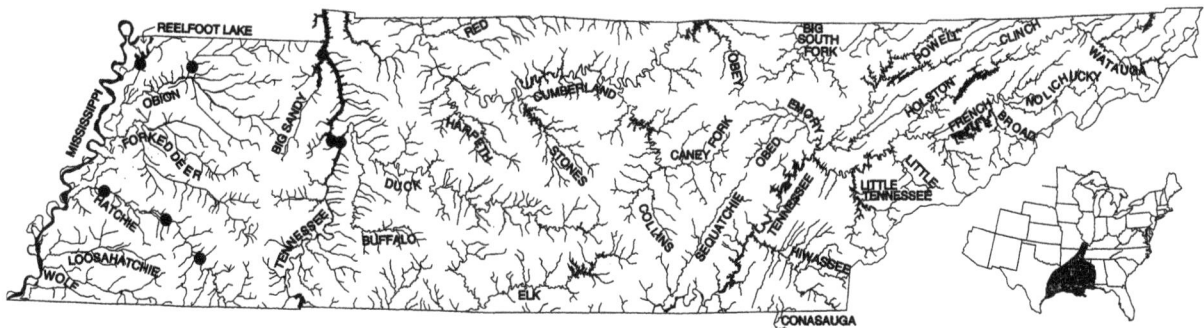

Range Map 75. *Plectomerus dombeyanus* (Valenciennes, 1827), Bankclimber.

Plate 76. *Plectomerus dombeyanus* (Valenciennes, 1827), Bankclimber.

Plethobasus cicatricosus (Say, 1829)
White Wartyback

RANGE MAP 76; PLATE 77

Synonymy:

Unio cyphia (Rafinesque, 1820); Conrad, 1834a:68 [in part]

Plethobasus cyphyus (Rafinesque, 1820); Ortmann and Walker, 1922:19 [in part]

Plethobasus pachosteus (Rafinesque, 1820); Morrison, 1969:24

Unio cicatricosus Say, 1829; Say, 1829:292

Pleurobema (Plethobasus) cicatricosum (Say, 1829); Simpson, 1900a:765

Quadrula cicatricosa (Say, 1829); Sterki, 1907:392

Pleurobema cicatricosa (Say, 1829); F. C. Baker, 1906:78

Plethobasus cicatricosa (Say, 1829); Ortmann, 1919:61

Unio varicosus Lea, 1829 non Lamarck, 1819; Lea, 1829:424; Lea, 1831:9, pl. 11, fig. 20

Margarita (Unio) varicosus (Lea, 1829); Lea, 1836:17

Margaron (Unio) varicosus (Lea, 1829); Lea, 1852c:23

Unio cicatricosus var. *varicosus* (Lea, 1829); Paetel, 1890:148

Unio cicatricoides Frierson, 1911; Frierson, 1911:53, pl. 2, upper fig.

Pleurobema cicatrioideum (Frierson, 1911); Frierson, 1927:45,46

Plethobasus cicatriodeus (Frierson, 1911); Morrison, 1969:24

Unio detectus Frierson, 1911; Frierson, 1911:52, pl. 2, lower fig., pl. 3, upper fig.

Plethobasus cicatricosus form *detectus* (Frierson, 1911); Warren, 1975:66

Type Locality: Wabash River.

General Distribution: Lower Wabash, Ohio, Cumberland, and Tennessee rivers. Stansbery (1972) identified archaeological specimens from the Buffalo site, West Virginia (Kanawha River).

Tennessee Distribution: There are few Tennessee records for *P. cicatricosus*. Lewis (1870) reported it from the Holston River, and Hinkley and Marsh collected the White Wartyback from the Cumberland River (Marsh, 1885). Wilson and Clark (1914) identified *Pleurobema aesopus (=Plethobasus cyphyus)* from the Cumberland River, but noted that there were questions concerning the identifications. Ortmann (1918) reported *Plethobasus cyphyus compertus* from the Holston, Clinch, and French Broad rivers. He noted that this was the species which Lewis had reported as *cicatricosus*. The specimens of Ortmann and Wilson and Clark may in fact be *P. cicatricosus*. The authors have identified numerous specimens of *P. cicatricosus* from archaeological sites along the Tennessee River in Rhea and Meigs counties (Parmalee et al., 1982).

is unrayed and dark brown to blackish. The nacre is a fairly uniform light to dark purple, sometimes mottled or streaked with white.

Life History and Ecology: In the Hatchie River in West Tennessee, the Bankclimber is most often encountered in a soft mud and sand substrate at depths varying from two to four feet. In streams and rivers of the bootheel region of Missouri, Oesch (1984:79) noted that "[i]t appears to be best suited to mud or mud-rock/gravel stream beds with moderate to sluggish current." *Plectomerus dombeyanus* is one of several species inhabiting the Hatchie River that exhibit a definite southern or Gulf Coast affinity (Manning, 1989). In spite of the wide range and local abundance of this mussel, life history studies, including information on its reproductive cycle and fish host for the glochidia, are wanting.

Status: Currently Stable (Williams et al., 1993:13).

Range Map 76. *Plethobasus cicatricosus* (Say, 1829), White Wartyback.

Description: Shells of the White Wartyback are thick, solid, and considerably inflated. Mature individuals may reach 100 mm in length. The general shell outline ranges from subtriangular to elongate or almost subquadrate with the beaks high, full, and turned forward over a lunule. The dorsal margin is almost straight, and the posterior and ventral margins are evenly rounded. The posterior ridge is low and narrowly rounded to almost truncated. The surface is marked by uneven concentric growth lines and a row of low, irregular knobs, beginning just behind the umbo and extending diagonally down across the disc of the shell to the ventral margin.

The left valve has two sculptured pseudocardinal teeth separated from two short, thick, slightly curved lateral teeth by a relatively broad interdentum. The right valve has a single triangular, sculptured pseudocardinal tooth and a single short, slightly curved lateral tooth. The long axis of the pseudocardinal tooth in the right valve is slightly angled off from being parallel with the lateral tooth. Anterior muscle scars are deep and sculptured, while posterior scars are shallow. The beak cavity is broad and shallow. The periostracum is a dull, rayless yellow or greenish yellow, becoming yellowish brown in old individuals. The nacre is silvery white and iridescent posteriorly.

Shell differences between *Plethobasus cyphyus* and *P. cicatricosus* are often subtle, and conceivably they could be the same animal (ecophenotypes?). Simpson (1914:807) commented that "[t]he species *[P. cyphyus]* differs from *cicatricosum* in being more angular at the base of the nodulous ridge, and in having a harder, smoother epidermis." Ortmann and Walker (1922) synonymized *P. cicatricosus* with *P. cyphyus,* yet other authors continued to recognize the White Wartyback as a species distinct from but closely related to the Sheepnose.

Life History and Ecology: Assuming *Plethobasus cicatricosus* is a valid species, it was one that inhabited shoals and riffles in big rivers like the Ohio and Tennessee. Little else is known about the life history of this mussel. Although unknown, the breeding season of the White Wartyback may be inferred as tachytictic; both *P. cyphyus* and *P. cooperianus* are short-term summer brooders.

Plate 77. *Plethobasus cicatricosus* (Say, 1829), White Wartyback.

Status: Endangered (Williams et al., 1993:13). *Plethobasus cicatricosus* is probably extirpated in Tennessee. The only remaining population is represented by a few nonreproducing individuals below Wilson Dam in northern Alabama (Isom, 1969; Stansbery, 1970). The species was last reported alive in 1987 (2 individuals) below Pickwick Landing Dam (S. A. Ahlstedt, per. comm., 1996). The U.S. Fish and Wildlife Service has developed a recovery plan for this species (U.S. Fish and Wildlife Service, 1984d) and has created a watershed implementation schedule for the recovery plan (U.S. Fish and Wildlife Service, 1989b).

Plethobasus cooperianus (Lea, 1834)
Orangefoot Pimpleback

RANGE MAP 77; PLATE 78

Synonymy:

Unio striatus (Rafinesque, 1820); Reeve, 1864:pl. 8, fig. 30 (changed in errata to *cooperianus*) [misidentification]

Quadrula striata striata (Rafinesque, 1820); Frierson, 1927:50 [misidentification]

Quadrula (Luteacarnea) striata (Rafinesque, 1820); Haas, 1969a:296 [misidentification]

Plethobasus striatus (Rafinesque, 1820); Morrison, 1969:24 [misidentification]

Unio cooperianus Lea, 1834; Lea, 1834:61, pl. 8, fig. 21

Margarita (Unio) cooperianus (Lea, 1834); Lea, 1836:16

Margaron (Unio) cooperianus (Lea, 1834); Lea, 1852c:22

Quadrula cooperianus (Lea, 1834); Simpson, 1900a:781

Pleurobema cooperianus (Lea, 1834); Ortmann, 1910:117

Plethobasus cooperianus (Lea, 1834); Ortmann, 1912a:261

Quadrula cooperiana (Lea, 1834); Walker, 1918a:166

Type Locality: Ohio River.

General Distribution: Ohio, Cumberland, and Tennessee River systems.

Tennessee Distribution: *Plethobasus cooperianus* formerly inhabited the lower Holston River, Knox County; the Clinch River, Roane and Anderson counties; and the French Broad, Sevier County (Lewis, 1870; Ortmann, 1918). The Orangefoot Pimpleback was recorded from the Tennessee River at Knoxville, Knox County, downstream to Marion County (Ortmann, 1918, 1919; van der Schalie, 1939; Isom, 1972). It formerly occurred throughout the lower Tennessee River from Hardin County downstream to Stewart County (Ortmann, 1925; van der Schalie, 1939). Isom and Yokley (1968a) reported *P. cooperianus* as a species new to the Duck River naiad fauna, Maury County. D. H. Stansbery said this was a misidentification (Ahlstedt, per. comm., 1996). The Orangefoot Pimpleback was common in the Cumberland River from Clay County to Stewart County (Marsh, 1885; Wilson and Clark, 1914).

Description: Nearly circular or subtriangular in outline, shells of *Plethobasus cooperianus* are solid, heavy, inequilateral, and only moderately inflated. Mature individuals will reach a length of 85–90 mm. Beaks are high, full, and directed forward. A posterior ridge, if present, is low and rounded. The surface is marked by concentric, irregular growth lines, and the posterior two-thirds of the shell is covered with numerous raised, irregular pustules.

The left valve has two triangular, roughened pseudocardinal teeth, a broad interdentum, and two short, straight lateral teeth. The right valve has a large triangular pseudocardinal tooth, often with a small, low tooth on either side, and a short lateral tooth. The beak cavity is deep and compressed. Anterior muscle scars are deep; posterior scars, although pronounced, tend to be shallow. The periostracum color varies from yel-

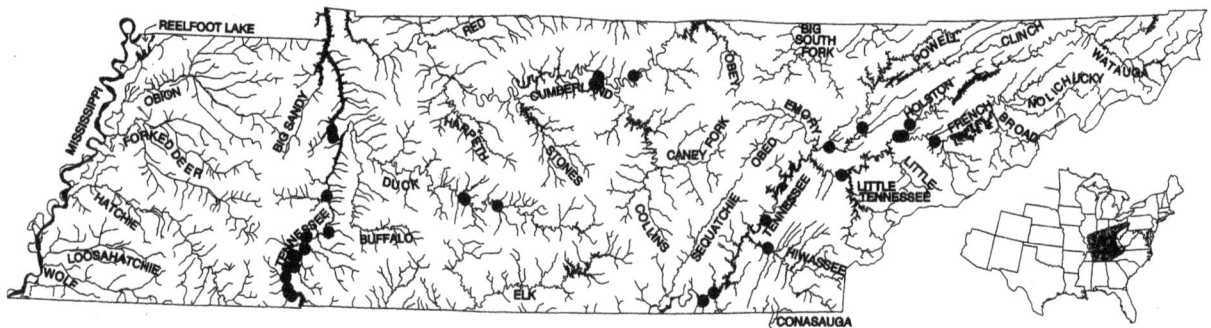

Range Map 77. *Plethobasus cooperianus* (Lea, 1834), Orangefoot Pimpleback.

Plate 78. *Plethobasus cooperianus* (Lea, 1834), Orangefoot Pimpleback.

lowish brown to a reddish brown; the posterior slope may be a lighter shade of brown or tan. Numerous dark green rays typical of juveniles become obliterated as the individual matures. The nacre color is white or varying intensities of pink inside the pallial line, becoming iridescent posteriorly.

Life History and Ecology: The Orangefoot Pimpleback, primarily a big river species, was fairly common locally in shoal areas of the Tennessee River as evidenced from prehistoric aboriginal midden deposits (e.g., Warren, 1975; Parmalee et al., 1982). It has now been extirpated or populations have been greatly reduced in much of its former range. Commercial shellers report encountering individuals at a depth of 12 to 18 feet, usually in a sand and coarse gravel substrate. On the basis of two gravid individuals collected in early June

by Wilson and Clark (1914) in the Cumberland River, *Plethobasus cooperianus* is considered to be tachytictic. Host fish for the glochidia of this mussel unknown.

Status: Endangered (Williams et al., 1993:13). Although Stansbery (1970, 1976) noted that the Orangefoot Pimpleback was restricted to the Tennessee River, Parmalee et al. (1980) reported that a small relict population of old individuals was still extant (1980) in the Cumberland River, Smith County, Tennessee. One relict individual was collected in 1978 below Fort Loudoun Dam, Knox County (Gooch et al., 1979). What appears to be a viable population occurs in the Tennessee River below Pickwick Landing Dam (about Tennessee River Mile 198.0), Hardin County, Tennessee. The U.S. Fish and Wildlife Service has developed a recovery plan for this species (U.S. Fish and Wildlife Service, 1984c) and has created a watershed implementation schedule for the recovery plan (U.S. Fish and Wildlife Service, 1989b).

Plethobasus cyphyus (Rafinesque, 1820)
Sheepnose

RANGE MAP 78; PLATE 79

Synonymy:

Obliquaria cypha Rafinesque, 1820; Rafinesque, 1820:305
Unio scyphius (Rafinesque, 1820); Küster, 1861:181, pl. 57, fig. 2 [misspelling]
Unio cyphius (Rafinesque, 1820); Say, 1834:no pagination
Unio cyphia (Rafinesque, 1820); Conrad, 1834a:68
Plethobasus cyphyus (Rafinesque, 1820); Ortmann, 1919:65. pl. 5, fig. 6, pl. 6, figs. 1–3
Pleurobema (Plethobasus) cyphyum (Rafinesque, 1820); Frierson, 1927:45
Plethobasus cyphyum (Rafinesque, 1820); Haas, 1969a:250
Unio aesopus Green, 1827; Green, 1827:46, fig. 3
Margarita (Unio) aesopus (Green, 1827); Lea, 1836:17
Margaron (Unio) aesopus (Green, 1827); Lea, 1852c:23
Unio cyphias var. *aesopus* Green, 1827; Paetel, 1890:150
Pleurobema (Plethobasus) aesopus (Green, 1827); Simpson, 1900a:764
Pleurobema aesopus (Green, 1827); F. C. Baker, 1905:255
Plethobasus aesopus (Green, 1827); Ortmann, 1912a:260, fig. 8
Unio compertus Frierson, 1911; Frierson, 1911:53, pl. 3, middle and lower figs.
Plethobasus cyphyus compertus (Frierson, 1911); Ortmann, 1918:544
Pleurobema (Plethobasus) compertus (Frierson, 1911); Simpson, 1914:809
Plethobasus compertus (Frierson, 1911); Haas, 1969a:251

Type Locality: Falls of the Ohio.

General Distribution: Ohio, Cumberland, and Tennessee River systems; upper Mississippi River north to Minnesota (Simpson, 1914).

Tennessee Distribution: Ortmann (1918:544) reported the distribution of *Plethobasus cyphyus* as "[i]n the larger rivers, Tennessee, Clinch, Powell, Holston, and French Broad, going up, in the Powell, to Bryant Shoals, Claiborne Co., Tenn.; in the Holston to the mouth of the North Fork at Rotherwood, Hawkins Co., Tenn." Ortmann (1918) also noted a "variety," *P. c. compertus* (Frierson, 1911), that occurred in the Tennessee, French Broad, and Holston rivers. Although reduced in numbers or entirely extirpated from much of its former range, extant local populations may be found in the lower Tennessee River (Kentucky Lake: Hardin County downstream), Cumberland River (Trousdale and Smith counties), Powell River (Claiborne and Hancock counties), Clinch River (Hancock County), and the Holston River (Knox, Grainger, and Jefferson counties). A relic specimen was collected in 1975 by S. A. Ahlstedt in the Hiwassee River, Polk County (Parmalee and Hughes, 1994).

Description: Shells of *Plethobasus cyphyus* are elongated, ovate, thick, and moderately inflated. The anterior end is rounded, but posterior end is somewhat bluntly pointed to truncate. Large, mature individuals may attain a length of 110–120 mm. Typically there is a row of large, broad tubercular swellings on the center of the valve, extending down from near the umbones to the ventral margin. A shallow, furrowlike depression between the row of tubercles and the rounded dorsal ridge is present in most specimens. Beaks are elevated, high, and placed near the anterior margin; sculpture consists of a few concentric ridges at the tip of the umbones, evident in only very young shells.

Plate 79. *Plethobasus cyphyus* (Rafinesque, 1820), Sheepnose.

The left valve has two heavy, erect, roughened, somewhat triangular, divergent pseudocardinal teeth; the two lateral teeth are low, heavy, finely serrated, and slightly curved. The right valve has a large, triangular, roughened pseudocardinal tooth, rarely with a smaller tubercular tooth on either side; the lateral tooth is heavy and serrated. Usually a second smaller lateral tooth is present on the inner side. The interdentum

Range Map 78. *Plethobasus cyphyus* (Rafinesque, 1820), Sheepnose.

varies from moderately wide to narrow, or absent; the beak cavity is shallow. The periostracum is rayless, light yellow to a dull, yellowish brown; concentric ridges resulting from rest periods are usually darker. The nacre is pearly white and iridescent posteriorly.

Life History and Ecology: In stretches of the unimpounded Clinch and Powell rivers and, for example, shoals of the Holston River (Knox County), the Sheepnose may be found in relatively fast current in less than two feet of water. In contrast, it occurs at depths of 12 to 15 feet in the Cumberland and Tennessee rivers (reservoirs). The most suitable substrate for this mussel is a mixture of coarse sand and gravel. *Plethobasus cyphyus* is tachytictic, with most reproduction taking place in early summer. Based on data from Surber (1913) and Wilson (1916), Fuller (1974) lists the sauger *(Stizostedion canadense)* as the fish host for glochidia of the Sheepnose.

Status: Threatened (Williams et al., 1993:13). Like so many species of freshwater mussels, the Sheepnose has been extirpated throughout much of its former range or reduced to local populations and relict individuals. Probably the most stable and viable populations of this mussel to be found in Tennessee occur in the upper Clinch River, Hancock County, and in the Tennessee River below Pickwick Landing Dam, Hardin County.

Pleurobema chattanoogaense (Lea, 1858)
Painted Clubshell

RANGE MAP 79; PLATE 80

Synonymy:

Unio chattanoogaensis Lea, 1858; Lea, 1858c:166; Lea 1859f:209, pl. 25, fig. 90

Margaron (Unio) chattanoogaensis (Lea, 1858); Lea, 1870:40

Pleurobema chattanoogaensis (Lea, 1858); Simpson 1900a:753

Pleurobema chattanoogaense (Lea, 1858); Walker, 1905:135

Pleurobema decisum chattanoogaensis (Lea, 1858); Frierson, 1927:41

Pleurobema decisum chattanoogaense (Lea, 1858); Haas, 1969a:253

Comments: Although shell characters of *Pleurobema chattanoogaense* are relatively consistent and distinct, those defining other closely related taxa found in the Conasauga River *(P. georgianum, P. hanleyanum, P. trochelianum)* are much less so. It is nearly impossible to identify and separate many individuals, especially juveniles, of the latter three species from one another because of close similarity in shell characteristics.

Type Locality: Chattanooga, Tennessee, Coosawattee, Etowah, and Oostanaula rivers, Georgia.

General Distribution: Alabama River system (Simpson, 1914).

Tennessee Distribution: Conasauga River, Bradley and Polk counties.

Description: The shell is solid, somewhat ovate or broadly elliptical, elongate in mature specimens, and moderately inflated. Adult individuals from the Conasauga River, Murray County, Georgia, reach a maximum length of about 65 mm. The anterior end is evenly rounded; the dorsal and ventral margins are straight to very slightly curved; the posterior end is rounded, sometimes bluntly pointed or biangulate. Beaks are fairly high, inflated, extending only slightly beyond the hinge line. The posterior ridge is typically low and rounded when present, but it is absent in the majority of individuals.

The left valve has two triangular, low, stumpy pseudocardinal teeth, typically widely separated and roughened; the two lateral teeth are low, long, and straight. The right valve has a moderately high, wedge-shaped, often serrated pseudocardinal tooth. The lateral tooth is straight, long, and fairly high; some specimens have a second low lateral tooth. Adductor muscle scars are deeply impressed; the pallial line is faint. The beak cavity is shallow, and the interdentum is short and fairly wide. The periostracum is a dull yellowish or greenish tan in juveniles, becoming a more uniform dark yellow tan, gray, or brown with age. Distinct, often wide dark concentric rest lines are present; some juveniles exhibit a few narrow green rays, primarily on the umbo. The nacre is a dull bluish white.

Life History and Ecology: The Painted Clubshell has been found in the Conasauga River living in a substrate composed of coarse gravel and sand on riffles in water less than two feet in depth. The breeding season and fish host for the glochidia of *Pleurobema chattanoogaense* are unknown.

Status: Endangered (Williams et al., 1993:13). Once common in many stretches of the Conasauga River in Tennessee and northern Georgia, this mussel is now rare and has been extirpated from much of its former

Range Map 79. *Pleurobema chattanoogaense* (Lea, 1858) Painted Clubshell

range. The U.S. Fish and Wildlife Service has developed a recovery plan for *Pleurobema decisum* and has included this species as a synonym (U.S. Fish and Wildlife Service, 1994).

Plate 80. *Pleurobema chattanoogaense* (Lea, 1858) Painted Clubshell

Pleurobema clava (Lamarck, 1819) Clubshell

RANGE MAP 80; PLATE 81

Synonymy:
Unio clava Lamarck, 1819; Lamarck, 1819:74
Margarita (Unio) clavus (Lamarck, 1819); Lea, 1836:22
Unio clavus Lamarck, 1819; Reeve, 1841:117, pl. 88, fig. 3
Pleurobema clava (Lamarck, 1819); Agassiz, 1852:49
Margaron (Unio) clavus (Lamarck, 1819); Lea, 1852c:26
Pleurobema clava (Lamarck, 1819); Simpson, 1900a:745
Pleurobema clava clava (Lamarck, 1819); Haas, 1969a:252
Pleurobema cuneata Rafinesque, 1820; Rafinesque, 1820:313
Unio cuneatus (Rafinesque, 1820); Say, 1834:no pagination
Unio (Aximedia) elliptica Rafinesque, 1820; Rafinesque, 1820:296
Obliquaria (Scalenaria) scalenia Rafinesque, 1820; Rafinesque, 1820:309, pl. 81, figs. 24, 25
Obliquaria scalenia Rafinesque, 1820; Johnson and Baker, 1973:169, pl. 1, fig. 6
Pleurobema mytiloides Rafinesque, 1820; Rafinesque, 1820:313, pl. 82, figs. 8–10
Unio mytiloides (Rafinesque, 1820); Lea, 1828:269
Margarita (Unio) mytiloides (Rafinesque, 1820); Lea, 1836:21
Margaron (Unio) mytiloides (Rafinesque, 1820); Lea, 1852c:25
Unio patulus Lea, 1829; Lea, 1829:441, pl. 12, fig. 20
Margarita (Unio) patulus (Lea, 1829); Lea, 1836:22
Cunicula patula (Lea, 1829); Swainson, 1840:378
Margaron (Unio) patulus (Lea, 1829); Lea, 1852c:26
Pleurobema mytiloides patulum (Lea, 1829); Frierson, 1927:40
Pleurobema clava patulum (Lea, 1829); Haas, 1969a:252
Unio consanguineus De Gregorio, 1814; De Gregorio, 1914:46 (non Lea, 1863)
Unio anaticulus var. *ohiensis* De Gregorio, 1914; De Gregorio, 1914:51

Type Locality: Lake Erie.

General Distribution: Ohio, Cumberland, and Tennessee River systems. Maumee Basin, western New York;

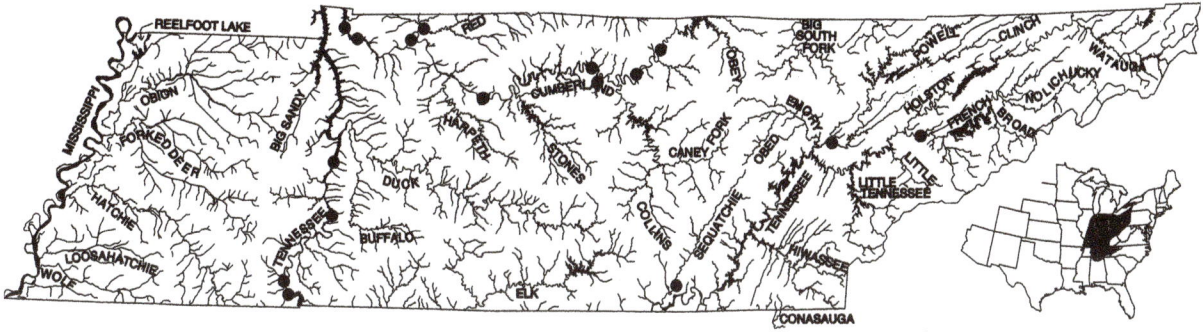

Range Map 80. *Pleurobema clava* (Lamarck, 1819), Clubshell.

Iowa River, Iowa; St. Peter's River, Minnesota, and from Nebraska (Simpson, 1914). Although Simpson (1914) also lists it from Ottawa, Canada, based on a record by Call, this species is not included in *The Freshwater Molluscs of Canada* (Clarke, 1981a).

Tennessee Distribution: Lower Tennessee River (Hardin County downstream) and the Cumberland River. The Clubshell formerly occurred in the Clinch and Sequatchie rivers (Starnes and Bogan, 1988).

Description: The shell is solid, thickest anteriorly, elongated, and generally triangular; beaks are high, full, placed at and projecting forward of the anterior end, usually being turned inward; beak sculpture consists of a few strong, irregular, often broken ridges that turn up behind (Simpson, 1914). The few individuals obtained from the Tennessee and Cumberland rivers in recent years averaged about 65 mm; shells of senile individuals from these rivers are greatly thickened and inflated. The anterior end is truncate, and the dorsal and ventral margins are nearly straight; the posterior end is pointed; the posterior ridge is rounded and prominent in most individuals, usually with a wide, shallow sulcus in front. The pseudocardinal teeth—two in the left valve (often united anteriorly), one in the right, somewhat compressed—are triangular, serrated, erect, and typically parallel to the hinge line. There are two lateral teeth in the left valve, one in the right, which are long, thin, elevated, straight, or slightly curved. The interdentum is narrow; the beak cavity is shallow. The periostracum is a dull yellowish or yellowish brown; dark green rays, often interrupted and forming irregular blotches, are usually prominent, especially on the beaks. The nacre is white and iridescent posteriorly.

Life History and Ecology: The Clubshell is a species typical of medium-sized and large rivers, and formerly it was fairly numerous in the shoals and riffles of the lower Cumberland and Tennessee rivers. *Pleurobema clava* is now nearly extirpated in the state, and most of the former populations in big river habitats are reduced

Plate 81. *Pleurobema clava* (Lamarck, 1819), Clubshell.

to an occasional relict individual, surviving since impoundment. The few individuals taken in the last few years were found at depths of 15 to 18 feet on a firm substrate of sand and gravel. In Pennsylvania, Ortmann (1919) reported finding gravid females in May, June, and July, the species apparently being a short-term, tachytictic breeder. Host fish for the glochidia unknown.

Status: Endangered (Williams et al., 1993:13). The status of viable populations in Tennessee, if such exist, is unknown. The U.S. Fish and Wildlife Service has developed a recovery plan for this species (U.S. Fish and Wildlife Service, 1993).

Pleurobema cordatum (Rafinesque, 1820)
Ohio Pigtoe

RANGE MAP 81; PLATE 82

Synonymy:
Unio obliqua Lamarck, 1819; Lamarck, 1819:72
Margarita (Unio) obliqua (Lamarck, 1819); Lea, 1836:20
Unio obliquus Lamarck, 1819; Hanley, 1842a:186
Margarita (Unio) obliquus (Lamarck, 1819); Lea, 1838a:17
Margaron (Unio) obliquus (Lamarck, 1819); Lea, 1852c:25
Quadrula obliqua (Lamarck, 1819); Simpson, 1900a:788
Pleurobema obliquum (Lamarck, 1819); Ortmann, 1912a:264
Obovaria cordata Rafinesque, 1820; Rafinesque, 1820:312
Comment: There is confusion surrounding the identification of the type of *Unio obliqua* Lamarck, 1819 (see Lea, 1834; Ortmann and Walker, 1922). Ortmann and Walker (1922) suggested the next available name, *Obovaria cordata,* should be used.
Obovaria cordata var. *rosea* Rafinesque, 1820; Rafinesque, 1820:312
Unio cordatus (Rafinesque, 1820); Conrad, 1836:48
Pleurobema obliquum cordatum (Rafinesque, 1820); Ortmann, 1918:548
Pleurobema cordatum (Rafinesque, 1820); Ortmann, 1924a:34
Fusconaia cordata (Rafinesque, 1820); Morrison, 1942:354

Quadrula (Obliquata) cordata (Rafinesque, 1820); Frierson, 1927:53
Quadrula (Obliquata) cordata cordata (Rafinesque, 1820); Haas, 1969a:297
Pleurobema cordatum cordatum (Rafinesque, 1820); Burch, 1973:15, fig. 38
Unio mytiloides (Rafinesque, 1820); Short and Eaton, 1831:74 [misidentification]

Type Locality: Ohio River.

General Distribution: Upper Mississippi River drainage and the St. Lawrence River drainage, from western New York west to Michigan, Wisconsin, Iowa, and Kansas, south to Arkansas and Alabama (Burch, 1975a).

Tennessee Distribution: Cumberland, Tennessee, Little Tennessee (prior to 1979), Holston, Nolichucky, Elk, Duck, Clinch, and Hatchie rivers. Formerly it occurred in the French Broad River (Starnes and Bogan, 1988), and valves of the Ohio Pigtoe have been reported from an aboriginal site along the Little Pigeon River, Sevier County (Parmalee, 1988).

Description: The shell is solid, inflated, and typically subtriangular in outline; beaks are full, elevated, projecting forward well above the hinge line, but turned inward; the sculpture consists of a few coarse, irregular ridges. The anterior margin is broadly rounded; the ventral margin is straight; the posterior ridge is distinct but low, most sharply angled behind the beaks; typically, there is a wide, shallow sulcus in front of the posterior ridge. Mature individuals from the Tennessee and Cumberland rivers average 80–90 mm in length; very old individuals may reach a length of 120 mm.

The left valve has two pseudocardinal teeth, the right valve has one; all are low, serrated, and heavy.

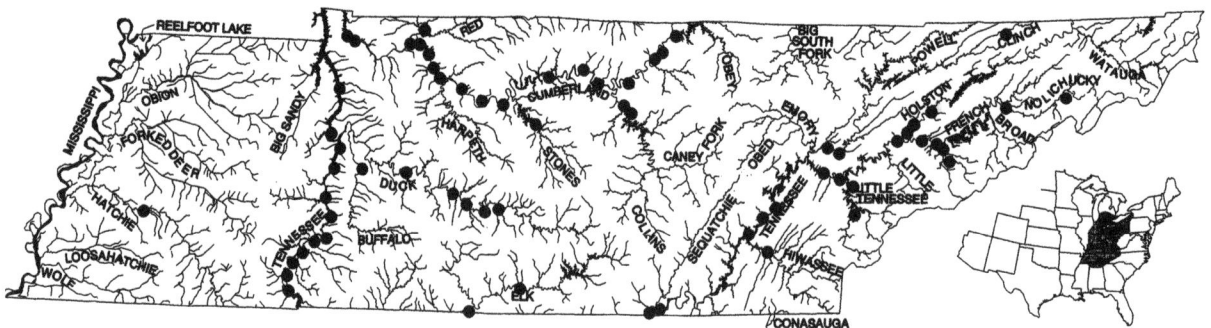

Range Map 81. *Pleurobema cordatum* (Rafinesque, 1820), Ohio Pigtoe.

Plate 82. *Pleurobema cordatum* (Rafinesque, 1820), Ohio Pigtoe.

The lateral teeth, two in the left valve, one in the right, are relatively short, low, and straight. Muscle scars are deep, the interdentum is wide, and the beak cavity is moderately deep. The surface is roughened by concentric rest lines. The periostracum is a dull reddish yellow or brown; juveniles are a lighter hue, often marked with dark green rays. The nacre is white, iridescent posteriorly; some individuals have a nacre of varying light (occasionally dark) shades of pink or salmon.

Life History and Ecology: The Ohio Pigtoe reaches its greatest abundance and individual size in large rivers such as the Cumberland and Tennessee, where it may occur at depths of up to 18 to 24 feet. It favors sections of the larger rivers with strong current and a firm substrate composed of sand and gravel. However, it appears not to have adapted well in the impounded reservoirs, even in sections where these habitat condi-

tions still exist. The bluegill *(Lepomis macrochirus)* and rosefin shiner *(Lythrurus ardens)* have been recorded as host fish for the glochidia of the Ohio Pigtoe (Fuller, 1974). *Pleurobema cordatum* is apparently tachytictic with glochidia most abundant in June or "approximately 3–4 weeks after the water temperature rises to 21° C" (Yokley, 1972:359).

Status: Special Concern (Williams et al., 1993:13). In the Tennessee River upstream from Chattanooga, the species exists as a relict population from pre- and post-impoundment of the river (Ahlstedt and McDonough, 1994).

Pleurobema georgianum (Lea, 1841)
Southern Pigtoe
RANGE MAP 82; PLATE 83

Synonymy:
Unio georgianus Lea, 1841; Lea, 1841a:31; Lea, 1842b:235, pl. 21, fig. 49
Margaron (Unio) georgianus (Lea, 1841); Lea, 1852c:27
Pleurobema georgiana (Lea, 1841); Simpson, 1900a:761
Quadrula (Pleuronaia) georgiana (Lea, 1841); Frierson, 1927:58
Unio favosus Lea, 1856; Lea, 1856b:262; Lea, 1858e:58, pl. 8 fig. 40
Margaron (Unio) favosus (Lea, 1856); Lea, 1870:38
Pleurobema favosa (Lea, 1856); Simpson, 1900a:761

Type Locality: Stump Creek, northwest Georgia.

General Distribution: Alabama River system (Burch, 1975a).

Tennessee Distribution: Conasauga River, Bradley, and Polk counties.

Description: The shell is solid and obovate in most individuals; it is moderately inflated in juveniles, more compressed and somewhat flattened (disc) in old adults. The anterior end is broadly rounded, and the ventral margin is straight or slightly curved; the posterior end has a blunt point, sometimes becoming truncated in mature specimens. The posterior ridge is low, rounded, often indistinct. Adult individuals from the Conasauga River, Bradley and Polk counties, Tennessee, attain a maximum length of about 65 mm. Beaks are relatively low, projecting only slightly beyond the hinge line; the beak cavity is very shallow, and the interdentum is wide in older specimens.

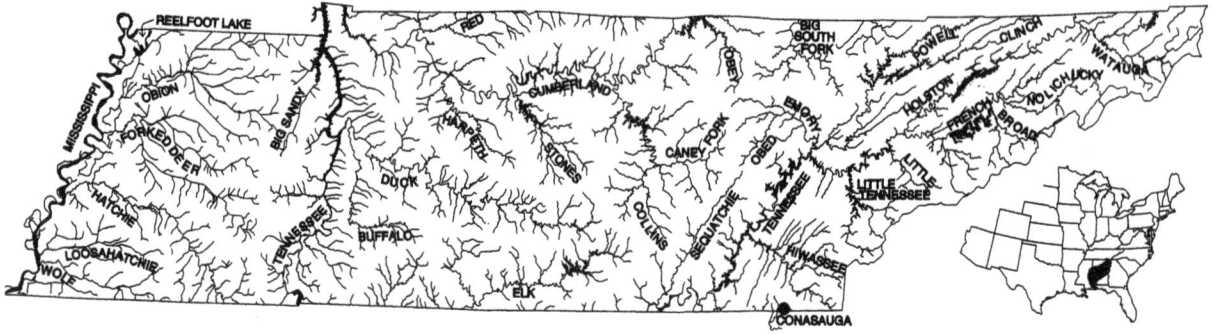

Range Map 82. *Pleurobema georgianum* (Lea, 1841), Southern Pigtoe.

Pseudocardinal teeth, two in the left valve, one in the right with occasionally a small tooth dorsally and anteriorly, are low, triangular, or linear, especially in the right valve. Lateral teeth, two in the left valve, one in the right, are low, thick, and slightly curved. The periostracum is a dull light yellowish or greenish brown in juveniles, usually becoming darker with age. Typi-

cally there is one wide dark green ray on the umbone and disc areas, usually broken and fading out ventrally. Dark rest lines are prominent on juvenile shells. The nacre is a dull bluish white.

Life History and Ecology: Like the Georgia Pigtoe, *Pleurobema georgianum* may be found in a coarse gravel and sand substrate in moderate current. It also may become fairly common locally under preferred habitat conditions that include shallow riffle areas less than two feet in depth. The breeding season and fish host for the glochidia of this mussel are unknown.

Status: Endangered (Williams et al., 1993:13). The U.S. Fish and Wildlife Service has developed a recovery plan for this species (U.S. Fish and Wildlife Service, 1994).

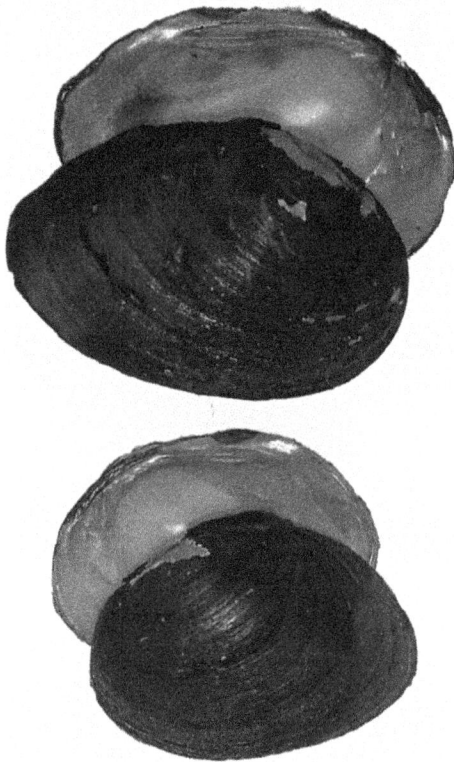

Pleurobema gibberum (Lea, 1838)
Cumberland Pigtoe

RANGE MAP 83; PLATE 84

Synonymy:
Margarita (Unio) gibber Lea, 1836; Lea, 1836:20 [nomen nudum]
Unio gibber Lea, 1838; Lea, 1838b:34, pl. 10, fig. 30
Margaron (Unio) gibber (Lea, 1838); Lea, 1852c:24
Pleurobema gibber (Lea, 1838); Simpson, 1900a:762
Pleurobema gibberum (Lea, 1838); Simpson, 1914:796

Type Locality: Caney Fork River, Tennessee.

General Distribution: Tributaries of the Cumberland River, Middle Tennessee.

Tennessee Distribution: Formerly in the Caney Fork River, now apparently restricted to Cane Creek, Van

Plate 83. *Pleurobema georgianum* (Lea, 1841), Southern Pigtoe.

Range Map 83. *Pleurobema gibberum* (Lea, 1838), Cumberland Pigtoe.

Buren County; Big Hickory Creek, Warren County; Calfkiller River, White County; and the Collins River, Grundy County. Possibly still present in other small tributaries of the Cumberland River in this region. A single adult specimen was collected on August 14, 1993, in Bradley Creek, Arnold Engineering Development Center, Coffee County, Tennessee. Bradley Creek, tributary to the Elk River, is part of the Tennessee River system; its occurrence there may be fortuitous.

Description: The shell is relatively solid, subrhomboid, and only slightly to moderately inflated; it is also small (maximum length of adult individuals from the Collins River, Grundy County, Tennessee, is about 55 mm). Beaks are moderately full, elevated, but not projecting beyond the hinge line. The anterior end is rounded, the ventral margin is straight, and the posterior margin ends in a blunt point; the posterior ridge is low and rounded. The left valve has two low but distinct, somewhat lamellar, finely serrated pseudocardinal teeth, typically projecting ventrally at a 90° angle to the hinge line. The lateral teeth are long, straight, thin, and elevated; the beak cavity is shallow, while the interdentum is wide. The periostracum lacks rays and varies from a dull light tan in juveniles to a dark reddish brown in adults; the surface is roughened with closely spaced, uneven concentric ridges. The nacre is a bluish white, dull bronze, or pale salmon.

Life History and Ecology: The Cumberland Pigtoe is a species typical of small to medium-sized rivers. It occurs in stretches with moderately strong current and a substrate of firm sand and small gravel, typically at depths of less than two feet. The breeding season and fish host for the glochidia are unknown.

Plate 84. *Pleurobema gibberum* (Lea, 1838), Cumberland Pigtoe.

Status: Endangered (Williams et al., 1993:13). The U.S. Fish and Wildlife Service has developed a recovery plan for this species (U.S. Fish and Wildlife Service, 1991c).

Family Unionidae 183

Pleurobema hanleyanum (Lea, 1852)
Georgia Pigtoe

RANGE MAP 84; PLATE 85

Synonymy:
Unio henleyianus Lea, 1852; Lea, 1852a:251
Unio hanleyianus Lea, 1852; Lea, 1852b:279, pl. 23, fig. 37
Margaron (Unio) hanleyianus (Lea, 1852); Lea, 1852c:26
Pleurobema hanleyana (Lea, 1852); Simpson, 1900a:759
Pleurobema aldrichianum Goodrich, 1931; Goodrich, 1931:2, pl. 1

Type Locality: Coosawattee River, Murray County, Georgia.

General Distribution: Alabama River system (Burch, 1975a).

Tennessee Distribution: Conasauga River, Bradley and Polk counties.

Description: The shell is solid, nearly equilateral, elliptical, and inflated; beaks are full and high. Mature individuals from the Conasauga River, Murray County, Georgia, attain a maximum length of about 65 mm. The posterior ridge is absent or, when discernible, low and evenly rounded. The anterior end is rounded, and the ventral margin is slightly curved; the posterior margin is rounded, bluntly pointed, or slightly biangulate in some individuals. The pseudocardinal teeth, two in the left valve, one in the right, are low, wide, triangular, and finely serrated; in some specimens a narrow, linear pseudocardinal tooth is present dorsally and anterior to the large tooth in the right valve. The lateral teeth, two in the left valve, one in the right, are long, straight, and moderately elevated. The beak cavity is shallow, and the interdentum is narrow or wanting. The periostracum is a dull yellowish tan, gray, or red-

Plate 85. *Pleurobema hanleyanum* (Lea, 1852), Georgia Pigtoe.

dish brown; some individuals are marked with one or a few dark green rays, sometimes broken or incomplete, most often on the umbonal area. The nacre is a dull bluish white.

Life History and Ecology: A substrate composed of coarse sand and gravel in stretches of rivers with good current provides the most suitable habitat for the Georgia Pigtoe. In the Conasauga River, *Pleurobema han-*

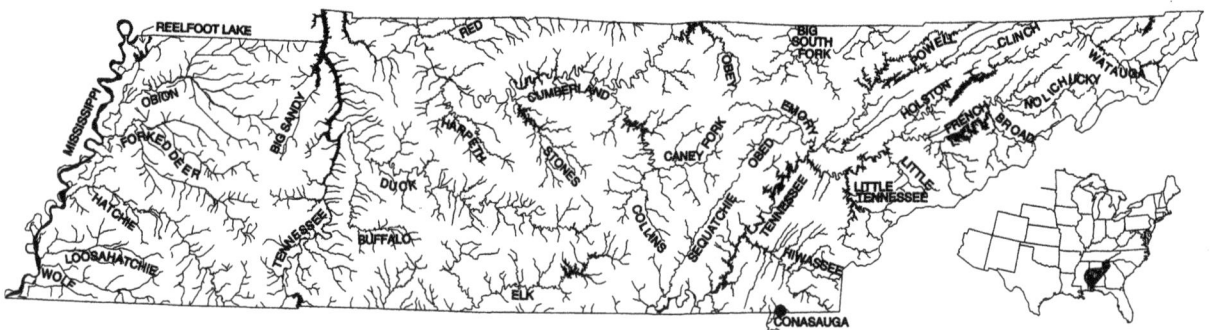

Range Map 84. *Pleurobema hanleyanum* (Lea, 1852), Georgia Pigtoe.

leyanum formerly was fairly common locally where it was most often encountered at normal depths of less than two feet. Neither the host fish for the glochidia of this species nor its breeding season is known.

Status: Endangered (Williams et al., 1993:13).

Pleurobema johannis (Lea, 1859)
Alabama Pigtoe

RANGE MAP 85; PLATE 86

Synonymy:
Unio johannis Lea, 1859; Lea, 1859d:171; Lea, 1860e:340, pl. 55, fig. 168
Margaron (Unio) johannis (Lea, 1859); Lea, 1870:41
Pleurobema johannis (Lea, 1859); Simpson, 1900a:759

Type Locality: Conasauga and Etowah rivers, Georgia.

General Distribution: Alabama River system (Simpson, 1914).

Tennessee Distribution: Conasauga River, Polk County.

Description: The shell is small, solid, inflated, and somewhat irregularly elliptical to subrhomboid. The average shell length of adult individuals is about 35 mm. Beaks are full and high, projecting slightly beyond the hinge line. The anterior end is somewhat narrowly rounded, the ventral margin is slightly curved, and the posterior end is bluntly pointed; the posterior ridge is distinct but low, slightly angled, broad, and rounded.

The pseudocardinal teeth, two in the left valve, one in the right, are erect, narrowly triangular, and finely serrated. There are two lateral teeth in the left valve, one in the right, and they are straight and fairly high. The beak cavity is shallow, while the interdentum is

Plate 86. *Pleurobema johannis* (Lea, 1859), Alabama Pigtoe.

wide. The periostracum varies from a uniform dull yellowish tan lacking rays (Cahaba River specimens) to wide rays or bands of dark green (Conasauga River specimens). Rest lines occasionally appear as dark brown bands. The nacre is a bluish white, and the posterior third is iridescent.

Life History and Ecology: Little habitat data are available for *Pleurobema johannis*. It has been collected in

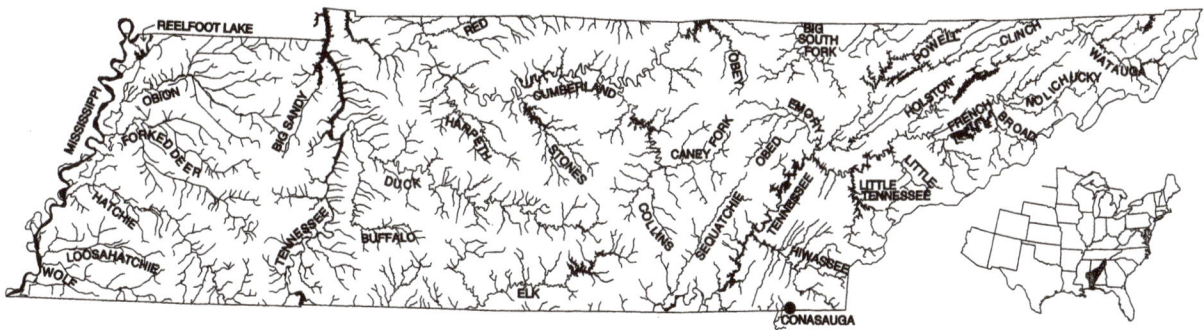

Range Map 85. *Pleurobema johannis* (Lea, 1859), Alabama Pigtoe.

stretches of the Conasauga River in which *Pleurobema hanleyanum* and *Pleurobema troschelianum* occur, where normal depth is about one to two feet and where there is moderately strong current and a substrate composed of coarse sand and gravel. Host fish for the glochidia of the Alabama Pigtoe unknown.

Status: Undetermined (Williams et al., 1993:13). If the Alabama Pigtoe still occurs in the Conasauga River in Tennessee, it is a rare shell.

Pleurobema oviforme (Conrad, 1834)
Tennessee Clubshell

RANGE MAP 86; PLATE 87

Synonymy:

Unio oviformis Conrad, 1834; Conrad, 1834a:46, 70, pl. 3, fig. 6

Margaron (Unio) oviformis (Conrad, 1834); Lea, 1852c:26

Pleurobema oviformis (Conrad, 1834); Simpson, 1900a:748

Pleurobema oviforme (Conrad, 1834); Goodrich, 1913:94

Unio ravenelianus Lea, 1834; Lea, 1834:32, pl. 3, fig. 5

Margarita (Unio) ravenelianus (Lea, 1834); Lea, 1836:22

Margaron (Unio) ravenelianus (Lea, 1834); Lea, 1852c:26

Quadrula (Pleuornaia) raveneliana (Lea, 1834); Frierson, 1927:60

Unio rudis Conrad, 1837; Conrad, 1837:76, pl. 43, fig. 1 [in part]

Unio decisus Lea, 1831; Küster, 1852:41, pl. 8, fig. 1 [in part]

? *Unio patulus* Lea, 1829; Conrad, 1838:92, pl. 50, fig. 2

Pleurobema cuneolus (Lea, 1840); Simpson, 1900a:748 [in part]

Unio holstonensis Lea, 1840; Lea, 1840:288; Lea, 1842b:212, pl. 15, fig. 27

Margaron (Unio) holstonensis (Lea, 1840); Lea, 1852c:25

Pleurobema holstonensis (Lea, 1840); Simpson, 1900a:746

Pleurobema oviforme holstonense (Lea, 1840); Ortmann, 1918:554

Unio argenteus Lea, 1841; Lea, 1841b:82; Lea, 1842b:242, pl. 25, fig. 57

Margaron (Unio) argenteus (Lea, 1841); Lea, 1852c:26

Pleurobema argentea (Lea, 1841); Simpson, 1900a:763

Pleurobema argenteum (Lea, 1841); Goodrich, 1913:94

Pleurobema oviforme argenteum (Lea, 1841); Ortmann, 1918:552

Unio mundus Lea, 1857; Lea, 1857b:83; Lea, 1866:40, pl. 14, fig. 38

Margaron (Unio) mundus (Lea, 1857); Lea, 1870:40

Quadrula (Pleuronaia) munda (Lea, 1857); Frierson, 1927:60

Unio lesleyi Lea, 1860; Lea, 1860c:306; Lea, 1860e:352, pl. 58, fig. 177

Margaron (Unio) lesleyi (Lea, 1860); Lea, 1870:40

Pleurobema lesleyi (Lea, 1860); Simpson, 1900a:748

Unio ornatus Lea, 1861 non Conrad, 1835; Lea, 1861a:41; Lea, 1862b:85, pl. 11, fig. 234

Margaron (Unio) ornatus (Lea, 1861); Lea, 1870:57

Pleurobema ornatus (Lea, 1861); Simpson, 1900a:749

Pleurobema ornatum (Lea, 1861); Simpson, 1914:746

Unio tesserulae Lea, 1861; Lea, 1861d:392; Lea, 1866:40, pl. 15, fig. 39

Margaron (Unio) tesserulae (Lea, 1861); Lea, 1870:36

Pleurobema tesserulae (Lea, 1861); Simpson, 1900a:749

Quadrula (Pleuronaia) tesserulae (Lea, 1861); Frierson, 1927:60

Unio striatissimus Anthony, 1865; Anthony, 1865:156, pl. 12, fig. 1

Quadrula (Pleuronaia) striatissima (Anthony, 1865); Frierson, 1927:58

Unio clinchensis Lea, 1867; Lea, 1867:81; Lea, 1868b:278, pl. 37, fig. 91

Margaron (Unio) clinchensis (Lea, 1867); Lea, 1870:38

Pleurobema clinchense (Lea, 1867); Simpson, 1914:743

Pleurobema fassinans (Lea, 1868); Ortmann, 1913a:310 non Lea

Unio planior Lea, 1868; Lea, 1868a:145; Lea, 1869:316, pl. 1, fig. 129

Margaron (Unio) planior (Lea, 1868); Lea, 1870:35

Pleurobema planior (Lea, 1868); Simpson, 1900a:763

Pleurobema planius (Lea, 1868); Simpson, 1914:802

Unio pattinoides Lea, 1871; Lea, 1871:193; Lea, 1874:16, pl. 4, fig. 12

Unio acuens Lea, 1871; Lea, 1871:190; Lea, 1874:27, pl. 8, fig. 24

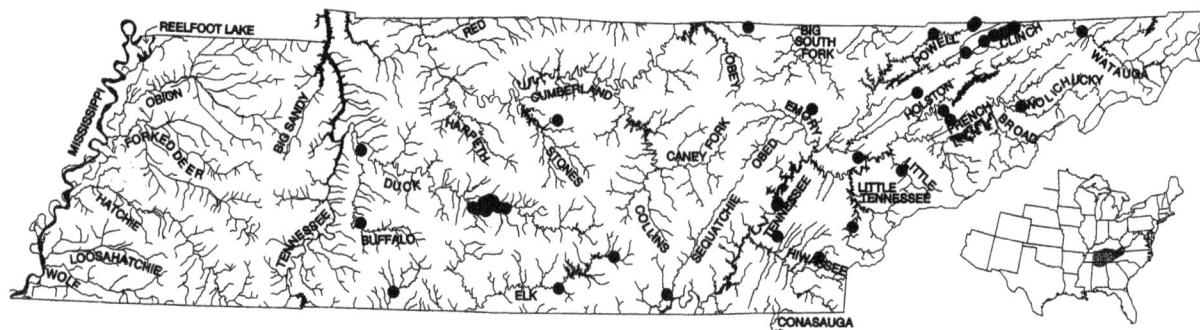

Range Map 86. *Pleurobema oviforme* (Conrad, 1834), Tennessee Clubshell.

Pleurobema acuens (Lea, 1871); Simpson, 1900a:749

Unio lawi Lea, 1871; Lea, 1871:189; Lea, 1874:8, pl. 2, fig. 4

Unio conasaugaensis Lea, 1872; Lea, 1872b:155; Lea, 1874:33, pl. 10, fig. 30

Pleurobema conasaugaensis (Lea, 1872); Simpson, 1900a:763

Pleurobema conasaugaense (Lea, 1872); Simpson, 1914:800

Unio bellulus Lea, 1872; Lea, 1872b:161; Lea, 1874:50, pl. 17, fig. 48

Unio brevis Lea, 1872; Lea, 1872b:157; Lea, 1874:35, pl. 12, fig. 32

Pleurobema brevis (Lea, 1872); Simpson, 1900a:763

Pleurobema breve (Lea, 1872); Simpson, 1914:800

Unio swordianus S. H. Wright, 1897; S. H. Wright, 1897:4

Pleurobema swordianus (S. H. Wright, 1897); Simpson, 1900b:81, pl. 4, fig. 4

Pleurobema swordianum (S. H. Wright, 1897); Simpson, 1914:757

Type Locality: Tennessee. Haas (1969a) records the Holston River as the type locality.

General Distribution: Tennessee and Cumberland river drainages (Ortmann, 1925).

Tennessee Distribution: Prior to 1960 the Tennessee Clubshell was widespread in East and Middle Tennessee and was known to have occurred in the Emory, Watauga, French Broad, Holston, Hiwassee, Obey, Buffalo, and Tennessee rivers (Starnes and Bogan, 1988). It is still found in the unimpounded stretches of the Clinch and Powell rivers in upper East Tennessee (Ahlstedt, 1991a, b), the Tellico River (Parmalee and Klippel, 1984), Elk River (Ahlstedt, 1983), Hiwassee River (Parmalee and Hughes, 1994), and the Duck, Little Pigeon, Big South Fork Cumberland, and Stones rivers.

Description: The shell is solid, obovate, elliptical to nearly rhomboid in outline, and slightly to moderately inflated. Ortmann (1925:341) commented that "[i]n the Tennessee at the Mussel Shoals it is represented by the more swollen *holstonense,* and in the headwaters it passes into the compressed *argenteum.*" Beaks are only moderately full, high, turned forward, and extend only slightly beyond the hinge line; the beak cavity is shallow, and the interdentum is wide. The anterior end is broadly rounded, and the ventral margin is slightly curved; the posterior end is bluntly pointed and sometimes somewhat truncated. The posterior ridge is low, rounded, and often indistinct. Old individuals may attain a length of 90 mm, although the majority of adults seldom exceed 70 mm in length.

Pseudocardinal teeth, two in the left valve, one with vestiges of two others in the right, are stout, triangular, deeply serrated, and erect; they project ventrally from the hinge line at nearly a 90° angle; the one in the right valve is directed more posterior-ventrally. Lateral teeth, two in the left valve, one in the right, are erect, heavy, long, and straight. Muscle scars are large and deeply impressed. The surface is roughened by evenly spaced, concentric, often darkened rest lines. The periostracum is a dull straw yellow, greenish yellow, or gray brown, typically patterned with narrow and/or wide broken green rays, primarily on the umbonal and disc areas. The nacre is a silvery or bluish white.

Life History and Ecology: Viable populations of *Pleurobema oviforme* in Tennessee typically may be found in small, shallow (less than two feet in depth) streams

Plate 87. *Pleurobema oviforme* (Conrad, 1834), Tennessee Clubshell.

and rivers with good current and a substrate of coarse gravel and sand. The breeding season of this species is unknown, although it may be tachytictic; Kitchel (1985) reported peak densities of glochidia during mid-July. Weaver et al. (1991) have shown, under both natural and laboratory conditions, that the whitetail shiner (*Cyprinella galactura*), common shiner (*Luxilus cornutus*), river chub (*Nocomis micropogon*), central stoneroller (*Campostoma anomalum*), and fantail darter (*Etheostoma flabellare*) may serve as host fish for glochidia of the Tennessee Clubshell.

Status: Special Concern (Williams et al., 1993:13).

Pleurobema perovatum (Conrad, 1834)
Ovate Clubshell
RANGE MAP 87; PLATE 88

Synonymy:
Unio perovatus Conrad, 1834; Conrad, 1834a:338, pl. 1, fig. 3.
Margarita (Unio) perovatus (Conrad, 1834); Lea, 1836:23
Margaron (Unio) perovatus (Conrad, 1834); Lea, 1852c:27
Pleurobema perovata (Conrad, 1834); Simpson, 1900a:755
Pleurobema perovatum (Conrad, 1834); Simpson, 1914:767
Unio nux Lea, 1852; Lea, 1852b:283, pl. 24, fig. 43
Margaron (Unio) nux (Lea, 1852); Lea, 1852c:31
Pleurobema nux (Lea, 1852); Simpson, 1900a:758
Unio cinnamomicus Lea, 1861; Lea, 1861a:39; Lea, 1862b:100, pl. 16, fig. 248
Margaron (Unio) cinnamominus (Lea, 1861); Lea, 1870:49 [misspelling]
Unio cinnamominus Lea, 1861; Sowerby, 1868:pl. 83, fig. 436 [misspelling]
Unio concolor Lea, 1861; Lea, 1861a:40; Lea, 1862b:89, pl. 12, fig. 237
Margaron (Unio) concolor (Lea, 1861); Lea, 1870:49
Unio flavidulus Lea, 1861; Lea, 1861a:39; Lea, 1862b:79, pl. 15, fig. 245
Margaron (Unio) flavidulus (Lea, 1861); Lea, 1870:40
Pleurobema flavidulus (Lea, 1861); Simpson, 1900a:759
Pleurobema flavidulum (Lea, 1861); Simpson, 1914:781
Unio pinkstonii B. H. Wright, 1897; B. H. Wright, 1897:136
Pleurobema pinkstonii (B. H. Wright, 1897); Simpson, 1900b:81, pl. 1 fig. 8

Type Locality: Prairie Creek, Marengo County, Alabama.

General Distribution: Alabama River system (Simpson, 1914).

Tennessee Distribution: *Pleurobema perovatum* is known only from southeastern Tennessee in the Conasauga River, Polk County.

Description: The Ovate Clubshell is roughly oval in outline and rather thick-shelled. The anterior end is rounded, the ventral margin curved, and the posterior-basal margin is drawn out posteriorly to a blunt point. The posterior ridge is narrowly rounded. Beaks are high, full, and inflated; sculpture has not been reported. The beak cavity is open and shallow. Maximum shell length is approximately 45 mm.

The left valve has two erect, narrowly triangular pseudocardinal teeth and two long, compressed, straight lateral teeth. The right valve has one erect, triangular pseudocardinal and one straight, compressed lateral tooth. Anterior adductor muscle scars are deep; posterior adductor muscle scars are shallow; the pallial line is impressed anteriorly. The periostracum is a dull yellow to olive green in color; some individuals have a few subtle green rays that may be either narrow or wide; in the latter case they appear as broken squares or blotches. The nacre color is white.

Life History and Ecology: In Tennessee this small clubshell was found inhabiting a sand and fine gravel sub-

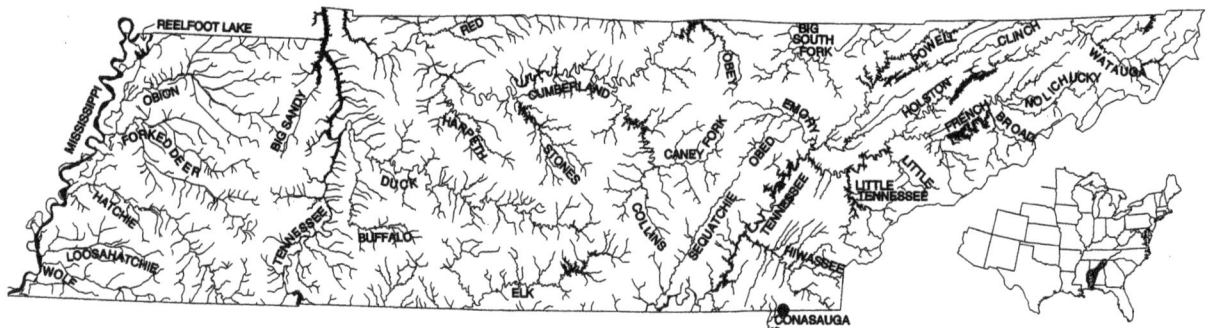

Range Map 87. *Pleurobema perovatum* (Conrad, 1834), Ovate Clubshell.

Plate 88. *Pleurobema perovatum* (Conrad, 1834), Ovate Clubshell.

strate in stretches of river with moderate current and typically at a depth of less than three feet. Like other closely related species of *Pleurobema*, the Ovate Clubshell is probably a summer breeder. Host fish for *Pleurobema perovatum* unknown. This mussel apparently no longer occurs in Tennessee stretches of the Conasauga River.

Status: Endangered (Williams et al. 1993:13). The U.S. Fish and Wildlife Service has developed a recovery plan for this species (U.S. Fish and Wildlife Service, 1994).

Pleurobema plenum (Lea, 1840)
Rough Pigtoe

RANGE MAP 88; PLATE 89
Synonymy:
Quadrula obliqua (Lamarck, 1819); Daniels, 1903:652 [in part]
Pleurobema obliquum (Lamarck, 1819); Ortmann, 1910:117 [in part]

Pleurobema obliquum cordatum (Rafinesque, 1820); Ortmann, 1918:548 [in part]
Quadrula cordata (Rafinesque, 1820); Vanatta, 1915:558 [misidentification]
Pleurobema cordatum (Rafinesque, 1820); Bickel, 1968:20 [in part]
Pleurobema premorsa (Rafinesque, 1820); Morrison, 1969:23 [misidentification]
Unio plenus Lea, 1840; Lea, 1840:286; Lea, 1842b:211, pl. 14, fig. 26
Margaron (Unio) plenus (Lea, 1840); Lea, 1852c:25
Quadrula plena (Lea, 1840); Simpson, 1900a:790
Pleurobema obliquum var. *plenum* (Lea, 1840); Utterback, 1915:188
Pleurobema plenum (Lea, 1840); Ortmann and Walker, 1922:23
Pleurobema cordatum var. *plenum* (Lea, 1840); Ortmann, 1925:339
Quadrula (Obliquata) cordata plena (Lea, 1840); Frierson, 1927:53
Fusconaia plena (Lea, 1840); Morrison, 1942:354
Quadrula (Obliquata) cordata var. *plena* (Lea, 1840); Haas, 1969a:298

Type Locality: Ohio River, Cincinnati, Ohio.

General Distribution: Ohio, Cumberland, and Tennessee River systems, southwest to Kansas and Arkansas (Simpson, 1914).

Tennessee Distribution: The Rough Pigtoe, prior to 1960, occurred in the French Broad, Holston, and Tennessee rivers (Starnes and Bogan, 1988). During his study of the Clinch River mussel fauna from 1978 to 1983, Ahlstedt (1991a) found *Pleurobema plenum* to be a rare species, reporting it from only three locations in Hancock County, Tennessee. It is a rare shell in stretches of the middle Cumberland River, primarily Smith and Trousdale counties, where commercial shellers inadvertently take an occasional individual. The species exists as a relict population in the Tennessee River above Chattanooga (Ahlstedt and McDonough, 1994).

Description: The shell is solid, inflated, and subtriangular in outline. Mature individuals reach about 75–80 mm in length. Beaks are full, high, turned forward, and projecting well beyond the hinge line; sculpture consists of a few irregular nodulous ridges. The anterior margin is sharply truncated, the dorsal margin is slightly curved, the ventral margin is rounded, and the posterior margin is nearly straight, often with a shallow sulcus before the posterior ridge. The posterior

Range Map 88. *Pleurobema plenum* (Lea, 1840), Rough Pigtoe.

ridge is narrowly rounded, ending in a blunt point. The median ridge is high, wide, and rounded, being separated from the posterior ridge by a radial depression. The surface is marked by irregular growth lines.

The left valve has two sculptured, radiate pseudocardinal teeth separated from two short, thick lateral teeth by a broad interdentum. The right valve has one large, striated pseudocardinal tooth, often with a small tooth before and behind. The lateral tooth in the right valve may be single or have a second reduced tooth. The beak cavity is moderately deep, wide, and open. Muscle scars are small and deeply impressed; anterior scars are sculptured. The periostracum is satinlike and yellowish brown to reddish brown in color. Shells may be unrayed or have a series of fine dark green lines over the posterior half of the shell or beak; these often become obliterated with age. The nacre color varies from white to pink and is iridescent posteriorly.

Life History and Ecology: Although *Pleurobema plenum* may become established in small rivers or in headwater stretches of medium-sized rivers, such as the upper Clinch River, it is a species most typical of large rivers such as the Cumberland. Individuals taken in the impounded stretches of the Cumberland River (Smith County, Tennessee) occurred at depths of 12 to 15 feet, and on a substrate composed of firmly packed gravel and sand. The Rough Pigtoe appears to be tachytictic, based on gravid females of *Pleurobema cordatum* found in May (Ortmann, 1919). Host fish for the glochidia of this species unknown.

Status: Endangered (Williams et al., 1993:13). Ortmann (1919) considered certain recognized "species" of *Pleurobema* (e.g., *catillus, sintoxia, rubrum*) as simply forms or varieties of *Pleurobema obliquum* (=*cordatum*); *P. plenum* would also be considered a part of this complex. The U.S. Fish and Wildlife Service has developed a recovery plan for this species (U.S. Fish and Wildlife Service, 1984e) and has created a watershed implementation schedule for the recovery plan (U.S. Fish and Wildlife Service, 1989b).

Plate 89. *Pleurobema plenum* (Lea, 1840), Rough Pigtoe.

Pleurobema rubellum (Conrad, 1834)
Warrior Pigtoe

RANGE MAP 89; PLATE 90

Synonymy:
Unio rubellus Conrad, 1834; Conrad, 1834a:38, pl. 6, fig. 2
Margarita (Unio) rubellus (Conrad, 1834); Lea, 1836:33
Margaron (Unio) rubellus (Conrad, 1834); Lea, 1852c:34
Pleurobema rubella (Conrad, 1834); Simpson, 1900a:757
Pleurobema rubellum (Conrad, 1834); Simpson, 1914:776
Unio rudis Conrad, 1837; Conrad, 1837:76, pl. 43, fig. 1 [in part]
Unio pulvinulus Lea, 1845; Lea, 1845:164; Lea, 1848:81, pl. 8, fig. 24
Margaron (Unio) pulvinulus (Lea, 1845); Lea, 1852c:26
Unio irrasus Lea, 1861; Lea, 1861a:38; Lea, 1862b:91, pl. 13, fig. 239
Margaron (Unio) irrasus (Lea, 1861); Lea, 1870:38
Pleurobema irrasa (Lea, 1861); Simpson, 1900a:756
Pleurobema irrasum (Lea, 1861); Simpson, 1914:771

Type Locality: Black Warrior River, near its source, among the mountains of Alabama.

General Distribution: Restricted to the Black Warrior, Coosa, and Cahaba River drainages in Alabama, Georgia, and Tennessee.

Tennessee Distribution: *Pleurobema rubellum* is known only from southeastern Tennessee in a short stretch of the Conasauga River, Polk County.

Description: Shells of the Warrior Pigtoe are small, inflated, and vary in outline from subtriangular to almost elliptical. The anterior end is slightly truncated above, rounded in the middle, and slopes quickly posteriorly; the ventral margin is well rounded, the dorsal margin is rounded, and the posterior margin is drawn out almost to a blunt point. The posterior ridge

Plate 90. *Pleurobema rubellum* (Conrad, 1834), Warrior Pigtoe.

is well developed and narrowly rounded. Beaks are high and full; sculpture has not been reported. The beak cavity is open and shallow. Maximum shell length is usually less than 35 mm.

The left valve has two triangular pseudocardinal teeth and two straight lateral teeth. The right valve has one main erect, narrowly triangular pseudocardinal, often with two smaller additional pseudocardinal teeth; there is one straight, high, thin lateral tooth. Adductor muscle scars are small and impressed. The pallial line is impressed anteriorly. The periostracum is mostly smooth, rayless, and reddish brown. The nacre color is whitish.

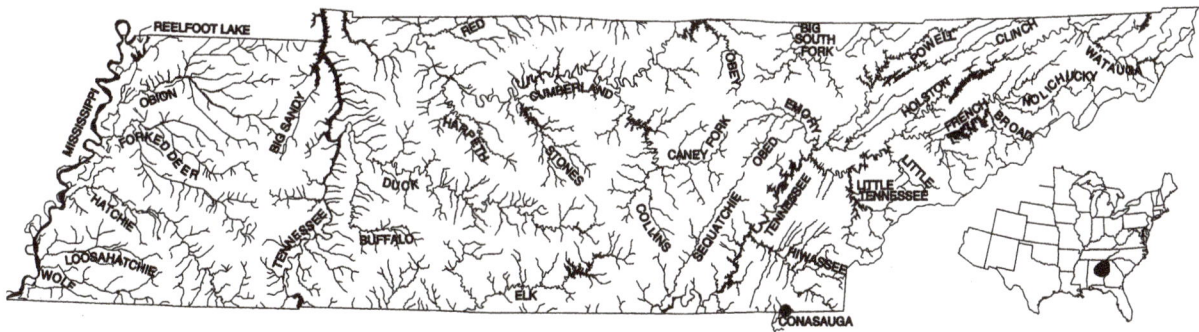

Range Map 89. *Pleurobema rubellum* (Conrad, 1834), Warrior Pigtoe.

Life History and Ecology: This small mussel most often occurs in shallow riffles, less than three feet in depth, in a firm substrate composed of coarse sand, cobbles, and silt. It is probably tachytictic, a summer breeder. Host fish for *Pleurobema rubellum* unknown.

Status: Endangered (Williams et al., 1993:13). The Warrior Pigtoe is probably extirpated in the stretch of Conasauga River flowing through Tennessee.

Pleurobema rubrum (Rafinesque, 1820)
Pyramid Pigtoe

RANGE MAP 90; PLATE 91

Synonymy:
Unio obliqua Lamarck, 1819; Wood, 1856:200 [misidentification]
Quadrula (Obliquata) obliquata (Rafinesque, 1820); Frierson, 1927:52 [misidentification]
Obliquaria rubra Rafinesque, 1820; Rafinesque, 1820:314
Pleurobema rubrum (Rafinesque, 1820); Stansbery, 1976a:50
Quadrula rubra (Rafinesque, 1820); Vanatta, 1915:557
Pleurobema obliquum rubrum (Rafinesque, 1820); Ortmann, 1918:550
Unio ruber (Rafinesque, 1820); Conrad, 1853:257
Unio mytiloides (Rafinesque, 1820); Deshayes, 1830:586, pl. 249, fig. 4 [misidentification]
Unio mytilloides Lea, 1829 (non Rafinesque); Frierson, 1927:52 [misspelling, misidentification, and wrong author and date cited by Frierson]
Unio cardiacea Guérin, 1828; Guérin, 1828:pl. 28, fig. 7
Unio pyramidatus Lea, 1834; Lea, 1834:109; pl. 16, fig. 39
Margarita (Unio) pyramidatus (Lea, 1834); Lea, 1836:21
Margaron (Unio) pyramidatus (Lea, 1834); Lea, 1852c:25
Unio mytiloides var. *pyramidatus* Lea, 1834; Paetel, 1890:160
Quadrula pyramidata (Lea, 1834); Simpson, 1900a:790
Pleurobema pyramidatum (Lea, 1834); Ortmann, 1912a:264
Pleurobema cordatum var. *pyramidatum* (Lea, 1834); Goodrich and van der Schalie, 1994:309
Pleurobema cordatum pyramidatum (Lea, 1834); Murray and Leonard, 1962:76

Type Locality: Ohio.

General Distribution: Ohio, Cumberland, and Tennessee River systems; southwest to Arkansas, west to Nebraska?; north in the Mississippi River to Prairie du Chien, Wisconsin (Simpson, 1914).

Tennessee Distribution: The Pyramid Pigtoe may be found locally in stretches of the Clinch, Tennessee, Cumberland, Duck, and Stones rivers. Prior to 1960 it was known to inhabit the French Broad, Holston, and Little Tennessee rivers (Starnes and Bogan, 1988), and prior to the closing of the Tellico Dam gates (1979) it occurred in the Little Tennessee River.

Description: The shell is solid, greatly inflated in the umbone area, and subtriangular in outline; beaks are full, high, and somewhat narrow toward the anterior end, projecting well forward of the hinge line and turned forward. Mature individuals attain a length of 80–90 mm. The anterior margin is rounded or somewhat truncated, the short ventral margin is straight, and the posterior end is bluntly pointed. The posterior ridge is distinct but low and rounded with a very shallow, wide sulcus anterior to the ridge toward the ventral margin.

There are two moderately low, triangular, roughened pseudocardinal teeth in the left valve; the two lateral teeth are straight, low, fairly short, and serrated. Right valve has a single, pronounced, roughened pseudocardinal tooth with a low "circular" tooth anterior and above the main pseudocardinal. The lateral tooth is high, straight, and finely serrated. The beak cavity is open and very shallow; the interdentum is wide. Muscle scars are deeply impressed. The periostracum is a light tan or brownish green in juveniles, usually

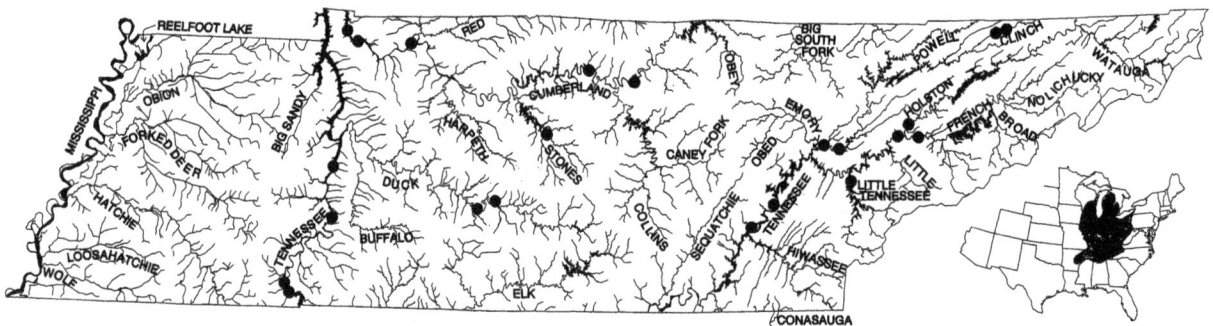

Range Map 90. *Pleurobema rubrum* (Rafinesque, 1820), Pyramid Pigtoe.

Plate 91. *Pleurobema rubrum* (Rafinesque, 1820), Pyramid Pigtoe.

with green rays; with maturity the rays become obscured and the periostracum becomes dark brown to blackish. The surface is roughened, with raised concentric rest lines. The nacre is white, more often various shades of pink, more rarely salmon.

Life History and Ecology: *Pleurobema rubrum* occurs in rivers with strong current and a substrate composed of firm sand and gravel. In the Duck River and unimpounded stretches of the Clinch, this species may be found at a depth of less than three feet, while individuals inhabiting sections of the impounded Cumberland River can occur at depths exceeding 20 feet. Since it is a closely related species (or form?) of *Pleurobema cordatum,* the Pyramid Pigtoe is probably tachytictic, a summer breeder. Host fish for the glochidia unknown.

Status: Threatened (Williams et al., 1993:13).

Pleurobema sintoxia (Rafinesque, 1820)
Round Pigtoe

RANGE MAP 91; PLATE 92

Synonymy:

Obliquaria sintoxia Rafinesque, 1820; Rafinesque, 1820:48

Pleurobema sintoxia (Rafinesque, 1820); Oesch, 1984:122

Unio rubens Menke, 1828; Menke, 1828:90

Unio catillus Conrad, 1836; Conrad, 1836:30, pl. 13, fig. 2

Unio catilus Conrad, 1836; B. H. Wright, 1888b:no pagination [misspelling]

Pleurobema obliquum catillus (Conrad, 1834); Utterback, 1915:190, pl. 20, figs. 62a, b

Pleurobema catillus (Conrad, 1834); Utterback, 1915:193, pl. 20, figs. 59a, b

Quadrula (Obliquata) catillus (Conrad, 1834); Frierson, 1927:54

Pleurobema coccineum var. *catillus* (Conrad, 1834); Baker, 1928a:117, pl. 55, figs. 1–3

Pleurobema cordatum var. *catillus* (Conrad, 1834); Goodrich, 1932:90

Pleurobema cordatum catillus (Conrad, 1834); Murray and Leonard, 1962:74

Unio coccineus Conrad, 1836; Conrad, 1836:29, pl. 13, fig. 1

Margarita (Unio) coccineus (Conrad, 1836); Lea, 1836:34

Margaron (Unio) coccineus (Conrad, 1836); Lea, 1852c:35

Quadrula coccinea (Conrad, 1836); F. C. Baker, 1898:79, pl. 14, fig. 1, pl. 19, fig. 3

Pleurobema coccineum (Conrad, 1836); Ortmann, 1912a:263

Pleurobema obliquum coccineum (Conrad, 1836); Utterback, 1915:191, pl. 20, figs. 61, a, b, c, d

Quadrula (Obliquata) coccinea (Conrad, 1836); Frierson, 1927:53

Pleurobema catillus var. *coccinea* (Conrad, 1836); Richardson, 1928:458

Pleurobema cordatum var. *coccineum* (Conrad, 1836); Goodrich, 1932:90

Pleurobema cordatum coccineum (Conrad, 1834); Murray and Leonard, 1962:71

Quadrula (Obliquata) coccinea coccinea (Conrad, 1836); Haas, 1969a:298

Margarita (Unio) solidus Lea, 1836; Lea, 1836:20 [nomen nudum]

Unio solidus Lea, 1838; Lea, 1838b:13, pl. 5, fig. 13

Margaron (Unio) solidus (Lea, 1838); Lea, 1852c:25

Quadrula (Fusconaia) solida (Lea, 1838); Simpson, 1900a:789

Fusconaja solida (Lea, 1838); F. C. Baker, 1920:382

Pleurobema coccineum solida (Lea, 1838); F. C. Baker, 1928a:118, pl. 52, pl. 53, fig. 6

Pleurobema catillus var. *solida* (Lea, 1838); Richardson, 1928:458

Quadrula solida (Lea, 1838); La Rocque, 1953:98

Pleurobema coccineum forma *solida* (Lea, 1838); Starrett, 1971:309

Unio gouldianus "Ward" Jay, 1839; Jay 1839:24 [nomen nudum]

Unio cardiacea Guérin, 1828; Deshayes, 1839:19, pl. 31, figs. 1, 2 [misidentification]

Unio cuneus Conrad, 1838; Conrad, 1838:No. 11, back cover; Conrad, 1840:105, pl. 58, fig. 1

Quadrula (Obliquata) coccinea var. *cuneus* (Conrad, 1838); Frierson, 1927:53

Unio fulgidus Lea, 1845; Lea, 1845:164; Lea, 1848:73, pl. 4, fig. 10

Margaron (Unio) fulgidus (Lea, 1845); Lea, 1852c:25

Unio obovalis (Rafinesque, 1820); Conrad, 1853:253 [misidentification]

Quadrula (Fusconaia) coccinea var. *paupercula* Simpson, 1900; Simpson, 1900a:789 (non Lea, 1861)

Quadrula coccinea paupercula Simpson, 1900; Sterki, 1907:392

Pleurobema obliquum pauperculum (Simpson, 1900); Ortmann, 1919:83, pl. 7, fig. 6

Pleurobema coccineum pauperculum (Simpson, 1900); Ortmann and Walker, 1922:24

Pleurobema cordatum pauperculum (Simpson, 1900); Goodrich 1932:90

Pleurobema missouriensis Marsh, 1901; Marsh, 1901:74; Walker, 1915:140, pl. 5, figs. 1, 2

Quadrula (Obliquata) missouriensis (Marsh, 1901); Frierson, 1927:54

Quadrula (Fusconaia) coccinea var. *magnalacustris* Simpson, 1914; Simpson, 1914:884 [replacement name for *paupercula* Simpson, 1900, perceived secondary homonym]

Quadrula (Obliquata) coccinea magnalacustris Simpson, 1914; Frierson, 1927:53

Pleurobema coccineum var. *mississippiensis* F. C. Baker, 1928; F. C. Baker, 1928a:121, pl. 53, figs. 1–5

Type Locality: Mahoning River near Pittsburgh, Pennsylvania.

General Distribution: Entire upper Mississippi River drainage (Simpson, 1914). Baker (1928a:115) states that "[i]ts range is apparently bounded by western New York on the east, Iowa and Kansas on the west, Michigan and Wisconsin on the north, and Alabama and Arkansas on the south." In Canada, the Round Pigtoe is reported to be rare, occurring in medium-sized to large rivers and in Lake Erie (Clarke, 1981a). It is the only species of *Pleurobema* reported from Missouri (Oesch, 1984).

Tennessee Distribution: The Round Pigtoe is now apparently restricted to the Cumberland, Big South Fork Cumberland, and Stones rivers. The "species" *Pleurobema catillus* reported as formerly (prior to 1960) in the Cumberland, Duck, Clinch, Holston, and Tennessee rivers (Starnes and Bogan, 1988) is probably only a medium-sized river form of *Pleurobema sintoxia*. Oesch (1984:124) provides a sagacious interpretation by David H. Stansbery of the taxonomic problem involving this *Pleurobema* complex:

> *Pleurobema coccineum [= sintoxia]* is a highly variable species since it is quite plastic and varies from a compressed, low-umbone, round, non-sulcate form upstream to exactly the opposite in big rivers.
>
> *P. coccineum* form *coccineum* (Con., 1836) headwaters
>
> *P. coccineum* form *catillus* (Con., 1836) medium rivers
>
> *P. coccineum* form *solida* (Lea, 1838) big rivers.
>
> These forms are related to habitat and may be more due to environmental differences than to genetic differences. Just where one form stops and another begins is arbitrary and taxonomically of no importance, since they are neither species nor sub-species.

Description: Shells characteristic of medium-sized rivers such as the Big South Fork Cumberland are compressed, flattened, solid, and somewhat rectangular, but often oval or elongated. Beaks are compressed, only slightly elevated, turned forward, and only slightly extending beyond the hinge line. However, beaks are full, elevated, and project forward well beyond the hinge line in the big river form of *P. sintoxia*. Sculpture con-

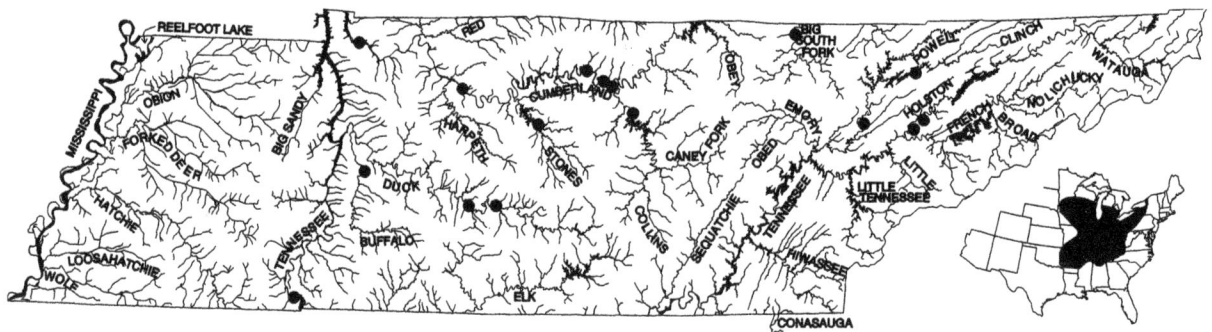

Range Map 91. *Pleurobema sintoxia* (Rafinesque, 1820), Round Pigtoe.

sists of a few coarse, irregular ridges, curving upward behind. The anterior end is rounded; the posterior end is squarely truncated; the posterior ridge is rounded, ending in a blunt point. Mature individuals from the Big South Fork Cumberland River, Scott County, Tennessee, attain a length of 110–120 mm. Shells of Cumberland River specimens are solid, inflated, and subquadrate; the anterior-ventral margin is broadly rounded; the posterior ridge is low and rounded, and some specimens have a wide but very shallow sulcus anterior to the posterior ridge.

There are two stout, rectangular, serrated pseudocardinal teeth in left valve, one in the right, with a low, roughened linear tooth anteriorly and dorsally. The lateral teeth—two in the left valve, one in the right— are straight, moderately high, and finely serrated. The interdentum is wide; the beak cavity is very shallow;

Plate 92. *Pleurobema sintoxia* (Rafinesque, 1820), Round Pigtoe.

muscle scars are deeply impressed. The surface is roughened with concentric rest lines. The periostracum in juveniles is a dull tan with distinct green rays; these become obliterated as the surface darkens to a deep reddish brown or black with age. The nacre is white or varying shades of pink.

Life History and Ecology: In Tennessee *Pleurobema sintoxia* is found most abundantly, and almost exclusively, in medium-sized (Big South Fork Cumberland) and big (Cumberland) rivers and in current on a firm substrate of coarse gravel and sand. It may occur at depths of less than three feet to more than 20 feet. Baker (1928a) stated that the Round Pigtoe is tachytictic, the breeding season lasting from early May to late July in Wisconsin. The fish host for glochidia of the Round Pigtoe was reported by Surber (1913) and Coker et al. (1921) to be the bluegill *(Lepomis macrochirus).* Hove (1995b), in laboratory studies, showed the spotfin shiner *(Cyprinella spiloptera),* bluntnose minnow *(Pimephales notatus),* and northern redbelly dace *(Phoxinus eos)* as host fish for the glochidia of the Round Pigtoe.

Status: Currently Stable (Williams et al., 1993:13).

Pleurobema troschelianum (Lea, 1852)
Alabama Clubshell

RANGE MAP 92; PLATE 93

Synonymy:

Unio troschelianus Lea, 1852; Lea, 1852a:252; Lea, 1852b:280, pl. 23, fig. 39.

Margaron (Unio) troschelianus (Lea, 1852); Lea, 1852c:26

Pleurobema troscheliana (Lea, 1852); Simpson, 1900a:756

Pleurobema troschelianum (Lea, 1852); Simpson, 1914:770

Pleurobema pulvinulum (Lea, 1845); Frierson, 1927:42 [in part]

Type Locality: Coosawattee River, Murray County, Georgia.

General Distribution: Alabama River system (Simpson, 1914).

Tennessee Distribution: Conasauga River, Bradley and Polk counties.

Description: The shell is relatively small (mature specimens from the Conasauga River, Bradley and Polk counties, Tennessee, measure 45–50 mm in length),

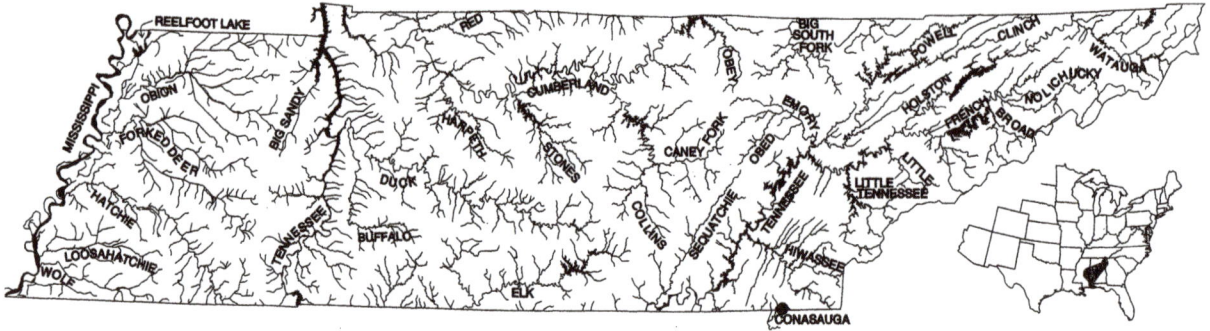

Range Map 92. *Pleurobema troschelianum* (Lea, 1852), Alabama Clubshell.

solid, and compressed. Beaks are full, projecting slightly beyond the hinge line but generally found eroded away. The anterior end is rounded, while the ventral margin is straight to very slightly curved; the posterior end is broadly pointed, sometimes ending in a slight biangulation in specimens exhibiting a distinct but rounded posterior ridge.

Plate 93. *Pleurobema troschelianum* (Lea, 1852), Alabama Clubshell.

The pseudocardinal teeth, two in the left valve, one in the right, are prominent but low, triangular to compressed, and slightly serrated to nearly smooth. There are two lateral teeth in the left valve, one in the right, and they are moderately elevated, thin, and straight. The beak cavity is shallow, and the interdentum is fairly wide. Muscle scars are deeply impressed. The periostracum varies from a dull straw yellow to yellowish brown. Some specimens are rayless, but others have a few or a series of dark green rays, sometimes broken and appearing as squarish blotches. The shell surface in old individuals is somewhat roughened by the raised, concentric, darkened rest lines. The nacre is a dull white to bluish white.

Life History and Ecology: In stretches of the Conasauga River flowing through Tennessee and northeastern Georgia (Murray County), *Pleurobema troschelianum* may be found in areas of moderate current, at a depth of three feet or less, and in a substrate composed of coarse sand and gravel. Populations of this species and those of several other naiads appear to have decreased during the past several decades as a result of siltation and various unfavorable agricultural practices affecting the river. The breeding season and host fish for the glochidia of this species are unknown.

Status: Endangered (Williams et al., 1993:13).

Potamilus alatus (Say, 1817)
Pink Heelsplitter
RANGE MAP 93; PLATE 94
Synonymy:
Unio alatus Say, 1817; Say, 1817:pl. 4, fig. 2
Unio alata Say, 1817; Lamarck, 1819:76

Metaptera alata (Say, 1817); Rafinesque, 1820:300

Mya alata (Say, 1817); Eaton, 1826:219

Symphynota alata (Say, 1817); Lea, 1829:448

Margarita (Unio) alatus (Say, 1817); Lea, 1836:11

Lymnadia alata (Say, 1817); Swainson, 1840:265, fig. 48

Mysca alata (Say, 1817); Swainson, 1841:28

Margaron (Unio) alatus (Say, 1817); Lea, 1852c:19

Metaptera alata (Say, 1817); Stimpson, 1851:14

Lampsilis alatus (Say, 1817); Baker, 1898:97

Lampsilis alata (Say, 1817); Scammon, 1906:299–300

Lampsilis (Proptera) alata (Say, 1817); Simpson, 1900a:567

Proptera alata (Say, 1817); Ortmann, 1912a:333

Potamilus alatus (Say, 1817); Morrison, 1969:24

Potamilus alata megapterus Rafinesque, 1820; Rafinesque, 1820:314–315

Metaptera megaptera Rafinesque, 1820; Rafinesque, 1820:300, pl. 80, figs. 20–22

Unio poulsoni Conrad, 1834; Conrad, 1834a:25, 71, pl. 1

Symphynota poulsoni (Conrad, 1834); Férussac, 1835:25

Lampsilis (Proptera) alatus var. *poulsoni* (Conrad, 1834); Simpson, 1900a:568

Lampsilis (Proptera) alata var. *poulsoni* (Conrad, 1834); Simpson, 1914:164

Type Locality: Unknown.

General Distribution: The Pink Heelsplitter ranges throughout the Mississippi River drainage from western Pennsylvania to Minnesota, west to Kansas and Nebraska, and south to Arkansas. In the St. Lawrence River system it occurs from Lake Huron to Lake Champlain, also in the Canadian Interior Basin in parts of the Red River of the North and the Winnipeg River.

Tennessee Distribution: *Potamilus alatus* is widespread throughout the Tennessee River system and its tributaries, including the French Broad, Holston, Powell, Clinch, Little Tennessee, Emory, Hiwassee, Sequatchie, Elk, and Duck rivers. It is found in the Cumberland River and its tributaries, including the Big South Fork

Cumberland, Obey, Caney Fork, Stones, Harpeth, and Red rivers. Interestingly, it has not been reported from Reelfoot Lake or any of the direct tributaries to the Mississippi River in western Tennessee.

Description: The Pink Heelsplitter is large, moderately thick, compressed, and ovate in outline, generally with a high posterior wing which often makes the shell triangular in appearance. There is often a smaller anterior wing. The shape of the dorsal margin will be affected by the degree of erosion of this wing. Maximum shell length is about 185 mm. The anterior end is evenly rounded, the ventral margin is slightly convex to almost straight posteriorly, and the dorsal margin ascends posteriorly to meet at a sharp angle with the posterior margin. The posterior margin is straight dorsally, becoming rounded ventrally. A posterior ridge is lacking, the area being evenly rounded. The posterior slope is compressed and blends into the posterior dorsal wing. The female shell is slightly more swollen, and the ventral margin is more broadly rounded posteriorly. Sexual dimorphism is slight and not always obvious. Beaks are low and only slightly elevated above the hinge line; sculpture consists of 3–4 narrow bars; the first is subconcentric, the rest double-looped. The beak cavity is open and shallow and marked by an irregular, almost vertical row of dorsal muscle scars.

The left valve has two knobby, erect pseudocardinal teeth and two slightly curved lateral teeth. There is no interdentum. The right valve has one erect, triangular, sculptured pseudocardinal, often with a small tooth anteriorly; there is one slightly curved lateral tooth. Anterior adductor muscle scar are impressed, while the posterior adductor muscle scar is shallow but distinct. The pallial line is impressed. The periostracum is dark greenish, often with green rays in young specimens that

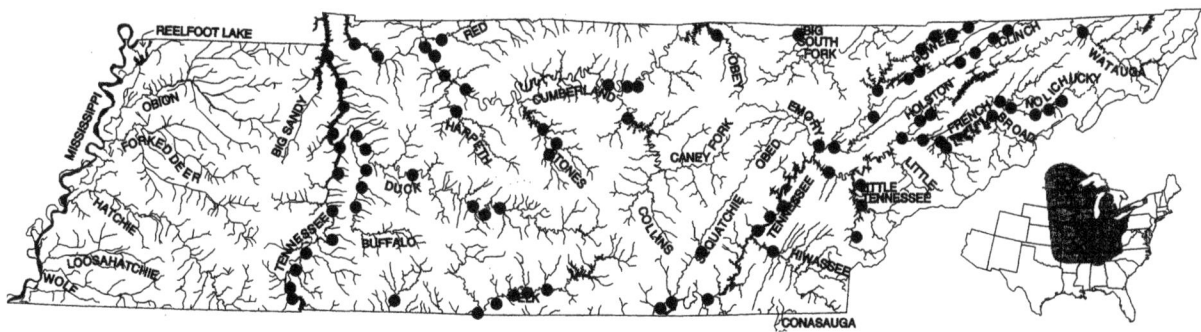

Range Map 93. *Potamilus alatus* (Say, 1817), Pink Heelsplitter.

Plate 94. *Potamilus alatus* (Say, 1817), Pink Heelsplitter.

become obscured as the epidermis color changes to dark brown or black with age. The nacre color is always a dark purple and is iridescent posteriorly.

Life History and Ecology: Ortmann (1919:254) reported that "the species seems to breed all the year round, and the breeding seasons appear to overlap in June and July." The Pink Heelsplitter is found in a variety of habitats, from sandy bottoms in shallow lakes and soft sandy river overbanks to coarse gravel in good current in areas up to about three feet in depth. Watters (1994a) and Weiss and Layzer (1995) list the freshwater drum *(Aplodinotus grunniens)* as the host for the glochidia of *Potamilus alatus*.

Status: Currently Stable (Williams et al., 1993:13).

Potamilus ohiensis (Rafinesque, 1820)
Pink Papershell

RANGE MAP 94; PLATE 95

Synonymy:

Anodonta (Lastena) ohiensis Rafinesque, 1820; Rafinesque, 1820:316–317

Unio ohioensis (Rafinesque, 1820); Say, 1834:no pagination

Symphynota ohioensis (Rafinesque, 1820); Férussac, 1835:25

Metaptera ohiensis (Rafinesque, 1820); Conrad, 1853:260

Potamilus ohiensis (Rafinesque, 1820); Morrison, 1969:24

Symphynota laevissima Lea, 1830; Lea, 1829:444, pl. 13, fig. 23

Unio laevissimus (Lea, 1830); Conrad, 1834a:70

Unio laevissima (Lea, 1830); Deshayes, 1835:558

Margarita (Unio) laevissimus (Lea, 1830); Lea, 1836:11

Margaron (Unio) laevissimus (Lea, 1830); Lea, 1852c:19

Lampsilis (Proptera) laevissima (Lea, 1830); Simpson, 1900a:574

Proptera laevissima (Lea, 1830); Ortmann, 1912a:334

Leptodea laevissima (Lea, 1830); Goodrich and van der Schalie, 1944:316

Unio discoideus Lea, 1834; Sowerby, 1866:pl. 53, fig. 275 [misidentification]

Type Locality: "l'Ohio et toutes les rivières adjacentes."

General Distribution: *Potamilus ohiensis* is widely distributed throughout the Mississippi River drainage, including the Tennessee and Cumberland River basins, and from western New York west to North Dakota and Nebraska and south to Louisiana and eastern Texas.

Tennessee Distribution: The Pink Papershell is of sporadic occurrence in Tennessee. In East Tennessee it is known from the main channel of the Tennessee River and the Little Tennessee and Little rivers. In Middle Tennessee it is found in the Tennessee River at the mouth of the Duck River and in the lower Harpeth River. *Potamilus ohiensis* occurs at widely scattered localities in the main channel of the Mississippi River and in the Hatchie River in West Tennessee.

Description: The shell is thin, generally elliptical to ovate in shape, but with a high, triangular posterior wing and usually a smaller anterior wing. The anterior end is sharply rounded, the dorsal margin is straight, the ventral margin is broadly curved, and the posterior margin is sharply rounded. The posterior ridge is low, gently rounded, and not prominent. The female shell is slightly fuller along the posterior ventral margin. Beaks are narrow, low, and subcompressed; sculpture is faint, composed of a few nodulous, broken,

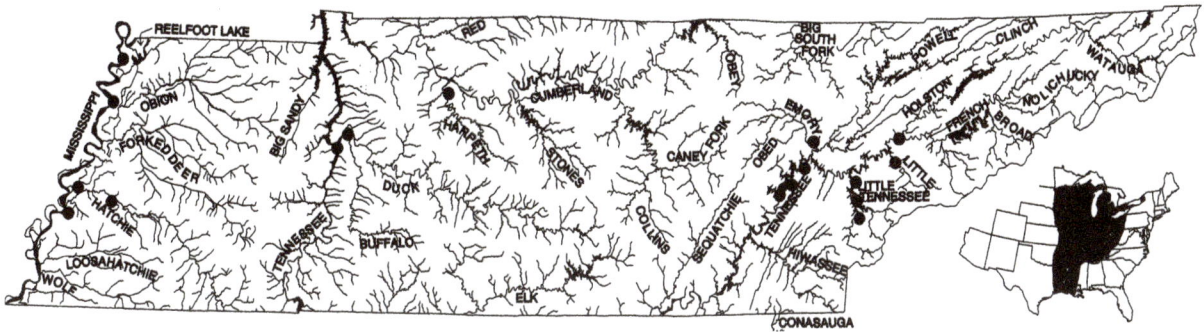

Range Map 94. *Potamilus ohiensis* (Rafinesque, 1820), Pink Papershell.

slightly looped lines. The beak cavity is shallow with an irregular row of dorsal muscle scars. Maximum shell length is about 140 mm.

The left valve has two thin, compressed pseudocardinal teeth and two fragile, curved lateral teeth. The right valve has one long, thin pseudocardinal and one short, high lateral tooth. Adductor muscle scars are large and shallow. The interdentum is absent. The pallial line is impressed anteriorly. The surface is marked by rest marks that form raised concentric ridges. The periostracum is shiny, and rest lines are usually a dark color; beaks are tan, while the rest of the shell is greenish to light brown, becoming dark brown with age. The nacre color varies from a light cream or gun blue to purplish and is iridescent throughout.

Life History and Ecology: The Pink Papershell inhabits substrates of sand, gravel, or mud in quiet water or in slow current at varying depths of usually three feet or less. This species may become established in lakes where it often becomes abundant. Surber (1912) reported glochidia in April, June, and September. Watters (1994a) lists the freshwater drum *(Aplodinotus grunniens)* and white crappie *(Pomoxis annularis)* as hosts for the glochidia of *Potamilus ohiensis*.

Status: Currently Stable (Williams et al., 1993:14).

Plate 95. *Potamilus ohiensis* (Rafinesque, 1820), Pink Papershell.

Potamilus purpuratus (Lamarck, 1819)
Bleufer
RANGE MAP 95; PLATE 96

Synonymy:
Unio purpurata Lamarck, 1819; Lamarck, 1819:71
Unio purpuratus Lamarck, 1819; Lea, 1834:199
Margarita (Unio) purpuratus (Lamarck, 1819); Lea, 1836:39
Margaron (Unio) purpuratus (Lamarck, 1819); Lea, 1852c:38
Lampsilis purpuratus (Lamarck 1819); Simpson, 1900a:568
Proptera purpurata (Lamarck, 1819); Ortmann, 1912a:234
Potamilus purpuratus (Lamarck, 1819); Valentine and
    Stansbery, 1971:26
Unio ater Lea, 1830 non Nilsson, 1822; Lea, 1829:426, p. 7, fig. 9

Unio atra Deshayes, 1830; Deshayes, 1830:582

Unio lugubris Say, 1832; Say, 1832:pl. 43, fig. 6 [replacement name for *Unio ater* Lea, 1830]

Unio poulsoni Conrad, 1834; Conrad, 1834a:25, pl. 1

Proptera purpurata poulsoni (Conrad, 1834); Frierson, 1927:87

Unio coloradoensis Lea, 1856; Lea, 1856a:103; Lea, 1858d:314, pl. 31, fig. 29

Margaron (Unio) coloradoensis (Lea, 1856); Lea, 1870:41

Lampsilis coloradoensis (Lea, 1856); Simpson, 1900a:568

Unio permiscens Lea, 1859; Lea, 1859a:112; Lea, 1862b:102, pl. 17, fig. 251

Margarita (Unio) permiscens (Lea, 1859); Lea, 1870:61

Lampsilis permiscens (Lea, 1859); Simpson, 1900a:569

Unio dolosus Lea, 1860; Lea, 1860c:307; Lea, 1862b:75, pl. 9, fig. 224

Margarita (Unio) dolosus (Lea, 1860); Lea, 1870:61

Type Locality: "Africa." Obviously a spurious locality.

General Distribution: The Bleufer is widespread across the Gulf Coast drainages from southern Georgia to eastern Texas. It ranges up the Mississippi Embayment as far north as the southern tip of Illinois, southeastern Missouri, and western Kentucky, and west into southeastern Kansas, eastern Arkansas, and western Tennessee.

Tennessee Distribution: In Tennessee the range of the Bleufer is restricted to the mainstem of the Mississippi River (occasional individuals) and direct tributaries including the Wolf, Loosahatchie, and Hatchie. It has also been reported from Reelfoot Lake.

Plate 96. *Potamilus purpuratus* (Lamarck, 1819), Bleufer.

Description: Shells of the Bleufer are large, ovate to oblong or quadrate, and greatly inflated with a hint of a wing anteriorly and a low, angled wing posterior to the beak area. It reaches a maximum shell length of about 170 mm. The shell is relatively thin for its size, but thick anteriorly. The anterior end is broadly to sharply rounded, and the dorsal margin is slightly curved; the ventral margin is straight in females and gently curved in males. The posterior end is rather sharply rounded in male shells, but angled and truncated in females. There are two or three low, radiating ridges on the posterior slope, which is broadly rounded. Females have a wide, rounded marsupial swelling at the posterior margin of the shell. Beaks are

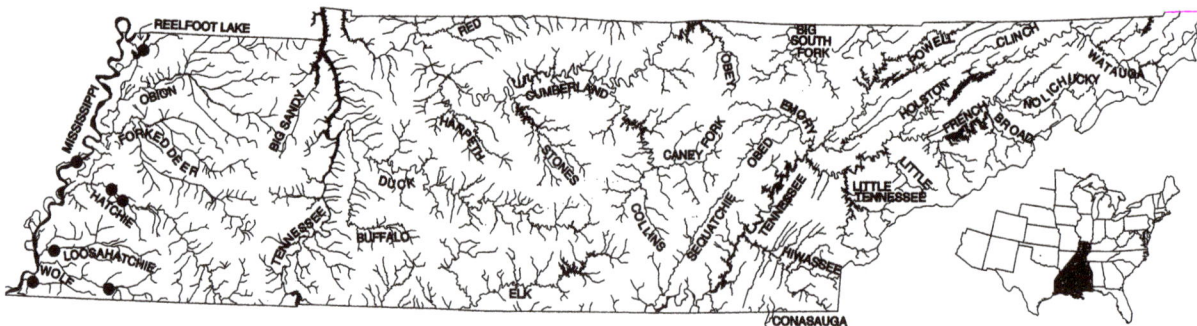

Range Map 95. *Potamilus purpuratus* (Lamarck, 1819), Bleufer.

high and full, moderately raised above the hinge line; they possess a faint corrugated sculpture. The beak cavity is broad and moderately deep, with a row of dorsal muscle scars running anterio-ventrally.

The left valve has two somewhat compressed, solid, erect, and roughened pseudocardinal teeth and two slightly curved, short lateral teeth widely separated from the pseudocardinal teeth. The right valve has a single pointed, subcompressed pseudocardinal, often with a low compressed tooth anteriorly. The lateral tooth is heavy, short, and curved. The interdentum is lacking. Anterior adductor muscle scars are deep and smooth, while the posterior adductor muscle scar is only slightly impressed. The pallial line is impressed anteriorly and ventrally. The periostracum is rayless, brown to black, often with a gun metal bluish sheen. The nacre color is dark purple throughout and iridescent posteriorly.

Life History and Ecology: The Bleufer is generally found in quiet or slow-moving waters in a mud or gravel bottom. In the Hatchie River, it occurs most commonly at depths of up to three feet, in slow current, and in a substrate composed of a stable mixture of silt and mud. Watters (1994a) lists the freshwater drum *(Aplodinotus grunniens)* as the fish host for glochidia of *Potamilus purpuratus.* The species is Bradytictic.

Status: Currently Stable (Williams et al., 1993:14).

Ptychobranchus fasciolaris (Rafinesque, 1820) Kidneyshell

RANGE MAP 96; PLATE 97

Synonymy:
Potamilus fasciolaris Rafinesque, 1818; Rafinesque, 1818a:355 [nomen nudum]
Plagiola fasciolaris Rafinesque, 1819; Rafinesque, 1819a:426 [nomen nudum]
Obliquaria (Ellipsaria) fasciolaris Rafinesque, 1820; Rafinesque, 1820:203
Unio fasciolaris (Rafinesque, 1820); Conrad, 1834a:69
Ellipsaria fasciolaris (Rafinesque, 1820); Frierson, 1914:7
Ptychobranchus fasciolare (Rafinesque, 1820); Ortmann and Walker, 1922:42
Ptychobranchus (Ptychobranchus) fasciolare (Rafinesque, 1820); Frierson, 1927:64
Ptychobranchus fasciolaris (Rafinesque, 1820); Goodrich and van der Schalie, 1944:319
Obliquaria (Ellipsaria) fasciolaris var. *interrupta* Rafinesque, 1820; Rafinesque, 1820:303

Obliquaria (Ellipsaria) fasciolaris var. *fuscata* Rafinesque, 1820; Rafinesque, 1820:304
Obliquaria (Ellipsaria) fasciolaris var. *obliterata* Rafinesque, 1820; Rafinesque, 1820:304
Obliquaria (Ellipsaria) fasciolaris var. *longa* Rafinesque, 1820; Rafinesque, 1820:304
Ptychobranchus (Ptychobranchus) fasciolare longum (Rafinesque, 1820); Frierson, 1927:64
Obliquaria (Plagiola) interrupta Rafinesque, 1820; Rafinesque, 1820:302
Comment: Bogan (1997) identified the syntype of *O. (Plagiola) interrupta* as a specimen of *P. fasciolaris.* Based on the correct identification of the type species, *Plagiola* Rafinesque, 1820 is the senior synonym of *Ptychobranchus* Simpson, 1900. However, to preserve nomenclatural stability the International Commission on Zoological Nomenclature has been asked to suppress *Plagiola.*
Unio phaseolus Hildreth, 1828; Hildreth, 1828:283
Margarita (Unio) phaseolus (Hildreth, 1828); Lea, 1836:38
Margaron (Unio) phaseolus (Hildreth, 1828); Lea, 1852c:38
Ptychobranchus phaseolus (Hildreth, 1828); Simpson, 1900a:612
Unio planulatus Lea, 1829; Lea, 1829:431, pl. 9, fig. 13
Unio planulata Lea, 1829; Deshayes, 1839:672
Unio camelus Lea, 1834; Lea, 1834:102, pl. 15, fig. 45
Margarita (Unio) camelus (Lea, 1834); Lea, 1836:19
Margaron (Unio) camelus (Lea, 1834); Lea, 1852c:24
Ptychobranchus phaseolus var. *camelus* (Lea, 1834); Simpson, 1914:335
Unio compressissimus Lea, 1845; Lea, 1845:163; Lea, 1848:81, pl. 8, fig. 23
Margaron (Unio) compressissimus (Lea, 1845); Lea, 1852c:24
? *Unio arquatus* Conrad, 1854; Conrad, 1854:297, pl. 26, fig. 8
Ptychobranchus (Ptychobranchus) fasciolare arquatum (Conrad, 1834); Frierson, 1927:64
Unio compressissimus var. *performosus* De Gregorio, 1914; De Gregorio, 1914:53–54
Unio lanceolatus var. *blandus* De Gregorio, 1914; De Gregorio, 1914:52
Unio imperitus De Gregorio, 1914; De Gregorio, 1914:45–46
Ptychobranchus fasciolaris var. *lacustris* F. C. Baker, 1928; F. C. Baker, 1928b:52

Type Locality: Muskingum River, Ohio.

General Distribution: Ohio, Tennessee, and Cumberland River systems; Lower Peninsula of Michigan, Kansas, Arkansas, Oklahoma, and Louisiana (Simpson, 1914); Pennsylvania west to Illinois, south to Tennessee.

Tennessee Distribution: In addition to local populations in the Tennessee and Cumberland rivers, the Kidneyshell occurs in small to medium-sized rivers such as the upper Clinch and Powell, the Big South Fork Cumberland,

Range Map 96. *Ptychobranchus fasciolaris* (Rafinesque, 1820), Kidneyshell.

Emory, Nolichucky, Elk, Duck, Harpeth, and Stones. In the compilation of distribution records that includes those prior to 1960, Starnes and Bogan (1988) list this species as formerly occurring in the French Broad, Holston, Little Tennessee, Obey, and Red rivers.

Description: The shell is elongate, elliptical, and compressed; it is solid, heavy, and thick in old individuals. The anterior end is rounded, and the posterior end is bluntly pointed. The posterior-dorsal ridge is prominent but rounded. Under apparently ideal habitat conditions, such as those found in the upper Clinch and Powell rivers as well as those in certain stretches of the Tennessee (Hardin County) and Cumberland (Smith County) rivers, individuals grow to exceptional length and thickness of shell. Senile individuals may reach a length of 150 mm, although the majority of mature individuals seldom exceed 120–130 mm. Beaks are flattened, compressed, and low; sculpture consists of several fine, indistinct, wavy ridges. The surface is marked with numerous, usually prominent, coarse rest lines.

The left valve has two low, thick, heavy, serrated triangular pseudocardinal teeth; the two lateral teeth are nearly straight, short, heavy, and usually widely separated. The right valve has a heavy, somewhat compressed and pyramidal elevated pseudocardinal tooth, sometimes with a low, roughened tubercular tooth on either side; the lateral tooth is wide, heavy, elongated, and serrated. The interdentum is long and wide; the beak cavity is shallow. The periostracum is yellow or yellowish green, becoming a dark chestnut brown in old shells; most individuals are patterned with dark green rays that are usually wide, often wavy, and usually interrupted or broken. Some individuals lack rays. The nacre is pearly white, and the posterior half or third is iridescent.

Life History and Ecology: The Kidneyshell appears tolerant of a variety of habitat conditions, although rivers with moderately strong current and a substrate of coarse gravel and sand provide the most suitable one. *Ptychobranchus fasciolaris* may be found at depths of less than three feet up to those as great as 18 to 24 feet in large rivers (reservoirs) such as the Tennessee and Cumberland. Ortmann (1919) records the species

Plate 97. *Ptychobranchus fasciolaris* (Rafinesque, 1820), Kidneyshell.

as bradytictic; the breeding season in Pennsylvania begins in August with the discharging of glochidia taking place in June through August. Host fish for the glochidia unknown.

Status: Currently Stable (Williams et al., 1993:14).

Ptychobranchus greeni (Conrad, 1834)
Triangular Kidneyshell

RANGE MAP 97; PLATE 98

Synonymy:
Unio greeni Conrad, 1834; Conrad, 1834a:32, pl. 4, fig. 1
Margarita (Unio) greeni (Conrad, 1834); Lea, 1836:24
Margaron (Unio) greeni (Conrad, 1834); Lea, 1852c:27
Ptychobranchus greeni (Conrad, 1834); Simpson, 1900a:614
Ptychobranchus (Ptychobranchus) greeni (Conrad, 1834); Frierson, 1927:65
Unio brumleyanus Lea, 1841; Lea, 1841b:82
Unio brumbyanus Lea, 1842; Lea, 1842b:245, pl. 25, fig. 62
Margaron (Unio) brumbyanus (Lea, 1842); Lea, 1852c:31
Pleurobema brumbyana (Lea, 1842); Simpson, 1900a:760
Unio foremanianus Lea, 1842; Lea, 1842a:224; Lea, 1842b:224, pl. 27, fig. 64
Margaron (Unio) foremanianus (Lea, 1842); Lea, 1852c:23
Ptychobranchus foremanianum (Lea, 1842); Simpson, 1900a:613
Ptychobranchus (Ptychobranchus) foremanianum (Lea, 1842); Frierson, 1927:65
Unio flavescens Lea, 1845; Lea, 1845:163; Lea, 1848:72, pl. 3, fig. 9
Margaron (Unio) flavescens (Lea, 1845); Lea, 1852c:27
Ptychobranchus flavescens (Lea, 1845); Simpson, 1900a:614
Unio simplex Lea, 1845; Lea, 1845:163; Lea, 1848:76, pl. 5, fig. 15
Margaron (Unio) simplex (Lea, 1845); Lea, 1852c:27
Unio velatus Conrad, 1854; Conrad, 1854:298, pl. 27, fig. 6
Unio woodwardius Lea, 1857; Lea, 1857f:170
Unio woodwardianus Lea, 1859; Lea, 1859f:199, pl. 23, fig. 82, pl. 29, fig. 103
Margaron (Unio) woodwardianus (Lea, 1859); Lea, 1870:36

Unio trinacrus Lea, 1861; Lea, 1861c:59; Lea, 1862b:86, pl. 12, fig. 235
Margaron (Unio) trinacrus (Lea, 1861); Lea, 1870:36
Ptychobranchus trinacrus (Lea, 1861); Simpson, 1900a:614

Type Locality: Headwaters, Black Warrior River, Alabama.

General Distribution: Alabama River drainage (Burch, 1975a).

Tennessee Distribution: Conasauga River, Polk and Bradley counties.

Description: The shell is subtriangular, rather elongated, and moderately inflated; the anterior margin is broadly rounded, the ventral margin is straight, and the posterior end is broadly pointed. Posterior ridge usually distinct, broadly rounded. Valves are fairly heavy and solid. Mature specimens from the Conasauga River, Polk and Bradley counties, Tennessee, average between 65 and 70 mm in length. Beaks are full and high in old individuals, but do not extend above the hinge line.

The two low, triangular pseudocardinal teeth in left valve and one in the right generally project downward at a 90° angle from the hinge line. The lateral teeth, two in the left valve and one in the right, are high, thin, and straight. The beak cavity is fairly shallow, and the interdentum is narrow; muscle scars are deeply impressed. The periostracum is a dull yellow, usually becoming a dull light brown with maturity. Some specimens lack rays; when present, they number from one to a few narrow, occasionally wide dark green rays situated primarily on the disc. The nacre is creamy or bluish white.

Life History and Ecology: *Ptychobranchus greeni* was a fairly common mussel in stretches of the Conasauga River flowing through Tennessee and northern Georgia.

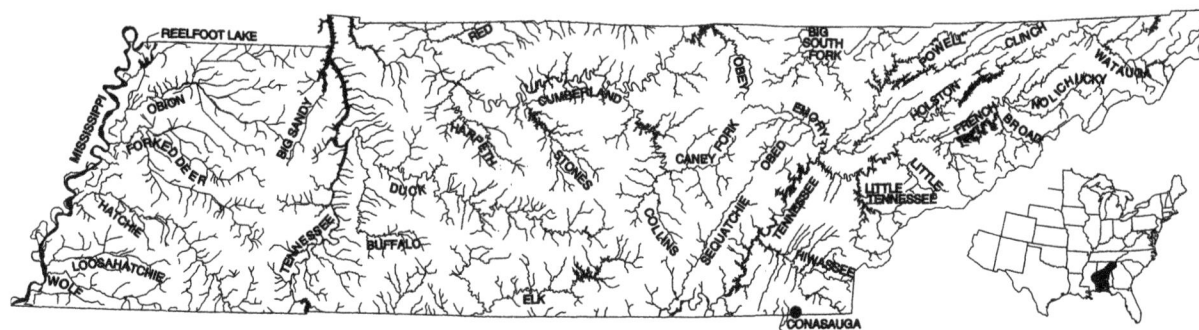

Range Map 97. *Ptychobranchus greeni* (Conrad, 1834), Triangular Kidneyshell.

Plate 98. *Ptychobranchus greeni* (Conrad, 1834), Triangular Kidneyshell.

It appears most prevalent in sections of the river less than three feet in depth and having a good current and a firm substrate composed of coarse gravel and sand. The breeding season and host fish for the glochidia are unknown. However, Haag and Warren (1996, 1997) list five species as potential hosts for the glochidia of the Triangular Kidneyshell: Warrior darter *(Etheostoma bellator)*, Tuskaloosa darter *(E. douglasi)*, redfin darter *(E. whipplei)*, blackbanded darter *(Percina nigrofasciata)*, and river darter *(P. shumardi)*.

Status: Endangered (Williams et al., 1993:14). It appears to be extirpated in the Tennessee stretch of the Conasauga River. The U.S. Fish and Wildlife Service has developed a recovery plan for this species (U.S. Fish and Wildlife Service, 1994).

Ptychobranchus subtentum (Say, 1825)
Fluted Kidneyshell

RANGE MAP 98; PLATE 99

Synonymy:
Unio subtentus Say, 1825; Say, 1825:130; Say, 1831a:pl. 15
Unio subtenta (Say, 1825); Deshayes, 1835:555
Margarita (Unio) subtentus (Say, 1825); Lea, 1836:13
Margaron (Unio) subtentus (Say, 1825); Lea, 1852c:21
Medionidus subtentus (Say, 1825); Simpson, 1900a:591
Ptychobranchus subtentus (Say, 1825); Ortmann, 1912a:308, fig. 15
Ptychobranchus subtentum (Say, 1825); Simpson, 1914:339
Ellipsaria subtenta (Say, 1825); Ortmann, 1918:564
Ptychobranchus (Subtentus) subtentum (Say, 1825); Frierson, 1927:65
Ptychobranchus subtenta (Say, 1825); Hickman, 1937:51, figs. 21a–b
Unio subteritus purcheornatus De Gregorio, 1914; De Gregorio, 1914:31, pl. 9, fig. 2

Type Locality: North Fork Holston River, Tennessee.

General Distribution: Tennessee and Cumberland River systems (Simpson, 1914).

Tennessee Distribution: The Fluted Kidneyshell is found throughout the headwater streams of the Cumberland and Tennessee rivers, including the Clinch, Powell, North Fork and South Fork Holston, Nolichucky, Big South Fork Cumberland, and Elk rivers. Now extirpated from numerous rivers where it once occurred, *Ptychobranchus subtentum* is no longer found in the main Holston and Tennessee rivers, Little Tennessee, Duck, Buffalo, Obey, Caney Fork, and Harpeth rivers (Starnes and Bogan, 1988). Prehistorically it occurred downstream in the mainstem of the Tennessee River as far as Decatur County.

Description: The shell is slightly obovate and subrhomboid in outline; the anterior end is broadly rounded, and the ventral margin is straight, although often recessed centrally in mature individuals, presenting a slightly pinched look; the hinge ligament is long. The posterior ridge is pronounced but rounded, sloping to a rounded margin; the posterior dorsal slope is radially plicate or corrugated in most individuals. Valves are solid, moderately inflated, and fairly large in adults. Very old individuals may attain a length of 120 mm, but shells of mature specimens will average somewhat less, 80–90 mm. The male shell is somewhat arcuate

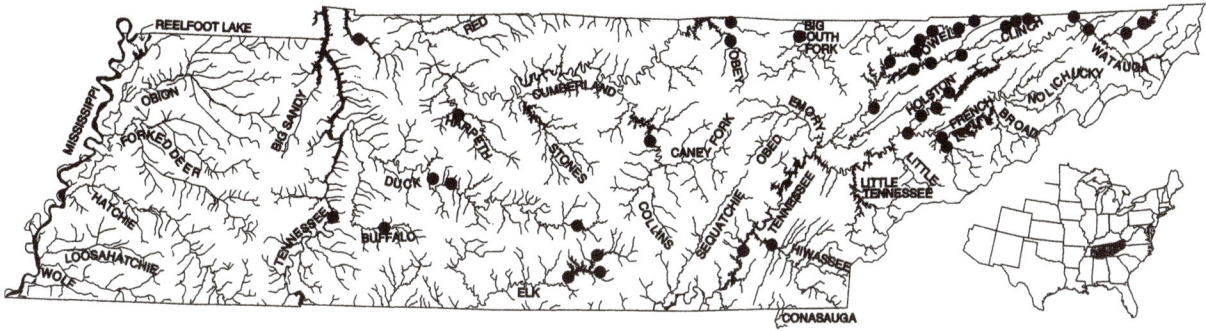

Range Map 98. *Ptychobranchus subtentum* (Say, 1825), Fluted Kidneyshell.

in older specimens and generally rounded posteriorly. The female shell has a slightly pronounced basal region with the rounded posterior point raised a little above the base line. Beaks are fairly compressed, barely extending beyond the hinge line.

The left valve has two stumpy to narrow triangular pseudocardinal teeth, a narrow curved interdentum, and two heavy lateral teeth. The right valve has one subtriangular pseudocardinal tooth and one large, wide lateral tooth. The beak cavity is shallow; muscle scars are well impressed; the pallial line is distinct and impressed anteriorly. The periostracum is smooth and subshiny in young individuals, often becoming duller and satinlike along the ventral margins in older ones. The periostracum is greenish yellow, often becoming a dull brown with age, and generally having a few to several pronounced broken, wide green rays; the rays often break into square spots or zigzag markings. The nacre color varies from bluish white, flesh color or yellowish, to dull white with a wash of salmon in the beak area.

Life History and Ecology: The Fluted Kidneyshell is primarily a stream and small river species, inhabiting a sand or sand and gravel substrate in riffles with fast current, usually at depths of two feet or less. Morrison (1942) noted that the shoal areas of big rivers (e.g., Muscle Shoals in the Tennessee River) were apparently similar enough to its typical small stream habitat that *Ptychobranchus subtentum* was able to survive in such big river situations. The species is bradytictic. Fish hosts for the glochidia, based on induced infections under laboratory conditions (Luo, 1993), include the rainbow darter *(Etheostoma caeruleum),* redline darter *(E. rufilineatum),* fantail darter *(E. flabellare),* barcheek darter *(E. obeyense),* and banded sculpin *(Cottus carolinae).*

Status: Special Concern (Williams et al., 1993:14). This mussel appears to be maintaining viable local populations in the upper Clinch River, Hancock County, Tennessee, but elsewhere in the state it has been extirpated or greatly reduced in numbers and range.

Plate 99. *Ptychobranchus subtentum* (Say, 1825), Fluted Kidneyshell.

Pyganodon grandis (Say, 1829)
Giant Floater

RANGE MAP 99; PLATE 100

Synonymy:

Anodonta fluviatilis of authors; Lapham, 1860:156 [misidentification]

Anodonta implicata Say, 1829 [of authors]; Chadwick, 1906:94 [misidentification]

Anodonta marginata Say, 1817 [of authors]; F. C. Baker, 1920:383 [misidentification]

Anodon edentulus (Say, 1829); Sowerby, 1867:pl. 17, fig. 60 [misidentification]

Anodonta grandis Say, 1829; Say, 1829:341

Margarita (Anodonta) grandis Say, 1829; Lea, 1836:52

Anodon grandis (Say, 1829); Catlow and Reeve, 1845:67

Anodonta (Pyganodon) grandis Say, 1829; Frierson, 1927:14

Anodonta (Anodonta) grandis Say, 1829; Clarke and Berg 1959:36, fig. 38

Anodonta grandis grandis Say, 1829; Oesch, 1984:40

Pyganodon grandis (Say, 1829); Hoeh, 1990:63

Anodonta lugubris Say, 1829; Say, 1829:340

Anodonta grandis lugubris Say, 1829; Strecker, 1931:10

Anodonta (Pyganodon) grandis lugubris Say, 1829; Frierson, 1927:15

Symphynota benedictensis Lea, 1834; Lea, 1834:104, pl. 16, fig. 48

Anodonta benedictensis (Lea, 1834); Férussac, 1835:25

Margarita (Anodonta) benedictensis (Lea, 1834); Lea, 1836:28

Anodon benedictensis (Lea, 1834); Catlow and Reeve, 1845:66

Margaron (Anodonta) benedictensis (Lea, 1834); Lea, 1852c:47

Margaron (Anodonta) benedictii Lea, 1870; Lea, 1870:75 [unjustified emendation]

Anodonta benedictii (Lea, 1870); Latchford, 1882:55

Anodonta grandis var. *benedictensis* (Lea, 1834); Simpson, 1900a:644

Anodonta (Pyganodon) grandis benedictensis (Lea, 1834); Frierson, 1927:15

Anodonta corpulenta Cooper, 1834; Cooper, 1834:154

Anodon corpulenta (Cooper, 1834); Sowerby, 1870:pl. 32, fig. 129

Margaron (Anodonta) corpulenta (Cooper, 1834); Lea, 1870:81

Anodonta grandis corpulenta Cooper, 1834; Starrett, 1971:319

Anodonta declivis Conrad, 1834; Conrad, 1834b:341, pl. 1, fig. 11

Anodonta gigantea Lea, 1834; Lea, 1834:1, pl. 1, fig. 1

Margarita (Anodonta) gigantea (Lea, 1834); Lea, 1836:52

Margaron (Anodonta) gigantea (Lea, 1834); Lea, 1852c:50

Anodon gigantea (Lea, 1834); Sowerby, 1867:pl. 8, fig. 18

Anodon giganteus (Lea, 1834); Sowerby, 1870:pl. 37, fig. 152

Anodonta grandis var. *gigantea* Lea, 1834; Simpson, 1900a:643

Anodonta grandis gigantea Lea, 1834; F. C. Baker, 1905:254

Anodonta palna Lea, 1834; Lea, 1834:48, pl. 7, fig. 18 [error for plana]

Margarita (Anodonta) plana (Lea, 1834); Lea, 1836:52

Anodon plana (Lea, 1834); De Kay, 1843:201, pl. 17, fig. 232

Margaron (Anodonta) plana (Lea, 1834); Lea, 1852c:50

Anodonta grandis plana (Lea, 1834); F. C. Baker, 1928a:155, pl. 60, fig. 5, pl. 61, figs. 3–6

Anodonta stewartiana Lea, 1834; Lea, 1834:47, pl. 6, fig. 17

Margarita (Anodonta) stewartiana (Lea, 1834); Lea, 1836:52

Anodon stewartiana (Lea, 1834); Catlow and Reeve, 1845:68

Anodon stewartianus (Lea, 1834); Sowerby, 1870:pl. 32, fig. 133

Margaron (Anodonta) stewartiana (Lea, 1834); Lea, 1852c:51

Margarita (Anodonta) decora Lea, 1836; Lea, 1836:52 [nomen nudum]

Anodonta decora Lea, 1838; Lea, 1838b:64, pl. 20, fig. 63

Anodon decora (Lea, 1838); Catlow and Reeve, 1845:66

Margaron (Anodonta) decora (Lea, 1838); Lea, 1852c:50

Anodonta grandis decora (Lea, 1838); Sterki, 1907:394

Margarita (Anodonta) ovata Lea, 1836; Lea, 1836:52 [nomen nudum]

Anodonta ovata Lea, 1838; Lea, 1838b:2, pl. 2, fig. 2

Anodon ovata (Lea, 1838); Catlow and Reeve, 1845:67

Margaron (Anodonta) ovata (Lea, 1838); Lea, 1852c:50

Anodonta ovata Lea, 1838; Sowerby, 1868:pl. 22, fig. 85

Anodonta pepinianus Lea, 1838; Lea, 1838b:96, pl. 16, fig. 51

Anodon pepinianus (Lea, 1838); Sowerby, 1870:pl. 36, fig. 150

Anodonta pepiniana Lea, 1838; Clessin, 1874:158, pl. 53, figs. 1, 2

Margarita (Anodonta) pepiniana (Lea, 1838); Lea, 1838a:30

Margaron (Anodonta) pepiniana (Lea, 1838); Lea, 1852c:49

Anodonta (Pyganodon) grandis pepiniana Lea, 1838; Frierson, 1927:15

Margarita (Anodonta) salmonia Lea, 1836; Lea, 1836:51 [nomen nudum]

Anodonta salmonia Lea, 1838; Lea, 1838b:45, pl. 14, fig. 41

Anodon salmonia (Lea, 1838); Catlow and Reeve, 1845:68

Margaron (Anodonta) salmonia (Lea, 1838); Lea, 1852c:50

Anodonta salmonea Lea, 1838; F. C. Baker, 1928a:155 [misspelling]

Anodonta footiana Lea, 1840; Lea, 1840:289; Lea, 1842b:225, pl. 20, fig. 44

Margaron (Anodonta) footiana (Lea, 1840); Lea, 1852c:49

Anodon footiana (Lea, 1840); Sowerby, 1867:pl. 14, fig. 48

Anodonta grandis var. *footiana* Lea, 1840; Simpson, 1900a:642

Anodonta grandis footiana Lea, 1840; Chadwick, 1906:94

Anodonta harpethensis Lea, 1840; Lea, 1840:289; Lea, 1842b:224, pl. 19 fig. 42

Margaron (Anodonta) harpethensis (Lea, 1840); Lea, 1852c:50

Anodon harpethensis (Lea, 1840); Sowerby, 1869:pl. 21, fig. 82

Anodonta maryattiana Lea, 1840; Lea, 1840:289

Anodonta maryattana Lea, 1842; Lea, 1842b:226, pl. 20, fig. 45 [emendation]

Margaron (Anodonta) maryattana (Lea, 1840); Lea, 1852c:51

Anodon marryattana (Lea, 1840); Sowerby, 1870:pl. 28, fig. 111 [misspelling]

Anodonta marryattiana (Lea, 1840); Simpson, 1900a:642 [misspelling]

Anodonta linneana Lea, 1852; Lea, 1852b:289, pl. 27, fig. 51

Margaron (Anodonta) linneana (Lea, 1852); Lea, 1852c:51

Anodon linneanus (Lea, 1852); Sowerby, 1870:pl. 35, fig. 144

Anodonta opaca Lea, 1852; Lea, 1852b:285, pl. 25, fig. 46

Margaron (Anodonta) opaca (Lea, 1852); Lea, 1852c:50

Anodonta virens Lea, 1852; Lea, 1852b:290, pl. 18, fig. 53

Margaron (Anodonta) virens (Lea, 1852); Lea, 1852c:51

Anodon virens (Lea, 1852); Sowerby, 1870:pl. 34, fig. 138

Anodonta nilssonii Küster, 1853; Küster, 1853:61, pl. 17, figs. 3, 4

Anodonta lewisii Lea, 1857; Lea, 1857c:84; Lea, 1860e:362, pl. 62, fig. 187

Anodon lewisii (Lea, 1857); Sowerby, 1870:pl. 35, fig. 142

Margaron (Anodonta) lewisii (Lea, 1857); Lea, 1870:70

Anodonta danielsii Lea, 1858; Lea, 1858b:139; Lea, 1860e:365, pl. 63, fig. 190

Margaron (Anodonta) danielsii (Lea, 1858); Lea, 1870:78

Margaritana danielsii (Lea, 1858); Paetel, 1890:173

Anodonta texasensis Lea, 1859; Lea, 1859a:113; Lea, 1860e:366, pl. 63, fig. 191

Anodon texasensis (Lea, 1859); Sowerby, 1870:pl. 36, fig. 146

Margaron (Anodonta) texasensis (Lea, 1859); Lea, 1870:81

Anodonta kennicottii Lea, 1861; Simpson, 1900a:647 [in part]

Anodonta simpsoniana Lea, 1861; Lea, 1861b:56; Lea, 1862c:212, pl. 32, fig. 281

Anodonta grandis simpsoniana Lea, 1861; Clarke, 1973:85, pl. 5, figs. 5–10, map 9

Anodonta leonensis Lea, 1862; Lea, 1862a:169; Lea, 1866:25, pl. 9, fig. 24

Margaron (Anodonta) leonensis (Lea, 1862); Lea, 1870:78

Anodonta grandis var. *leonensis* (Lea, 1862); Simpson, 1900a:643

Anodonta bealei Lea, 1863; Lea, 1863:194; Lea, 1866:26, pl. 9, fig. 25

Margaron (Anodonta) bealei (Lea, 1863); Lea, 1870:81

Anodonta (Pyganodon) grandis bealei Lea, 1863; Frierson, 1927:15

Anodonta grandis bealei Lea, 1863; Strecker, 1931:10

Anodonta bealii Lea, 1863; Johnson, 1980:92 [misspelling]

Anodonta dallasiana Lea, 1863; Lea, 1863:190; Lea, 1866:29, pl. 11, fig. 28

Margaron (Anodonta) dallasiana (Lea, 1863); Lea, 1870:78

Anodon imbricata Anthony, 1865; Anthony, 1865:159, pl. 14, fig. 1

Anodonta imbricata (Anthony, 1865); B. H. Wright, 1888b:no pagination

Anodon micans Anthony, 1865; Anthony, 1865:162, pl. 16, fig. 1

Anodonta micans (Anthony, 1865); B. H. Wright, 1888b:no pagination

Margaron (Anodonta) micans (Anthony, 1865); Lea, 1870:78

Anodon opalina Anthony, 1865; Anthony, 1865:159, pl. 14, fig. 2

Anodon subangulata Anthony, 1865; Anthony, 1865:158, pl. 13, fig. 1

Anodonta subangulata (Anthony, 1865); B. H. Wright, 1888b:no pagination

Anodon subinflata Anthony, 1865; Anthony, 1865:160, pl. 15, fig. 1

Anodonta subinflata (Anthony, 1865); B. H. Wright, 1888b:no pagination

Anodon inornata Anthony, 1866; Anthony, 1866:145

Anodon mcnielii Anthony, 1866; Anthony, 1866:144, pl. 6, fig. 1

Anodon subgibbosa Anthony, 1866; Anthony, 1866:144, pl. 6, fig. 2

Anodonta subgibbosa (Anthony, 1866); B. H. Wright, 1888b:no pagination

Anodon subgibbosus Anthony, 1866; Sowerby, 1870:pl. 28, fig. 107

Margaron (Anodonta) subglobosa (Anthony, 1866); Lea, 1870:81 [misspelling?]

Anodonta lewisii Lea, 1857; Lea, 1857c:84; Lea, 1860e:362, pl. 62, fig. 187

Anodon lewisii (Lea, 1857); Sowerby, 1870:pl. 35, fig. 142

Anodonta sulcata Küster, 1873; Küster, 1873:62, pl. 18, fig. 1

Anodonta hockingensis "Moores" Call, 1880; Call, 1880:530 [nomen nudum]

Anodonta somersii "Moores" Call, 1880; Call, 1880:530 [nomen nudum]

Anodonta houghtonensis Currier, 1868; Currier, 1868:11 [nomen nudum]

Anodonta houghtonensis Currier in DeCamp, 1881; DeCamp, 1881:14, pl. 1, fig. 2

Anodonta dakota Frierson, 1910; Frierson, 1910:113, pl. 10

Anodonta (Pyganodon) grandis dakota Frierson, 1910; Frierson, 1927:15

Anodonta dakotana Utterback, 1915; Utterback, 1915:265, pl. 24, figs. 77a, b [error for *Anodonta dakota* Frierson, 1910 ?]

Anodonta (Pyganodon) grandis dakotana Utterback, 1915; Frierson, 1927:15 [non *Anodonta dakota* Frierson, 1910]

Type Locality: Fox River of the Wabash River, Indiana.

General Distribution: This common mussel is found throughout the Mississippi and Missouri River drainages, the St. Lawrence drainage and the Canadian Interior Basin from western Ontario to Alberta (Burch, 1975a), in the Gulf of Mexico drainage area of Louisiana and Texas (Clarke, 1973), and in the Red River drainage, Texas and Oklahoma.

Tennessee Distribution: The Giant Floater may be found throughout the state in medium-sized rivers, such as the upper Powell and Clinch in East Tennessee, the Elk, Harpeth, and Duck in Middle Tennessee, and the Obion and Hatchie in West Tennessee. It occurs throughout the impounded stretches of the Tennessee and Cumberland rivers, in overflow ponds and sloughs, and in Reelfoot Lake (Hoff, 1943; Najarian, 1955).

Description: The shell is variable, usually elongated, ovate, often somewhat elliptical or rhomboid, thin to moderately solid, and inflated to swollen in the beak

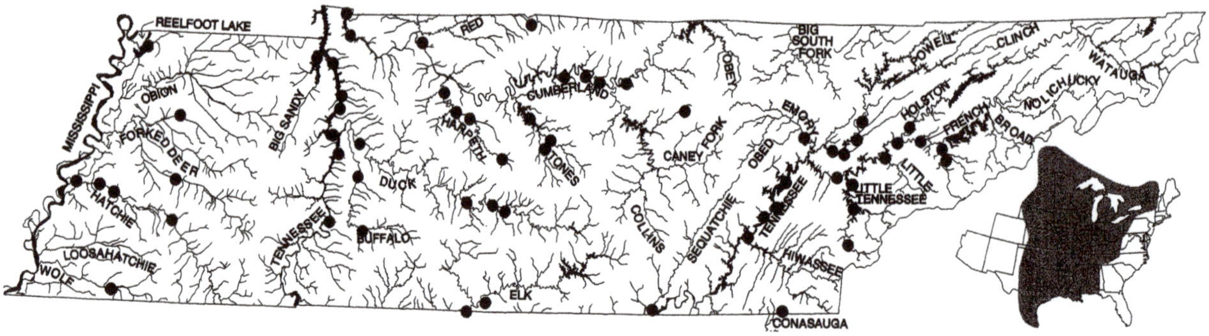

Range Map 99. *Pyganodon grandis* (Say, 1829), Giant Floater.

area. The anterior end is broadly rounded, and the posterior end is rather bluntly pointed; the dorsal margin usually forms a sharp angle with the posterior end. Beaks are swollen, typically flush with the hinge line or moderately elevated; sculpture consists of 4–5 heavy bars, the first two concentric, the rest strongly double-looped. Individuals may become extremely large under ideal habitat conditions; mature specimens may reach a length of 130–140 mm. Both valves are edentulous; the hinge line is usually slightly thickened. The beak cavity is shallow. There are pronounced fine concentric ridges on the surface indicative of rest periods. The periostracum is yellowish green and occasionally faintly rayed in young shells and dark green or brown to black in old shells, often with the umbone area a lighter ash brown. The nacre is a dull white, somewhat iridescent, and occasionally tinged or washed with cream, pink, or salmon.

Life History and Ecology: *Pyganodon grandis* does occur in Tennessee rivers possessing a substrate of sand and gravel, but it, like all closely related species within the genera *Anodonta* and *Utterbackia* that are found in the state, reaches its greatest abundance and individual size in reservoirs, lakes, and ponds having a mud bottom with little or no current. The diversity of fish hosts for the glochidia of this floater (e.g., gar, *Lepisosteus* sp.; catfish, *Ictalurus* sp.; sunfish, *Lepomis* sp.; freshwater drum, *Aplodinotus grunniens*, according to Oesch, 1984; Watters, 1994a) may well be the primary factor contributing to its extensive geographical range, abundance, and adaptability. Baker (1928a) reported the bradytictic reproductive period for *Pyganodon* (=*Anodonta*) *grandis* in Wisconsin as extending from August to April or May.

The greatly inflated shell and elevated beaks that are more or less centrally located on the dorsal margin of *Anodonta corpulenta* are characters that appear to clearly distinguish this form or subspecies from the somewhat more compressed and elongated shell of

Plate 100. *Pyganodon grandis* (Say, 1829), Giant Floater.

Anodonta g. grandis. Some researchers (e.g., van der Schalie and van der Schalie, 1950) considered these naiads to be distinct species, and, although the shells of these two subspecies appear quite different, certain local populations suggest a cline or intermediate form on the basis of shell obesity and position and elevation of the beaks. Stream gradient (current) and type of substrate may be the primary factors in influencing shell characters of these subspecies.

The Giant Floater attains a maximum size and abundance in impounded sections of large rivers such as the Cumberland and Tennessee, and in sloughs and floodplain lakes such as those associated with West Tennessee rivers. Quiet, mud-bottomed sections of rivers and lakes provide ideal habitat for this floater; it may be found at depths that vary from one foot to 20 feet or more.

Status: Currently Stable (Williams et al., 1993:14).

Quadrula apiculata (Say, 1829)
Southern Mapleleaf

RANGE MAP 100; PLATE 101

Synonymy:
Unio apiculatus Say, 1829; Say, 1829:309; Say, 1834:pl. 52
Margarita (Unio) apiculatus (Say, 1829); Lea, 1836:15
Margaron (Unio) apiculatus (Say, 1829); Lea, 1852c:22
Quadrula (Quadrula) apiculata (Say, 1829); Simpson, 1900a:778
Quadrula (Quadrula) quadrula apiculata (Say, 1829); Haas, 1969a:290
Quadrula quadrula (Rafinesque, 1820); Murray and Roy, 1968:29 [in part]
Unio nobilis Conrad, 1854; Conrad, 1854:297, pl. 27, fig. 2 [in part]

Type Locality: New Orleans, Louisiana.

General Distribution: Southern portion of the Mississippi Interior, eastern and western Gulf drainages (Vidrine, 1993). Introduced into the Tennessee River; lower Kentucky Lake, Kentucky.

Tennessee Distribution: Tennessee River (Kentucky Lake) below Pickwick Landing Dam, Hardin County, downstream locally to the Tennessee–Kentucky border.

Description: Shells of the Southern Mapleleaf, which reach an average length of 100 mm at maturity, are subrhomboid; some individuals are nearly square in outline while others are elongated posteriorly. Valves are solid; beaks are high and full. The posterior ridge is well developed, sharply angled posterior to the beaks, and becomes rounded, low, and indistinct near the posterior-ventral margin. Some individuals have a broad, shallow sulcus anterior to the posterior ridge. The anterior end is broadly rounded; the dorsal and posterior margins are straight; the ventral margin is usually slightly convex. The surface is covered with low, fine to slightly enlarged dense pustules; the posterior slope is covered with distinct, closely spaced parallel plications that angle upward from the dorsal margin to the posterior ridge.

The left valve has two heavy, erect, widely separated pseudocardinal teeth; the anterior tooth is the largest, and the space between the teeth is roughly serrated. The two lateral teeth are long and straight. The right valve has a large, triangular serrated pseudocardinal tooth, usually with a small, low, but erect tooth dorsal and anterior to it. The lateral tooth is high and long and sometimes has a short, low tooth ventral to it. Anterior muscle scars are deeply impressed; the interdentum is short or lacking; the beak cavity is moderately deep. The periostracum is greenish in young

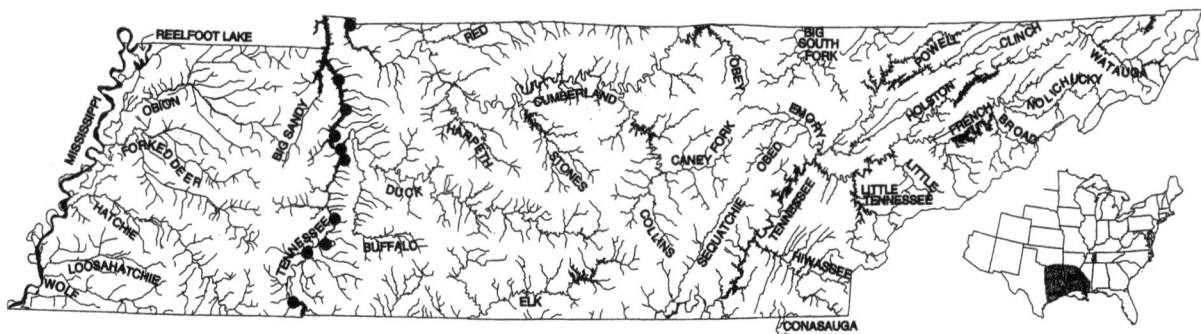

Range Map 100. *Quadrula apiculata* (Say, 1829) Southern Mapleleaf.

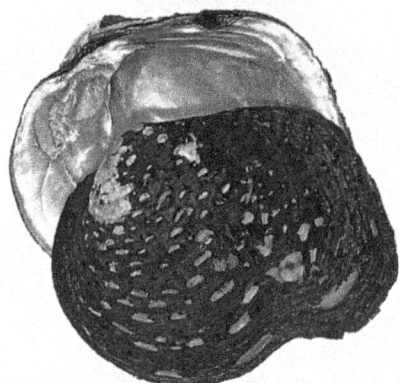

Plate 101. *Quadrula apiculata* (Say, 1829), Southern Mapleleaf.

individuals and unrayed, becoming reddish to ashy brown in old specimens. The nacre is white and iridescent behind.

Life History and Ecology: *Quadrula apiculata* appears to be primarily a big river mussel inhabiting unimpounded rivers with good current as well as lakes and reservoirs (for example, Lay Lake [Coosa River], Alabama). Neck (1986) reported it to be abundant in Lake Tawakoni, Sabine River, Texas. Vidrine (1993:53) noted that "[t]his is one of the species which is able to often coexist with *Rangia cuneata* and reside in the southern (coastal) reaches of Louisiana streams." Since its apparent intentional introduction into the Tennessee River (Kentucky Lake) in the 1980s (pers. comm., S. A. Ahlstedt, 1994), it has become well established locally, where it is taken by commercial shellers from a coarse sand and gravel substrate at depths of 10 to 15 feet or more.

Like other species of *Quadrula* to which *Q. apiculata* is most closely related (e.g., *Q. quadrula*), the Southern Mapleleaf is probably tachytictic, being gravid during the summer months. Although unreported for *Q. apiculata,* one or more species of catfish (Ictaluridae) and/or bass and sunfish (Centrarchidae) may well serve as hosts for the glochidia. Representatives of these two families of fishes have been recorded as the principal hosts for other taxa in the genus *Quadrula.*

Status: Currently Stable (Williams et al., 1993:14).

Quadrula cylindrica (Say, 1817)
Rabbitsfoot

RANGE MAP 101; PLATE 102

Synonymy:
Unio cylindricus Say, 1817; Say, 1817:pl. 4, fig. 3
Unio (Eurynia) solenoides var. *cylindrica* Say, 1817;
    Rafinesque, 1820:298
Mya cylindricus (Say, 1817); Eaton, 1826:219
Margarita (Unio) cylindricus (Say, 1817); Lea, 1836:17
Unio (Theliderma) cylindrica (Say, 1817); Swainson, 1840:271,
    fig. 54c
Margaron (Unio) cylindricus (Say, 1817); Lea, 1852c:23
Orthonymus cylindricus (Say, 1817); Agassiz, 1852:48
Quadrula cylindrica (Say, 1817); Lewis, 1870:218
Quadrula (Quadrula) cylindrica (Say, 1817); Simpson,
    1900a:773
Quadrula cylindrica cylindrica (Say, 1817); Oesch, 1984:91
Unio naviformis Lamarck, 1819; Lamarck, 1819:75
Unio rugosus (Barnes, 1823); Chenu, 1859:138 [misidentification]
Unio cylindricus var. *strigillatus* B. H. Wright, 1898; B. H.
    Wright, 1898c:6; Johnson, 1967b:9, pl. 3, fig. 2
Quadrula (Orthonymus) cylindrica strigillata (Wright, 1898);
    Frierson, 1927:51
Quadrula cylindrica strigillata (Wright, 1898); Goodrich,
    1913:93

Type Locality: Wabash River.

General Distribution: Entire Ohio, Cumberland, and Tennessee River systems; south to Arkansas and Oklahoma (Simpson, 1914). Lower Mississippi River drainage: Louisiana north to Missouri, west to Kansas.

Tennessee Distribution: Two subspecies of the Rabbitsfoot occur in Tennessee: *Quadrula cylindrica cylindrica* (Rabbitsfoot) and *Q. c. strigillata* (Rough Rabbitsfoot). The Rough Rabbitsfoot is found primarily in upper East Tennessee in the headwaters of the Tennessee River. It

Range Map 101. *Quadrula cylindrica* (Say, 1817), Rabbitsfoot.

occurs in the Powell River, Claiborne, and Hancock counties; North Fork Holston River, Hawkins County and the Clinch River, Hancock County. This form apparently intergraded with typical *Q. c. cylindrica* in the lower sections of these rivers prior to impoundment (Ortmann, 1918). The species has also been found in the Elk River, Lincoln County (Ahlstedt, 1983). Specimens from the Duck River (Marshall County) and East Fork Stones River (Rutherford County) in Middle Tennessee, and the Paint Rock River, Alabama, appear tuberculate like typical *Q. c. strigillata,* but are somewhat more compressed. Those found in the Tennessee River (Kentucky Lake) below Pickwick Landing Dam, Hardin County, downstream are *Q. c. cylindrica.*

Description: The shell is elongate, being rhomboid or rectangular in outline. Valves are thick and solid, ranging from compressed when young to inflated as adults and becoming almost cylindrical in cross section. Large mature individuals may reach a length of 120 mm. Beaks are moderately elevated; sculpture consists of a few irregular strong ridges or coarse folds, ending or continuing on the umbonal area as small tubercles. The posterior ridge is full and rounded, extending diagonally from the umbo to the posterior ventral margin, while above there is a wide radial impression that may end in a slight sinus behind. In *Q. c. strigillata* the posterior ridge is low and less distinct. The anterior end of the shell is rounded, while the posterior end is usually squared or obliquely truncated. The surface of the shell is marked by raised annual growth lines and possesses a row of knobs extending down the posterior ridge. In the Rough Rabbitsfoot, the shell surface is more or less covered with tear-shaped nodules and plications. Many specimens of *Quadrula c. strigillata* are devoid

of large tubercles, but are thickly covered with small ones. *Quadrula c. cylindrica,* on the other hand, is typically devoid of tubercles except occasionally for a few large, rounded, low ones on the posterior slope. The dorsal area behind the beaks is compressed and covered with several to many broadly rounded ridges curving up toward the dorsal margin.

The left valve has two low, triangular, radially split pseudocardinal teeth and two long, straight lateral teeth. The right valve has a single low triangular, deeply

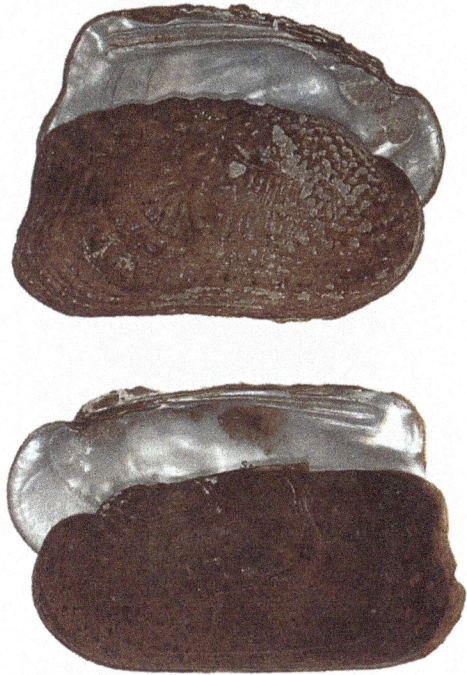

Plate 102. *Quadrula cylindrica* (Say, 1817), Rabbitsfoot.

serrated pseudocardinal tooth, often with a smaller elongated tubercular tooth on either side; the single lateral tooth is long and straight. Anterior muscle scars and the pallial line are deeply impressed. The interdentum is either narrow or absent; the beak cavity is moderately deep. The periostracum is straw-colored, light yellowish or greenish, becoming darker and yellowish-brown in old shells; numerous dark green streaks, chevrons, or triangular spots, with the point directed ventrally, appear over most of the surface. The nacre is white and iridescent. The shell is much thinner posteriorly. Wright (1898c) noted that the nacre in *Q. c. strigillata* is often pink. The beak cavity and often much of the area within the pallial line may be a shade of blue-gray.

Life History and Ecology: The Rough Rabbitsfoot is found in small to medium-sized rivers such as the upper Clinch and Powell, living in clear, shallow water in a sand and gravel substrate. Shoal and riffle areas near the banks seem to provide the most suitable habitat. In the case of *Q. c. cylindrica,* individuals are encountered most frequently in current at depths of 9 to 12 feet. The genus *Quadrula* is tachytictic (Heard and Guckert, 1970), so the brooding period of the Rabbitsfoot may be inferred to be from May to July. Yeager and Neves (1986) identified three species of fish that serve as hosts for the glochidia of the subspecies *Quadrula cylindrica strigillata:* whitetail shiner *(Cyprinella galactura),* spotfin shiner *(C. spiloptera),* and bigeye chub *(Hybopsis amblops).*

Status: Williams et al. (1993:14) list *Quadrula cylindrica cylindrica* as Threatened and *Q. c. strigillata* as Endangered.

Quadrula fragosa (Conrad, 1835)
Winged Mapleleaf

RANGE MAP 102; PLATE 103

Synonymy:
Unio fragosus Conrad, 1835; Conrad, 1835b:12, pl. 6, fig. 2
Margarita (Unio) fragosus (Conrad, 1835); Lea, 1836:14
Unio tragosus Conrad, 1835; Hanley, 1843:178, pl. 20, fig. 40 [misspelling]
Unio fragosa Conrad, 1835; Catlow and Reeve, 1845:59
Margaron (Unio) fragosus (Conrad, 1835); Lea, 1852c:22
Quadrula (Quadrula) fragosa (Conrad, 1835); Simpson, 1900a:778

Quadrula quadrula fragosa (Conrad, 1835); Ortmann, 1925:335
Quadrula (Quadrula) quadrula fragosa (Conrad, 1835); Frierson, 1927:47

Type Locality: Scioto River, Ohio.

General Distribution: Ohio, Cumberland, and Tennessee River systems; westward probably to Minnesota, Nebraska and Kansas (Simpson, 1914). Murray and Leonard (1962:53) conclude that *"Q. fragosa* as reported from Kansas is actually *Q. quadrula."* Cummings and Mayer (1992:28) state that the "only known extant population in the Midwest occurs in Wisconsin."

Tennessee Distribution: Starnes and Bogan (1988) list *Quadrula fragosa* for the Duck and Harpeth rivers prior to 1960, and from the Cumberland and lower Tennessee rivers. A relic right valve (105 mm in length) was collected from the Harpeth River near Pegram, Davidson County, August 1988.

Description: Similar to *Quadrula quadrula,* shells of the Winged Mapleleaf are squarish in outline, solid, and moderately inflated. Maximum shell length is about 120 mm. The most distinguishing character that separates these two species is the pronounced wing or expanded posterior slope, posterior to the beak, in *Q. fragosa.* Beaks are high, full, and turned forward over a well-developed lunule. The anterior end is rounded, and the posterior end is almost squarely truncated. The posterior ridge is single above, double below, with a radical depression or sulcus in front of and behind it (Simpson, 1914). There are two rows of several large, raised tubercles, separated by a wide sulcus, that radiate from the beak to the ventral margin. Small, scattered pustules may be present on the beaks and posterior wing.

The left valve has two strong, triangular, serrated pseudocardinal teeth; the two lateral teeth are short, erect, and roughened. The right valve has a single but somewhat divided heavy pseudocardinal tooth; the lateral tooth is broad and finely serrated. The beak cavity is fairly deep and open. The periostracum varies from yellowish green to light brown, becoming dark brown in mature individuals. Some have a few light green rays. The nacre is pearly white and iridescent posteriorly.

Life History and Ecology: *Quadrula fragosa* is a species characteristic of medium-sized rivers, such as the

Range Map 102. *Quadrula fragosa* (Conrad, 1835), Winged Mapleleaf.

Duck and Harpeth, and large rivers like the Cumberland and Tennessee. Possibly because of its similarity in shell characteristics to *Q. quadrula,* the Winged Mapleleaf may have gone unrecognized in recent decades. Cummings and Mayer (1992) noted that it occurs in mud, sand, or gravel. The species is probably tachytictic, gravid during summer months. Fish host for the glochidia unknown.

Plate 103. *Quadrula fragosa* (Conrad, 1835), Winged Mapleleaf.

Status: Endangered (Williams et al., 1993:14). Possibly *Quadrula fragosa* still occurs in the lower Tennessee and Cumberland rivers, but elsewhere in the state it is apparently extirpated.

Quadrula intermedia (Conrad, 1836) Cumberland Monkeyface

RANGE MAP 103; PLATE 104

Synonymy:
Unio intermedius Conrad, 1836; Conrad, 1836:63, pl. 35, fig. 2
Margaron (Unio) intermedius (Conrad, 1836); Lea, 1852c:22
Quadrula (Quadrula) intermedia (Conrad, 1836); Simpson, 1900a:775
Orthonymus intermedius (Conrad, 1836); Haas, 1969a:310
Unio kleinianus (Lea, 1852); Küster, 1861:191 [misidentification]
Unio tuberosus perlobatus De Gregorio, 1914; De Gregorio, 1914:9

Type Locality: Nolichucky River, Tennessee.

General Distribution: Tennessee River system, from extreme southwestern Virginia (Powell River, Lee County: Neves, 1991) to Muscle Shoals, Alabama (prehistorically).

Tennessee Distribution: Ortmann (1918) collected *Quadrula intermedia* from the Holston River, Hawkins County, and Stansbery (1973) recorded it from the Clinch River. The authors obtained specimens from a well-established population in the Powell River, Claiborne, and Hancock counties, and have recovered archaeological specimens from the lower Clinch River, Roane County, and from the Tennessee River, Rhea and Meigs counties, and downstream to Perry and Decatur counties in Middle Tennessee. The Cumberland Monkeyface has been found in the Elk River, Lincoln and

Family Unionidae 213

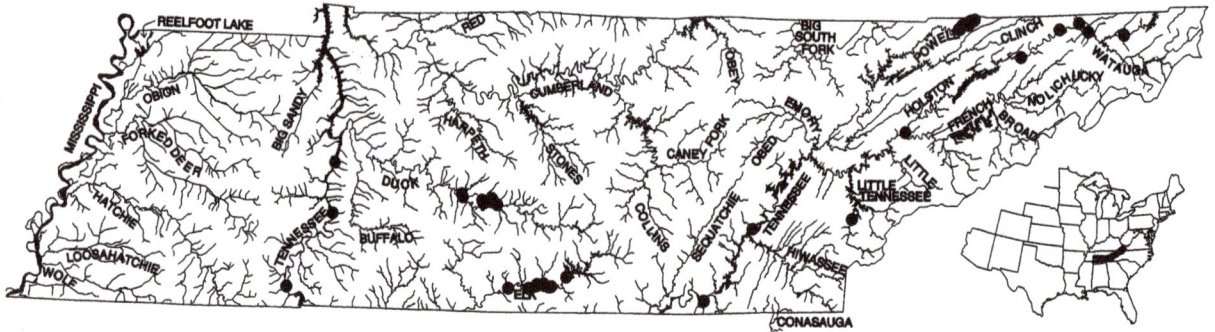

Range Map 103. *Quadrula intermedia* (Conrad, 1836), Cumberland Monkeyface.

Franklin counties, and in the Duck River, Maury County (Ortmann, 1924a; 1925; Isom and Yokley, 1968a; Isom et al., 1973; van der Schalie, 1973; Stansbery, 1976g; Ahlstedt, 1983; 1991b).

Description: Shells of the Cumberland Monkeyface are solid, only slightly inflated, and subelliptical, suborbicular, or subquadrate in outline. Maximum length of mature individuals is about 75–80 mm. Beaks are moderately high but not full. The anterior end and ventral margin of the shell are rounded; the posterior-dorsal surface is rounded but interrupted by a sinus formed by the radial depression. The posterior ridge is only slightly elevated above the outline of the shell anteriorly, but in females it is marked by a deep, wide, radial depression along the posterior-dorsal margin. This depression ends in a deep, rounded sinus. The surface is usually well covered with large, elevated warts or tubercles, except on the anterior third of the shell.

The right valve has one large radially striate pseudocardinal tooth, typically with a smaller low tooth or raised serrated surface on either side. The lateral tooth is low, broad, and roughened. The beak cavity is deep and compressed; the interdentum is wide. The left valve has two blunt, erect radially striate pseudocardinal teeth and two short, slightly curved lateral teeth. The inner lateral tooth is usually wider, more serrated, and shorter than the outer (dorsal) lateral. Mussel scars are deeply impressed. The periostracum is greenish yellow, darkening with age in some individuals, and variegated with fine, angular green spots, chevrons, or zigzags, sometimes with broken green rays. The nacre is white and often tinted or shaded with salmon, usually in posterior areas of the valve.

Life History and Ecology: Based on valves recovered from aboriginal shell middens, *Quadrula intermedia* inhabited the shoals and riffles of the Tennessee River from the confluence of the French Broad and Holston rivers at Knoxville, Tennessee, to Muscle Shoals, Alabama, and downstream to at least TRM 120 (Perry

Plate 104. *Quadrula intermedia* (Conrad, 1836), Cumberland Monkeyface.

County). Individuals in remaining populations, primarily in the upper Powell River, may be found living in a substrate of coarse sand and gravel, typically in current at a depth of less than two feet.

Like other species of *Quadrula,* the Cumberland Monkeyface has been shown to be tachytictic, spawning in spring and becoming gravid in May and June. Females lay on the substrate surface when spawning. Laboratory-induced infestations of glochidia identified the streamline chub *(Erimystax dissimilis)* and the blotched chub *(E. insignis)* as possible natural fish hosts (Tennessee Valley Authority, 1986; Neves, 1991; Yeager and Saylor, 1995).

Status: Endangered (Williams et al., 1993:14). Stansbery (1970) noted that there were scattered populations in the Powell, Clinch, and Duck rivers and, until 1980, a population in the Elk River (Lincoln County), Tennessee (Ahlstedt, 1983). Although the authors found only relic specimens (in 1978) of *Quadrula intermedia* in the Duck River, Ahlstedt (1991b) collected six live specimens from four sites on the Duck River, Maury County, in 1979. Ahlstedt (1991a) failed to find *Q. intermedia* during his mussel survey of the Clinch River, but noted that R. Neves collected a fresh dead specimen in 1983 at Pendleton Island, Virginia. Except for a possible small population in the Duck River (Marshall County) and Elk River (Lincoln County), the last remaining local populations of the Cumberland Monkeyface are to be found in the upper Powell River. The U.S. Fish and Wildlife Service has developed a recovery plan for this species (U.S. Fish and Wildlife Service, 1982) and has created a watershed implementation schedule for the recovery plan (U.S. Fish and Wildlife Service, 1989b).

Quadrula metanevra (Rafinesque, 1820) Monkeyface

RANGE MAP 104; PLATE 105

Synonymy:

Obliquaria (Quadrula) metanevra Rafinesque, 1820; Rafinesque, 1820:305, pl. 81, figs. 15, 16

Unio metanever (Rafinesque, 1820); Short and Eaton, 1831:76

Unio metaneverus (Rafinesque, 1820); Say, 1834:no pagination

Margarita (Unio) metanever (Rafinesque, 1820); Lea, 1836:15

Unio (Theliderma) metanevra (Rafinesque, 1820); Swainson, 1840:268, figs. 50, 54b

Unio meteniver (Rafinesque, 1820); Catlow and Reeve, 1845:61 [misspelling]

Unio metaneurus (Rafinesque, 1820); Küster, 1852:50, pl. 10, fig. 4 [misspelling]

Margaron (Unio) metanevrus (Rafinesque, 1820); Lea, 1852c:22

Quadrula (Quadrula) metanevra (Rafinesque, 1820); Simpson, 1900a:774

Quadrula metanevra (Rafinesque, 1820); Ortmann, 1912a:255, figs. 6, 6a

Quadrula (Orthonymus) metanevra (Rafinesque, 1820); Frierson, 1927:51

Orthonymus metanevrus metanevrus (Rafinesque, 1820); Haas, 1969a:309

Unio nodosus Barnes, 1823; Barnes, 1823:124, pl. 6, figs. 7, 7a, 7b

Mya nodosa (Barnes, 1823); Eaton, 1826:216

Unio tuberosus Lea, 1840; Lea, 1840:286; Lea, 1842b:210, pl. 14, fig. 25

Margaron (Unio) tuberosus (Lea, 1840); Lea, 1852c:22

Quadrula (Quadrula) tuberosa (Lea, 1840); Simpson, 1900a:774

Quadrula (Orthonymus) metanevra tuberosa (Lea, 1840); Frierson, 1927:52 [in part]

Orthonymus metanevrus tuberosus (Lea, 1840); Haas, 1969a:310

Unio wardii Lea, 1861; Lea, 1861d:392; Lea, 1862c:187, pl. 24, fig. 257

Margaron (Unio) wardii (Lea, 1861); Lea, 1870:33

Quadrula (Quadrula) metanevra var. *wardii* (Lea, 1861); Simpson, 1900a:774

Quadrula (Orthonymus) metanevra wardii (Lea, 1861); Frierson, 1927:52

Quadrula metanevra wardii (Lea, 1861); La Rocque, 1967:136

Orthonymus metanevrus wardii (Lea, 1861); Haas, 1969a:310

Comment: In our view there is some question as to the taxonomic status of the Rough Rockshell, *Quadrula tuberosa* (Lea, 1840), a species described from specimens collected in the Caney Fork and Cumberland rivers, Tennessee (type localities). Simpson (1914:836), in commenting on differences between it and *Quadrula metanevra,* stated that *Q. tuberosa* was "[g]enerally a little shorter in proportion than *metanevra,* not knobbed on the posterior ridge, smaller, less brightly painted, and having a more decided projection of the posterior ridge. The lateral of the right valve is more completely double than it is in *metanevra.*" Further, in discussing *Quadrula sparsa* (which he considered a variety of *Q. tuberosa*), Simpson (1914:837) made the following comment: "After critically comparing large series of this [*Q. sparsa*] and *tuberosa,* it does not seem to me that it can take any higher rank than that of a variety of the latter. It is smaller and generally a little less nodulous, but the coloring, texture, form and general characters are practically the same. I am very strongly inclined to believe that both should be considered mere varieties of *metanevra.*"

Although some malacologists consider *Q. tuberosa* a distinct or valid species, we are of the opinion at this time that it is a variety or form (ecophenotype) of the Monkeyface (*Quadrula metanevra*) and have listed *Q. tuberosa* as a synonym of it.

Type Locality: Kentucky River.

General Distribution: Upper Mississippi River drainage south to the Tennessee and Arkansas rivers (Simpson, 1914). Also found in the Cumberland River (Tennessee and Kentucky) and Tombigbee River (Alabama and Mississippi systems). Cahaba River, Alabama. Northern Louisiana (Vidrine, 1993).

Tennessee Distribution: The Monkeyface is present throughout major stretches of the main Tennessee and Cumberland rivers. It is fairly common in the middle Cumberland River (Smith and Trousdale counties) and in the Tennessee River from below Pickwick Landing Dam (Hardin County, Tennessee) downstream to lower Kentucky Lake, and below Watts Bar Dam (Rhea/Meigs counties) (Ahlstedt and McDonough, 1994). Starnes and Bogan (1988) recorded it from the Holston and Obey rivers prior to 1960 and from the Little Tennessee, Elk, and Clinch rivers since that date. In the Tennessee stretch of the Little Tennessee River, however, *Quadrula metanevra* has been extirpated as a result of completion of the Tellico Dam in 1979, and Ahlstedt (1991a) noted that it is a rare species in the Clinch River, found only in the lower reaches.

Description: Shells of the Monkeyface are thick, solid, inflated, and rather squared or rhomboid in outline. Mature individuals attain a maximum length of about 110 mm. Beaks are slightly to moderately elevated; sculpture consists of a few short, pronounced ridges, becoming nodulous on the posterior part. The anterior end of the shell is broadly rounded; the posterior end is squared or sharply truncated with a sinus above the prominent posterior ridge. The dorsal area behind the umbones is flattened and appears alate. The wide, elevated posterior ridge extends diagonally from the umbones toward the posterior ventral margin and typically is characterized by several large, high, elongated knobs. The remaining surface, except the lower anterior third, is covered with numerous rounded tubercles. Several specimens taken from the Cumberland River, Smith and Trousdale counties, however, are only moderately inflated, possess only a few low pustules, and lack the characteristic large knobs on the posterior ridge. They resemble *Quadrula sparsa,* which some consider an ecomorph of *Q. metanevra* (e.g., Bates and Dennis, 1978), more than the typical Monkeyface.

The left valve has two heavy, triangular, divergent pseudocardinal teeth, often so deeply serrated or sulcated as to give the appearance of three pseudocardinals; the two lateral teeth are short, wide, heavy, straight, and finely striated. The right valve has a large triangular, serrated pseudocardinal tooth, sometimes with a large tubercular tooth in front (giving the appearance of two large pseudocardinals) and a smaller one behind on the wide interdentum; the lateral tooth is short, wide, straight, and finely striated, with an occasional partial lateral tooth evident on the inner side. The interdentum is wide and flat; the beak cavity is compressed and deep. The periostracum varies from a plain straw yellow to a yellowish green or brown with numerous rays and triangular or chevron-shaped markings of dark green, more common on the umbone area. Some individuals are rayless. The nacre is white and iridescent posteriorly.

Life History and Ecology: In Tennessee the Monkeyface is found in medium-sized rivers such as the Elk as well as in the Tennessee and Cumberland rivers (reservoirs). In Missouri, Oesch (1984:96) noted that "[t]his is a typical riffle species, preferring clear water, a swift current and a gravel stream bottom." In Tennessee, however, *Quadrula metanevra* has adapted well to im-

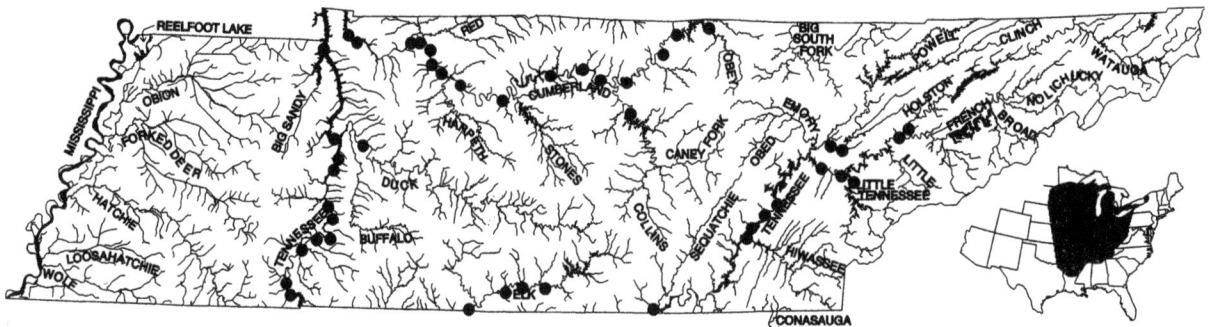

Range Map 104. *Quadrula metanevra* (Rafinesque, 1820), Monkeyface.

Plate 105. *Quadrula metanevra* (Rafinesque, 1820), Monkeyface.

poundment where it occurs in the Cumberland and Tennessee rivers (reservoirs) at depths of 12 to 15 feet, in current, and in a substrate of coarse sand and gravel.

The species is tachytictic, being gravid from May to July. Based on the work of Surber (1913), Coker et al. (1921), and others, Fuller (1974) recorded the green sunfish *(Lepomis cyanellus)*, bluegill *(L. macrochirus)*, and sauger *(Stizostedion canadense)* as host fish for the glochidia.

Status: Currently Stable (Williams et al., 1993:10). Viable populations of the Monkeyface appear to be limited to sections of the Tennessee and Cumberland rivers (reservoirs) in Tennessee.

Quadrula nodulata (Rafinesque, 1820)
Wartyback

RANGE MAP 105; PLATE 106

Synonymy:

Obovaria (Quadrula) nodulata Rafinesque, 1820; Rafinesque, 1820:307

Unio nodulatus (Rafinesque, 1820); Say, 1834:no pagination

Quadrula nodulata (Rafinesque, 1820); Utterback, 1915:139

Quadrula (Bullata) nodulata (Rafinesque, 1820); Frierson, 1927:49

Quadrula (Pustulosa) nodulata (Rafinesque, 1820); Haas, 1969a:295

Unio pustulatus Lea, 1834; Lea, 1834:79, pl. 7, fig. 9

Margarita (Unio) pustulatus (Lea, 1834); Lea, 1836:15

Margaron (Unio) pustulatus (Lea, 1834); Lea, 1852c:22

Unio nodulatus var. pustulatus (Lea, 1834); Paetel, 1890:161

Quadrula (Quadrula) pustulata (Lea, 1834); Simpson, 1900a:781

Quadrula pustulata (Lea, 1834); F. C. Baker, 1905:256

Type Locality: Kentucky River.

General Distribution: The entire Ohio, Cumberland, and Tennessee River systems; Mississippi River drainage from southeastern Minnesota south to Louisiana, west to southeastern Kansas, and northeastern Texas (Murray and Leonard, 1962).

Tennessee Distribution: The Wartyback is known to occur only in the "lower" Tennessee River (Starnes and Bogan, 1988) and in the Hatchie River, Lauderdale County (Manning, 1989).

Description: The shell of *Quadrula nodulata*, which seldom exceeds 70 mm in length, is circular to somewhat quadrate in outline, solid, often thick, and moderately inflated. Beaks are elevated, moderately swollen, and turned forward; sculpture consists of only 1–2 coarse bars or ridges. These continue down the surface of the shell as two diverging, more or less straight rows of widely separated, prominent elongated tubercles; one row is on the angled posterior ridge. Typically there are no more than 5–6 pronounced tubercles in each row. Some individuals develop a few small pustules elsewhere on the shell, most often on the posterior slope. The anterior end is rounded; the posterior end is squared or truncate. Anterior muscle scars and the pallial line are distinct but not deeply impressed.

The left valve has two high, pyramidal, divergent, roughened pseudocardinal teeth; the two lateral teeth

Family Unionidae 217

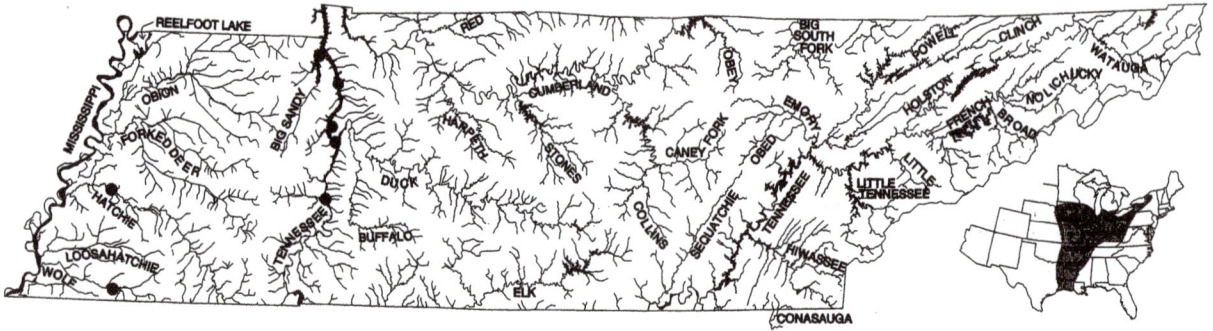

Range Map 105. *Quadrula nodulata* (Rafinesque, 1820), Wartyback.

are short, wide, and straight. The right valve has a large, triangular pseudocardinal tooth that is deeply serrated, often with a small tubercular tooth on either side; the lateral tooth is wide, heavy, and finely striated and occasionally has a small, incomplete inner lateral. The interdentum is moderately wide; the beak cavity

Plate 106. *Quadrula nodulata* (Rafinesque, 1820), Wartyback.

is fairly deep but open. The periostracum is yellowish green to light brown; it is rarely faintly rayed, often shiny, but roughened with pronounced concentric rest lines. The nacre is white and iridescent posteriorly.

Life History and Ecology: In Tennessee the Wartyback appears locally common in stretches of the lower Tennessee River (Kentucky Lake) where it reportedly occurs at depths of 15–18 feet on a sand and mud substrate. Its presence in the Hatchie River is based on a single specimen, now in the McClung Museum collections, that Manning (1989:14) classified as relict: "R-found at only one station or represented by only one or two specimens." The occurrence of *Quadrula nodulata* in the Hatchie River may well be fortuitous.

A variety of fish are known to serve as hosts for the glochidia of *Q. nodulata*: channel catfish *(Ictalurus punctatus)*, flathead catfish *(Pylodictis olivaris)*, bluegill *(Lepomis macrochirus)*, largemouth bass *(Micropterus salmoides)*, white crappie *(Pomoxis annularis)*, and black crappie *(P. nigromaculatus)* (Fuller, 1978). The Wartyback is tachytictic, being gravid in June and July.

Status: Currently Stable (Williams et al., 1993:14).

Quadrula pustulosa (Lea, 1831)
Pimpleback

RANGE MAP 106; PLATE 107

Synonymy:

Obliquaria bullata Rafinesque, 1820; Rafinesque, 1820:307

Comment: Vanatta (1915:557) recognized this species as the same as *Quadrula pustulosa pernodosa* (Lea, 1845) and would have priority over *Unio pustulosus* Lea, 1831. We have chosen to use the long-established name *pustulosa* Lea, 1831.

Unio bullatus (Rafinesque, 1820); Conrad, 1834a:68
Quadrula bullata (Rafinesque, 1820); Morrison, 1969:24
Unio pustulosus Lea, 1831; Lea, 1831:76, pl. 7, fig. 7
Margarita (Unio) pustulosus (Lea, 1831); Lea, 1836:15
Margaron (Unio) pustulosus (Lea, 1831); Lea, 1852c:22
Quadrula pustulosa (Lea, 1831); Baker, 1898:86, pl. 25, fig. 2, pl. 28, fig. 13
Quadrula (Quadrula) pustulosa (Lea, 1831); Simpson, 1900a:779
Quadrula (Bullata) pustulosa (Lea, 1831); Frierson, 1927:48
Quadrula (Pustulosa) pustulosa pustulosa (Lea, 1831); Haas, 1969a:292
Unio verrucosa Barnes, 1823; Valenciennes, 1827:231, pl. 53, fig. 2 [misidentification]
Unio nodulosus Say, 1834; Say, 1834:no pagination
Unio prasinus Conrad, 1834; Conrad, 1834a:44, 71, pl. 3, fig. 1
Unio bullatus var. *prasinus* Conrad, 1834; Paetel, 1890:146
Quadrula pustulosa prasina (Conrad, 1834); Ortmann and Walker, 1922:16
Quadrula (Bullata) pustulosa prasina (Conrad, 1834); Frierson, 1927:48
Quadrula (Pustulosa) pustulosa prasina (Conrad, 1834); Haas, 1969a:292
Unio schoolcraftensis Lea, 1834; Lea, 1834:37, pl. 3, fig. 9
Margarita (Unio) schoolcraftensis (Lea, 1834); Lea, 1836:15
Margaron (Unio) schoolcraftensis (Lea, 1834); Lea, 1852c:22
Unio bullatus var. *schoolcraftensis* (Lea, 1834); Paetel, 1890:146
Quadrula pustulosa var. *schoolcraftensis* (Lea, 1834); Simpson, 1914:850
Quadrula pustulosa schoolcraftensis (Lea, 1834); Utterback, 1915:132
Margaron (Unio) schoolcraftii Lea, 1870; Lea, 1870:33 [unjustified emendation]
Unio schoolcraftii (Lea, 1870); B. H. Wright, 1888b:no pagination
Unio mortoni Conrad, 1836; Conrad, 1836:11, pl. 6, fig. 1
Quadrula (Quadrula) mortoni (Conrad, 1836); Simpson, 1900a:781
Quadrula pustulosa mortoni (Conrad, 1836); Ortmann, 1926a:88
Quadrula (Bullata) pustulosa mortoni (Conrad, 1836); Frierson, 1927, 49
Quadrula (Pustulosa) pustulosa mortoni (Conrad, 1836); Haas, 1969a:293
Margarita (Unio) turgidus Lea, 1836; Lea, 1836:16 [nomen nudum]
Unio turgidus Lea, 1838; Lea, 1838b:11, pl. 5, fig. 11
Margaron (Unio) turgidus (Lea, 1838); Lea, 1852c:22
Unio dorfeuillianus Lea, 1838; Lea, 1838b:73, pl. 17, fig. 54
Margarita (Unio) dorfeuillianus (Lea, 1838); Lea, 1838a:15
Margaron (Unio) dorfeuillianus (Lea, 1838); Lea, 1852c:22
Unio pernodosus Lea, 1845; Lea, 1845:163; Lea, 1848:71, pl. 3, fig 8
Margaron (Unio) pernodosus (Lea, 1845); Lea, 1870:34
Quadrula (Quadrula) pustulosa var. *pernodosa* (Lea, 1845); Simpson, 1900a:780

Quadrula pustulosa pernodosa (Lea, 1845); Vanatta, 1915:557
Quadrula (Pustulosa) pustulosa pernodosa (Lea, 1845); Haas, 1969a:293
Quadrula pustulosa asperatus (Lea, 1861); Utterback, 1915:134, pl. 17, figs. 42a, 42b [misidentification]
Quadrula (Bullata) pustulosa asperata (Lea, 1861); Frierson, 1927:49 [in part]

Type Locality: Ohio; Alabama River.

General Distribution: The entire Mississippi River drainage, from New York and Pennsylvania west to the Dakotas, and south to eastern Texas and Louisiana. Tombigbee and Cahaba rivers, Alabama.

Tennessee Distribution: The Pimpleback occurs in the majority of medium-sized to large rivers throughout the state, from the Hatchie River in West Tennessee to the upper Clinch, Powell, Holston, and Nolichucky rivers in East Tennessee. Starnes and Bogan (1988) list it as formerly (prior to 1960) occurring in Reelfoot Lake and the Loosahatchie, Harpeth, and French Broad rivers.

Description: Shells of this common mussel, which seldom exceed 80 mm in length, are rounded to somewhat quadrate, solid, and moderately to greatly inflated. Beaks are high, full, and turned forward; sculpture consists of 3–4 coarse ridges. The anterior end is rounded, while the posterior end is squarish or sharply truncated. The posterior ridge is prominent and rounded. The posterior two-thirds of the surface is usually densely covered with rounded tubercles or pustules, occasionally with a few low, narrow ridges on the posterior slope. Some individuals may be totally lacking in pustules or possess only a very few.

The pseudocardinal teeth, two in the left valve and one in right, are triangular, elevated, and roughened (divergent in the right). In the left valve, the anterior pseudocardinal tooth is bladelike, more elevated, and the larger of the two. The lateral teeth, one in the right valve and two in the left, are short, slightly curved, and heavy. Muscle scars and the pallial line are deeply impressed. The beak cavity is deep and somewhat compressed. The periostracum is yellowish green or brown in young shells, often with a broad broken green ray extending from the umbone toward the ventral margin; it is dark brown to black in old shells. The nacre is white and iridescent posteriorly.

Range Map 106. *Quadrula pustulosa* (Lea, 1831), Pimpleback.

Life History and Ecology: Like the Mapleleaf, *Quadrula pustulosa* is rather generalized in habitat preference and can maintain abundant and viable populations in shallow to deep sections of large reservoirs, such as Watts Bar Lake (Tennessee River), as well as in small to medium-sized free-flowing rivers. It is usually found in a substrate consisting of coarse gravel, sand, and silt. The Pimpleback is tachytictic, being gravid from about mid-June to mid-August. The shovelnose sturgeon *(Scaphirhynchus platorynchus),* black bullhead *(Ameiurus melas),* brown bullhead *(A. nebulosus),* channel catfish *(Ictalurus punctatus),* flathead catfish *(Pylodictis olivaris),* and the white crappie *(Pomoxis annularis)* are listed by Fuller (1974) as host fish for glochidia of this mussel.

Status: Currently Stable (Williams et al., 1993:14).

Quadrula quadrula (Rafinesque, 1820) Mapleleaf

RANGE MAP 107; PLATE 108

Synonymy:

Obliquaria (Quadrula) quadrula Rafinesque, 1820; Rafinesque, 1820:305

Unio quadrulus (Rafinesque, 1820); Say, 1834:no pagination

Quadrula quadrula (Rafinesque, 1820); Utterback, 1915:135

Quadrula (Quadrula) quadrula (Rafinesque, 1820); Frierson, 1927:47

Quadrula (Quadrula) quadrula quadrula (Rafinesque, 1820); Haas, 1969a:289

Unio quadratus (Rafinesque, 1820); Reeve, 1864:pl. 6, fig. 24 [misspelling]

Unio rugosus Barnes, 1823; Barnes, 1823:126, pl. 8, fig. 9

Unio lacrymosus Lea, 1828; Lea, 1828:272, pl. 6, fig. 8

Margarita (Unio) lacrymosus (Lea, 1828); Lea, 1836:14

Margaron (Unio) lacrymosus (Lea, 1828); Lea, 1852c:21

Quadrula lachrymosa (Lea, 1828); Baker, 1898:83, pl. 25, fig. 1, pl. 12, fig. 2 [misspelling]

Quadrula (Theliderma) lachrymosa (Lea, 1828); Simpson, 1900a:776 [misspelling]

Unio asperrimus Lea, 1831; Lea, 1831:71, pl. 5, fig. 3

Margarita (Unio) asperrimus (Lea, 1831); Lea, 1836:14

Plate 107. *Quadrula pustulosa* (Lea, 1831), Pimpleback.

Margaron (Unio) asperrimus (Lea, 1831); Lea, 1852c:21
Tritogonia nobilis (Conrad, 1854); Simpson, 1914:321 [in part]
Unio lunulatus Pratt, 1876; Pratt, 1876:167, pl. 31, fig. 1
Quadrula (Theliderma) lachrymosa var. *lunulata* (Pratt, 1876); Simpson, 1900a:777
Quadrula quadrula contraryensis Utterback, 1915; Utterback 1915:138, pl. 18, figs. 47a, 47b
Quadrula quadrula contrayensis Utterback, 1915; Johnson, 1969b:132 [misspelling]
Quadrula (Quadrula) quadrula var. *bullocki* F. C. Baker, 1928; F. C. Baker, 1928a:87, pl. 46, figs. 1–3
Quadrula biangulata Morrison, 1942; Morrison, 1942:356

Type Locality: Ohio River.

General Distribution: The entire Mississippi River drainage; various localities in the St. Lawrence Basin; Red River of the North; southwest into eastern Texas (Simpson, 1914) and southeast to Louisiana (Vidrine, 1993).

Tennessee Distribution: Starnes and Bogan (1988) list *Quadrula quadrula* as presently (since 1960) occurring in the lower Tennessee, Cumberland, Stones, Elk, Duck, and Hatchie rivers. Prior to 1960 it was reported from the North Fork Obion and Loosahatchie rivers and Reelfoot Lake (Hoff, 1943; Starnes and Bogan, 1988); however, a population of this mussel still occurs in Reelfoot Lake and is locally uncommon in the Mississippi River, Lake County. Specimens were collected in 1973 in the Harpeth River, Davidson County, where it is still present but rare.

Description: Shells of the Mapleleaf are typically squarish in outline, quadrate, compressed to slightly inflated, and solid. Mature specimens may reach 120 mm in length. Beaks are elevated and only slighted inflated; sculpture consists of double loops or zigzag bars with radiating extensions behind which continue diagonally downward on the shell as two rows of raised, elongated, or rounded tubercles. There is a row of tubercles on the elevated posterior ridge; the two rows of tubercles are separated by a wide sulcus or depression, gradually becoming progressively wider from the beak to the ventral margin. Tubercles on the beaks are small and crowded, and there are usually a few pustules present on dorsal surface behind the beaks. Some individuals exhibit a scattering of small, low pustules over much of the shell's surface. The anterior end is rounded; the posterior end is square or sharply truncated. The surface of the valves is often roughened by concentric rest lines, which, in individuals from lakes and river embayments, may be quite distinct (see fig. 8).

The left valve has two vertically elongated, solid, divergent, and roughened pseudocardinal teeth, sometimes joined above; the two lateral teeth are wide, rather long, and nearly straight. The right valve has a thick, heavy, roughened, triangular pseudocardinal tooth, often with a small tubercular tooth on either side. The lateral tooth is high, straight, and finely serrated; some shells have an incomplete, low second lateral. The interdentum is usually wide, and the beak cavity is fairly deep but open. The periostracum is yellowish green in young shells, occasionally marked with faint rays; old shells are brownish to black, such as those from Reelfoot Lake and the Hatchie River. The nacre is pearly white and iridescent posteriorly.

Life History and Ecology: *Quadrula quadrula* is an adaptable species and does well in shallow lakes (Klippel and Parmalee, 1979) and big river embayments, such as those found at Cuba Landing, Tennessee River, Humphreys County, or in deep (15–18 feet) reservoir impoundments. The most suitable substrate is one composed of sand and fine gravel.

Range Map 107. *Quadrula quadrula* (Rafinesque, 1820), Mapleleaf.

Plate 108. *Quadrula quadrula* (Rafinesque, 1820), Mapleleaf.

Like other species of *Quadrula* for which the breeding season is known, the Mapleleaf is tachytictic, the gravid period lasting from May to August (Baker, 1928a). Thus far, only one fish, the flathead catfish *(Pylodictis olivaris),* has been shown to serve as a host for the glochidia of this mussel (Howard and Anson, 1922).

Status: Currently Stable (Williams et al., 1993:14). Populations of the Mapleleaf continue to remain viable in the Tennessee River (Kentucky Lake) below Pickwick Landing Dam and locally in the middle Cumberland River. Populations of this mussel in the Hatchie River and Reelfoot Lake also appear stable.

Quadrula sparsa (Lea, 1841)
Appalachian Monkeyface

RANGE MAP 108; PLATE 109

Synonymy:

Quadrula intermedia (Conrad, 1836); Ortmann, 1918:541 [in part]

Unio sparsus Lea, 1841; Lea, 1841b:82; Lea, 1842b:80, pl. 25, fig. 58

Margaron (Unio) sparsus (Lea, 1841); Lea, 1852c:22

Quadrula (Quadrula) sparsa (Lea, 1841); Simpson, 1900a:775

Quadrula (Quadrula) tuberosa var. *sparsa* (Lea, 1841); Simpson, 1914:837

Quadrula intermedia form *tuberosa* (Lea, 1840); Ortmann, 1925:336 [in part]

Orthonymus metanevrus var. *tuberosus* (Lea, 1840); Haas, 1969a:310 [in part]

Type Locality: Holston River, Tennessee.

General Distribution: Upper Tennessee River drainage. Ortmann (1918) combined *Quadrula tuberosa, Q. sparsa,* and *Q. intermedia* together under *Q. intermedia,* which tends to confuse the distribution records. If *Q. tuberosa* and *Q. sparsa* are synonymous, then the form Ortmann (1912a) called *Q. sparsa* was also in the headwaters of the Cumberland River.

Tennessee Distribution: The only confirmed records of this species are from the Holston River, the type locality, the upper Powell, Claiborne and Hancock counties, and the unimpounded portion of the Clinch River above Norris Dam (Stansbery, 1973, 1976d; Bates and Dennis, 1978; Ahlstedt, 1991a). H. D. Athearn, Cleveland, Tennessee, reportedly collected a specimen from the Tellico River, a tributary to the Little Tennessee River, Monroe County. Archaeological specimens referred to *Quadrula sparsa* have been identified from the lower Clinch River, Roane County (Parmalee and Bogan, 1986) and the Hiwassee River, Bradley County (Parmalee and Hughes, 1994).

Description: The shell is triangular to irregularly rhomboidal in outline, solid, and rather sparsely tuberculate. Maximum shell length attained by mature specimens of *Quadrula sparsa* is about 80 mm. Beaks are elevated and situated in about the middle of the shell. Valves are somewhat inflated and subequilateral, the anterior margin of the shell being broadly rounded. The hinge ligament is very short and thick. The dorsal slope is obliquely truncated, and the sulcus above

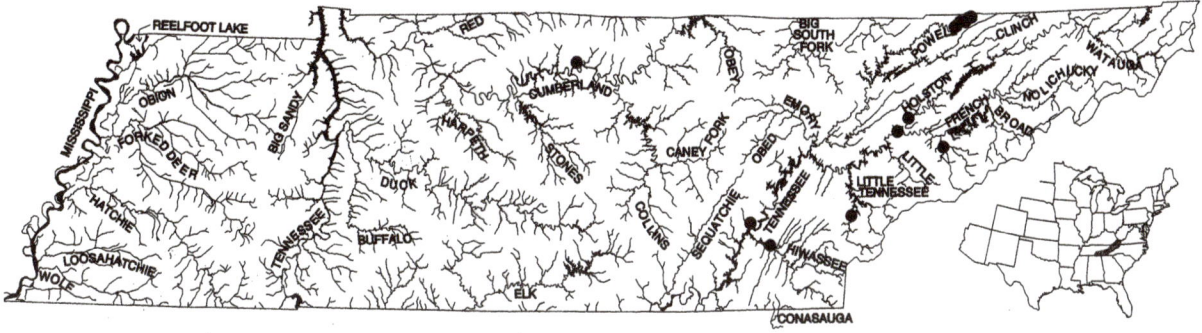

Range Map 108. *Quadrula sparsa* (Lea, 1841), Appalachian Monkeyface.

the posterior ridge ends in a shallow sinus. The posterior ridge is rounded, slightly elevated, and typically lacking pustules; it is well defined in many individuals by a posterior and anterior sulcus. The surface is only sparsely covered with small lachrymose knobs or tubercles; valves lack large or pronounced knobs on the posterior ridge. Tubercles are lacking on the anterior third of the shell.

The left valve has two rugged pseudocardinal teeth, a broad interdentum, and two short, straight lateral teeth. The right valve has two or three rough pseudocardinals, the middle triangular tooth being the largest, the others quite reduced; the lateral tooth is short and straight. The beak cavity is deep and compressed, and the interdentum is wide. The periostracum is yellowish, yellowish green, or brownish, subshiny to dull, often with small, greenish chevrons or triangles. The nacre color is white and iridescent, although in some specimens it may be salmon-colored posteriorly.

Life History and Ecology: The Appalachian Monkeyface has been found inhabiting a sand and gravel substrate in riffles and shallow shoal areas with moderate current. This mussel is presumed to be tachytictic, breeding from May to July, judging by the reproductive patterns of other species in the genus (Heard and Guckert, 1970). Individuals come out of the substrate and lay on their side while spawning, a trait also observed for *Q. intermedia* (Ahlstedt, 1991a). Information on the host fish and other life history data are lacking.

Status: Endangered (Williams et al., 1993:14). *Quadrula sparsa* has been extirpated from nearly all of its former range and is now reduced to one population in the upper Powell River, Claiborne and Hancock counties, Tennessee (Bates and Dennis, 1978) and one popula-

tion in the upper Clinch River in Virginia (Ahlstedt, 1991a). It has now (1996) nearly disappeared from this stretch of the Powell River. The U.S. Fish and Wildlife Service has developed a recovery plan for this species (U.S. Fish and Wildlife Service, 1983e) and has created a watershed implementation schedule for the recovery plan (U.S. Fish and Wildlife Service, 1989b).

Plate 109. *Quadrula sparsa* (Lea, 1841), Appalachian Monkeyface.

Simpsonaias ambigua (Say, 1825)
Salamander Mussel

RANGE MAP 109; PLATE 110

Synonymy:

Alasmodonta ambigua Say, 1825; Say, 1825:131

Margaritana ambigua (Say, 1825); Küster, 1862:300, pl. 99, fig. 7

Simpsoniconcha ambigua (Say, 1825); Simpson, 1900a:673

Simpsonaias ambigua (Say, 1825); Frierson, 1914:7

Hemilastena ambigua (Say, 1825); Simpson, 1900a:673

Unio hildrethianus Lea, 1834; Lea, 1834:36, pl. 3, fig. 8

Margarita (Unio) hildrethianus (Lea, 1834); Lea, 1836:28

Margaron (Margaritana) hildrethianus (Lea, 1834); Lea, 1852c:43

Strophitus hildrethiana (Lea, 1834); Conrad, 1853:263

Baphia hildrethiana (Lea, 1834); H. and A. Adams, 1857:499

Margaritana hildrethiana (Lea, 1834); B. H. Wright, 1888b:no pagination

Alasmodonta dubia Férussac, 1835; Férussac, 1835:26

Type Locality: Northwestern Territory.

General Distribution: For Canada, Clarke (1981a) recorded one location from the Lake St. Clair drainage (Sydenham River) in southern Ontario. Simpson (1914) listed it for the Ohio River system, north to Michigan, west to Iowa, south to Arkansas, and east to Tennessee.

Tennessee Distribution: Although Simpson (1914) reported the Salamander Mussel as occurring "east to Tennessee" and this distribution was repeated by Baker (1928a), the actual former occurrence and location of this species in Tennessee is uncertain. There are four lots of *Simpsonaias ambigua,* two from the East Fork Stones River, Rutherford County, in the collections of the Museum of Zoology, The Ohio State University, Columbus, obtained in 1965 by David H. Stansbery.

Plate 110. *Simpsonaias ambigua* (Say, 1825), Salamander Mussel.

We are unaware of other Tennessee records; in all probability, the species is now extirpated in the state.

Description: The shell is thin, fragile, considerably elongated, and inflated, especially along the broadly rounded posterior ridge, which is somewhat more swollen in females. This is a small species—individuals rarely exceed 50 mm in length. Beaks are somewhat compressed and only slightly elevated above the hinge

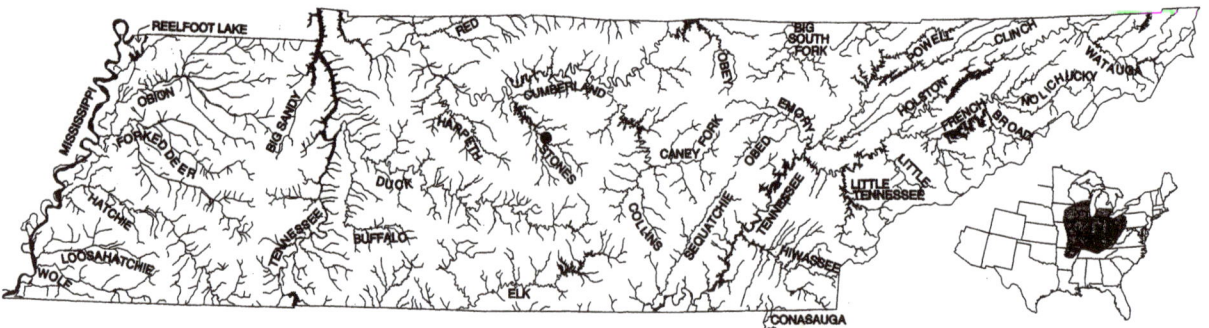

Range Map 109. *Simpsonaias ambigua* (Say, 1825), Salamander Mussel.

line; sculpture consists of 4–5 fine but prominent ridges drawn up in the middle. The anterior and posterior ends are rounded; the dorsal and ventral margins are nearly straight and parallel. A small, thin, flattened pseudocardinal tooth is present in each valve; lateral teeth are represented by a swelling of the hinge line. There is no interdentum; the beak cavity is shallow. The periostracum is a rayless, dull yellowish tan. The nacre is a dull bluish white; the posterior half is iridescent.

Life History and Ecology: *Simpsonaias ambigua* is the only North American freshwater mussel known to utilize a salamander, the mudpuppy *(Necturus maculosus),* as a host for the developing glochidia (which parasitize the mudpuppy's external gills) (Howard, 1951). This mussel is typically encountered living under flat rocks in a sandy substrate, although Clarke (1981a) comments that it also occurs in mud or on gravel. The undersides of rocks provide a hiding place for the mudpuppy, so this close association of the mussel and its host permits a greater opportunity for completion of the mussel's life cycle. Baker (1928a) indicated that the breeding season was not known, but suggested that the animal was bradytictic.

Status: Special Concern (Williams et al., 1993:14).

Strophitus connasaugaensis (Lea, 1858)
Alabama Creekmussel

RANGE MAP 110; PLATE 111

Synonymy:
Margaritana connasaugaensis Lea, 1858; Lea, 1858b:135; Lea, 1859f:229, pl. 32, fig. 113
Unio connasaughensis (Lea, 1858); Sowerby, 1868:pl. 88, fig. 474 [misspelling]
Margaron (Margaritana) connasaugaensis (Lea, 1858); Lea, 1870:69
Margaritana connesaugensis (Lea, 1858); Clessin, 1875:269 [misspelling]
Strophitus connasaugaensis (Lea, 1858); Simpson, 1900a:618
Strophitus (Strophitus) connasaugaensis (Lea, 1858); Haas, 1969a:375
Margaritana alabamensis Lea, 1861; Lea, 1861a:41; Lea, 1862b:104, pl. 16, fig. 249
Unio alabamensis (Lea, 1861); Sowerby, 1868:pl. 82, fig. 433
Margaron (Margaritana) alabamensis (Lea, 1861); Lea, 1870:68
Strophitus (Strophitus) alabamensis (Lea, 1861); Haas, 1969a:376

Type Locality: Conasauga River, Gilmer County, Georgia (Simpson, 1914).

General Distribution: Alabama River system.

Tennessee Distribution: Conasauga River, Polk County.

Description: The shell of the Alabama Creekmussel is thin, but slightly more solid in the beak cavity area; it is slightly to moderately inflated. Mature adults reach a maximum length of between 100 and 120 mm. Beaks are full, moderately high, and have sculpture consisting of a few strong ridges running parallel with the growth lines; the posterior ridge is full but widely rounded. The anterior margin is rather sharply rounded, while the ventral and posterior margins are nearly straight, making generally oval shells somewhat rhomboid in outline. The surface has irregular, slightly raised concentric growth lines.

Both valves possess an irregular, compressed, only slightly elevated pseudocardinal tooth at or immediately anterior to the beak. Lateral teeth are nearly wanting, represented by a slightly raised, rounded ridge along the hinge line. The beak cavity is open and shallow; muscle scars are large and slightly impressed. The periostracum is yellowish green, becoming a dull brown with age; rarely it is patterned with subtle dark green rays, then mostly on the posterior half of the valve. The nacre color is a bluish gray, the beak cavity–lateral tooth area sometimes has a light salmon or flesh-colored wash.

Life History and Ecology: *Strophitus connasaugaensis* is a small to medium-sized river species, or one that inhabits shallow embayments of larger rivers. It is most often found in substrates composed of fine gravel, sand, and silt, typically in stretches with some current in less than two feet of water. Little else is known of its life history.

The Creeper, *Strophitus undulatus,* is bradytictic, gravid from July to April and May (Baker, 1928a). This mussel may complete its growth with or without the glochidia passing a parasitic stage on the gills of fish. Whether this is true of all species in the genus *Strophitus* is unknown. Several fish taxa have been reported as hosts for the glochidia of the Creeper, but those utilized by glochidia of *S. connasaugaensis* are unknown.

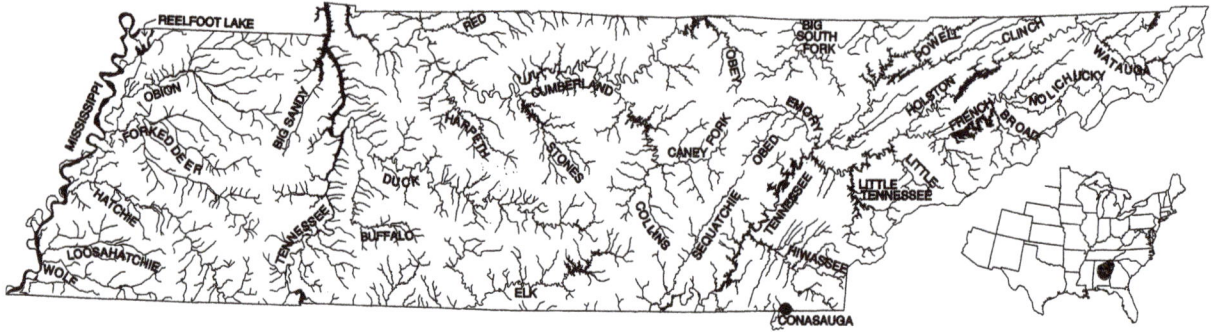

Range Map 110. *Strophitus connasaugaensis* (Lea, 1858), Alabama Creekmussel.

Status: Special Concern (Williams et al., 1993:14). Hurd (1974) recorded the Alabama Creekmussel from the Conasauga River, Polk County, Tennessee; it has been collected there periodically since that time, but it now appears uncommon to rare.

Plate 111. *Strophitus connasaugaensis* (Lea, 1858), Alabama Creekmussel.

Strophitus undulatus (Say, 1817)
Creeper (formerly Squawfoot)

RANGE MAP 111; PLATE 112

Synonymy:

Anodonta undulata Say, 1817; Say, 1817:pl. 3, fig. 5

Margarita (Anodonta) undulata (Say, 1817); Lea, 1836:50

Anodon undulatus (Say, 1817); Sowerby, 1836:316

Anodon undulata (Say, 1817); Gould, 1841:120, fig. 79

Alasmodonta undulata (Say, 1817); C. B. Adams, 1842:165

Strophitus undulatus (Say, 1817); Stimpson, 1851:15

Margaron (Anodonta) undulata (Say, 1817); Lea, 1852c:49

Unio undulata (Say, 1817); Deshayes, 1853:217, pl. 30, figs. 8, 9

Strophitus (Strophitus) undulatus undulatus (Say, 1817); Haas, 1969a:372

Strophitus undulatus undulatus (Say, 1817); Oesch, 1984:49

Anodonta pensylvanica [*sic*] Lamarck, 1819; Lamarck, 1819:86

Anodonta pennsylvanica Lamarck, 1819; Stark, 1828:89

Strophitus undulatus pennsylvanicus (Lamarck, 1819); Frierson, 1927:22

Strophitus (Strophitus) undulatus pennsylvanicus (Lamarck, 1819); Haas, 1969a:373

Anodon rugosus Swainson, 1822; Swainson, 1822:pl. 96

Anodonta rugosus (Swainson, 1822); Hanley, 1843:217

Strophitus rugosus (Swainson, 1822); Ortmann and Walker, 1922:40

Strophitus undulatus rugosus (Swainson, 1822); Frierson, 1927:22

Strophitus (Strophitus) undulatus rugosus (Swainson, 1822); Haas, 1969a:373

Alasmodonta edentula Say, 1829; Say, 1829:340

Anodonta edentula (Say, 1829); Férussac, 1835:25

Margarita (Anodonta) edentula (Say, 1829); Lea, 1836:50

Anodon edentula (Say, 1829); Catlow and Reeve, 1845:66

Uniopsis edentula (Say, 1829); Agassiz, 1852:49

Margaron (Anodonta) edentula (Say, 1829); Lea, 1852c:49

Strophitus edentulus (Say, 1829); Conrad, 1853:263

Strophitus (Strophitus) undulatus edentulus (Say, 1829); Haas, 1969a:374

Anodon areolatus Swainson, 1829; Swainson, 1829:pl. 18

Anodonta areolatus (Swainson, 1829); Férussac, 1835:25

Anodonta marginata Say, 1817; Férussac, 1835:25 [misidentification]

Anodonta ferussaciana Lea, 1834; Férussac, 1835:25 [misidentification]

Anodonta pavonia Lea, 1836; Lea, 1836:78, pl. 21, fig. 65

Margarita (Anodonta) pavonia (Lea, 1836); Lea, 1838a:30

Anodon pavonia (Lea, 1836); De Kay, 1843:203, pl. 40, fig. 358

Margaron (Anodonta) pavonia (Lea, 1836); Lea, 1852c:50

Strophitus pavonius (Lea, 1836); F. C. Baker, 1898:68, pl. 3, fig. 5, pl. 5, fig. 3

Strophitus edentulus var. *pavonius* (Lea, 1836); Simpson, 1900a:617

Strophitus rugosus pavonius (Lea, 1836); F. C. Baker, 1928a:202, pl. 73, figs. 1–5, pl. 74, figs. 7, 9, 10, pl. 67, figs. 1, 2

Margarita (Anodonta) wardiana Lea, 1836; Lea, 1836:50 [nomen nudum]

Anodonta wardiana Lea, 1838; Lea, 1838b:46, pl. 14, fig. 42

Anodon wardiana (Lea, 1838); Sowerby, 1867:pl. 28, fig. 114

Anodonta virgata Conrad, 1836; Conrad, 1836:cover of No. 5

Strophitus virgatus (Conrad, 1836); Conrad, 1853:263

Strophitus undulatus virgatus (Conrad, 1836); Frierson, 1927:22

Strophitus (Strophitus) undulatus virgatus (Conrad, 1836); Haas, 1969a:373

Anodon unadilla De Kay, 1843; De Kay, 1843:199, pl. 15, fig. 228

Strophitus unadilla (De Kay, 1843); Conrad, 1853:263

Anodonta tetragona Lea, 1845; Lea, 1845:165; Lea, 1848:82, pl. 8, fig. 25

Margaron (Anodonta) tetragona (Lea, 1845); Lea, 1852c:51

Strophitus tetragona (Lea, 1845); Conrad, 1853:263

Anodonta arkansasensis Lea, 1852; Lea, 1852b:293, pl. 29, fig. 56

Margaron (Anodonta) arkansasensis (Lea, 1852); Lea, 1852c:50

Strophitus arkansasensis (Lea, 1852); Conrad, 1853:262

Anodonta shaefferiana Lea, 1852; Lea, 1852b:288, pl. 26, fig. 50

Margaron (Anodonta) shaefferiana (Lea, 1852); Lea, 1852c:51

Strophitus shaefferiana (Lea, 1852); Conrad, 1853:263

Anodonta shefferiana Lea, 1852; Clessin, 1873:243, pl. 17, figs. 5–7 [misspelling]

Anodon shaefferianus (Lea, 1852); Sowerby, 1870:pl. 35, fig. 143

Anodonta shafferiana Lea, 1852; Currier, 1868:12 [misspelling]

Anodonta showalterii Lea, 1860; Lea, 1860c:307; Lea, 1862c:215, pl. 33, fig. 284

Anodon showalterii (Lea, 1860); Sowerby, 1870:pl. 27, fig. 104

Margaron (Anodonta) showalterii (Lea, 1860); Lea, 1870:79

Alasmodon rhombica Anthony, 1865; Anthony, 1865:158, pl. 12, fig. 5

Margaritana rhombica (Anthony, 1865); Paetel, 1890:183

Anodon papyraceus Anthony, 1865; Anthony, 1865:161

Anodonta papyracea (Anthony, 1865); B. H. Wright, 1888b:no pagination

Margaron (Anodonta) papyracea (Anthony, 1865); Lea, 1870:81

Anodon annulatus Sowerby, 1867; Sowerby, 1867:pl. 18, fig. 67

Anodon quadriplicatus Sowerby, 1867; Sowerby, 1867:pl. 28, fig. 110

Strophitus undulatus var. *quadriplicatus* (Sowerby, 1867); Simpson, 1914:350

Anodonta salmonea Lea, 1838; Clessin, 1873:91, pl. 24, figs. 1, 2 [misspelling and misidentification]

Strophitus undulatus ovatus Frierson, 1927; Frierson, 1927:22

Strophitus undulatus tennesseensis Frierson, 1927; Frierson, 1927:22

Strophitus (Strophitus) undulatus tennesseensis Frierson, 1927; Haas, 1969a:374

Strophitus rugosus pepinensis F. C. Baker, 1928; F. C. Baker, 1928a:204, pl. 74, fig. 8

Strophitus rugosus winnebagoeinsis F. C. Baker, 1928; F. C. Baker, 1928a:205, pl. 74, figs. 1–6

Strophitus rugosus lacustris F. C. Baker, 1928; F. C. Baker, 1928a:207, pl. 75, figs. 6–8

Type Locality: None given for *Anodonta undulata* Say, 1817; For *Alasmodonta edentula* Say, 1829: Wabash River, Indiana.

General Distribution: Canadian Interior Basin in the Red River–Nelson River system from western Ontario to eastern Saskatchewan, and throughout the Great Lakes–St. Lawrence northern drainage system (Clarke, 1981a). Entire Mississippi River drainage from Minnesota to central Texas, Pennsylvania to Tennessee; Atlantic coastal drainage from Nova Scotia to the upper Savannah River system of South Carolina.

Tennessee Distribution: Statewide from the upper Clinch and Powell rivers, the Watauga, Little, Elk, Big South Fork Cumberland, Stones, and Harpeth in East and Middle Tennessee, to the Hatchie and other river systems of West Tennessee. Formerly in the French Broad, Holston, Tennessee, Buffalo, Cumberland, Obey, and Red rivers (Starnes and Bogan, 1988). Localized populations of this species may still occur in these and other rivers in Tennessee.

Description: The shell is elliptical, somewhat rhomboid, solid, compressed, and thin when young, moderately inflated and thick in mature and old individuals. The anterior end is rounded, and the posterior end is bluntly pointed and often obliquely truncated. The posterior ridge is broadly rounded and either compressed or quite pronounced (especially in old shells). One specimen in the McClung Museum collections from the Big South Fork Cumberland River, Scott County, Tennessee, measures 115 mm in length. However, most shells of mature individuals from Tennessee seldom exceed 85–90 mm in length. Beaks are depressed, only slightly

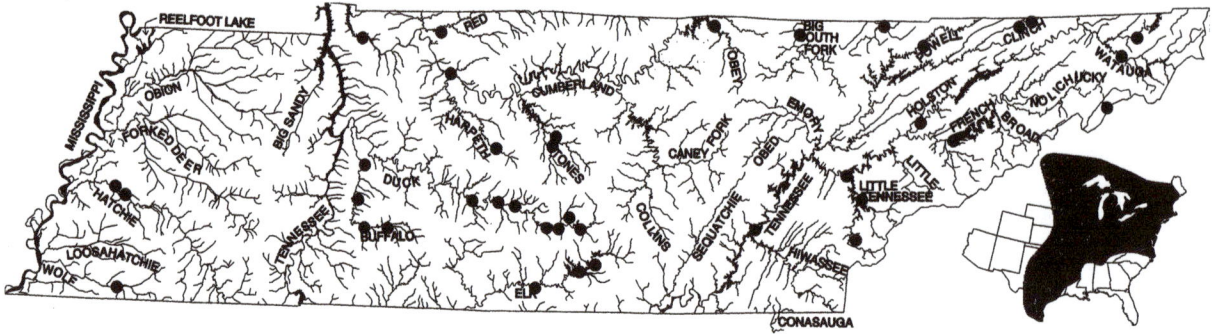

Range Map 111. *Strophitus undulatus* (Say, 1817), Creeper.

elevated above the hinge line; sculpture consisting of 3–4 heavy concentric bars, somewhat oblique to the hinge line, rounded anteriorly, and angled posteriorly.

The pseudocardinal tooth in the left valve is represented by an elongated, low thickening of the hinge line below the beak; the pseudocardinal tooth in the right valve appears as a low, thick swelling anterior to the beak. Lateral teeth are absent or suggested by a thickened hinge line. The beak cavity is shallow. The periostracum is yellowish or greenish, marked by greenish, often wavy rays; old shells are dark brown or black and usually rayless. The nacre is white or bluish white and iridescent around the margins, the center, and occasionally along the pallial line where it may also be cream-colored or salmon.

Life History and Ecology: *Strophitus undulatus* appears adaptable to a variety of aquatic habitats, from the high-gradient streams of East Tennessee to the meandering or channelized rivers of West Tennessee. It has been shown experimentally that the glochidia of this species may develop on the fins and skin of the largemouth bass *(Micropterus salmoides)* and the creek chub *(Semotilus atromaculatus)* (Baker, 1928a) and that the Rio Grande killifish *(Fundulus zebrinus)* and green sunfish *(Lepomis cyanellus)* may serve as natural hosts for the glochidia of this mussel (Fuller, 1978). In addition, Hove (1995a) has identified seven fish as definite hosts for the glochidia of the Creeper: spotfin shiner *(Cyprinella spiloptera),* fathead minnow *(Pimephales promelas),* yellow bullhead *(Ameiurus natalis),* black bullhead *(Ameiurus melas),* bluegill *(Lepomis macrochirus),* largemouth bass *(Micropterus salmoides),* and walleye *(Stizostedion vitreum).* However, the Creeper is one of the few freshwater mussels able to

complete its life cycle without a fish host; the glochidia undergo a complete development in the female before being expelled (Lefevre and Curtis, 1910). Having direct development of the glochidia, resulting in the elimination of the parasitic stage on fish, is probably a primary factor in the species's wide distribution and local abundance. It is bradytictic, the reproductive period extending from July to April and May (Baker, 1928a).

Plate 112. *Strophitus undulatus* (Say, 1817), Creeper.

In Tennessee, *Strophitus undulatus* reaches maximum growth in small to medium-sized rivers with current and a substrate of fine sand and mud. It typically inhabits depths of no more than three or four feet.

Status: Currently Stable (Williams et al., 1993:14).

Toxolasma cylindrellus (Lea, 1868)
Pale Lilliput

RANGE MAP 112; PLATE 113

Synonymy:

Carunculina parva (Barnes, 1823); Johnson, 1967a:128 [in part]

Unio glans, Lea, 1831; Call, 1896:113 [in part]

Carunculina glans (Lea, 1831); Haas, 1969a:429 [in part]

Toxolasma lividum Rafinesque, 1831; Rafinesque, 1831:2 [in part]

Carunculina moesta (Lea, 1841); Ortmann and Walker, 1922:54 [in part]

Unio cylindrellus Lea, 1868; Lea, 1868a:144; Lea, 1868c:308, pl. 48, fig. 121

Margaron (Unio) cylindrellus (Lea, 1868); Lea, 1870:49

Lampsilis (Carunculina) cylindrella (Lea, 1868); Simpson, 1900a:565

Carunculina cylindrella (Lea, 1868); Ortmann, 1924a:44

Carunculina moesta cylindrella (Lea, 1868); Ortmann, 1925:353

Carunculina glans cylindrella (Lea, 1868); Frierson, 1927:88

Toxolasma cylindrellus (Lea, 1868); Stansbery, 1971:12, fig. 39

Type Locality: Duck Creek, Tennessee; Swamp Creek, Whitfield County, Georgia; North Alabama.

General Distribution: The Pale Lilliput is a Cumberlandian species restricted to tributaries of the Tennessee River. One exception to its Cumberlandian distribution is a single record from Swamp Creek, Whitfield County, Georgia, which is in the Mobile River system (Lea, 1868a).

Tennessee Distribution: The Pale Lilliput has a restricted range in Tennessee. It is absent from the headwaters of the Tennessee River in East Tennessee and from the Cumberland River system. H. D. Athearn (pers. comm., 1984) has collected it from the Sequatchie River and a tributary of the Little Sequatchie River, both in Marion County. Ortmann (1925) obtained specimens from the Elk River in Franklin County. *Toxolasma cylindrellus* has been collected in the Duck River and a tributary, Big Rock Creek, from Hickman County upstream to Bedford County and from the Buffalo River in Perry County upstream to Lewis County (Ortmann, 1924a; 1925; Isom and Yokley, 1968a; van der Schalie, 1973; Stansbery, 1976c).

Description: The shell is small, adults reaching about 35 mm in length, rather solid, elongate, and elliptical, becoming almost cylindrical. The dorsal and ventral margins are nearly parallel, the anterior end being rounded and the posterior margin sloping and rounded toward the base. The posterior ridge is low or wanting. Valves are subinflated with a somewhat full beak. Beak sculpture has not been recorded. Shells of the female are only slightly distinct from the male, having a faint marsupial swelling near the posterior part of the base.

The left valve has two short, stumpy pseudocardinal teeth, a narrow interdentum, and two short laterals. The right valve has one triangular pseudocardinal tooth and a single lateral, which sometimes has a slight ridge at its base. Beak cavities are shallow. Muscle scars are well impressed; the pallial line distinct. The periostracum is clothlike in appearance and is tawny or yellowish green in color and rayless. The surface of the shell is smooth. The nacre is whitish outside the pallial line, while inside it is a rich coppery color. The shell becomes thinner and more iridescent behind. This mussel may

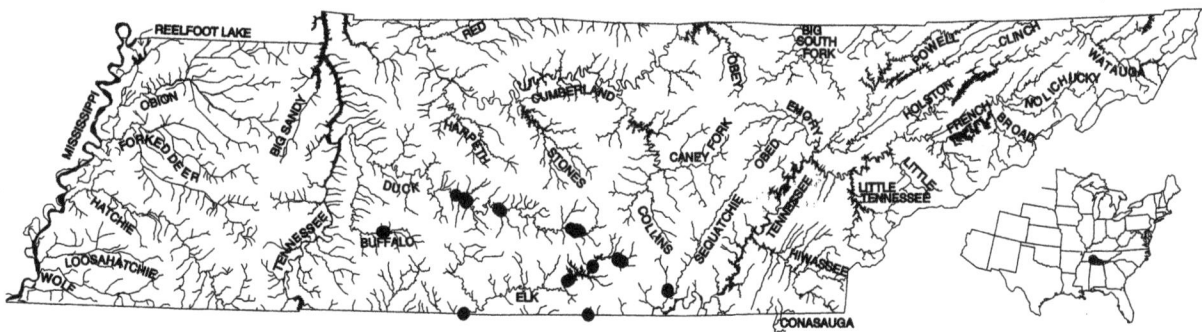

Range Map 112. *Toxolasma cylindrellus* (Lea, 1868), Pale Lilliput.

Plate 113. *Toxolasma cylindrellus* (Lea, 1868), Pale Lilliput.

be distinguished from closely related species by the elongate elliptical shell, yellowish periostracum, a usually lighter colored nacre overall, and the even lighter shade of the nacre outside the pallial line. Where *Toxolasma lividus* and *T. cylindrellus* occur together, *T. cylindrellus* is faster growing and reaches a larger size (Ortmann, 1924a; Stansbery, 1976c). Stansbery (1976c) found *Toxolasma cylindrellus* and *Toxolasma lividus* living together in the Paint Rock River, Alabama.

Life History and Ecology: This tiny naiad is found only in small tributary rivers and streams of the Tennessee River. The genus is bradytictic, but no other specific data on the host fish or life history was encountered. It may be found in a gravel and sand substrate, slow to moderate current, and in less than three feet of water.

Status: Endangered (Williams et al., 1993:14). Big Rock Creek, a tributary to the Duck River (Marshall County), appears to be one of the few creeks, if not the only one, still supporting a population of this mussel in Tennessee. It is known to still occur in the Paint Rock River and Hurricane Creek in northern Alabama (Ahlstedt,

1991b). The U.S. Fish and Wildlife Service has developed a recovery plan for this species (U.S. Fish and Wildlife Service, 1984f) and has created a watershed implementation schedule for the recovery plan (U.S. Fish and Wildlife Service, 1989b).

Toxolasma lividus Rafinesque, 1831
Purple Lilliput
RANGE MAP 113; PLATE 114

Synonymy:
Toxolasma lividus Rafinesque, 1831; Rafinesque, 1831:2
Toxolasma livida Rafinesque, 1831; Morrison, 1969:24
Toxolasma lividum Rafinesque, 1831; Ortmann, 1918:573
Toxolasma lividum lividum Rafinesque, 1831; Stansbery, 1972:46
Toxolasma lividus lividus Rafinesque, 1831; Stansbery, 1976a:48
Unio glans Lea, 1831; Lea, 1831:82, pl. 18, fig. 12
Margarita (Unio) glans (Lea, 1831); Lea, 1836:28
Margaron (Unio) glans (Lea, 1831); Lea, 1852c:31
Lampsilis (Carunculina) glans (Lea, 1831); Simpson, 1900a:565
Eurynia (Carunculina) glans (Lea, 1831); Ortmann, 1912a:339
Carunculina glans (Lea, 1831); Ortmann, 1910:119
Carunculina glans glans (Lea, 1831); Stansbery, 1970:18 ,
Toxolasma glans (Lea, 1831); Valentine and Stansbery, 1971:29
Toxolasma glans glans (Lea, 1831); Stansbery, 1971:14
Toxolasma lividus glans (Lea, 1831); Stansbery, 1976a:48
Unio moestus Lea, 1841; Lea, 1841b:82; Lea, 1842b:244, pl. 26, fig. 60
Margaron (Unio) moestus (Lea, 1841); Lea, 1852c:31
Lampsilis moestus (Lea, 1841); Simpson, 1900a:565
Lampsilis moesta (Lea, 1841); Simpson, 1914:156
Carunculina moesta (Lea, 1841); Ortmann, 1921:89
Carunculina glans moesta (Lea, 1841); Stansbery, 1970:18

Type Locality: Rockcastle River [Kentucky].

General Distribution: The Purple Lilliput is known from the Ohio River Basin west of Pennsylvania, including Ohio, Michigan, Illinois, Indiana, and Kentucky. It occurred in the Tennessee and Cumberland River drainages in Tennessee, Virginia, and Kentucky. Oesch (1984) reported the Purple Lilliput from streams in southern Missouri. Also known from the Arkansas River in Arkansas and Oklahoma.

Tennessee Distribution: *Toxolasma lividus* was found throughout the upper Tennessee River system, including the Powell, Clinch, Emory, Holston, French Broad, Tellico, Little Pigeon, and Little rivers, as well as the main channel of the Tennessee River below Knoxville. It was also found in the Duck and Elk rivers. In the

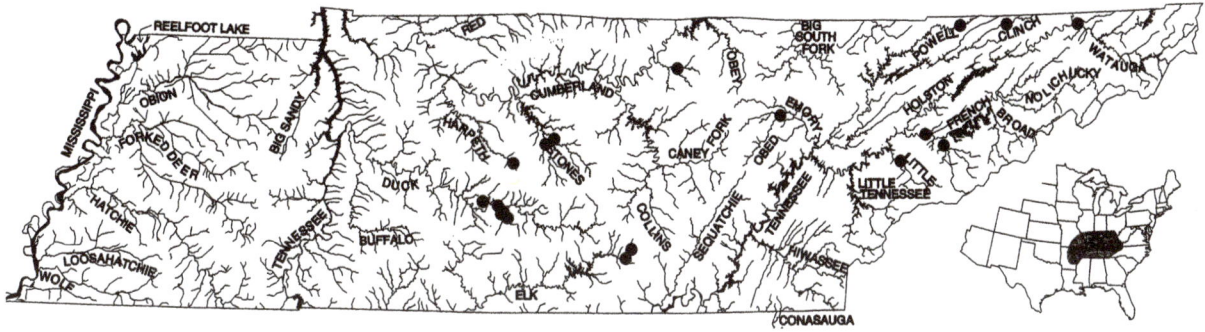

Range Map 113. *Toxolasma lividus* Rafinesque, 1831, Purple Lilliput.

Cumberland River system in Tennessee it occurred in the Caney Fork, Stones, and Harpeth rivers and numerous tributary creeks.

Description: The Purple Lilliput is short, solid, inflated, and elliptical in outline, with a distinct lunule. Shell length seldom exceeds 35 mm. The anterior end is uniformly rounded; the dorsal margin is nearly straight; the ventral margin is straight; the posterior margin has a sharp point above and is broadly rounded below. The posterior ridge is low and rounded. The male shell is short and elliptical to almost rhomboidal in outline with a bluntly rounded posterior margin. The female shells are shorter and somewhat more inflated and have an angular marsupial swelling at the posterior ventral margin. Beaks are full and elevated; sculpture consists of irregular ridges which curve upward behind and become nodulous at the posterior ridge. The beak cavity is open and shallow, and the interdentum is narrow or absent.

The left valve has two erect, triangular, compressed pseudocardinal teeth and two long, nearly straight lateral teeth. The right valve has a large elongate pseudocardinal tooth, often with a vestige of another above, and a long lateral tooth often with a vestigial lateral or small shelf below it. Adductor muscle scars are impressed; the pallial line is impressed anteriorly. The periostracum has irregular growth lines; color varies from dark brown to black and is rayless. The nacre color is usually a deep purple, lighter beyond the pallial line, but sometimes it is a creamy white, becoming iridescent posteriorly.

Life History and Ecology: Ahlstedt (in Neves, 1991) reported the Purple Lilliput from small to medium-sized rivers in mud, sand, and gravel substrates. This species has also been found on shallow, rocky gravel points or sandbars in Wheeler Reservoir, Alabama. Ortmann (1921) reported gravid females in May and July. Watters (1994a) listed the green sunfish *(Lepomis cyanellus)* and longear sunfish *(L. megalotis)* as fish hosts for the glochidia of *Toxolasma lividus*.

Status: Special Concern (Williams et al., 1993:14).

Plate 114. *Toxolasma lividus* Rafinesque, 1831, Purple Lilliput.

Toxolasma parvus (Barnes, 1823)
Lilliput

RANGE MAP 114; PLATE 115

Synonymy:
Unio parvus Barnes, 1823; Barnes, 1823:pl. 13, fig. 18
Mya parva (Barnes, 1823); Eaton, 1826:222
Margarita (Unio) parvus (Barnes, 1823); Lea, 1836:28
Margaron (Unio) parvus (Barnes, 1823); Lea, 1852c:31
Lampsilis (Corunculina) parvus (Barnes, 1823); Baker, 1898:109, pl. 13, fig. 3
Lampsilis (Carunculina) parvus (Barnes, 1823); Simpson, 1900a:564
Eurynia (Carunculina) parva (Barnes, 1823); Ortmann, 1912a:338
Carunculina parva (Barnes, 1823); Wheeler 1914:75; Utterback, 1916a:396
Toxolasma parvum (Barnes, 1823); Ortmann, 1919:260
Toxolasma parva (Barnes, 1823); Valentine and Stansbery, 1971:29
Carunculina parva cahni F. C. Baker, 1928; Baker, 1928a:253–254; pl. 105, figs. 14–18

Type Locality: Fox River, Wisconsin.

General Distribution: Generally throughout the Mississippi River drainage from western New York to Minnesota, and from southern Canada southward to central Texas.

Tennessee Distribution: The Lilliput has been collected from the lower Holston, Clinch, Tennessee, and Little Tennessee rivers in East Tennessee. In the Cumberland River drainage it is found in the Cumberland and Stones rivers. It is also reported from Reelfoot Lake, tributaries to the Hatchie River, and lower Tennessee River.

Description: The Lilliput is a small species with a maximum shell length of about 30 mm. The shell is rather solid and subelliptical in outline and often quite inflated. The anterior and posterior ends are evenly rounded; the ventral margin is nearly straight to very slightly curved and almost parallel with the dorsal margin. A posterior ridge is lacking; the posterior slope is flattened and somewhat compressed. Male shells are less swollen than those of the females, and the posterior end is narrower, sometimes coming almost to a point. The female shell is more swollen than the male shell, and the posterior end is more broadly rounded and very blunt. Beaks are more or less inflated but only slightly elevated above the hinge line. Sculpture consists of 5–6 relatively strong, subconcentric, distinct bars; latter bars become subangular posteriorly. The beak cavity is open and shallow; there is no interdentum.

The left valve has two erect, compressed, triangular pseudocardinal teeth and two moderately strong, straight lateral teeth. The right valve has one compressed, erect triangular pseudocardinal and a single straight lateral tooth. The anterior adductor muscle scar is deep, and the posterior adductor muscle scar is shallow. Dorsal muscle scars are present. The pallial line is impressed anteriorly. The periostracum is usually dark greenish or gray, sometimes dark brown to black, and rayless; growth lines may be coarse and elevated. The nacre color is typically silvery white, iridescent posteriorly, often with a tinge of pale yellow within the beak cavities.

Life History and Ecology: The Lilliput is found typically in the shallows of lakes, ponds, and reservoirs, as well as in small to large rivers, where it lives in mud, sand, or fine gravel. Ortmann (1919) and Baker (1928a) reported gravid females in May, June, and July with eggs present in August, the species probably being bradytictic. Watters (1994a) lists the following fish

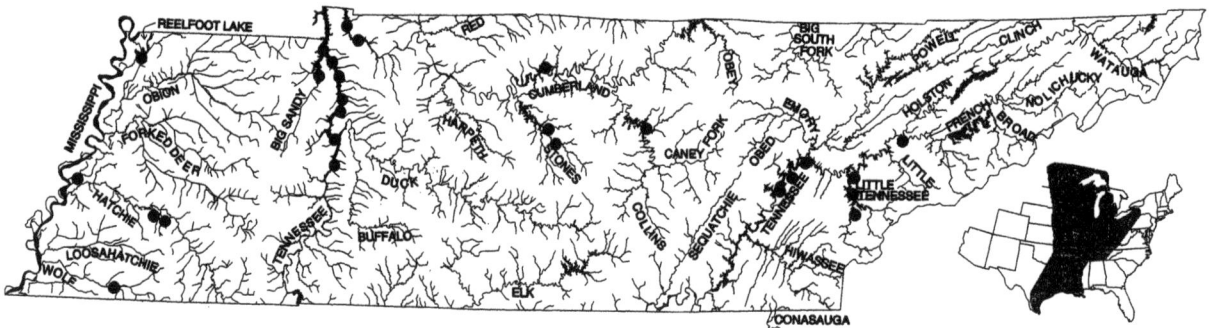

Range Map 114. *Toxolasma parvus* (Barnes, 1823), Lilliput.

Plate 115. *Toxolasma parvus* (Barnes, 1823), Lilliput.

hosts for the glochidia of *Toxolasma parvus:* bluegill *(Lepomis macrochirus),* green sunfish *(L. cyanellus),* orangespotted sunfish *(L. humilis),* warmouth *(L. gulosus),* and white crappie *(Pomoxis annularis).* Hove (1995c) confirmed the green sunfish as a suitable host.

Status: Currently Stable (Williams et al., 1993:14).

Toxolasma texasensis (Lea, 1857)
Texas Lilliput
RANGE MAP 115; PLATE 116

Synonymy:
Unio texasiensis Lea, 1857; Lea, 1857c:84
Unio texasensis Lea, 1860; Lea, 1860e:359, pl. 61, fig. 84
Margaron (Unio) texasensis (Lea, 1857); Lea, 1870:49
Lampsilis texasensis (Lea, 1857); Simpson, 1900a:563
Eurynia (Carunculina) texasensis (Lea, 1860); Ortmann, 1912a:339
Carunculina texasensis (Lea, 1857); Ortmann, 1915:141
Carunculina parva texasensis (Lea, 1857); Strecker, 1931:45
Toxolasma texasensis (Lea, 1857); Valentine and Stansbery, 1971:29
Unio texasianus (Lea, 1857); Sowerby, 1866:pl. 40, fig. 218 [misspelling]

Unio bairdianus Lea, 1857; Lea, 1857e:102; Lea, 1860e:361, pl. 61, fig. 186
Margaron (Unio) bairdianus (Lea, 1857); Lea, 1870:49
Unio bealei Lea, 1862; Lea, 1862a:169; Lea, 1862c:204, pl. 30, fig. 273
Margaron (Unio) bealei (Lea, 1862); Lea, 1870:49
Lampsilis texasensis var. *compressus* Simpson, 1900; Simpson, 1900a:564; Johnson, 1975:11, pl. 1, fig. 512
Carunculina parva compressa (Simpson, 1900); Strecker, 1931:47
Comment: The taxon *Unio haleianus* Lea, 1842 (Lea, 1842a:224; Lea, 1842b:247, pl. 27, fig. 63) has been placed near *Toxolasma texasensis* but there are questions. If *haleianus* is the same as *texasensis* it will have priority.

Type Locality: DeWitt County, Texas.

General Distribution: The species ranges from the southern coast of Texas, east to western Mississippi, up the Mississippi River embayment through Louisiana, eastern Arkansas, and Missouri to southern Illinois, Indiana, and western Tennessee.

Tennessee Distribution: The Texas Lilliput has been found only in the Hatchie River and adjoining backwaters and at Reelfoot Lake, all West Tennessee localities.

Description: The Texas Lilliput is almost elliptical in outline, varying from subcompressed to inflated. Maximum shell length is usually less than 50 mm. The anterior end is evenly rounded, and the ventral margin is nearly straight. The posterior ridge is well defined. Male shells are inflated with a nearly straight ventral margin, although it is often slightly pinched or slightly angular in the center, and the posterior margin is evenly rounded to slightly biangulate behind. The female shell has a well-developed, almost angular marsupial swelling located somewhat anterior of the posterior margin, which is truncated. Adult female shells are generally smaller than those of males. Beaks are low but full in more inflated specimens; sculpture consists of 7–9 ridges, which are curved and turn upward suddenly behind the posterior ridge, where they turn dorsally. The beak cavity is open and shallow or entirely wanting; there is no interdentum.

The left valve has two short compressed, ragged triangular pseudocardinal teeth and two slightly curved lateral teeth. The right valve has one short stubby pseudocardinal and one curved lateral tooth. The anterior adductor muscle scar is deep, but the posterior

Family Unionidae   233

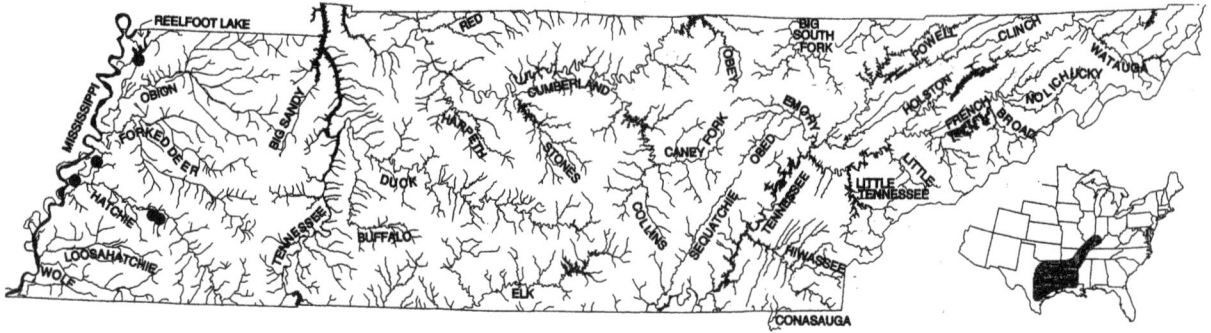

Range Map 115. *Toxolasma texasensis* (Lea, 1857), Texas Lilliput.

adductor muscle scar is shallow. The pallial line is impressed anteriorly. The periostracum appears thick and rayless and is brownish to blackish in color, often with a lighter brown tint in the beak area. The nacre color is bluish white to salmon and iridescent posteriorly.

Plate 116. *Toxolasma texasensis* (Lea, 1857), Texas Lilliput.

Life History and Ecology: *Toxolasma texasensis* is found in streams and in small to medium-sized rivers and sloughs, living in sand and silty bottoms usually in areas with little or no current and normally in water less than three feet deep. Watters (1994a) lists the bluegill *(Lepomis macrochirus)* and warmouth *(L. gulosus)* as fish hosts for the glochidia of *Toxolasma texasensis*.

Status: Currently Stable (Williams et al., 1993:14).

Tritogonia verrucosa (Rafinesque, 1820) Pistolgrip

RANGE MAP 116; PLATE 117

Synonymy:

Obliquaria verrucosa Rafinesque, 1820; Rafinesque, 1820:48, pl. 81, figs. 10–12

Unio verrucosus (Rafinesque, 1820); Say, 1834:no pagination

Tritogonia verrucosa (Rafinesque, 1820); Agassiz, 1852:48

Tritigonia verrucosa (Rafinesque, 1820); Smith, 1899:291, pl. 81 [misspelling]

Quadrula verrucosa (Rafinesque, 1820); Ortmann and Walker, 1922:18

Quadrula (Tritogonia) verrucosa (Rafinesque, 1820); Frierson, 1927:48

Unio tuberculatus Barnes, 1823; Barnes, 1823:125, pl. 7, figs. 8a, 8b

Mya tuberculata (Barnes, 1823); Eaton, 1826:217

Margarita (Unio) tuberculatus (Barnes, 1823); Lea, 1836:17

Margaron (Unio) tuberculatus (Barnes, 1823); Lea, 1852c:23

Tritogonia tuberculata (Barnes, 1823); Simpson, 1900a:608

Quadrula tuberculata (Barnes, 1823); Ortmann, 1912a:254

Unio (Theliderma) pustulata Swainson, 1840; Swainson, 1840:271, fig. 54d non Lea, 1834

Unio gigas Swainson, 1824; Sowerby, 1867:pl. 56, fig. 287 [misidentification]

Unio conjugans B. H. Wright, 1899; B. H. Wright, 1899:89

Tritogonia conjugans (B. H. Wright, 1899); Simpson, 1900b:79, pl. 4, fig. 1

Quadrula (Quadrula) quadrula conjugans (B. H. Wright, 1899); Frierson, 1927:48

Quadrula tritogonia Ortmann, 1909; Ortmann, 1909b:101 [new name for *Tritogonia tuberculata* Barnes]

Tritogonia tuberculata var. *obesa* Simpson, 1900; Simpson, 1900a:609

Quadrula obesa (Simpson, 1900); Vanatta, 1910:102

Quadrula parkeri Geiser, 1911; Geiser, 1911:15 [new name for *Tritogonia tuberculata* Barnes]

Type Locality: Ohio River.

General Distribution: Generally throughout the Mississippi River drainage, from western Pennsylvania west to southern Minnesota, south and west to Oklahoma and Texas; the Cumberland, Tennessee, and Alabama River systems (Ortmann, 1919).

Tennessee Distribution: Statewide, from the Hatchie River in southwestern Tennessee eastward through much of the lower stretches of the Cumberland and Tennessee River drainages including the Big South Fork Cumberland, Duck, Buffalo (Ahlstedt, 1991b), Elk, and Harpeth rivers. It is present in the Conasauga River, southeastern Tennessee. Prior to 1960 the Pistolgrip was known to have occurred in the North Fork Obion, Loosahatchie, Wolf, Hiwassee, Obey, and Red rivers (Starnes and Bogan, 1988).

Description: The shell is solid, elongate and rhomboid in outline, rather compressed but with a distinct, elevated, and rounded posterior ridge. In the same population, shells of old males may reach a length of 160 mm, while those of females may reach 120 mm. Shell size varies greatly in this species and is apparently dependent upon local habitat conditions. The anterior end is broadly rounded, and the ventral margin is slightly curved; the posterior end is squarely or obliquely trun-

cated in the male, more compressed and expanded into a broad wing in the female. Although sexual dimorphism is usually apparent in the shells of mature specimens, it is not well defined in many individuals. Ortmann (1919:45) commented that "the female shell is on the average more flattened and compressed than that of the male." The entire shell surface, with the occasional exception of the posterior ridge and slope, is densely covered with rather small, low tubercles. The posterior slope is usually sculptured with several parallel elevated ridges or plications. Beaks are compressed to slightly swollen, barely elevated above the hinge line.

The left valve has two solid, triangular, divergent, roughened pseudocardinal teeth; the two lateral teeth are long and straight. The right valve has a large, heavy, triangular, serrated pseudocardinal, with a small tubercular tooth on either side. The lateral tooth is solid, high, straight, and finely striated. The interdentum is narrow; the beak cavity is moderately deep. The anterior pallial line and muscle scars are deeply impressed, the posterior scars weakly so. The periostracum is a dark olive or yellowish tan, unrayed, becoming brown to black in old shells. The nacre is white and iridescent posteriorly. Some individuals in local populations—for example, those in the Hatchie River, Lauderdale County, and in the Tennessee River, Hardin County—possess a nacre color of light purple or lavender.

Life History and Ecology: Because of its apparent adaptability to a variety of habitat conditions, *Tritogonia verrucosa* may be found living at river depths of one foot up to 20 feet and in a substrate composed of coarse gravel, sand, and/or mud. Under favorable conditions, including a stable substrate and moderate current, the Pistolgrip may become locally numerous. The species is tachytictic, and the reproductive period occurs from

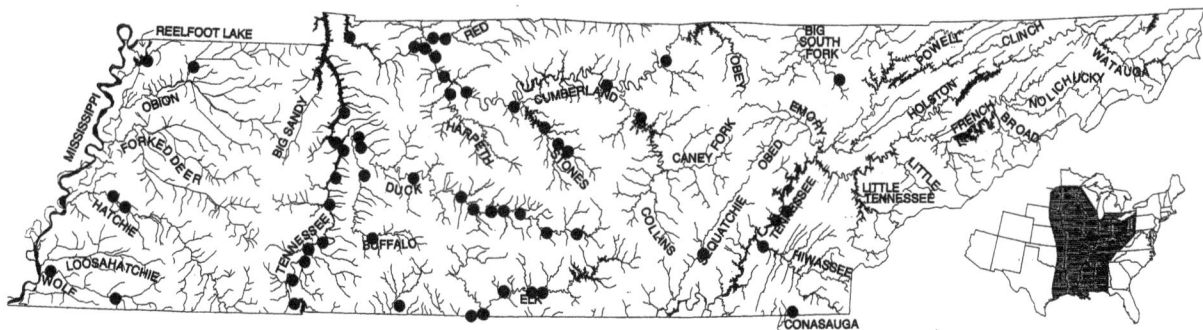

Range Map 116. *Tritogonia verrucosa* (Rafinesque, 1820), Pistolgrip.

Plate 117. *Tritogonia verrucosa* (Rafinesque, 1820), Pistolgrip.

April to August (Ortmann, 1919; Utterback, 1915–1916a). Howells (1996) reported the flathead catfish *(Pylodictis olivaris)* to be a suitable fish host for this species. Pepi and Hove (1997) have added the yellow bullhead *(Ameiurus natalis)* to the list of host fish for the glochidia of the Pistolgrip.

Status: Currently Stable (Williams et al., 1993:14).

Truncilla donaciformis (Lea, 1828)
Fawnsfoot
RANGE MAP 117; PLATE 118
Synonymy:
Unio donaciformis Lea, 1828; Lea, 1828:267, pl. 4, fig. 3
Margarita (Unio) donaciformis (Lea, 1828); Lea, 1836:18
Margaron (Unio) donaciformis (Lea, 1828); Lea, 1852c:24
Plagiola donaciformis (Lea, 1828); Baker, 1898:91, pl. 13, fig. 4
Plagiola (Amygdalonaias) donaciformis (Lea, 1828); Simpson, 1900a:605

Amygdalonaias donaciformis (Lea, 1828); Ortmann, 1914:67
Truncilla donaciformis (Lea, 1828); Ortmann and Walker, 1922:50
Unio zigzag Lea, 1829; Lea, 1829:440, pl. 12, fig. 19
Margarita (Unio) zigzag (Lea, 1829);Lea, 1836:18
Margaron (Unio) zigzag (Lea, 1829); Lea, 1852c:24
Unio nervosa (Rafinesque, 1820); Conrad, 1834a:70 [misidentification]
Unio nervosus (Rafinesque, 1820); Say, 1834:no pagination [misidentification]

Type Locality: Ohio.

General Distribution: Mississippi river drainage; western Pennsylvania west to Kansas, north to Minnesota, south to eastern Texas (Baker, 1928a). Kentucky, Tennessee, and Alabama. In Canada it is known only from Lake Erie and the Grand River of southern Ontario (Clarke, 1981a).

Tennessee Distribution: The Fawnsfoot occurs locally in the main Tennessee and Cumberland rivers as well as in streams and smaller rivers, such as the impounded lower Clinch River, and the Emory, Elk, Duck, and Stones rivers in Middle Tennessee. It has been reported as having occurred prior to 1960 in the Harpeth River (Starnes and Bogan, 1988).

Description: The shell is small—mature individuals from Tennessee rivers seldom exceed 40 mm in length—elongated, somewhat ovate, relatively thin but strong, and compressed to moderately inflated. Relict individuals surviving under impoundment conditions become greatly inflated and thick-shelled and may reach 50 mm in length. The anterior end is rounded, while the posterior end is bluntly pointed. The posterior-dorsal ridge is prominent and often sharply angled with the dorsal surface flattened. Beaks are full and elevated slightly above the hinge line; sculpture consists of 3–4 fine bars, the first is concentric, and the others are double-looped.

The left valve has two thin, compressed, divergent, elevated, and serrated pseudocardinal teeth; the posterior one located directly below the beak is more erect and flared upward. The right valve has a flattened, somewhat triangular, elevated pseudocardinal tooth. The lateral teeth, two in the left valve, one in the right, are thin, nearly straight, elevated, slightly roughened, and long. There is no interdentum; the beak cavity is shallow. The anterior pallial line and muscle scars are

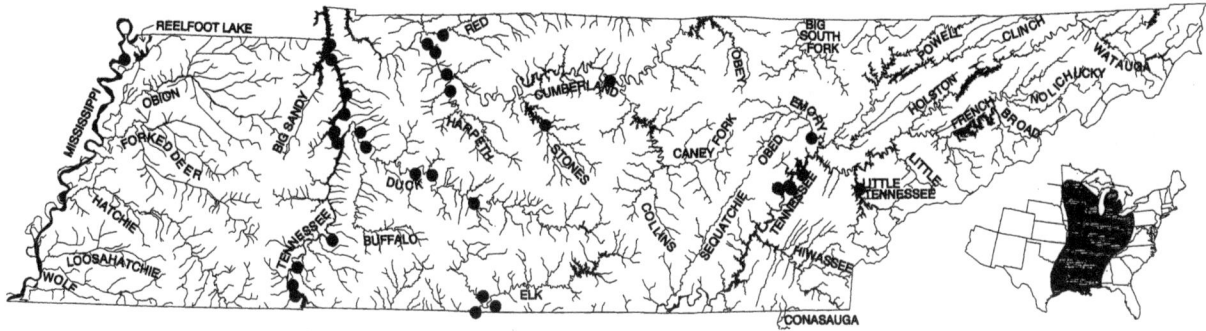

Range Map 117. *Truncilla donaciformis* (Lea, 1828), Fawnsfoot.

deeply impressed in mature individuals. The periostracum is a dull yellow or greenish, patterned with numerous, rather striking dark green rays which are interrupted or broken up into zigzag, triangular, or arrowhead lines. The nacre is silvery or bluish white and iridescent posteriorly.

Plate 118. *Truncilla donaciformis* (Lea, 1828), Fawnsfoot.

Life History and Ecology: One of the smallest and most beautifully marked naiads found in the state, *Truncilla donaciformis* may occur in both large and medium-sized rivers at normal depths varying from less than three feet up to 15 to 18 feet in big rivers such as the Tennessee. A substrate of either sand or mud is suitable for this species, and, although typically found in stretches of river with current, the Fawnsfoot can adapt to a lake or embayment environment lacking current. Baker (1928a) indicated that the breeding season is not clear, but that it is probably similar to *Truncilla truncata* (bradytictic). The freshwater drum *(Aplodinotus grunniens)* is the primary fish host for the glochidia of this mussel, rarely the sauger *(Stizostedion canadense)* (Surber, 1913; Wilson, 1916).

Status: Currently Stable (Williams et al., 1993:14).

Truncilla truncata Rafinesque, 1820
Deertoe

RANGE MAP 118; PLATE 119

Synonymy:

Truncilla truncata Rafinesque, 1819; Rafinesque, 1819a:426 [nomen nudum]

Truncilla truncata Rafinesque, 1820; Rafinesque, 1820:301

Truncilla truncata var. *fusca* Rafinesque, 1820; Rafinesque, 1820:301

Truncilla truncata var. *vermiculata* Rafinesque; 1820, Rafinesque, 1820:301

Truncilla truncata vermiculata Rafinesque, 1820; Frierson, 1927:89

Unio truncatus (Rafinesque, 1820); Say, 1834:no pagination

Amygdalonaias truncata (Rafinesque, 1820); Utterback, 1916a:348

Truncilla truncata Rafinesque, 1820; Ortmann and Walker, 1922:49

Unio elegans Lea, 1831; Lea, 1831:83, pl. 9 fig. 13

Margarita (Unio) elegans (Lea, 1831); Lea, 1836:18

Margaron (Unio) elegans (Lea, 1831); Lea, 1852c:23

Unio truncatus var. *elegans* Lea, 1831; Paetel, 1890:170

Plagiola elegans (Lea, 1831); Baker, 1898:91, pl. 21, fig. 1

Plagiola (Amygdalonaias) elegans (Lea, 1831); Simpson, 1900a:604

Amygdalonajas elegans (Lea, 1831); Ortmann, 1912a:328 [misspelling]

Truncilla truncata var. *lacustris* F. C. Baker, 1928; F. C. Baker 1928a:227, p. 78, figs. 1, 2

Type Locality: Ohio River.

General Distribution: Throughout the Mississippi River drainage, from western Pennsylvania to Michigan and Minnesota, south to Iowa, Kansas, Texas, Louisiana, Tennessee, and northern Alabama (Baker, 1928a). In Canada, southern Ontario in the Lake Erie and Lake St. Clair drainages (Clarke, 1981a).

Tennessee Distribution: Statewide, from the Clinch, Nolichucky, Elk, Duck, Cumberland, Stones, Harpeth, and Tennessee rivers in East and Middle Tennessee to the Hatchie in West Tennessee. Prior to 1960 it was known from the North Fork Obion River, Reelfoot Lake, Holston River, Caney Fork River, and the Red River (Starnes and Bogan, 1988).

Description: The shell is somewhat triangular in outline, solid, and moderately inflated; the anterior end is rounded, while the posterior end is pointed and obliquely truncated. The posterior ridge is prominent, sharply angled, and often has a shallow sulcus in front; the dorsal surface behind the beaks is wide and flattened. Individuals inhabiting large impoundments, such as the Tennessee and Cumberland River reservoirs, may become extremely thick-shelled and attain 65–70 mm in length; 40–50 mm is more typical for mature adults. Beaks are full, curved inward, and elevated well above the hinge line; sculpture consists of 3–4 fine ridges, the first being concentric, the others double-looped. Dark growth lines are often prominent, raised as low ridges, indicating rest periods.

The left valve has two strong, somewhat compressed, elevated, serrated, triangular pseudocardinal teeth; the two lateral teeth are moderately long, thin, elevated, and roughened. The right valve has a heavy, triangular, serrated, erect pseudocardinal tooth, sometimes with a low tubercular tooth in front. The lateral tooth is high, compressed, and roughened, occasionally with an incomplete, smaller inner lateral tooth developed. The lateral teeth are slightly curved and directed ventrally. The interdentum is narrow or absent; the beak cavity is fairly shallow. Muscle scars are lightly to fairly deeply impressed. The periostracum is yellow, yellowish brown, or greenish; it is sometimes rayless, especially in old shells, but usually with numerous distinct green rays of varying widths, often marked with darker zigzag or wavy blotches. The nacre is silvery white and iridescent posteriorly. An occasional specimen will have a pale pink nacre.

Life History and Ecology: *Truncilla truncata* is rather generalized in the type of substrate—usually a composite of fine gravel mixed with sand and mud—and the size of rivers it inhabits. It appears to be more common in medium-sized rivers, such as the Harpeth, although locally the Deertoe may become numerous in large rivers, such as the Cumberland River, where it has been encountered at depths of 12 to 18 feet. Although typically a river species, *Truncilla truncata* will establish viable populations in lakes lacking current. Ortmann (1919) indicated that this mussel was bradytictic as evidenced by gravid females with glochidia be-

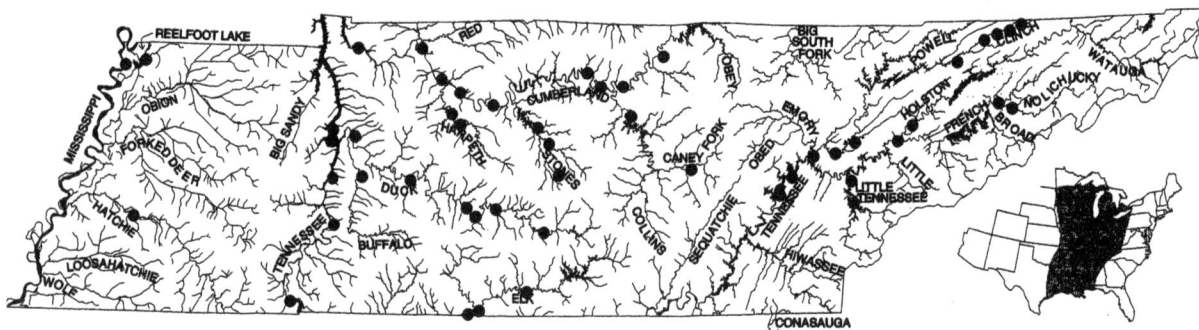

Range Map 118. *Truncilla truncata* Rafinesque, 1820, Deertoe.

Plate 119. *Truncilla truncata* Rafinesque, 1820, Deertoe.

ing found in May and July. Wilson (1916) recorded the sauger *(Stizostedion canadense)* and freshwater drum *(Aplodinotus grunniens)* as fish hosts for the glochidia of the Deertoe (Fuller, 1974).

Status: Currently Stable (Williams et al., 1993:14).

Uniomerus declivis (Say, 1831)
Tapered Pondhorn

RANGE MAP 119; PLATE 120
Synonymy:
Unio declivis Say, 1831; Say, 1831c:527
Margarita (Unio) declivis (Say, 1831); Lea, 1836:32
Margaron (Unio) declivis (Say, 1831); Lea, 1852c:33
Unio (Uniomerus) tetralasmus var. *declivis* Say, 1831; Simpson, 1900a:740
Elliptio (Uniomerus) declivis (Say, 1831); Frierson, 1927:34
Elliptio declivis (Say, 1831); Strecker, 1931:16
Uniomerus tetralasmus (Say, 1831); Utterback, 1915:244, pl. 21, figs. 69a, b [in part]
Uniomerus tetralasmus declivus (Say, 1831); Murray and Roy, 1968:29

Uniomerus declivis (Say, 1831); Morrison, 1976:10
Unio geometricus Lea, 1834; Lea, 1834:38, pl. 4, fig. 10
Unio declivis geometricus Lea, 1834; Frierson, 1903:49, pl. 3, middle fig.

Type Locality: Bayou Teche, Louisiana.

General Distribution: Simpson (1914:708), in considering *declivis* a variety of *Unio tetralasmus,* recorded this mussel as occurring in the "[l]ower part of the Gulf States from Alabama to Louisiana." However, Morrison (1976:11) stated that "[t]he species *declivis* Say is recorded from the Lake Erie drainage, Ohio and Indiana; from Tennessee; from the Coosa River system in Alabama, and southwest across Texas to the south side of the Rio Grande system in Chihuahua, Mexico."

Tennessee Distribution: The Tapered Pondhorn appears to be restricted to the Hatchie River system in West Tennessee (Manning, 1989): Bear Creek, Haywood County, and various, often temporary, overflow ponds, sloughs, and ditches.

Description: The shell is solid, somewhat inflated, and subrhomboid. The largest specimens obtained from Bear Creek (Hatchie River system), Haywood County, Tennessee, measures between 80 and 90 mm in length. Beaks are elevated, projecting only slightly above hinge line; the sculpture is similar to that of *Uniomerus tetralasmus,* consisting of 5–6 fairly concentric ridges that appear to radiate from a central point. The anterior end is rounded; the ventral margin is straight; the posterior end terminates in a rather distinct, sharp point. The posterior ridge is high and narrowly rounded; the dorsal-posterior slope is wide, appearing slightly depressed or as a shallow radial furrow.

The left valve has two low, moderately thin to stubby, finely serrated pseudocardinal teeth; the two lateral teeth are low, fairly short (for the length of the shell), and slightly curved. The right valve has a low, heavy, triangular pseudocardinal tooth; the single lateral tooth is thin, fairly low, and slightly curved. The interdentum is narrow to wanting, and the beak cavity is shallow; anterior muscle scars are moderately deep. The periostracum, somewhat roughened, is a dark tan or brown becoming considerably darkened in old individuals. The nacre is a dull bluish white, sometimes with a slight tint of purple.

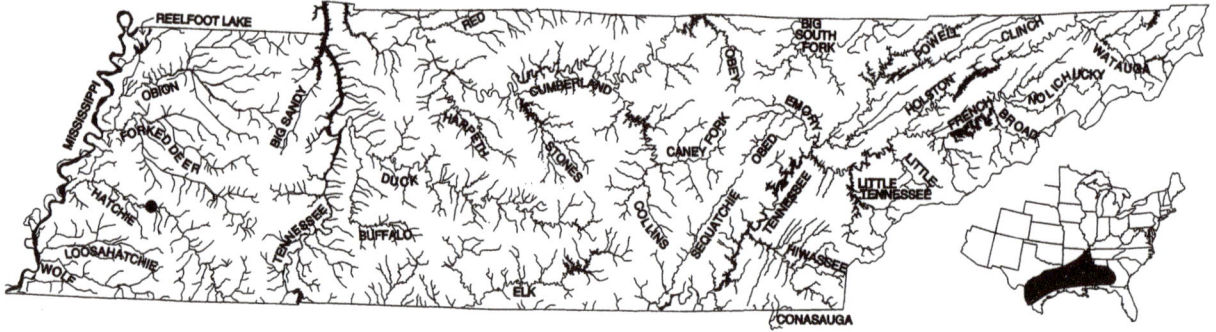

Range Map 119. *Uniomerus declivis* (Say, 1831), Tapered Pondhorn.

Life History and Ecology: Like the Pondhorn, this species may be encountered in shallow, quiet, or slow-moving water at depths seldom exceeding two feet. It is typically found buried in a substrate of fine sand and mud in shallow sloughs and ditches, and it is a species tolerant of adverse habitat conditions, surviving for periods of weeks or even months buried in the bottoms or banks of dried-up ponds. No information is available on the reproductive period of this mussel or about which species of fish serve as a host for its glochidia.

Status: Currently Stable (Williams et al., 1993:14).

Uniomerus tetralasmus (Say, 1831)
Pondhorn

RANGE MAP 120; PLATE 121

Synonymy:

Unio tetralasmus Say, 1831; Say, 1831a:pl. 13

Margarita (Unio) tetralasmus (Say, 1831); Lea, 1836:30

Margaron (Unio) tetralasmus (Say, 1831); Lea, 1852c:32

Unio (Uniomerus) tetralasmus Say, 1831; Simpson, 1900a:739

Uniomerus tetralasmus (Say, 1831); Ortmann, 1912a:272

Elliptio (Uniomerus) tetralasmus (Say, 1831); Frierson, 1927:34

Uniomerus tatralasmus (Say, 1831); F. C. Baker, 1928a:494 (index) [misspelling]

Uniomerus tetralasmus (Say, 1831); Strecker, 1931:14

Unio camptodon Say, 1832; Say, 1832:pl. 42

Margarita (Unio) camptodon (Say, 1832); Lea, 1836:30

Margaron (Unio) camptodon (Say, 1832); Lea, 1852c:32

Unio (Uniomerus) tetralasmus var. *camptodon* Say, 1832; Simpson, 1900a:740

Uniomerus tetralasma comptodon (Say, 1832); Utterback, 1915:48; Utterback, 1915:247 [misspelling]

Elliptio (Uniomerus) tetralasmus camptodon (Say, 1832); Frierson, 1927:34

Uniomerus tetralasmus camptodon (Say, 1832); Murray and Roy, 1968:29

Unio declivis Say, 1831; Conrad, 1836:45, pl. 23 fig. 1 [non Say]

Unio rivularis Conrad, 1853; Conrad, 1853:257 [replacement name for *Unio declivis* fide Conrad, 1836 non Say, 1831]

Unio excultus Conrad, 1838; Conrad, 1838:99, pl. 54, fig. 1

Unio sayanus Conrad, 1838; Conrad, 1838:102, pl. 56, fig. 2

Unio sayi Ward in Tappan, 1839; Tappan, 1839:268, pl. 3, fig. 1

Unio (Uniomerus) tetralasmus var. *sayi* Ward, 1839; Simpson, 1900a:741

Unio tetralasmus sayi Ward, 1839; Sterki, 1907:393

Plate 120. *Uniomerus declivis* (Say, 1831), Tapered Pondhorn.

Elliptio (Uniomerus) tetralasmus var. *sayi* (Ward, 1839);
    Frierson, 1927:34
Uniomerus tetralasmus sayi (Ward, 1839); F. C. Baker, 1928a:137
Unio parallelus Conrad, 1841; Conrad, 1841:20
Unio symmetricus Lea, 1845; Lea, 1845:164; Lea, 1848:73,
    pl. 4, fig. 11
Margaron (Unio) symmetricus (Lea, 1845); Lea, 1852c:32
Unio porrectus Conrad, 1854; Conrad, 1854:296, pl. 26, fig. 7
Unio subcroceus Conrad, 1854; Conrad, 1854:297, pl. 27, fig. 1
Unio manubius Gould, 1856; Gould, 1856:229
Margaron (Unio) manubius (Gould, 1856); Lea, 1870:54
Unio manubrius Gould, 1856; Paetel, 1890:158 [misspelling]
Unio (Uniomerus) tetralasmus var. *manubius* Gould, 1856;
    Simpson, 1900a:741
Unio (Uniomerus) tetralasmus var. *manubus* Gould, 1856;
    Simpson, 1914:708 [misspelling]
Elliptio (Uniomerus) tetralasmus manubius (Gould, 1856);
    Frierson, 1927:34
Uniomerus tetralasmus manubius (Gould, 1856); Strecker,
    1931:16
Uniomerus tetralasmus manubius (Gould, 1856); Murray and
    Roy, 1968:29
Unio jamesianus Lea, 1857; Lea, 1857c:84; Lea, 1858e:53, pl.
    6, fig. 35
Margaron (Unio) jamesianus (Lea, 1857); Lea, 1870:50
Unio electrinus Reeve, 1865; Reeve, 1865:pl. 25, fig. 121

Type Locality: Bayou St. John, New Orleans, Louisiana.

General Distribution: This mussel is found throughout much of the central and lower Mississippi River drainages. It occurs from Indiana west through Iowa and Missouri to Colorado, western Oklahoma, and Texas.

Tennessee Distribution: This species occurs in the Hatchie and Forked Deer rivers of West Tennessee, including various borrow pits, drainage ditches and backwater sloughs associated with these and other local river systems, as well as those in the Reelfoot Lake area (Najarian, 1955). Starnes and Bogan (1988:29) list the Pondhorn from the "Lower Tennessee River." Brown and Pardue (1980) recorded it from Hurricane Creek, a small tributary of the Tennessee River, Henderson and Hardin counties.

Description: The shell is somewhat elliptical to trapezoidal, usually moderately inflated, and thin to fairly solid. Mature individuals from a borrow pit (North Fork, Forked Deer River) along Tenn. Rt. 104, Gibson County, Tennessee, measure 125–135 mm in length. Beaks are somewhat full and elevated; sculpture consists of 6–7 prominent concentric ridges rounded up sharply behind, appearing to radiate from a central point. The anterior end is rounded; the ventral margin is straight; the posterior end is bluntly pointed; the posterior ridge is long and widely rounded. Usually a few concentric, slightly elevated rest period ridges are on the surface.

The left valve has two elevated (sometimes compressed), thin to somewhat stout, widely divergent pseudocardinal teeth; the two lateral teeth are long, narrow, and straight. The right valve has an elongated, sharply pointed, finely serrated pseudocardinal tooth, occasionally with a low, thin, ridgelike tooth anteriorly; the lateral tooth is long, thin, and straight. There is no interdentum; the beak cavity is shallow. Muscle scar impressions are moderately deep in mature individuals. The periostracum is rayless, yellowish brown or ashy to dark brown in color, and sometimes banded with lighter shades that are often shiny. The nacre is a dull white.

Life History and Ecology: The Pondhorn is a species that typically inhabits the quiet or slow-moving, shallow waters of sloughs, borrow pits, ponds, ditches, and meandering streams typical of West Tennessee. This mussel can become quite numerous locally, where it

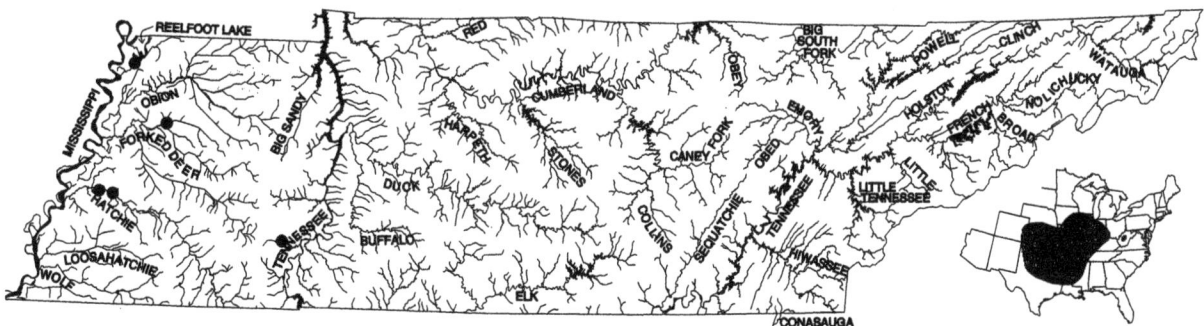

Range Map 120. *Uniomerus tetralasmus* (Say, 1831), Pondhorn.

Plate 121. *Uniomerus tetralasmus* (Say, 1831), Pondhorn.

may be found well buried in a substrate composed of fine sand, silt, and/or mud. *Uniomerus tetralasmus* has been known to survive for extended periods of time when the pond or slough in which it is living has temporarily dried up by burying itself deep into the substrate. This species may be bradytictic, judging from comments expressed by Utterback (1915–1916a). The fish host for the glochidia of specimens collected in Louisiana was identified as the golden shiner *(Notemigonus crysoleucas)* by Stern and Felder (1978).

Status: Currently Stable (Williams et al., 1993:14).

Utterbackia imbecillis (Say, 1829)
Paper Pondshell

RANGE MAP 121; PLATE 122

Synonymy:

Lastena ohiensis (Rafinesque, 1820); Utterback, 1915:260, pl. 23, figs. 74a, b [misidentification]

Anodonta ohiensis Rafinesque, 1820; Ortmann, 1919:162, pl. 11, fig. 4 [misidentification]

Anodonta (Lastena) ohiensis Rafinesque, 1820; Frierson, 1927:17 [misidentification]

Anodonta imbecillis Say, 1829; Say, 1829:355

Anodonta imbecilis Say, 1829; Lapham, 1852:370 [misspelling]

Margaron (Anodonta) imbecillis (Say, 1829); Lea, 1852c:50

Anodon imbecillis (Say, 1829); Sowerby, 1870:pl. 27, fig. 102

Utterbackia imbecillis (Say, 1829); F. C. Baker, 1928a:172, pl. 68, figs. 6–8

Anodonta (Utterbackia) imbecilis Say, 1829; Clarke and Berg, 1959:41, fig. 40 [misspelling]

Anodonta incerta Lea, 1834; Lea, 1834:36, pl. 6, fig. 16

Margarita (Anodonta) incerta (Lea, 1834); Lea, 1836:51

Anodon incerta (Lea, 1834); Catlow and Reeve, 1845:67

Anodon incertus (Lea, 1834); Sowerby, 1867:pl. 17, fig. 59

Anodon horda Gould, 1855; Gould, 1855:229

Anodonta hordeum (Gould, 1855); Paetel, 1890:180

Anodonta (Lastena) ohiensis horda (Gould, 1855); Frierson, 1927:17

Utterbackia imbecillis fusca F. C. Baker, 1927; F. C. Baker, 1927:222

Type Locality: Wabash River.

General Distribution: Mississippi River and Great Lakes drainages; from southern Michigan south to Georgia and northern Florida (Clench and Turner, 1956), west to Kansas, Oklahoma, and Texas, and southwest to extreme northeastern Mexico.

Tennessee Distribution: The Paper Pondshell occurs throughout the state from Watauga Lake in extreme northeastern Tennessee to the Hatchie River and Reelfoot Lake in West Tennessee. Although apparently absent from most high-gradient streams in East Tennessee and the Cumberland Plateau, it may be found in nearly all stretches of impounded rivers, as well as in lakes, ponds, sloughs, and other permanent standing bodies of water.

Description: The shell is thin, oblong, and inflated. Juveniles, however, are greatly compressed. In especially favorable habitat, individuals may exceed 100 mm in length and become extremely inflated, almost circular in cross section at the beaks. The posterior

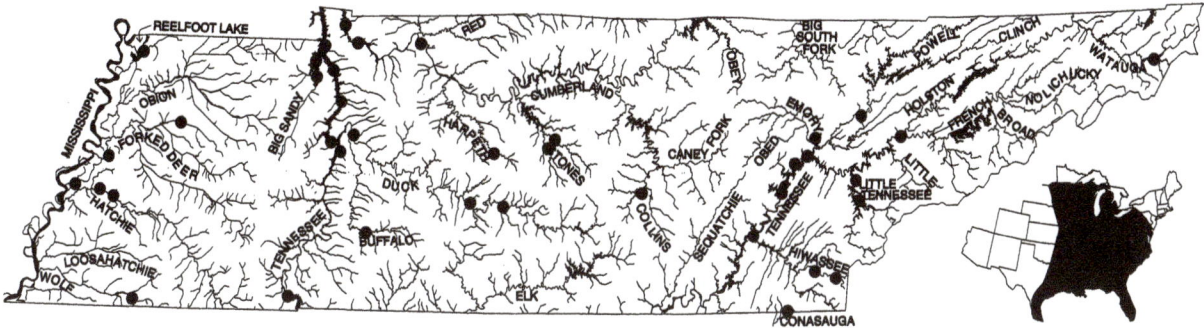

Range Map 121. *Utterbackia imbecillis* (Say, 1829), Paper Pondshell.

ridge is moderately angled; the dorsal and ventral margins are nearly straight and parallel. The anterior end is rounded; the posterior end is rather pointed. Beaks are flattened and usually flush with the hinge line; sculpture consists of 5–6 fine, irregular, often broken, somewhat concentric ridges, which are somewhat wavy, forming indistinct double loops. Rest periods are usually marked by distinct concentric ridges, edged with black. Both valves are edentulous; the hinge line is only very slightly thickened. The periostracum is yellowish or greenish with numerous fine green rays. The nacre is bluish white or silvery, and the outside margins and posterior end are iridescent; the shallow beak cavities are often cream or light yellowish brown.

Life History and Ecology: *Utterbackia imbecillis* is a species characteristic of nearly all impounded rivers in Tennessee where it inhabits the shallow bank and bay areas; it, like other closely related taxa *(Anodonta* and *Pyganodon)* occurring in Tennessee, thrives in a mud and fine sand substrate. Once it becomes established in a farm pond, borrow pit, or drainage canal, the Pond Mussel may become quite numerous. The species is probably bradytictic; Ortmann (1909b) suggested that it is an autumn breeder in Pennsylvania, noting gravid individuals in May and June. According to Sterki (1898), this species is hermaphroditic, and Baker (1928a) indicated that the parasitic developmental stage of glochidia on fish is often omitted. However, Tucker (1927) listed the green sunfish *(Lepomis cyanellus)* as a host for the Paper Pondshell, and Fuller (1978) reported the creek chub *(Semotilus atromaculatus)* as another. Stern and Felder (1978) also recorded the western mosquitofish *(Gambusia affinis),* warmouth *(Lepomis gulosus),* blue-gill *(L. macrochirus),* and dollar sunfish *(L. margina-*

tus) as host fish for the Paper Pondshell in Louisiana. Watters (1994a), citing Trdan and Hoeh (1982) and others, added the banded killifish *(Fundulus diaphanus),* largemouth bass *(Micropterus salmoides),* pumpkinseed *(Lepomis gibbosus),* rockbass *(Ambloplites rupestris),* and yellow perch *(Perca flavescens)* to the list of host fish. Hove et al. (1995), based on laboratory experiments, added the spotfin shiner *(Cyprinella spiloptera)* and black crappie *(Pomoxis nigromacula-*

Plate 122. *Utterbackia imbecillis* (Say, 1829), Paper Pondshell.

tus) to the list of fishes parasitized by glochidia of the Paper Pondshell. Watters (1997) had identified 26 exotic fish species, the tadpoles of the bullfrog and northern leopard frog, adult African clawed frogs, and larval tiger salamanders as surrogate hosts for the glochidia of *Utterbackia imbecillis*.

Status: Currently Stable (Williams et al., 1993:14).

Villosa fabalis (Lea, 1831)
Rayed Bean

RANGE MAP 122; PLATE 123

Synonymy:
Unio fabalis Lea, 1831; Lea, 1831:86, pl. 10, fig. 16
Margarita (Unio) fabalis (Lea, 1831); Lea, 1836:28
Margaron (Unio) fabalis (Lea, 1831); Lea, 1852c:31
Eurynia (Micromya) fabalis (Lea, 1831); Ortmann, 1912a:339
Micromya fabalis (Lea, 1831); Call, 1900:458
Micromya fabale (Lea, 1831); Wilson and Clark, 1912:51
Lemiox fabalis (Lea, 1831); Frierson, 1927:93
Villosa fabalis (Lea, 1831); Stein, 1963:19
Unio capillus Say, 1831; Say, 1831c:528
Unio lapillus Say, 1832; Say, 1832:pl. 41
Unio donacopsis De Gregorio, 1914; De Gregorio, 1914:60–61, pl. 10, fig. 5a–d

Type Locality: Ohio River.

General Distribution: *Villosa fabalis* is widely but discontinuously distributed; found in the Rogue River, Michigan, and in the St. Lawrence, Ohio, Duck, and upper Tennessee River drainages.

Tennessee Distribution: Ortmann (1918) found this small naiad chiefly in the headwaters of the Tennessee River in Tennessee. It was collected from the Tennessee River, Knox County; the Powell River from Campbell County upstream to Claiborne County; and the Clinch River in Claiborne County. *Villosa fabalis* also occurred in the South Fork Holston River, Sullivan County; the North Fork Holston River, Hawkins County; and from the Holston River, Knox County, upstream to Hawkins County (Ortmann, 1918; Stansbery, 1973). Specimens were also obtained from the Nolichucky River, Greene County (H. D. Athearn records). It formerly inhabited the Elk River in Lincoln County and the Duck River in Maury and Marshall counties (Ortmann, 1924a; 1925; Isom et al., 1973; van der Schalie, 1973). Two live specimens were collected in 1982 in the Duck River (Lillard Mill Dam), Marshall County (S. A. Ahlstedt, pers. comm., 1996). This species has not been recorded from the Cumberland River drainage.

Description: The shell of this small species, which only reaches about 38 mm in maximum length, is very solid and elongate, ovate, or elliptical in outline. Valves vary from subinflated to inflated. Male shells are usually long, ovate, and subinflated, while those of females are generally elliptical and somewhat more inflated but with the marsupial swelling only slightly pronounced and not distinct from the rest of the shell. Females are generally smaller and more inflated than males. Beaks are somewhat elevated, having a double-looped, subnodulous sculpture.

The left valve has two low, triangular pseudocardinal teeth, a short, narrow to wide interdentum, and two short, heavy lateral teeth. The right valve has a low triangular pseudocardinal tooth, occasionally with a smaller tooth before and behind. The lateral tooth is short, heavy, and somewhat elevated, occasionally with a vestigial tooth ventrally in some individuals. Adductor muscle scars are small and impressed; the pallial line is impressed. The periostracum is an ashy

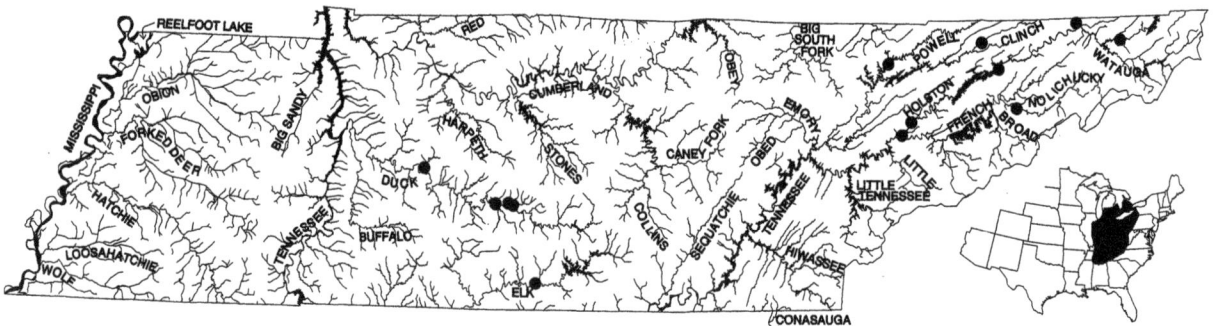

Range Map 122. *Villosa fabalis* (Lea, 1831), Rayed Bean.

green with wavy dark brown or greenish brown rays. The surface is marked by irregular growth lines. The nacre color is white or bluish and iridescent posteriorly.

Life History and Ecology: In West Virginia, Ortmann (1919) observed *V. fabalis* in and near riffles, generally in water weeds, and deeply buried in sand and gravel bound together by roots. La Rocque (1967) noted the habitat of *V. fabalis* as being sand among roots of aquatic vegetation in shallow water with current, a sand and gravel bottom, and in lakes with water up to four feet in depth.

This naiad is bradytictic (Ortmann, 1909b; van der Schalie, 1938). Gravid females were collected during mid- to late May, but it is not known when breeding commences (Ortmann, 1919). Host fish of *Villosa fabalis* unknown.

Status: Special Concern (Williams et al. 1993:14). It is apparently extirpated from all Tennessee rivers, except possibly the Duck River.

Plate 123. *Villosa fabalis* (Lea, 1831), Rayed Bean.

Villosa iris (Lea, 1829)
Rainbow

RANGE MAP 123; PLATE 124

Synonymy:
Unio iris Lea, 1829; Lea, 1829:439, pl. 11, fig. 18
Margarita (Unio) iris (Lea, 1829); Lea, 1836:37
Margaron (Unio) iris (Lea, 1829); Lea, 1852c:38
Lampsilis iris (Lea, 1829); Baker, 1898:105, pl. 13, fig. 1; pl. 14, fig. 2
Eurynia (Micromya) iris (Lea, 1829); Ortmann, 1912a:341, fig. 23
Micromya iris (Lea, 1829); Ortmann, 1924a:46
Eurynia iris (Lea, 1829); Ortmann, 1919:234
Eurynia (Micromya) iris (Lea, 1829); Ortmann, 1912a:341, fig. 23
Lampsilis iris (Lea, 1829); Simpson, 1900a:552
Villosa iris (Lea, 1829); Clarke and Berg, 1959:53, fig. 45
Villosa iris iris (Lea, 1829); Starrett, 1971:333
Lampsilis nervosa (Rafinesque, 1820); Frierson, 1927:77 [in part]
Ligumia nervosa (Rafinesque, 1820); Haas, 1969a:445–446 [in part]
Unio subrostratus Say, 1831; Küster, 1861:203, pl. 67, fig. 3 [misidentification]
Eurynia nebulosa (Conrad, 1834); Ortmann, 1913a:311 [in part]
Eurynia (Micromya) nebulosa (Conrad, 1834); Ortmann, 1915:64 [in part]
Micromya nebulosa (Conrad, 1834); Ortmann, 1924a:45–46 [in part]
Lampsilis (Ligumia) nebulosa (Conrad, 1834); Frierson, 1927:78 [in part]
Ligumia nebulosa (Conrad, 1834); Haas, 1969a:446–447 [in part]
Villosa nebulosa (Conrad, 1834); Burch, 1973:23 [in part]
Margarita (Unio) creperus Lea, 1836; Lea, 1836:28 [nomen nudum]
Unio creperus Lea, 1838; Lea, 1838b:33, pl. 10, fig. 28
Margarita (Unio) creperus (Lea, 1838); Lea, 1838a:20
Unio cresserus Lea, 1838; Hanley, 1842a:196 [misspelling]
Margaron (Unio) creperus (Lea, 1838); Lea, 1852c:31
Margarita (Unio) cumberlandianus Lea, 1836; Lea, 1836:27 [nomen nudum]
Unio cumberlandicus Lea, 1838; Lea, 1838b:25–26, pl. 7, fig. 19
Margaron (Unio) cumberlandianus (Lea, 1836); Lea, 1852c:30
Margaron (Unio) cumberlandicus (Lea, 1838); Lea, 1870:48
Margarita (Unio) glaber Lea, 1836; Lea, 1836:28 [nomen nudum]
Unio glaber Lea, 1838; Lea, 1838b:34, pl. 10, fig. 29
Margaron (Unio) glaber (Lea, 1838); Lea, 1852c:31
Margarita (Unio) mühlfeldianus Lea, 1836; Lea, 1836:27 [nomen nudum]
Unio mühlfeldianus Lea, 1838; Lea, 1838:41, pl. 12, fig. 36
Margarita (Unio) mühlfeldianus (Lea, 1838); Lea, 1838a:20
Margaron (Unio) mühlfeldianus (Lea, 1838); Lea, 1852c:30
Lampsilis mühlfeldianus (Lea, 1838); Simpson, 1900a:555
Unio muehlfeldianus Lea, 1838; Ortmann, 1918:578
Margarita (Unio) notatus Lea, 1836; Lea, 1836:26 [nomen nudum]

Unio notatus Lea, 1838; Lea, 1838b:28, pl. 8, fig. 22
Margarita (Unio) notatus (Lea, 1838); Lea, 1852c:29
Unio novi-eboraci Lea, 1838; Lea, 1838b:104, pl. 24, fig. 114
Margarita (Unio) novi-eboraci (Lea, 1838); Lea, 1838a:19
Margaron (Unio) novi-eboraci (Lea, 1838); Lea, 1852c:27
Lampsilis novi-eboraci (Lea, 1838); Sterki, 1907:389
Eurynia (Micromya) iris novi-eboraci (Lea, 1838); Ortmann, 1919:268, pl. 16, figs. 8, 9
Ligumia iris novi-eboraci (Lea, 1838); Baker, 1928a:260, pl. 86, figs. 8–12
Micromya iris novi-eboraci (Lea, 1838); Ortmann, 1924a:115
Margarita (Unio) obscurus Lea, 1836; Lea, 1836:26 [nomen nudum]
Unio obscurus Lea, 1838; Lea, 1838b:7, pl. 3, fig. 7
Margarita (Unio) obscurus (Lea, 1838); Lea, 1838a:20
Margaron (Unio) obscurus (Lea, 1838); Lea, 1852c:29
Lampsilis obscurus (Lea, 1838); Simpson, 1900a:549
Lampsilis obscura (Lea, 1838); Simpson, 1914:107
Margarita (Unio) simus Lea, 1836; Lea, 1836:29 [nomen nudum]
Unio simus Lea, 1838; Lea, 1838b:26, pl. 8, fig. 20
Margarita (Unio) simus (Lea, 1838); Lea, 1838a:21
Margaron (Unio) simus (Lea, 1838); Lea, 1852c:31
Lampsilis simus (Lea, 1838); Simpson, 1900a:556
Venustaconcha sima (Lea, 1838); Gordon, 1995:55–60
Margarita (Unio) zeiglerianus Lea, 1836; Lea, 1836:26 [nomen nudum]
Unio zeiglerianus Lea, 1838; Lea, 1838b:32, pl. 10, fig. 27
Margarita (Unio) zeiglerianus (Lea, 1838); Lea, 1838a:20
Margaron (Unio) zeiglerianus (Lea, 1838); Lea, 1852c:29
Margaron (Unio) zieglerianus (Lea, 1838); Lea, 1870:45 [misspelling]
Lampsilis zeiglerianus (Lea, 1838); Frierson, 1927:75
Ligumia zeigleriana (Lea, 1838); Haas, 1969a:439–440
Unio amoenus Lea, 1840; Lea, 1840:286; Lea, 1842b:200, pl. 10, fig. 12
Margaron (Unio) amoenus (Lea, 1840); Lea, 1852c:29
Lampsilis amoenus (Lea, 1840); Simpson, 1900a:555
Unio dactylus Lea, 1840; Lea, 1840:287; Lea, 1842b:196, pl. 9, fig. 7
Margaron (Unio) dactylus (Lea, 1840); Lea, 1852c:36
Unio fatuus Lea, 1840; Lea, 1840:287; Lea, 1842b:201, pl. 11, fig. 14
Margarita (Unio) fatuus (Lea, 1840); Lea, 1852c:38
Lampsilis fatua (Lea, 1840); Simpson, 1900a:553
Unio tener Lea, 1840; Lea, 1840:286; Lea, 1842b:198, pl. 10, fig. 10
Margaron (Unio) tener (Lea, 1840); Lea, 1852c:28
Lampsilis tener (Lea, 1840); Simpson, 1900a:555
Lampsilis tenera (Lea, 1840); Simpson, 1914:122–123
Ligumia tenera (Lea, 1840); Haas, 1969a:445
Unio regularis Lea, 1841; Lea, 1841b:82; Lea, 1842b:243, pl. 25, fig. 59
Margaron (Unio) regularis (Lea, 1841); Lea, 1852c:29
Unio puniceus Haldeman, 1842; Haldeman, 1842:201
Margaron (Unio) puniceus (Haldeman, 1842); Lea, 1870:46
Lampsilis puniceus (Haldeman, 1842); Simpson, 1900a:548
Lampsilis punicea (Haldeman, 1842); Simpson, 1914:104
Unio radiatus Gmelin, 1791; De Kay, 1843:189, pl. 17, fig. 236 [misidentification]
Unio spatulatus Lea, 1845; Sowerby, 1868:pl. 65, fig. 328 [misidentification]
Unio proximus Lea, 1852; Küster, 1861:248, pl. 83, fig. 4 [misidentification]
Unio discrepans Lea, 1860; Lea, 1860a:92; Lea, 1860e:340, pl. 55, fig. 165
Margaron (Unio) discrepans (Lea, 1860); Lea, 1870:48
Unio perpictus Lea, 1860; Lea, 1860c:306; Lea, 1860e:350, pl. 58, fig. 175
Margaron (Unio) perpictus (Lea, 1860); Lea, 1870:44
Unio planicostatus Lea, 1860; Lea, 1860a:92; Lea, 1860e:354, pl. 59, fig. 179
Margaron (Unio) planicostatus (Lea, 1860); Lea, 1870:43
Lampsilis planicostatus (Lea, 1860); Simpson, 1900a:553
Eurynia planicostata (Lea, 1860); Ortmann, 1913a:311
Unio scitulus Lea, 1860; Lea, 1860a:93; Lea, 1860e:342, pl. 55, fig. 167
Margaron (Unio) scitulus (Lea, 1860); Lea, 1870:45
Unio opalina Anthony, 1866; Anthony, 1866:146, pl. 7. fig. 2
Lampsilis fasciata opalina (Anthony, 1866); Frierson, 1927:72
Unio dispansus Lea, 1871; Lea, 1871:191; Lea, 1874:19, pl. 6, fig. 16

Comment: Simpson, 1914:106 put *dispansus* with *vanuxemensis* but Ortmann, 1918:578 placed it here.

Eurynia dispansa (Lea, 1871); Ortmann, 1913a:311

Comment: The name *Unio nebulosus* Conrad, 1834 has been variously used and confused with the *Villosa iris* complex. We have chosen to list all of the described taxa from the Ohio, Tennessee, and Cumberland River systems as synonyms of *Villosa iris* and restrict the use of *Villosa nebulosa* to the species occurring in the headwaters of the Mobile Bay Basin. We recognize that *Villosa iris* is a species complex and is probably composed of several valid species. However, this complex will not be resolved on conchological grounds (as attempted by Gordon, 1995, for example) but only with the combined application of detailed anatomy, behavior and biochemical techniques.

Type Locality: Ohio.

General Distribution: The Rainbow is found throughout the Tennessee, Cumberland, and Ohio River basins, the upper Mississippi River, and the St. Lawrence River system from Lake Huron to Lake Ontario including their tributaries (Burch, 1975a; Clarke, 1981a).

Tennessee Distribution: *Villosa iris* occurs throughout the upper Tennessee River drainage including the Powell, Clinch, Holston, Watauga, French Broad, Nolichucky, Little Pigeon, Little, Obed, Little Tennessee, Hiwassee, and Sequatchie rivers. In the lower Tennessee River

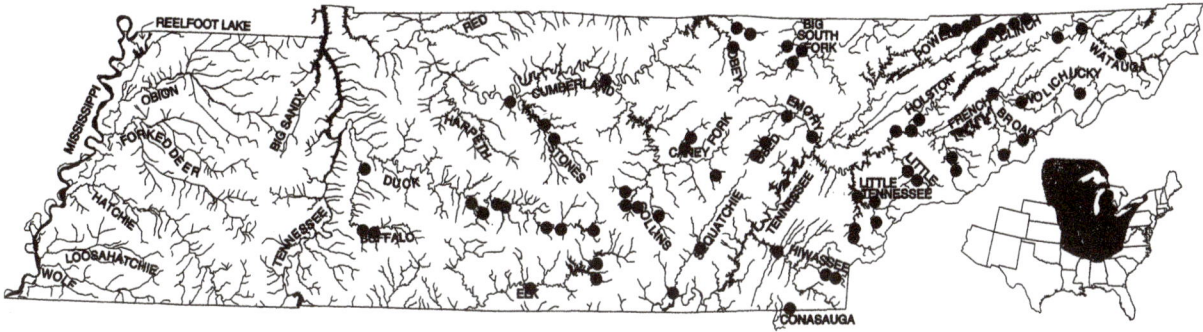

Range Map 123. *Villosa iris* (Lea, 1829), Rainbow.

system it may be found in numerous tributaries including the Elk, Buffalo, and Duck rivers. In the Cumberland River system, the Rainbow is found sporadically in the Big South Fork Cumberland, Obey, Caney Fork, Collins, and Stones rivers, but rarely in the mainstem of the Cumberland River (Starnes and Bogan, 1988). *Villosa nebulosa* (Conrad, 1834), the southern counterpart of the *Villosa iris* complex, occurs in the Conasauga River in southeastern Tennessee and is plotted with *Villosa iris* on Range Map 123.

Description: The Rainbow shell outline is elongate elliptical to long ovate, compressed to somewhat inflated; the shell is thicker anteriorly, becoming quite thin posteriorly. The anterior end is evenly rounded, the dorsal margin is almost straight to slightly convex, and the ventral margin is also almost straight to broadly curved. The posterior ridge is low and rounded. The male shells are more sharply pointed posteriorly, while the female shell is expanded posteriorly, producing a marsupial swelling and thus becoming more broadly rounded. Beaks are low and compressed; sculpture consists of 4–6 bars, the first concentric, the rest irregular, interrupted ridges, tending to become double-looped. The beak cavity is open and shallow with a few dorsal muscle scars. The maximum shell length attained is about 75 mm.

The left valve has two triangular, somewhat compressed, slightly sculptured pseudocardinal teeth which run parallel with the hinge line and two long, straight, thin lateral teeth. The right valve has one subcompressed pseudocardinal and one long, straight, and thin lateral tooth. The anterior adductor muscle scar is well impressed; the posterior adductor muscle scar is quite shallow. The pallial line is impressed anteriorly. The shell surface is covered with faint growth lines. The periostracum is yellowish to greenish yellow with numerous dark green rays varying from narrow to wide, complete or interrupted. The nacre color varies from white to salmon, pink and purple, and is iridescent posteriorly.

Life History and Ecology: Ortmann (1919) reported *Villosa iris* as bradytictic with glochidia from July to May. The Rainbow lives in riffles and along the edges

Plate 124. *Villosa iris* (Lea, 1829), Rainbow.

Family Unionidae 247

of emerging vegetation, such as *Justicia* beds, in gravel and sand in moderate to strong current. It becomes most numerous in clean, well-oxygenated stretches at depths of less than three feet. Watters (1994a) lists the following fish hosts for the glochidia of *Villosa iris*: largemouth bass *(Micropterus salmoides)*, smallmouth bass *(M. dolomieu)*, spotted bass *(M. punctulatus)*, Suwanee bass *(M. notius)*, rockbass *(Ambloplites rupestris)*, and western mosquitofish *(Gambusia affinis)*.

Status: Currently Stable (Williams et al., 1993:14).

Villosa lienosa (Conrad, 1834)
Little Spectaclecase
RANGE MAP 124; PLATE 125

Synonymy:
Unio lienosus Conrad, 1834; Conrad, 1834b:339, pl. 1, fig. 4
Margarita (Unio) lienosus (Conrad, 1834); Lea, 1836:26
Margaron (Unio) lienosus (Conrad, 1834); Lea, 1852c:29
Lampsilis lienosus (Conrad, 1834); Simpson, 1900a:547
Eurynia (=Micromya) lienosa (Conrad, 1834); Ortmann, 1912a:340–341
Lampsilis lienosa (Conrad, 1834); Simpson, 1914:100–101
Villosa (=Micromya) lienosa (Conrad, 1834); Parmalee, 1967:76
Ligumia lienosa (Conrad, 1834); Haas, 1969a:438
Villosa lienosa lienosa (Conrad, 1834); Oesch, 1984:204–206
Unio saxeus Conrad, 1838; Conrad, 1838:no. 11 (back cover). [Nomen dubium]
Unio saxeus Conrad, 1840; Conrad, 1840:109, pl. 60, fig. 1
Margaron (Unio) saxeus (Conrad, 1840); Lea, 1852c:27
Unio caliginosus Lea, 1845; Lea, 1845:165; Lea, 1848:79, pl. 7, fig. 21
Margaron (Unio) caliginosus (Lea, 1845); Lea, 1852c:29
Unio gouldii Lea, 1845; Lea, 1845:165; Lea, 1848:76, pl. 6, fig. 16
Margaron (Unio) gouldii (Lea, 1845); Lea, 1852c:20
Lampsilis gouldii (Lea, 1845); Simpson, 1900a:569
Unio nigerrimus Lea, 1852; Lea, 1852b:268, pl. 18, fig, 23
Unio niggerimus Lea, 1852; Frierson, 1927:74 [misspelling]

Margaron (Unio) nigerrimus (Lea, 1852); Lea, 1852c:31
Lampsilis nigerrimus (Lea, 1852); Simpson, 1900a:551
Lampsilis nigerrima (Lea, 1852); Simpson, 1914:113–114
Unio fuligo Reeve, 1856; Reeve, 1856:pl. 30, fig. 159
Unio apicinus Lea, 1857; Lea, 1857a:32; Lea, 1858e:76, pl. 14, fig. 56
Margaron (Unio) apicinus (Lea, 1857); Lea, 1870:44
Lampsilis apicinus (Lea, 1857); Simpson, 1900a:551
Lampsilis apicina (Lea, 1857); Simpson, 1914:112
Unio concestator Lea, 1857; Lea, 1857a:31; Lea, 1858e:66, pl. 12, fig. 48
Margaron (Unio) concestator (Lea, 1857); Lea, 1870:45
Lampsilis concestator (Lea, 1857); Simpson, 1900a:548
Unio fallax Lea, 1857; Lea, 1857a:32; Lea, 1858e:79, pl. 15, fig. 59
Margaron (Unio) fallax (Lea, 1857); Lea, 1870:45
Unio intercedens Lea, 1857; Lea, 1857a:32; Lea, 1858e:77, pl. 15, fig. 57
Margaron (Unio) intercedens (Lea, 1857); Lea, 1870:45
Unio prattii Lea, 1858; Lea, 1858c:166; Lea, 1859f:206, pl. 24, figs. 88, 88a
Margaron (Unio) prattii (Lea, 1858); Lea, 1870:45
Lampsilis prattii (Lea, 1858); Simpson, 1900a:550
Lampsilis pratti (Lea, 1858); Frierson, 1927:75 [misspelling]
Unio contiguus Lea, 1861; Lea, 1861d:392; Lea, 1862c:199, pl. 28, fig. 268
Margaron (Unio) contiguus (Lea, 1861); Lea, 1870:45
Unio porphyreus Lea, 1861; Lea, 1861c:60; Lea, 1862b:80, pl. 10, fig. 228
Unio bi-caelatus Reeve 1865; Reeve, 1865:pl. 26, fig. 130
Unio fontanus Conrad, 1866; Conrad, 1866b:279, pl. 15, fig. 13
Unio unicostatus B. H. Wright, 1899; B. H. Wright, 1899:69
Lampsilis lienosus var. *unicostata* (B. H. Wright, 1899); Simpson, 1900a:547
Lampsilis lienosa var. *unicostata* (B. H. Wright, 1899); Simpson, 1914:101–102

Type Locality: Small streams in South Alabama.

General Distribution: The Little Spectaclecase occurs in Gulf Coast rivers from the Suwannee River system in Florida west to Texas and Oklahoma, up the Mis-

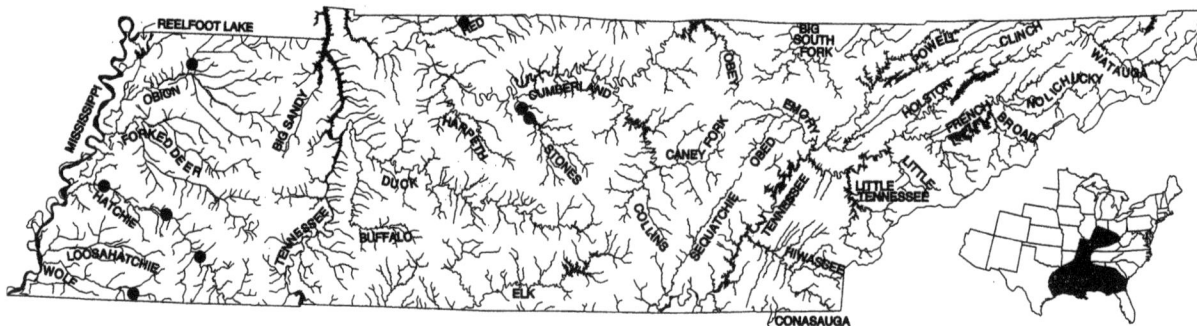

Range Map 124. *Villosa lienosa* (Conrad, 1834), Little Spectaclecase.

sissippi River to southern Missouri, up the Ohio and lower Wabash rivers, and in small streams in southern Ohio and West Virginia (Clench and Turner, 1956; Valentine and Stansbery, 1971; Burch, 1975a; Jenkinson and Kokai, 1977; Starnes and Bogan, 1988).

Tennessee Distribution: *Villosa lienosa* is found in direct tributaries to the Mississippi River in western Tennessee including the Wolf, Hatchie, and Obion rivers (Starnes and Bogan, 1988). Jenkinson and Kokai (1977) include the Stones and Red rivers in Tennessee on their distribution map, based on collections in the Museum of Zoology, The Ohio State University.

Description: Shells of the Little Spectaclecase are subelliptical, oval, or oblong in outline, often thin but strong and stout; shells of the males are only moderately inflated, while females are greatly inflated. The anterior end is evenly rounded; the dorsal margin is straight; the ventral margin is straight to slightly curved; the posterior margin in males is bluntly pointed; female shells are subangulate dorsally and truncated ventrally. The posterior ridge is low and rounded, and the posterior slope is flattened. Beaks are broad, but only slightly elevated above the hinge line; sculpture consists of 4–6 bars pulled up in the middle. The beak cavity is moderately deep. Maximum shell length is usually less than 70 mm.

The left valve has two erect, divergent, sculptured triangular pseudocardinal teeth and two short, curved lateral teeth. The right valve has one large, erect, sculptured pseudocardinal with a thin, elevated denticle anteriorly, and one short, curved lateral tooth. The interdentum is absent; the beak cavity is wide and shallow. The anterior adductor muscle scar is well defined, but the posterior adductor muscle scar is shallow and poorly defined. The pallial line is impressed anteriorly. The periostracum is yellowish brown to dark brown or black; young specimens may be olive green, and some may have faint green rays. The nacre color varies from white, salmon, or pink to deep purple and is iridescent posteriorly.

Life History and Ecology: Ortmann (1912a) reported a gravid female during November, making *Villosa lienosa* bradytictic. The Little Spectaclecase is found typically in shallow mud-bottomed creeks and small rivers, usually in slow current. Host fish for this species unknown.

Plate 125. *Villosa lienosa* (Conrad, 1834), Little Spectaclecase.

Status: Currently Stable (Williams et al., 1993:14). The Stones River population is now extirpated and its status in other known river locations in Tennessee is uncertain.

Villosa perpurpurea (Lea, 1861)
Purple Bean

RANGE MAP 125; PLATE 126

Synonymy:

Unio perpurpureus Lea, 1861; Lea, 1861a:41; 1866:46, pl. 16, fig. 44

Margaron (Unio) perpurpureus (Lea, 1861); Lea, 1870:48

Lampsilis perpurpureus (Lea, 1861); Simpson, 1900a:558

Eurynia (Micromya) perpurpurea (Lea, 1861); Ortmann, 1915:63

Eurynia perpurpurea (Lea, 1861); Goodrich, 1913:94

Micromya perpurpurea (Lea, 1861); Ortmann, 1924a:58

Villosa trabalis perpurpurea (Lea, 1861); Stansbery, 1971:14

Villosa perpurpurea (Lea, 1861); Bickel, 1968:22

Unio troostensis Lea, 1834; Sowerby, 1868:pl. 79, fig. 415 [misidentification]

Type Locality: Tennessee.

General Distribution: *Villosa perpurpurea* was found historically in the upper Tennessee River drainage in Tennessee and Virginia (Clinch River).

Tennessee Distribution: The Purple Bean was apparently quite rare in the Tennessee River drainage in Tennessee. *Villosa perpurpurea* was found (but is now extirpated) in the North Fork Holston River, Hawkins County, and in the Emory River, Roane County. It is also known from a few other localities, including Beech Creek, Hawkins County, and the Obed River, Cumberland County. Stansbery (1973) noted the occurrence of *V. perpurpurea* from unimpounded sections of the Clinch River above Norris Lake, but does not give exact localities in either Tennessee or Virginia. S. A. Ahlstedt (pers. comm., 1996) reported live specimens present in 1994 in the upper Clinch, Tazewell County, Virginia.

Description: The shell is elongate with slightly inflated, inequilateral and irregularly oval valves. The anterior end is rounded; the ventral margin is slightly rounded to straight, converging with the posterior-dorsal surface in a rounded point. The posterior ridge is low, somewhat full, and rounded. Male shells are slightly narrowed at their center and drawn out posteriorly. This elongation is obliquely truncated above and ends in a rounded point below. Female shells are higher and more evenly ovate and only slightly truncated behind and above the posterior ridge. The ventral margin is rather evenly curved. Beaks are fairly high and situated near the anterior end, where the shell is thickest, and sculptured with a few coarse, double-looped ridges. Female shells are often slightly larger than males, reaching a maximum of about 55 mm in length.

The left valve has two solid triangular pseudocardinal teeth, a narrow but long interdentum, and two long, straight, relatively heavy lateral teeth. The right valve has a large, sculptured triangular pseudocardinal tooth with a small tooth in front and behind. The lateral tooth is long, sometimes with a vestige of a second tooth below. The beak cavity is shallow with a few dorsal scars. Adductor muscle scars are deeply impressed; the pallial line is impressed anteriorly. The periostracum is a dingy olive green with numerous faint wavy green rays. The surface is marked by irregular growth lines. The nacre color exhibits varying intensities of purple with some iridescence posteriorly.

Shells of *Villosa perpurpurea* differ from those of *V. trabalis,* the Cumberland Bean, mainly by its purple nacre which may grade out to almost white. The shell is slightly more compressed than some specimens of *V. trabalis,* but this is a poor character. Eroded specimens of *V. trabalis* and *V. perpurpurea* are often difficult to separate (Simpson, 1914; Ortmann, 1925). Ortmann (1912a; 1915) described the anatomy of *V. trabalis* and *V. perpurpurea* and considered each a distinct species. Later, however, Ortmann (1918) noted that the only difference between *perpurpurea* and *trabalis* was the nacre color. This species has been viewed for many years as either falling within the variation of *Villosa trabalis* or as a subspecies or variety of *V. trabalis.*

Life History and Ecology: The Purple Bean is typically encountered in a substrate of coarse sand and gravel that includes some silt, in moderate to strong current, and at depths of less than three feet. It also occurs in rock piles and under large, flat rocks. This species is bradytictic, releasing glochidia in the spring and early summer. Suitable host fish for *Villosa perpurpurea* in-

Range Map 125. *Villosa perpurpurea* (Lea, 1861), Purple Bean.

Plate 126. *Villosa perpurpurea* (Lea, 1861), Purple Bean.

clude sculpin (*Cottus* sp.), greenside darter *(Etheostoma blennioides)*, and fantail darter *(E. flabellare)* (Watson and Neves, 1996).

Status: Endangered (Williams et al., 1993:15).

Villosa taeniata (Conrad, 1834)
Painted Creekshell

RANGE MAP 126; PLATE 127

Synonymy:
Unio taeniatus Conrad, 1834; Conrad, 1834a:26, 72, pl. 4, fig. 2
Margarita (Unio) taeniatus (Conrad, 1834); Lea, 1836:24
Lampsilis taeniata (Conrad, 1834); Simpson, 1900a:541
Ligumia taeniata (Conrad, 1834); Haas, 1969a:449
Unio pictus Lea, 1834; Lea, 1834:73, pl. 11, fig. 32
Margarita (Unio) pictus (Lea, 1834); Lea, 1836:24
Margaron (Unio) pictus (Lea, 1834); Lea, 1852c:27
Lampsilis picta (Lea, 1834); Simpson, 1900a:542
Eurynia (Micromya) picta (Lea, 1834); Ortmann, 1912a:342
Ligumia picta (Lea, 1834); Haas, 1969a:450

Unio menkianus Lea, 1838; Lea, 1838b:76, pl. 19, fig. 59
Margarita (Unio) menkianus (Lea, 1838); Lea, 1838a:19
Margaron (Unio) menkianus (Lea, 1838); Lea, 1852c:27
Lampsilis menkiana (Lea, 1838); Frierson, 1927:79
Ligumia menkiana (Lea, 1838); Haas, 1969a:449
Margarita (Unio) pulcher Lea, 1836; Lea, 1836:25 [nomen nudum]
Unio pulcher Lea, 1838; Lea, 1838b:6, pl. 3, fig. 6
Margaron (Unio) pulcher (Lea, 1838); Lea, 1852c:28
Margarita (Unio) interruptus Lea, 1836; Lea, 1836:24 [nomen nudum]
Unio interruptus Lea, 1838, non Rafinesque, 1820 [of authors], non Say, 1831; Lea, 1838b:15, pl. 6, fig. 15
Margaron (Unio) interruptus (Lea, 1838); Lea, 1852c:27
Unio latiradiatus Conrad, 1838; Conrad, 1838:96–97, pl. 53 [Replacement name for *Unio interruptus* Lea, 1838.]
Unio tennesseensis Lea, 1840; Lea, 1840:288; Lea, 1842b:199, pl. 10, fig. 11
Margaron (Unio) tennesseensis (Lea, 1840); Lea, 1852c:27
Unio camelopardilis Lea, 1860; Lea, 1860a:92; Lea, 1860e:355, pl. 59, fig. 180
Margaron (Unio) camelopardilis (Lea, 1860); Lea, 1870:53
Lampsilis camelopardilis (Lea, 1860); Simpson, 1900a:542
Unio fucatus Lea, 1860; Lea, 1860a:92; Lea, 1860e:353, pl. 59, fig. 178
Margaron (Unio) fucatus (Lea, 1860); Lea, 1870:43
Unio lindsleyi Lea, 1860; Lea, 1860c:306; Lea, 1860e:351, pl. 58, fig. 176
Margaron (Unio) lindsleyi (Lea, 1860); Lea, 1870:43
Unio perdix of authors non Lea, 1827; Reeve, 1864:pl. 18, fig. 82
Unio punctatus Lea, 1865; Lea, 1865:89; Lea, 1868b:261, pl. 32, fig. 76
Margaron (Unio) punctatus (Lea, 1865); Lea, 1870:43
Lampsilis punctatus (Lea, 1865); Simpson, 1900a:542

Type Locality: Flint River, Morgan County, Alabama.

General Distribution: The Painted Creekshell is restricted to the Tennessee and Cumberland River drainages in Kentucky, Tennessee, and Alabama.

Tennessee Distribution: In the Tennessee River system, *Villosa taeniata* is found only in the Elk, Buffalo, and Duck rivers; formerly it occurred in the lower main channel of the Tennessee River. In the Cumberland River Basin, the Painted Creekshell once inhabited the mainstem of the Cumberland River, but in recent times it is restricted to the Big South Fork Cumberland, Obey, Roaring, Caney Fork, Stones, Harpeth, and Red rivers (Starnes and Bogan, 1988).

Description: Shells of the Painted Creekshell are elongate oval to elliptical, solid, and rather compressed.

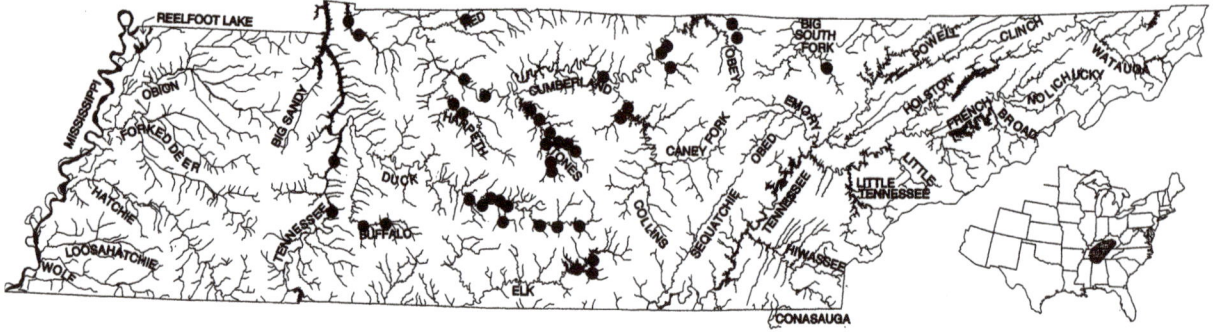

Range Map 126. *Villosa taeniata* (Conrad, 1834), Painted Creekshell.

The anterior end is evenly rounded, and the ventral margin is nearly straight to very broadly curved. The posterior margin in male shells is bluntly pointed at about the center of the posterior margin, while female shells have a wide marsupial swelling which ends in a broadly rounded posterior-ventral margin. A posterior ridge is lacking. Beaks are low, only slightly elevated above the hinge line; sculpture consists of several distinct, double-looped ridges, sometimes interrupted. The beak cavity is open and shallow; the interdentum is narrow or lacking. Maximum shell length seldom exceeds 80 mm.

The left valve has two equal-sized, erect, and sometimes slightly compressed pseudocardinal teeth and two long thin lateral teeth. The right valve has one erect pseudocardinal, often with a vestigial tooth posterior-dorsal to it and a low, narrow one anteriorly, and one long, thin, compressed lateral tooth. Anterior adductor muscle scars are well impressed, posterior adductor muscle scars are distinct. The pallial line is weakly impressed. The periostracum is sculptured with growth rings, its color being a greenish yellow with broad, slightly wavy green rays usually broken or interrupted. Juvenile shells are brightly colored, becoming dull with age. The nacre color is bluish white, often with a light salmon wash in the beak cavity area; there is some iridescence posteriorly.

Life History and Ecology: This mussel is presumed to be bradytictic like other species in the genus (e.g., Baker, 1928a). The Painted Creekshell is found in a substrate of mixed sand and gravel with good current in less than three feet of water. The rockbass *(Ambloplites rupestris)* has been identified as a host fish for the glochidia of *Villosa taeniata* (Gordon et al., 1994).

Status: Currently Stable (Williams et al., 1993:15).

Villosa trabalis (Conrad, 1834)
Cumberland Bean
RANGE MAP 127; PLATE 128
Synonymy:
Unio trabalis Conrad, 1834; Conrad, 1834a:27, 72 pl. 3, fig. 5.
Lampsilis trabalis (Conrad, 1834); Simpson, 1900a:558
Lampsilis (Venusta) trabalis (Conrad, 1834); Frierson, 1927:81

Plate 127. *Villosa taeniata* (Conrad, 1834), Painted Creekshell.

Lampsilis (Venustaconcha) trabalis (Conrad, 1834); Haas, 1969a:475
Eurynia (Micromya) trabalis (Conrad, 1834); Ortmann, 1915:340
Micromya trabalis (Conrad, 1834); Ortmann, 1924a:58
Villosa trabalis (Conrad, 1834); Stansbery, 1964:27
Unio troostensis Lea, 1834; Lea, 1834:71, pl. 10 fig. 30
Margarita (Unio) troostensis (Lea, 1834); Lea, 1836:21
Margaron (Unio) troostensis (Lea, 1834); Lea, 1852c:25
Unio troostensis Lea, 1834; Sowerby, 1868:pl. 79, fig. 415
Unio troostii Lea, 1866; Lea, 1866:47 [substitute name for *troostensis* Lea, 1834]
Margaron (Unio) troostii (Lea, 1866); Lea, 1870:39
Unio vanuxemensis Lea, 1838; Sowerby, 1866:pl. 39, fig. 216 [misidentification]

Type Locality: Flint River, Alabama.

General Distribution: The small Cumberland Bean is restricted to the upper Cumberland River system in Kentucky, formerly the main channel of the Tennessee River upstream from Muscle Shoals, Alabama, and in streams of the upper Tennessee River drainage in Tennessee and Virginia.

Tennessee Distribution: The Cumberland Bean was supposedly quite rare in the Tennessee River drainage in Tennessee. *Villosa trabalis* was reported from the Hiwassee River, Polk County (Ortmann, 1918; 1925); a viable population still exits there. It occurs with *V. perpurpurea* in Beech Creek, Hawkins County, and in the Obed River, Cumberland County, which suggests that they might be the same species. Stansbery (1973) noted the occurrence of *V. trabalis* from unimpounded sections of the Clinch River above Norris Lake, but does not give exact localities for either Tennessee or Virginia. S. A. Ahlstedt (pers. comm., 1996) reported finding it at Coytee Springs, Little Tennessee River (Loudon County) just prior to impoundment, and relic specimens from Little Chuckey Creek

(Greene County). The only modern record for *V. trabalis* from the Cumberland River system in Tennessee is from the lower Obey River, Clay County (Neel and Allen, 1964). As recently as 1995, three valves were recovered in a large archaeological shell midden sample from the Hogan Site (40SW24), Stewart County.

Description: Shells of the Cumberland Bean are solid and elongate with inflated, inequilateral, and irregularly oval valves. The anterior end is rounded, and the ventral margin slightly rounded to straight, converging with the posterior-dorsal surface in a rounded point. The posterior ridge is somewhat full and rounded. The male shell is slightly narrowed at its center and is drawn out posteriorly; this elongation is obliquely truncated above and ends in a rounded point below. Female shells are higher and more evenly ovate and only slightly truncated behind and above the posterior ridge. The ventral margin is rather evenly curved. Beaks are high, situated near the anterior end where the shell is thickest, and sculptured with a few coarse, double-looped ridges. Female shells reach a slightly larger size than males, attaining a maximum length of about 55 mm.

The left valve has two solid triangular pseudocardinal teeth, a narrow interdentum, and two long, straight, relatively heavy lateral teeth. The right valve has three pseudocardinals: the central tooth is large, sculptured, and triangular, while the anterior and posterior teeth are much reduced. The single lateral tooth is long, sometimes with a vestige of a second tooth below. The beak cavity is shallow with a few dorsal scars. Adductor muscle scars and the pallial line are well impressed. The periostracum is a dingy olive green with numerous faint wavy green rays. The surface is marked by irregular growth lines. The nacre color is a bluish white or white with a bluish iridescence posteriorly.

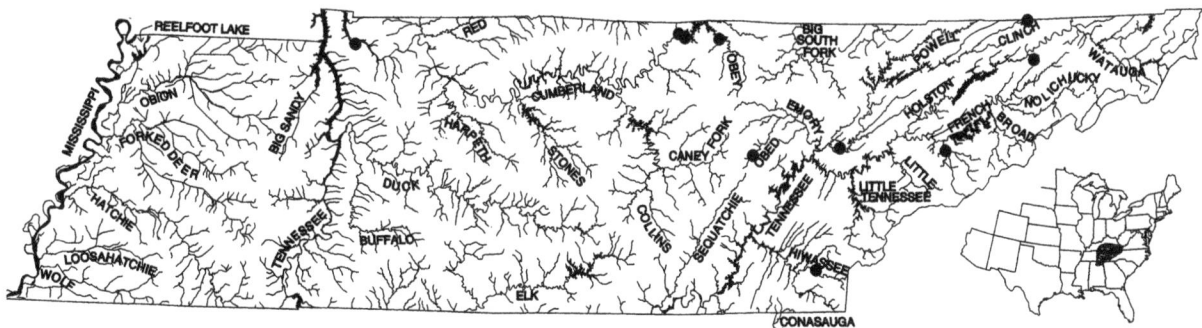

Range Map 127. *Villosa trabalis* (Conrad, 1834), Cumberland Bean.

Plate 128. *Villosa trabalis* (Conrad, 1834), Cumberland Bean.

Life History and Ecology: The Cumberland Bean has been collected in small rivers and streams in a typically gravel or sand and gravel substrate with fast current in riffle areas. The animal is bradytictic. Probable host fish for *Villosa trabalis* have been determined, based on laboratory experiments, as the arrow darter *(Etheostoma sagitta),* barcheek darter *(E. obeyense),* fantail darter *(E. flabellare),* johnny darter *(E. nigrum),* rainbow darter *(E. caeruleum),* snubnose darter *(E. simoterum atripinne),* sooty darter *(E. olivaceum),* striped darter *(E. virgatum),* and stripetail darter *(E. kennicotti).* However, the arrow darter is not found within the known range of *Villosa trabalis* and the johnny darter and rainbow darter both produced very few juveniles per fish (Layzer and Anderson, 1991, 1992, J. B. Layzer, pers. comm., 1997).

Status: Endangered (Williams et al., 1993:15). Populations of *Villosa trabalis* appear to be localized and restricted to a very few streams and rivers. Bogan and Parmalee (1983) listed this species as still occurring in the upper Cumberland River and its tributaries in Kentucky. A viable population is still extant in the Hiwassee River, Polk County (Parmalee and Hughes, 1994). The U.S. Fish and Wildlife Service has developed a recovery plan for this species (U.S. Fish and Wildlife Service, 1984g) and has created a watershed implementation schedule for the recovery plan (U.S. Fish and Wildlife Service, 1989b).

Villosa vanuxemensis (Lea, 1838)
Mountain Creekshell

RANGE MAP 128; PLATE 129

Synonymy:

Margarita (Unio) vanuxemensis (Lea, 1838); Lea, 1836:26 [nomen nudum]

Unio vanuxemensis Lea, 1838; Lea, 1838b:36, pl. 11, fig. 31

Margarita (Unio) vanuxemensis (Lea, 1838); Lea, 1838a:19

Margaron (Unio) vanuxemensis (Lea, 1838); Lea, 1852c:29

Lampsilis vanuxemensis (Lea, 1838); Simpson, 1900a:549

Eurynia (Micromya) vanuxemensis (Lea, 1838); Ortmann, 1912a:342

Eurynia vanuxemensis (Lea, 1838); Ortmann, 1913a:311

Ligumia vanuxemensis (Lea, 1838); Haas, 1969a:441

Micromya vanuxemensis (Lea, 1838); Isom and Yokley, 1968b:192

Villosa (Micromya) vanuxemensis (Lea, 1838); van der Schalie, 1973:46

Villosa vanuxemensis (Lea, 1838); Isom, 1968:516

Unio vanuxemii Lea, 1858; Lea, 1858e:83 [unjustified emendation]

Margaron (Unio) vanuxemii (Lea, 1858); Lea, 1870:46

Villosa vanuxemi (Lea, 1858); Isom and Yokley, 1968a:35 [misspelling]

Unio nitens Lea, 1840; Lea, 1840:288; Lea, 1842b:205, pl. 12, fig. 19

Margaron (Unio) nitens (Lea, 1840); Lea, 1852c:29

Unio caliginosus Lea, 1845; Pilsbry and Rhoads, 1896:501 [misidentification]

Unio tenebricus Lea, 1871; Lea, 1857f:171; Lea, 1858e:83, pl. 17, fig. 63

Margaron (Unio) tenebricus (Lea, 1857); Lea, 1870:45

Unio pybasii Lea, 1858; Lea, 1858a:40; Lea, 1862b:67, pl. 6, fig. 216

Margaron (Unio) pybasii (Lea, 1858); Lea, 1870:46

Unio fabaceus Lea, 1861; Lea, 1861a:38; Lea, 1862b:90, pl. 13, fig. 238

Margaron (Unio) fabaceus (Lea, 1861); Lea, 1870:46

Unio copei Lea, 1868; Lea, 1868a:144; Lea, 1868c:307, pl. 47, fig. 120

Margaron (Unio) copei (Lea, 1868); Lea, 1870:45

Nephronaias copei (Lea, 1868); Ortmann, 1913a:311

Type Locality: Cumberland River, Tennessee.

General Distribution: *Villosa vanuxemensis* is a Cumberlandian species restricted to the Tennessee and central Cumberland River basins (Ortmann, 1918, 1924a). Upper Coosa River system (Conasauga River, northern Georgia).

Tennessee Distribution: The Mountain Creekshell is found primarily in tributaries of the Tennessee River, including the Powell, Clinch, Holston, Watauga, Nolichucky, French Broad, Little Pigeon, Little, Little Tennessee, Emory, Hiwassee, Sequatchie, Elk, Buffalo, and Duck rivers. There are records of *Villosa vanuxemensis* in the Cumberland River system from the Stones and lower Cumberland rivers (Ortmann, 1918; Starnes and Bogan, 1988). *Villosa vanuxemensis umbrans* (Lea, 1857) is the recognized subspecies occurring in the Conasauga River of the Coosa River Basin in Polk County. Recognition of this subspecies is based solely on conchological characters. The true relationships among the species of *Villosa* need to be addressed with detailed anatomical and biochemical analyses.

Description: The Mountain Creekshell varies in outline from elliptical to somewhat obovate, the shell being rather solid and inflated. The anterior end is broadly rounded, the dorsal margin is straight to convex, and the ventral margin is broadly rounded to almost straight. The posterior ridge is slightly developed. Male shells are elongated elliptical in outline with the posterior margin rather sharply pointed, the point occurring in about the middle of the posterior margin. The female shell is marked by a prominent marsupial swelling along the posterior ventral margin; the shell is truncated beyond this to a point about two-thirds the way

Plate 129. *Villosa vanuxemensis* (Lea, 1838), Mountain Creekshell.

up the posterior margin. Shells in old females have a strong constriction posterior to the marsupial swelling. Beaks are low; sculpture consists of several ridges drawn up in the middle. The beak cavity is open and shallow. Maximum shell length rarely exceeds 70 mm.

The left valve has two short, compressed triangular pseudocardinal teeth and two slightly curved, thin

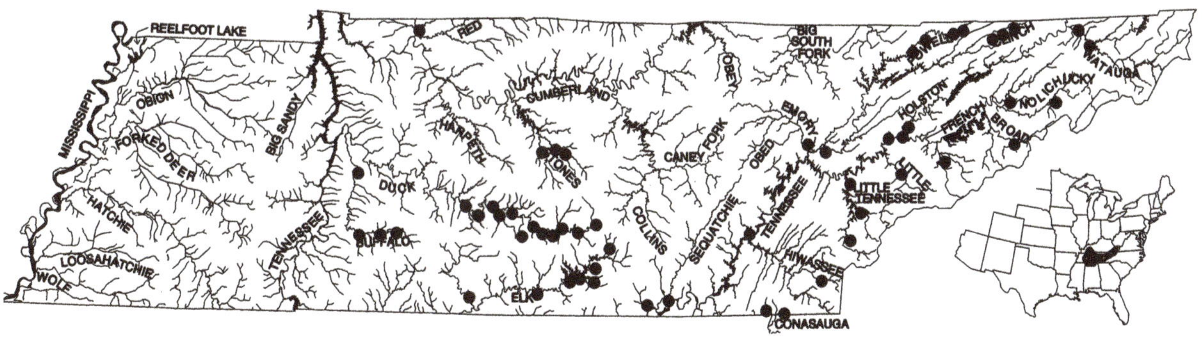

Range Map 128. *Villosa vanuxemensis* (Lea, 1838), Mountain Creekshell.

lateral teeth. The right valve has one short, compressed pseudocardinal and one curved, thin lateral tooth. The anterior adductor muscle scar is well impressed, while the posterior adductor muscle scar is shallow. The pallial line is impressed anteriorly. The periostracum varies from a tan or olive to dark brown, becoming black with age; rays are indistinct or absent. The nacre color varies from a light lavender or pinkish purple, to shades of copper or very dark purple.

Life History and Ecology: The Mountain Creekshell is found in gravel and sand substrates in riffles and along the edges of *Justicia* beds in very clean water at depths of less than three feet. It, along with *V. iris,* is a species most often encountered in small headwater creeks and streams throughout East and Middle Tennessee. *Villosa vanuxemensis* is bradytictic, holding glochidia from September to their release in May (Ortmann, 1921). Watters (1994a) lists the following fish hosts for the glochidia of *Villosa vanuxemensis:* banded sculpin *(Cottus carolinae),* black sculpin *(C. baileyi),* mottled sculpin *(C. bairdi),* and the slimy sculpin *(C. cognatus).*

Status: Special Concern (Williams et al., 1993:15).

Villosa vibex (Conrad, 1834)
Southern Rainbow

RANGE MAP 129; PLATE 130

Synonymy:
Unio vibex Conrad, 1834; Conrad, 1834a:31, pl. 4, fig. 3
Margarita (Unio) vibex (Conrad, 1834); Lea, 1836:27
Margaron (Unio) vibex (Conrad, 1834); Lea, 1852c:30
Lampsilis vibex (Conrad, 1834); Simpson, 1900a:559
Villosa vibex (Conrad, 1834); Clench and Turner, 1956:209, pl. 4, fig. 4
Ligumia vibex (Conrad, 1834); Haas, 1969a:442
Unio modioliformis Lea, 1834; Lea, 1834:97, pl. 13, fig. 40
Margarita (Unio) modioliformis (Lea, 1834); Lea, 1836:34
Margaron (Unio) modioliformis (Lea, 1834); Lea, 1852c:39
Lampsilis modioliformis (Lea, 1834); Simpson, 1900a:559
Ligumia modioliformis (Lea, 1834); Haas, 1969a:443
Unio exiguus Lea, 1840; Lea, 1840:287; Lea, 1842b:191, pl. 7 fig. 1
Margaron (Unio) exiguus (Lea, 1840); Lea, 1852c:27
Unio stagnalis Conrad, 1849; Conrad, 1849:152; Conrad, 1850:275, pl. 37, fig. 2
Margaron (Unio) stagnalis (Conrad, 1849); Lea, 1852c:27
Ligumia stagnalis (Conrad, 1849); Haas, 1969a:442
Margaron (Unio) prevostianus Lea, 1852; Lea 1852c:29 [nomen nudum]
Unio prevostianus Lea, 1852; Lea, 1852b:269, pl. 19, fig. 24

Margaron (Unio) nigrinus Lea, 1852; Lea, 1852c:39 [nomen nudum]
Unio nigrinus Lea, 1852; Lea, 1852b:284, pl. 24, fig. 44
Lampsilis vibex var. *nigrinus* (Lea, 1852); Simpson, 1900a:560
Unio gracilior Lea, 1856; Lea, 1856b:262; Lea, 1858e:56, pl. 8, fig. 38
Margaron (Unio) gracilior (Lea, 1856); Lea, 1870:45
Lampsilis gracilior (Lea, 1856); Simpson, 1900a:559
Ligumia gracilior (Lea, 1856); Haas, 1969a:448
Unio rutilans Lea, 1856; Lea, 1856b:262; Lea, 1858e:59, pl. 9, fig. 41
Margaron (Unio) rutilans (Lea, 1856); Lea, 1870:45
Unio subellipsis Lea, 1856; Lea, 1856b:262; Lea, 1858e:62, pl. 10 fig. 44
Margaron (Unio) subellipsis (Lea, 1856); Lea, 1870:45
Unio sudus Lea, 1857; Lea, 1857f:170; Lea, 1859f:194, pl. 21, fig. 77
Margaron (Unio) sudus (Lea, 1857); Lea. 1870:46
Lampsilis sudus (Lea, 1857); Simpson, 1900a:561
Unio obfuscus Lea, 1857; Lea, 1857f:172; Lea, 1859f:197, pl. 22, fig. 80
Margaron (Unio) obfuscus (Lea, 1857); Lea, 1870:45
Unio dispar Lea, 1860; Lea, 1860b:305; Lea, 1860e:327, pl. 51, fig. 153
Margaron (Unio) dispar (Lea, 1860); Lea, 1870:45
Lampsilis dispar (Lea, 1860); Simpson, 1900a:561
Unio averillii B. H. Wright, 1888; B. H. Wright, 1888a:115, pl. 3 fig. 4

Type Locality: Black Warrior River, south of Blount's Springs [Blount County], Alabama.

General Distribution: Gulf Coast drainages from the Pearl River, Mississippi, east across peninsular Florida, north to the Savannah River system, and the upper Coosa River drainage.

Tennessee Distribution: The Southern Rainbow's distribution in Tennessee is restricted to the Conasauga River, Polk County, the Hatchie River, Haywood and Tipton counties (Manning, 1989), and the Wolf River, Fayette County.

Description: Shells vary from thin to subsolid, being elliptical to elongate obovate in outline. The anterior and posterior margins are evenly rounded, and the ventral margin is straight to slightly curved in males and often arcuate in females. The dorsal margin is straight. The shell varies from slightly compressed to inflated. Male shells are often rhomboid with a bluntly pointed posterior margin, while female shells are slightly inflated with a broadly rounded posterior margin. The posterior ridge is broadly rounded. Beaks are only mod-

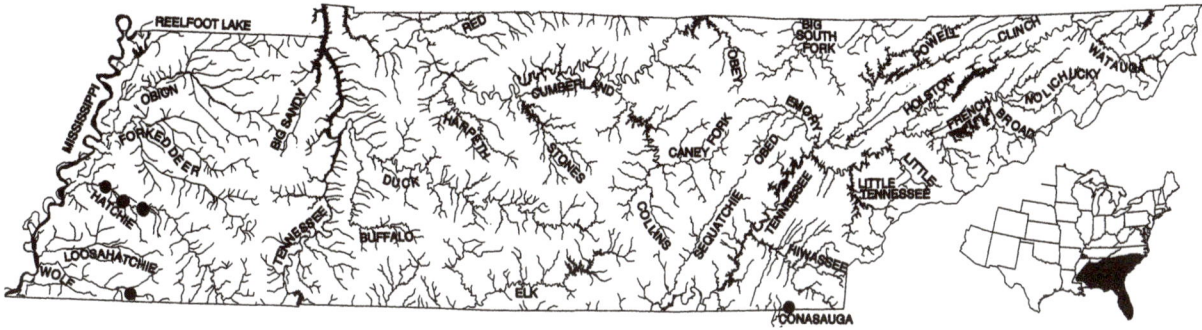

Range Map 129. *Villosa vibex* (Conrad, 1834), Southern Rainbow.

erately inflated and slightly elevated above the hinge line; sculpture consists of a few double-looped ridges. The shell length of adults averages about 60 mm but may reach 100 mm.

The left valve has two slightly compressed pseudocardinal teeth, the anterior tooth being longer and higher; the two lateral teeth are rather short and delicate. The right valve has a single pseudocardinal tooth, sometimes with a dorsal vestigial tooth and a single short lateral tooth. The beak cavity is fairly shallow and open. Adductor muscle scars are shallow, not impressed; the pallial line is lightly impressed. The periostracum is smooth and shiny, but interrupted by irregular growth lines. Color varies from a greenish yellow to olive brown, the surface covered with rather broad, unbroken to slightly wavy dark green rays over the entire surface. Some individuals have the rays restricted to the posterior area or are occasionally rayless. The nacre color is a bluish white, often becoming iridescent posteriorly.

Life History and Ecology: Johnson (1972:238) reported that *Villosa vibex* "[l]ives in small rivers, creeks, and lakes, in mud or soft sand, particularly where rich in vegetable detritus." In the Conasauga River in Tennessee, the Southern Rainbow occurs at depths of less than three feet, usually in stretches with moderate current. Haag et al. (1997) have shown through laboratory fish host identification experiments that the redeye bass *(Micropterus coosae),* spotted bass *(M. punctulatus),* and largemouth bass *(M. salmoides)* may serve as hosts for the glochidia of this mussel. No information on the breeding season for this species was encountered, but it is assumed to be bradytictic, holding glochidia from September to May, as is the case in other members of the genus *Villosa.*

Status: Currently Stable (Williams et al., 1993:15). *Villosa vibex* populations in the portion of the Conasauga River in Tennessee appear to be declining and may have already been extirpated.

Plate 130. *Villosa vibex* (Conrad, 1834), Southern Rainbow.

Family Corbiculidae

Corbicula fluminea (Müller, 1774)
Asian Clam
PLATE 131

General Distribution: The Asian Clam appears to have been introduced into North America sometime during or before the 1920s (Counts, 1986). It was first collected in the United States along the banks of the Columbia River in Pacific County, Washington, in 1938 (Burch, 1944), and since then it has invaded nearly every major river system in the country.

Tennessee Distribution: Discovery of a populations of *Corbicula fluminea* in 1959 below Pickwick Dam (Sinclair and Isom, 1961) appears to be the first official accounting of its presence in the state. With the exception of a few high-gradient creeks and streams in the upper Tennessee River drainage and on the Cumberland Plateau, the Asian Clam may be found in nearly every river, reservoir, lake, and pond in Tennessee.

Description: The shell is fairly small, seldom exceeding 50 mm in length, very solid, ovate when young, and triangular in outline when mature. Beaks are high, full, directed inward, and elevated well above the hinge line, and centrally located. Growth periods are indicated by thin, prominent concentric rings. There are three cardinal teeth directly below the beaks in each valve, with two straight to slightly curved lateral teeth on each side in the right valve and one on each side in the left valve. The lateral teeth are serrated, a character distinguishing *Corbicula* from the Sphaeriidae. The beak cavity is deep. The periostracum is a light yellowish olive to cream-colored in immature clams, changing with age to tan, olive, and, finally, dark brown to black in old individuals. Very young individuals possess a characteristic dark stripe or band on the anterior slope of the valves. The nacre is white to a shiny light purple, darkest along the lateral teeth and in the beak cavity. The entire inner surface of adults is a very light purple and white, appearing highly polished outside the pallial line.

Life History and Ecology: Unlike our native freshwater mussels, the juvenile or larva (called a veliger) of the Asian Clam is free swimming and does not require a host for partial development. Oesch (1984) noted that in Missouri the spawning time of *Corbicula* generally is between May and September. The period of growth of the free-swimming veliger lasts about 7–10 days.

In Tennessee the Asian Clam reaches its greatest population densities in a substrate of almost pure sand or one of mixed sand, silt, and mud. Although it thrives in rivers with slow to moderate current, typically at

Plate 131. *Corbicula fluminea* (Müller, 1774), Asian Clam.

depths of less than three feet, *C. fluminea* may become abundant and grow to a large size in the quiet waters of small ponds. This small clam is highly resistant to desiccation and can survive for weeks in damp sand or mud.

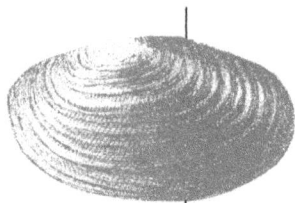

Family Dreissenidae

Dreissena polymorpha (Pallas, 1771)
Zebra Mussel

RANGE MAP 130; PLATE 132A, 132B

General Distribution: An invader from Europe's Black and Caspian seas, the Zebra Mussel entered North America about 1985 or 1986, probably in the larval (veliger) stage in jettisoned ballast water from an oceangoing freighter. It is believed this event took place on Lake St. Clair. Within the decade following its introduction, *Dreissena polymorpha* spread throughout the Great Lakes, the entire St. Lawrence River, the Hudson and Mohawk rivers, and the Illinois, Ohio, and Mississippi River systems in the Midwest south to the lower Tennessee and Cumberland rivers in the Midsouth. It has also been reported from the Arkansas River west to eastern Oklahoma and in the lower Mississippi River as far as New Orleans, Louisiana.

Tennessee Distribution: The Zebra Mussel has been reported from most sections of the entire lower Tennessee River, including all navigation locks from Kentucky Dam upstream to Ft. Loudoun Dam, Knox County. Its most upstream limit has been documented as the lower half mile of the French Broad River, which is the limit for navigation (barge traffic). It has also invaded the lower Cumberland River and has been found upstream as far as Nashville, Davidson County.

Description: *Dreissena polymorpha* is a small mussel, reaching a maximum length of about 40 mm. The shell is thin but strong, elongated, and inflated with the outer margin from the beak to the posterior end evenly rounded. Cardinal and lateral teeth are wanting; valves are hinged together by a short ligament. There is a small septum present at the anterior end of each valve for the attachment of the anterior adductor muscle; this forms a moderately deep beak cavity. The ventral surface is flat with a sharply angled high ridge running from the beaks to the posterior end. The periostracum varies from an unmarked cream color to one patterned with irregular, usually parallel, wide, curved brown or black bands. The nacre is a bluish white; the shell is nearly transparent in juveniles.

Life History and Ecology: The Zebra Mussel has adapted well to environmental conditions in our northern impounded rivers and the Great Lakes. In western Lake Erie, for example, as many as 700,000 mussels have been reported in one square meter of lake bottom. They may occur at depths of less than three feet to more than 20 feet. The free-swimming veliger attaches itself to any solid object by means of a glue and byssal threads. They grow as a crust, in clumps, or as a thick blanket covering any hard surface. When they occur in dense masses in municipal water intake pipes, for

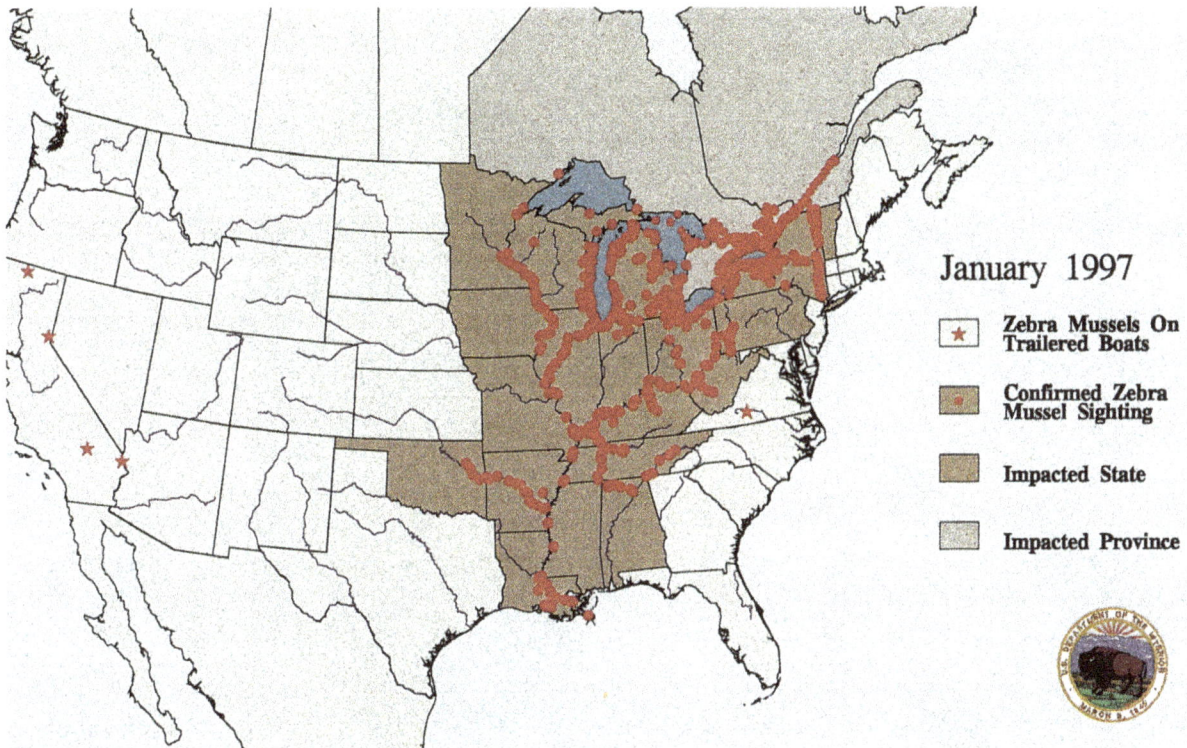

Range Map 130. *Dreissena polymorpha* (Pallas, 1771), Zebra Mussel distribution in North America as of January 1997. Reprinted with the permission of the U.S. Geological Survey, Biological Resources, Florida Caribbean Science Center in Gainesville, Florida. Non-indigenous aquatic species database.

example, eradication becomes a very costly operation. Although *Dreissena polymorpha* is widespread throughout much of the Tennessee and Cumberland River systems in Tennessee, by 1996 it had not become abundant. Only time will tell if this invader will become a detriment to the state's native aquatic fauna and an economic burden to its citizens.

Comments: There is another member of the Family Dreissenidae which is occasionally found in freshwater, the Dark Falsemussel, *Mytilopsis leucophaeata* (Conrad, 1831) (pl. 133). A few specimens of the Dark Falsemussel, a species most often inhabiting brackish waters, have been found recently in the Tennessee, Mississippi, and Ohio rivers (Koch, 1989; W. Pennington, pers. comm., 1995). *Mytilopsis* can be easily distinguished from *Dreissena* by the presence of a small, downward-projecting tooth on the edge of the shelf near the beak. The shell of *Mytilopsis* typically has a brown periostracum without bands, and the shell is long and narrow, rounded, and lacks the sharp posterior ridge characteristic of *Dreissena*. Abbott (1974) reported this species living in brackish to freshwater near rivers. Its occurrence in Tennessee (Tennessee River near New Johnsonville, TRM 100, Humphreys County), from an unknown source, is certainly fortuitous.

Family Dreissenidae 261

A

B

Plate 132A and 132B. *Dreissena polymorpha* (Pallas, 1771), Zebra Mussel.

Plate 133. *Mytilopsis leucophaeata* (Conrad, 1831),
Dark Falsemussel.

Glossary

Adductor muscle scar: large impressions in the nacreous layer of the shell near the anterior and posterior ends of the valve, the attachment area of the adductor muscle.

Alate: shells that have an anterior or posterior winglike projection of the valves that extends dorsally above the hinge line.

Angular, subangulate: having either the anterior or posterior margins forming a relatively acute (sharp) angle.

Anterior: front or forward.

Arcuate: bent in a bow or arched.

Beak: the raised portion of the dorsal margin of a shell; formed by the embryonic shell around which the rest of the shell develops distally in a concentric manner.

Beak cavity: the hollow on the inside of each valve leading into the beak, under the interdentum.

Beak sculpture: raised ridges or undulations on the umbo.

Biangulate: having two angles.

Bradytictic: mussels which are long-term breeders; females retain glochidia in their gills typically over the winter, as opposed to tachytictic.

Byssus, byssal threads: a bundle of tough threads secreted by the byssal gland in the foot of a bivalve, used to anchor the bivalve to some hard substrate.

Calcareous: composed of calcium, calcium carbonate.

Cardinal teeth: teeth located between the two sets of lateral teeth as in Corbiculidae and Sphaeriidae.

Chevron: shaped like a wide-angled V; appearing as lines or rays on the epidermis (periostracum).

Cilia, ciliary action: small, hairlike processes extending from cells, usually characterized by regular or rhythmic beating, resulting in movement of water and small particles.

Clade: a natural monophyletic group of two or more species.

Clinal variation: the graded variation in morphology over space, as exhibited by some species of mollusks from headwater areas to the mouth of the highest order stream.

Compressed, subcompressed: flattened out or pressed together.

Concentric: having a common center, such as ridges or loops radiating from the beak of a mussel valve.

Congeneric: belonging to the same group, e.g., the same genus.

Conglutinate: a mass of glochidia bound together in a gelatinous/mucous mass, often resembling prey items of potential host fish such as aquatic insect larvae, larval fish, fish eggs, or various worms.

Conspecific: pertaining to individuals or populations of the same species.

Corrugated: marked by wrinkles or ridges and grooves.

Crescentic: shaped like the figure of the crescent moon with a convex and a concave edge.

Demibranch: one-half of one of the paired gills of a freshwater bivalve.

Denticle, denticulate, denticulation: a small swelling, toothlike projection or minor tooth on the hinge line usually anterior to the major pseudocardinal tooth or teeth.

Disc: the middle or central portion of the exterior of a valve; distinct from the posterior slope or other areas immediately adjacent to the margin of the valve.

Discoidal: round and flat like a disc.

Dorsal: the top or back; in mussels, the hinge area.

Ecomorph (ecophenotype): a shell shape, form, or morphology which is determined or induced by nongenetic environmental factors.

Edentulous: lacking both pseudocardinal and lateral teeth.

Electrophoresis: a separation technique by which biological molecules such as proteins and nucleic acids are separated from one another on the basis of their electric charge and molecular weight.

Elliptical, subelliptical: elongated, having the form of an ellipse.

Elongate: long or extended.

Emarginate: having a shallow notching at the margin.

Endangered: this status at the state level includes peripheral forms which may be common in another part of their range, but whose continued existence within the political boundaries of the state is in danger of extirpation. At the national level, this status means the organism is in danger of extinction and is included on or being considered for the U.S. List of Endangered Fauna and Endangered and Threatened Plant Species of the United States, under the Endangered Species Act of 1973.

Epidermis: exterior or outside (corneous) layer of the shell (periostracum).

Extinct: a species that has no living representatives; all individuals of that species are no longer extant.

Extirpated: the elimination of a species from a portion of its total range; the loss of a local population.

Fluted: valves in which the posterior margin or slope is corrugate, the corrugations (usually parallel) opening onto the margin of the shell.

Form: an animal with questionable taxonomic status; that is, one exhibiting variation but the extent or degree is not well enough known to determine whether it is a species, subspecies, or simply an individual or population variation.

Fusiform: tapering toward each end.

Gills: a thin, platelike paired structure within the mantle cavity which serves as a respiratory organ in aquatic mollusks, and in female unionoids all four of the gills or certain portions of the outer gills serve as the marsupium.

Globose: globelike, spherical.

Glochidium (pl. glochidia): the bivalved larvae of freshwater mussels in the superfamily Unionoidea which are generally parasitic on the gills of fish.

Gravid female: a female that has embryos/glochidia in the marsupium.

Growth lines: compact lines of temporarily arrested growth or rest periods appearing on the epidermis of the shell as a raised or darker concentric line.

Hinge ligament: an elastic, elongate, corneous structure that unites the two valves dorsally along the hinge plate.

Hinge plate: the dorsal area of the unionoid mussel shell, including the pseudocardinal and lateral teeth and the interdentum, if present.

Holotype: single specimen designated as the "type" by the author in the publication of a new species-level taxon.

Inequilateral: in a bivalve, having the two ends unequal, i.e., one end is wider or thicker than the other.

Inflated, subinflated: moderately to greatly swollen.

Interdentum: a flattened area of the hinge plate between the pseudocardinal and lateral teeth.

Iridescent: showing lustrous colors like those of a rainbow.

Labial palps: a pair of thin, flattened structures extending from either side of the mouth region of bivalve mollusks to lie against the outer surface of the outer demibranch and the inner surface of the inner demibranch; they function to receive filtered particles from the surfaces of the demibranchs and sort them into those particles accepted for ingestion and those particles rejected as pseudofeces (Thorpe and Covich 1991:866).

Lachrymose: term describing teardrop-shaped pustules.

Lamellate: thin, leaf- or bladelike structure, lamellar.

Lateral teeth: the elongated, raised, and interlocking structures along the hinge line of the valve.

Lotic: running water environments including rivers, creeks, and springs.

Lunule: depressed area immediately anterior to the umbo.

Mainstem: main channel of a river, excluding tributaries.

Marsupial swelling: a section of the posterior ventral margin of certain female unionoid shell which is enlarged or inflated to provide space for expansion of the marsupium with the development of the glochidia.

Marsupium, marsupial pouch: in unionoids, a brood pouch for eggs and developing glochidia, formed by a restricted portion of the outer gill, the complete outer gill or all four gills.

Monophyletic, monophyly: a group that contains a common ancestor and all descendant taxa.

Muscle scar: the area of attachment of a muscle to the inside of the shell; e.g., the anterior adductor muscle scar is the location of attachment for the anterior adductor muscle.

Nacre: the interior iridescent, thin layer of a mussel shell.

Naiad: formerly a tribe of mollusca nearly equivalent taxonomically to the family Unionidae, often used as a synonym of unionoid.

Nodule, subnodulous: a small rounded mass of irregular shape.

Oblique: slanting; angled, but not horizontal or vertical.

Obovate, subobovate: ovate.

Orbicular, suborbicular: having the form of an orb; circular or nearly circular in outline.

Oval, Ovate, subovate: egg-shaped, broadly elliptical.

Overbank: areas of the floodplain, beyond the first natural levee (river bank), under water.

Pallial line: an indented groove or line approximately parallel with the ventral margin of a bivalve shell which marks the line of muscles attaching the mantle to the shell.

Paratype: each specimen of a type series other than the holotype designated in the original publication of the taxon.

Periostracum: exterior or outside (corneous) layer of the shell (epidermis).

Periphery: the external boundary on a surface, edge.

Phylogeny, phylogenetic: the complete evolutionary history of a group of organisms, descent with modification.

Plications: parallel ridges on the surface of the shell.

Posterior: hind or rear.

Posterior ridge: a ridge on the exterior of a mussel shell, extending from the umbo to the posterior margin.

Posterior slope: the area across the dorsal portion of the valve extending from the umbo to the posterior margin, often above or behind the posterior ridge.

Pseudocardinal teeth: triangular-shaped hinge teeth near the anterior-dorsal margin of the shell.

Pustule: small, raised structure on the external or outside surface of the shell (see also tubercle).

Quadrate, subquadrate: square or nearly square in outline.

Radial furrow: a groove or depression; in naiads a groove running from the umbo area toward the shell margin.

Radiating: proceeding outward from a central point.

Rare: seldom appearing, occurring widely separated in space; extremely few in number.

Rectangular: a shape with four sides possessing four right angles.

Rest mark: see growth lines.

Retractor muscle scar: smaller of the two impressions in the nacreous layer of the shell near the posterior and anterior ends of the valve, the attachment area of the pedal retractor muscles.

Rhomboid, subrhomboid: having generally four distinct sides, two sides being longer than the others.

Riverine: found in or characteristic of rivers.

Semicircular: a partial or incomplete circle.

Serrated: notched or grooved, like saw blade teeth.

Sexual dimorphism: a condition in which males and females of the same species are morphologically different, usually indicated in unionoids by an expanded posterior marsupial area in the female in contrast to a more pointed or bluntly rounded area in the male.

Sinuous: having a wavy or curved margin.

Sinus: a character of some unionoids which have a depression above or below the posterior ridge.

Solid, subsolid: shells which are thick and heavy.

Special Concern: this status covers cases where the organism exists in small populations over a broad range, may be over exploited which may pose a threat, the organism is especially vulnerable to specific pressures, or any other reasons identified by experienced researchers (Cooper et al. 1977:x).

Species: group of interbreeding natural populations that are reproductively isolated from all other such groups.

Striae: impressed or raised lines on a shell.

Striate: having striae.

Subspecies: a geographically defined aggregate of local populations within a species which differ morphologically and/or physiologically from other aggregations of local populations within that species.

Sulcus (pl. sulci): a longitudinal furrow or depression.

Sympatric: pertaining to populations of two or more closely related species which occupy identical or broadly overlapping geographical areas.

Syntype: one of a series of specimens of the same taxon which formed the material studied by the original author to describe a new species level taxon, from which no type specimen (holotype) was designated.

Tachytictic: mussels that are short-term breeders; i.e., when glochidia are found in the gills of the female only during the summer, as opposed to bradytictic.

Taxon (pl. taxa): any formal taxonomic unit or category of an organism; e.g., a species or genus.

Threatened: this status at the state level includes forms which are likely to become Endangered in the foreseeable future if certain conditions are not met. This includes forms which exhibit a considerable decrease in numbers beyond normal population fluctuations or a documented range contraction, but are not yet considered Endangered. At the national level this applies to the Endangered Species Act of 1973.

Torulus: twisted, gnarled, protuberant, swollen, as in tubercles of *Epioblasma torulosa torulosa*.

Trapezoid, subtrapezoid: a shape having four distinct sides with two sides parallel.

Triangular, subtriangular: a shape having three sides and three angles, like a triangle.

Truncate, subtruncate: having an end squared off.

Tubercle, tuberculate: small, raised, rounded knob on the outside of the shell.

Type: a designated specimen or specimens of an organism that serves as the basis for the original name and description of any species-level taxon.

Umbo, umbone: the dorsally raised, inflated area of the bivalve shell, centrally or anteriorly placed along the dorsal margin of the valve.

Undulation: pattern with waves; raised ridges or bars.

Unionoids: refers to any member of the freshwater bivalve mollusks that belong to the superfamily Unionoidea and by definition has glochidial larvae.

Valve: the right or left half of a mussel (or unionoid) shell.

Ventral: the underside or bottom.

Literature Cited

Abbott, R. T. 1974. *American Seashells. The marine Mollusca of the Atlantic and Pacific coasts of North America.* 2d ed. Van Nostrand Reinhold Co. New York. 663 pp.

Adams, C. B. 1842. Fresh-water and land shells of Vermont. Pp. 151–169 in *Thompson's History of Vermont,* Part 1. 169 pp.

Adams, H., and A. Adams. 1857. *The genera of Recent Mollusca; arranged according to their organization.* John van Vorst, Paternoster Row, London. 2:489–511.

Agassiz, L. 1852. Ueber die Gattungen unter den nordamerikanischen Najaden. *Archiv für Naturgeschichte* 18(1):41–52.

Ahlstedt, S. A. 1983. The molluscan fauna of the Elk River in Tennessee and Alabama. *American Malacological Bulletin* 1:43–50.

Ahlstedt, S. A. 1986. A status survey of the little-wing pearly mussel *Pegias fabula* (Lea, 1838). Endangered Species Field Office, U.S. Fish and Wildlife Service, Asheville, North Carolina. 38 pp.

Ahlstedt, S. A. 1991a. Twentieth century changes in the freshwater mussel fauna of the Clinch River (Tennessee and Virginia). *Walkerana* 5(13):73–122.

Ahlstedt, S. A. 1991b. Cumberlandian Mollusk Conservation Program. Activity 1: mussel surveys in six Tennessee Valley streams. *Walkerana* 5(13):123–160.

Ahlstedt, S. A., and T. A. McDonough. 1994. Summary of preoperational monitoring of the mussel fauna in upper Chickamauga Reservoir in the vicinity of the Watts Bar Nuclear Plant. Tennessee Valley Authority, Clean Water Initiative, Norris, Tennessee 37828. 30 pp.

Anthony, J. G. 1865. Descriptions of new species of North American Unionidae. *American Journal of Conchology* 1(2):155–164, 5 pls.

Anthony, J. G. 1866. Descriptions of new American freshwater shells. *American Journal of Conchology* 2(2):144–147, pls. 6, 7.

Apgar, A. C. 1887. The muskrat and the Unio. *Journal of the Trenton Natural History Society* 1:58–59.

Athearn, H. D. 1970. Discussion of Dr. Heard's paper. Symposium on Endangered Mollusks. *Malacologia* 10(1):28–31.

Baker, F. C. 1898. The Mollusca of the Chicago area, Part I: The Pelecypoda. *Bulletin of the Chicago Academy of Science* 3(1):1–130, 27 pls.

Baker, F. C. 1901. Some interesting molluscan monstrosities. *Transactions of the Academy of Science of St. Louis* 11:143–146, pl. 11.

Baker, F. C. 1905. The molluscan fauna of McGregor, Iowa. *Transactions of the Academy of Science of St. Louis* 15(3):249–258.

Baker, F. C. 1906. A catalogue of the Mollusca of Illinois. *Bulletin of the Illinois State Laboratory of Natural History* 7(6):53–136, 1 map.

Baker, F. C. 1920. *The life of the Pleistocene or glacial period as recorded in the deposits laid down by the great ice sheets.* University of Illinois Bulletin 17(41):xiv–476, 57 pls.

Baker, F. C. 1922a. New species and varieties of Mollusca from Lake Winnebago, Wisconsin, with new records from this state. *The Nautilus* 35(4):130–133; 36(1):19–21.

Baker, F. C. 1922b. The molluscan fauna of the Big Vermilion River, Illinois, with special reference to its modification as the result of pollution by sewage and manufacturing wastes. *Illinois Biological Monographs* 7(2):105–224, 15 pls.

Baker, F. C. 1923a. A new Anodontoides from Wisconsin. *The Nautilus* 36(4):123–125.

Baker, F. C. 1923b. The use of molluscan shells by the Cahokia mound builders. *Transactions of the Illinois State Academy of Science* 16:328–334.

Baker, F. C. 1924. The fauna of the Lake Winnebago region. A quantitative and qualitative survey, with special reference to the Mollusca. *Transactions of the Wisconsin Academy of Sciences, Arts and Letters* 21:109–146, 1 map.

Baker, F. C. 1926. The naiad fauna of the Rock River system: a study of the law of stream distribution. *Transactions of the Illinois State Academy of Science* 19:102–112.

Baker, F. C. 1927. On the division of the Sphaeriidae into two subfamilies: and the description of a new genus of Unionidae, with descriptions of new varieties. *The American Midland Naturalist* 10(7):220–223.

Baker, F. C. 1928a. The fresh water Mollusca of Wisconsin. Part II. Pelecypoda. *Bulletin of the Wisconsin Geological and Natural History Survey. University of Wisconsin.* 70(2):i–vi + 1–495.

Baker, F. C. 1928b. The Mollusca of Chautauqua Lake, New York, with descriptions of a new variety of *Ptychobranchus* and of *Helisoma*. *The Nautilus* 42(2):48–60.

Baker, F. C. 1941. A study of ethnozoology of the prehistoric Indians of Illinois. *Transactions of the American Philosophical Society* 32(1, Part 2):51–77.

Baker, H. B. 1964a. Some of Rafinesque's unionid names. *The Nautilus* 77(4):140–142.

Baker, H. B. 1964b. Dromus not a homonym. *The Nautilus* 77(4):142.

Ball, G. H. 1922. Variation in fresh-water mussels. *Ecology* 3(2):93–121, figs 1–6, tables i–iv.

Barnes, D. H. 1823. On the genera *Unio* and *Alasmodonta*; with introductory remarks. *American Journal of Science and Arts* 6(1):107–127; 6(2):258–280, 13 pls.

Barnes, D. H. 1828. Reclamation of Unios. *American Journal of Science and Arts* 13(2):358–364. Reprinted in *Sterkiana* 8(1962):35–38.

Barnhart, C., and A. Roberts 1996. When clams go fishing. *Missouri Conservationist* 57(2):22–25.

Barnhart, M. C., A. D. Roberts, and A. P. Farnsworth. 1995. Fish hosts of four unionids from Missouri and Kansas. *Triannual Unionid Report* 7:22.

Barr, W. C., S. A. Ahlstedt, G. D. Hickman, and D. M. Hill. 1993–1994. Cumberlandian mollusks Conservation Program. Activity 8: analysis of macrofauna factors. *Walkerana* 7(17/18):159–224.

Bates, J. M., and S. D. Dennis. 1978. The mussel fauna of the Clinch River, Tennessee and Virginia. *Sterkiana* 69–70:3–23.

Bickel, D. 1968. Checklist of the Mollusca of Tennessee. *Sterkiana* 31:15–39.

Binney, W. G. 1885. A manual of American land shells. *Bulletin of the U.S. National Museum* No. 28:1–528.

Black, D. W., and R. H. Whitehead. 1988. Prehistoric shellfish preservation and storage on the Northeast Coast. *North American Archaeologist* 9(1):17–30.

Boepple, J. F., and R. E. Coker. 1912. Mussel resources of the Holston and Clinch rivers of eastern Tennessee. *Reports and Special Papers of the U.S. Bureau of Fisheries.* Issued separately as U.S. Bureau of Fisheries Document 765. 1911:1–13.

Bogan, A. E. 1977. Analysis of faunal sample from Site FN-3 (Table 1). *In:* Archaeological and historical sites survey of the proposed Fuqua Energy, Inc., Coal Processing Facilities, Marion County, Tennessee. Report prepared for Fuqua Energy, Inc., Atlanta, Georgia, and the U.S. Army Corps of Engineers, Nashville District.

Bogan, A. E. 1978. Faunal remains from the Wiser-Stephens I Site (40CF81), Coffee County, Tennessee. Pp. 427–434. Appendix A to Excavations at the Wiser-Stephens Site by R. P. Stephen Davis, Jr., *In:* Sixth Report of the Normandy Archaeological Project, edited by M. C. R. McCollough and C. H. Faulkner. *The University of Tennessee, Department of Anthropology, Reports of Investigations No. 21.*

Bogan, A. E. 1983. Faunal remains from the historic Cherokee occupation at Citico (40MR7), Monroe County, Tennessee. *Tennessee Anthropologist* 8(1):28–49.

Bogan, A. E. 1987. Chapter II. Faunal analysis: a comparison of Dallas and Overhill Cherokee subsistence strategies. With additional bone and shell artifact analyses by R. R. Polhemus. Vol. 2. Pp. 971–1111. *In:* The Toqua Site: A late Mississippian Dallas Phase town. By Richard R. Polhemus, *University of Tennessee Department of Anthropology Reports of Investigations No. 41 and Tennessee Valley Authority Publications in Anthropology No. 44.*

Bogan, A. E. 1988. A bibliographic history of C. S. Rafinesque's work on North American freshwater bivalves. *Archives of Natural History* 15(2):149–154.

Bogan, A. E. 1990. Stability of recent unionid (Mollusca: Bivalvia) communities over the past 6000 years. Pp. 112–136. *In:* W. Miller III, editor. Paleocommunity Temporal Dynamics: The Long-Term Development of Multispecies Assemblages. *The Paleontological Society Special Publication No. 5.*

Bogan, A. E. 1997. A resolution of the nomenclatural confusion surrounding *Plagiola* Rafinesque, *Epioblasma* Rafinesque, and *Dysnomia* Agassiz (Mollusca: Bivalvia: Unionidae). *Malacological Review* 30: [in press]

Bogan, A. E., and C. M. Bogan. 1985. Faunal Remains, Chapter 7. Pp. 369–410. *In:* Archaeological Contexts and Assemblages at Martin Farm by Gerald F. Schroedl, R. P. Stephen Davis, Jr., and C. Clifford Boyd, Jr. *University of Tennessee, Department of Anthropology, Reports of Investigations No. 39, Tennessee Valley Authority Publications in Anthropology, No. 37.*

Bogan, A. E., and P. W. Parmalee. 1977. Comments on the faunal material from test excavations at 40RH62, Rhea County, Tennessee, pp. 35–37; 63–64. *In:* Phase II intensive survey of the proposed Tenna-Tech, Inc., Industrial Site in Rhea County, Tennessee. By Douglas Prescott. Report prepared for Tenna-Tech, Inc., Federal Antiquities Act Permit 77-TN-078.

Bogan, A. E., and P. W. Parmalee. 1983. *Tennessee's rare wildlife. Volume II: The Mollusks.* Tennessee Wildlife Resources Agency and Tennessee Department of Conservation. Nashville, Tennessee. 123 pp.

Bogan, A. E., and L. B. Starnes. 1983. The Little River unionid fauna and its implications for unionid biogeography (Abstract). *American Malacological Bulletin* 1:93–94.

Bosc, L. A. G. 1801–1804. *Histoire naturelle des coquilles, contenant leur description, les moeurs des animaux qui les habitent et leurs usages. Avec figures dessinées d'après nature.* Deterville, Paris. Five volumes.

Breitburg, E. n.d.a. Freshwater mussels from the Stone Site (40SW23), Stewart County, Tennessee. Species list on file, Tennessee Division of Archaeology, Department of Environment and Conservation, Nashville, Tennessee.

Breitburg, E. n.d.b. Freshwater mussels from the Meeks Site (40MT37), Montgomery County, Tennessee. Species list on file, Tennessee Division of Archaeology, Department of Environment and Conservation, Nashville, Tennessee.

Breitburg, E. n.d.c. Freshwater mussels from Dunbar Cave (40MT43), Cumberland/ Red rivers at Clarksville, Montgomery County, Tennessee. Species list on file, Tennessee Division of Archaeology, Department of Environment and Conservation, Nashville, Tennessee.

Breitburg, E. n.d.d. Freshwater mussels and vertebrates from the Archaic Cockrill Bend Site (40DV68), Cumberland River, Davidson County, Tennessee. Species list on file, Tennessee Division of Archaeology, Department of Environment and Conservation, Nashville, Tennessee.

Breitburg, E. 1983a. An analysis of faunal remains recovered from the Duncan Tract Site (40TR27), Trousdale County, Tennessee. *In:* The Duncan Tract Site (40TR27), Trousdale County, Tennessee. By C. H. McNutt and G. G. Weaver. *Tennessee Valley Authority Publications in Anthropology* No. 33.

Breitburg, E. 1983b. Paleoenvironmental exploitation strategies: the faunal data. *In:* P. A. Cridlebaugh, Penitentiary Branch: A late Archaic Cumberland River shell midden in Middle Tennessee. Contract Report to the Tennessee Division of Archaeology, Tennessee Department of Conservation, Contract #FA9234; Allotment 327.12, Nashville, Tennessee.

Brown, R. N., and W. J. Pardue. 1980. *Uniomerus tetralasmus* (Say, 1830) (Peleycypoda: Unionidae), a new distribution record for a freshwater mussel collected from the lower Tennessee River drainage. *Journal of the Tennessee Academy of Science* 55(3):108.

Bruenderman, S. A. 1989. Life history of the endangered fine-rayed pearly mussel, *Fusconaia cuneolus* (Lea, 1840), in the Clinch River, Virginia. Unpublished M.S. Thesis. Virginia Polytechnic Institute and State University. Blacksburg, Virginia. 114 pp.

Bruenderman, S. A., and R. J. Neves. 1993. Life history of the endangered fine-rayed pigtoe, *Fusconaia cuneolus* (Bivalvia: Unionidae) in the Clinch River, Virginia. *American Malacological Bulletin* 10(1):83–91.

Burch, J. B. 1973. Freshwater Unionacean clams (Mollusca: Pelecypoda) of North America. *Biota of Freshwater Ecosystems. Identification Manual* 11. U.S. Environmental Protection Agency. 176 pp.

Burch, J. B. 1975a. *Freshwater Unionacean clams (Mollusca: Pelecypoda) of North America.* Rev. ed. Malacological Publications. Hamburg, Michigan. 204 pp.

Burch, J. B. 1975b. *Freshwater sphaeriacean clams (Mollusca: Pelecypoda) of North America.* Malacological Publications, Hamburg, Michigan. xi + 96 pp.

Burch, J. Q. 1944. Checklist of west American Mollusks. *Minutes. Conchology Club of Southern California* 38:18.

Butler, R. S. 1989. Distributional records for freshwater mussels (Bivalvia: Unionidae) in Florida and south Alabama, with zoogeographic and taxonomic notes. *Walkerana* 3(10):239–261.

Calkins, W. W. 1874. The land and fresh water shells, of LaSalle County, Ills. *Proceedings of the Ottawa Academy of Sciences* 48 pp., 1 pl.

Call, R. E. 1880. Polymorphous anodontae. *American Naturalist* 14:529–530.

Call, R. E. 1885. A geographic catalogue of the Unionidae of the Mississippi valley. *Bulletin of the Des Moines Academy of Science* 1(1):5–57.

Call, R. E. 1895. A study of the Unionidae of Arkansas, with incidental reference to their distribution in the Mississippi valley. *Transactions of the Academy of Science of St. Louis* 7(1):1–65, 21 pls.

Call, R. E. 1896. A revision and synonymy of the parvus group of Unionidae. *Proceedings of the Indiana Academy of Science for 1895*:109–119, 6 pls.

Call, R. E. 1900. A descriptive illustrated catalogue of the Mollusca of Indiana. Indiana *Department of Geology and Natural Resources, 24th Annual Report* 1899:335–535, 1013–1017 [index], pls. 1–78.

Call, S. M., and P. W. Parmalee 1981. The discovery of extant populations of *Alasmidonta atropurpurea* (Rafinesque) (Bivalvia: Unionidae) in the Cumberland River basin. *Bulletin of the American Malacological Union, Inc., for 1981:* 42–43.

Casey, J. L. 1986. The prehistoric exploitation of unionacean bivalve molluscs in the lower Tennessee–Cumberland–Ohio River valleys in Western Kentucky. Unpublished M.S. Thesis. Simon Fraser University. Burnby, British Columbia. 176 pp.

Catlow, A., and L. Reeve. 1845. *The Conchologists Nomenclator.* A catalogue of all the Recent species of shells, etc. London. 326 pp.

Chadwick, G. H. 1906. Notes on Wisconsin Mollusca. *Bulletin of the Wisconsin Natural History Society* 4(3):67–99.

Chamberlain, T. K. 1934. The glochidial conglutinates of the Arkansas fanshell, *Cyprogenia aberti* (Conrad). *Biological Bulletin* (Woods Hole) 66:55–61.

Chenu, J. C. 1859–1862. *Manuel de Conchyliologie et de paléontologie conchyliologique.* 2 Volumes. Paris. 1(159):1–508; 2(1862):1–327.

Cicerello, R. R., M. L. Warren, Jr., and G. A. Schuster. 1991. A distributional checklist of the freshwater unionids (Bivalvia: Unionoidea) of Kentucky. *American Malacological Bulletin* 8(2):113–129.

Claassen, C. 1994. Washboards, pigtoes, and muckets: historic musseling in the Mississippi watershed. *Historical Archaeology* 28(2):1–145.

Clarke, A. H. 1973. The freshwater molluscs of the Canadian Interior Basin. *Malacologia* 13(1–2):1–509.

Clarke, A. H. 1981a. *The freshwater molluscs of Canada.* National Museum of Natural Sciences, National Museum of Canada, Ottawa, Canada. 446 pp.

Clarke, A. H. 1981b. The tribe Alasmidontini (Unionidae: Anodontinae), Part I. *Pegias, Alasmidonta,* and *Arcidens. Smithsonian Contributions to Zoology* No. 326. iii + 101 pp.

Clarke, A. H. 1985. The tribe Alasmidontini (Unionidae: Anodontinae), Part II: *Lasmigona* and *Simpsonaias. Smithsonian Contributions to Zoology* No. 399. iii + 75 pp.

Clarke, A. H., and C. O. Berg. 1959. The freshwater mussels of central New York with an illustrated key to the species of northeastern North America. *Memoir Cornell University Agricultural Experiment Station,* New York State College of Agriculture, Ithaca, New York 367:1–79.

Clench, W. J. 1959. Mollusca. Pp. 1117–1160. *In:* W. T. Edmondson, editor. *Freshwater Biology.* 2d ed. John Wiley and Sons, Inc. New York. 1248 pp.

Clench, W. J., and R. D. Turner. 1956. Freshwater mollusks of Alabama, Georgia, and Florida from the Escambia to the Suwannee River. *Bulletin of the Florida State Museum, Biological Sciences* 1(3):97–239, pls. 1–9.

Clessin, S. 1873–1876. Anodonta. *In:* Küster, H. C., editor. *Systmatiches Conchylien-Cabinet von Martini und Chemnitz.* Küster edition. 9(Part 1):288 pp., pl. 1–87, A. 11.

Coker, R. E. 1919. Fresh-water mussels and mussel industries of the United States. *Bulletin of the U.S. Bureau of Fisheries* 36(1917–1918):13–89, 46 pls. Issued separately as U.S. Bureau of Fisheries Document 865.

Coker, R. E., A. F. Shira, H. W. Clark, and A. D. Howard. 1921. Natural history and propagation of fresh-water mussels. *Bulletin of the Bureau of Fisheries* 37(1919–1920):77–181, 17 pls. Issued separately as U.S. Bureau of Fisheries Document 893.

Conrad, T. A. 1831. Descriptions of fifteen new species of recent and three of fossil shells, chiefly from the coast of the United States. *Journal of the Academy of Natural Sciences of Philadelphia* 6:256–268, pl. 11.

Conrad, T. A. 1834a. *New freshwater shells of the United States, with coloured illustrations; and a monograph of the genus Anculotus of Say; also a synopsis of the American naiades.* J. Dobson, 108 Chestnut Street, Philadelphia, Pennsylvania. 1–76, 8 pls.

Conrad, T. A. 1834b. Descriptions of some new species of fresh water shells from Alabama, Tennessee, etc. *American Journal of Science and Arts* 25(2):338–343, 1 pl. Reprinted in *Sterkiana* 9(1963):49–50.

Conrad, T. A. 1835a. Additions to, and corrections of, the Catalogue of species of American Naiades, with descriptions of new species and varieties of Fresh Water Shells. Pp. 1–8, pl. 9. Appendix to: *Synoptical table to New freshwater shells of the United States, with coloured illustrations; and a monograph of the genus Anculotus of Say; also a synopsis of the American naiades.* J. Dobson, 108 Chestnut Street, Philadelphia, Pennsylvania.

Conrad, T. A. 1835b–1840. *Monography of the Family Unionidae, or naiades of Lamarck, (fresh water bivalve shells) or North America, illustrated by figures drawn on stone from nature.* J. Dobson, 108 Chestnut Street, Philadelphia, Pennsylvania. 1835, 1:1–12, [pp. 13–16 not published], pls. 1–5; 1836, 2:17–24, pls. 6–10; 1836, 3:25–32, pls. 11–15; 1836, 4:33–40, pls. 16–19; 1836, 5:41–48, pls. 21–25; 1836, 6:49–56, pls. 26–30; 1836, 7:57–64, pls. 32–36; 1837, 8:65–72, pls. 36–40; 1837, 9:73–80, pls. 41–45; 1838, 10:81–94, 2i, pls. 46–51; 1838, 11:95–102, pls. 52–57; 1840, 12:103–110, pls. 58–60; [1840], 13:111–118, pls. 61–65.

Conrad, T. A. 1841. [Descriptions of three new species of Unio from the rivers of the United States.] *Proceedings of the Academy of Natural Sciences of Philadelphia* 1(2):19–20.

Conrad, T. A. 1849. Descriptions of new fresh water and marine shells. *Proceedings of the Academy of Natural Sciences of Philadelphia* 4(7):152–155.

Conrad, T. A. 1850. Descriptions of new fresh water and marine shells. *Journal of the Academy of Natural Sciences of Philadelphia* 1[New Series]:275–280, pls. 37–39.

Conrad, T. A. 1853. A synopsis of the family of Naïades of North America, with notes, and a table of some of the genera and sub-genera of the family, according to their geographical distribution, and descriptions of genera and sub-genera. *Proceedings of the Academy of Natural Sciences of Philadelphia* 6(7):243–269.

Conrad, T. A. 1854. Descriptions of new species of Unio. *Journal of the Academy of Natural Sciences of Philadelphia* 2(4)[New Series]:295–298, pls. 26, 27.

Conrad, T. A. 1855. Descriptions of three new species of Unio. *Proceedings of the Academy of Natural Sciences of Philadelphia* 7(7):256.

Conrad, T. A. 1866a. Description of a new species of Unio. *American Journal of Conchology* 2(2):107, pl. 10.

Conrad, T. A. 1866b. Descriptions of American fresh-water shells. *American Journal of Conchology* 2(3):278–279, pl. 15.

Cooper, J. E., S. S. Robinson, J. B. Funderburg, editors. 1977. *Endangered and threatened plants and animals of North Carolina*. North Carolina State Museum of Natural Sciences, Raleigh. xvi + 444 pp.

Cooper, W. 1834. List of shells collected by Mr. Schoolcraft in western and northwestern territory. Pp. 153–156. Appendix to *Narrative of Expedition through Upper Missouri to Itasca Lake, etc., under the direction of Henry B. Schoolcraft*. Harper & Bros., New York.

Counts, C. L. III. 1986. The zoogeography and history of the invasion of the United States by *Corbicula fluminea* (Bivalvia: Corbiculidae). *American Malacological Bulletin Special Edition* No. 2:7–39.

Cragin, F. W. 1887. A new species of Unio from Indian Territory. *Bulletin of the Washburn College Laboratory of Natural History* 1(8):6.

Crouch, E. A. 1827. *An illustrated introduction to Lamarck's Conchology, contained in his Histoire Naturelle des Animaux sans Vertebres, etc*. London. 22 colored pls.

Cummings, K. S., and J. M. K. Berlocher. 1990. The naiades or freshwater mussels (Bivalvia: Unionidae) of the Tippecanoe River, Indiana. *Malacological Review* 23(1–2):83–98.

Cummings, K. S., and C. A. Mayer. 1992. Field guide to freshwater mussels of the Midwest. *Illinois Natural History Survey, Manual 5*. 194 pp.

Cummings, K. S., and C. A. Mayer. 1993. Distribution and host species of the federally endangered freshwater mussel, *Potamilus capax* (Green, 1832) in the Lower Wabash River, Illinois and Indiana. *Illinois Natural History Survey, Center for Biodiversity Technical Report* 1993(1):1–29.

Currier, A. O. 1868. List of the shell-bearing Mollusca of Michigan, especially of Kent and adjoining counties. *Kent Scientific Institute. Miscellaneous Publications* 1:1–12.

Cvancara, A. M. 1963. Clines in three species of *Lampsilis* (Pelecypoda: Unionidae). *Malacologia* 1(2):215–225.

Danglade, E. 1922. The Kentucky River and its mussel resources. *U.S. Bureau of Fisheries Document* No. 934:1–8.

Daniels, L. E. 1902. A new species of Lampsilis. *The Nautilus* 16(2):13–14, pl. 2.

Daniels, L. E. 1903. A check list of Indiana Mollusca, with localities. *27th Annual Report of the Indiana Geological Survey, 1902*, 629–652.

Davis, G. M. 1983a. Relative roles of molecular genetics, anatomy, morphometrics and ecology in assessing relationships among North American Unionidae (Bivalvia). Pp. 193–222. *In*: G. S. Oxford and D. Rollinson, editors. Protein polymorphism: Adaptive and taxonomic significance. *Systematics Association Special Volume* No. 24, Academic Press.

Davis, G. M. 1983b. Genetic relationships among North American Pleurobemini and Amblemini (Bivalvia: Unionidae) with emphasis on *Elliptio, Uniomerus, Elliptoideus*, and *Quincuncina*. (Abstract). *American Malacological Bulletin* 1:109–110.

Davis, G. M., and S. L. H. Fuller. 1981. Genetic relationships among recent Unionacea (Bivalvia) of North America. *Malacologia* 20(2):217–253, 2 apps.

Davis, G. M., and M. Mulvey. 1993. Species status of Mill Creek *Elliptio*. *Savannah River Site Publication SRO-NERP-22*. 58 pp.

Davis, G. M., W. H. Heard, S. L. H. Fuller, and C. Hesterman. 1981. Molecular genetics and speciation in *Elliptio* and its relationships to other taxa of North American Unionidae (Bivalvia). *Biological Journal of the Linnean Society* 15(2):131–150.

De Gregorio, A. 1914. Su taluni molluschi di acqua dolce di America (American fresh water shells of America). *Il Naturalista Siciliano* 22(2–3):31–72, pl. 3–12.

De Kay, J. E. 1843. *Zoology of New York, or the New York Fauna; comprising detailed descriptions of all the animals hitherto observed within the state of New York; with brief notices of those occasionally found near its borders: and accompanied by appropriate illustrations*. Part 5. Mollusca. Carroll and Cook, Printers to the Assembly (Albany). 1–271, 40 pls.

DeCamp, W. H. 1881. List of the shell-bearing Mollusca of Michigan. *Kent Scientific Institute. Miscellaneous publication* (5):1–15, pl. 1.

Delessert, B. 1841. *Recueil des Coquilles decrites par Lamarck, dans son Histoire naturelle des Animaux sans Vertébres et non encore figurées*. 40 folios, colored pls.

Dennis, S. D. 1987. An unexpected decline in populations of the freshwater mussel *Dysnomia capsaeformis (= Epioblasma capsaeformis)* in the Clinch River of Virginia and Tennessee. *Virginia Journal of Science* 38(4):281–288.

Dennis, S. D. 1989. Status of the freshwater mussel fauna, Pendleton Island Mussel Preserve, Clinch River, Virginia. *Sterkiana* 72:19–27.

Deshayes, D. P. 1830–1832. *Encyclopédie Méthodique, Histoire des Vers par Bruguière et Lamarck, complétée par Deshayes*.

Deshayes, D. P. 1835. *Histoire naturelle des Animaux sans vertébres de Lamarck*. 2d ed. Vol. 6:iv + 600 pp.

Deshayes, G. P. 1839–1858. *Traité élémentaire de conchyliologie, avec les applications de cette science à la géologie*. Two volumes, Paris. 824 pp.

Eaton, A. 1826. *Zoological text-book, comprising Cuvier's four grand divisions of Animals: also Shaw's improved Linnean genera, arranged according to the classes and orders of Cuvier and Latreille*. Websters and Skinners, Albany. 282 pp.

Etnier, D. A., and W. C. Starnes. 1993. *The fishes of Tennessee*. University of Tennessee Press, Knoxville. 681 pp.

Fenneman, N. M. 1938. *Physiography of Eastern United States*. McGraw-Hill Book Company, Inc. New York. 714 pp.

Ferriss, J. H. 1900. A new Lampsilis from Arkansas. *The Nautilus* 14(4):38–39.

Férussac, A. E. de. 1835. Observations adressées en forme de lettre a M. M. Th. Say, C. S. Rafinesque, Is. Lea, S. P. Hildreth, T. A. Conrad, et C. A. Poulson, Sur la Synonomie de Coquilles bivalves de l'Amerique Septentrionale, et Essai d'une Table de Concordance à ce sujet. *Magasin de Zoologie* 5(59–60):1–36.

Fleming, J. 1828. *History of British animals*. Two volumes.

Frierson, L. S. 1902. Collecting Unionidae in Texas and Louisiana. *The Nautilus* 16(4):37–40.

Frierson, L. S. 1903. The specific value of Unio declivus, Say. *The Nautilus* 17(5):49–51, pl. 3.

Frierson, L. S. 1904. Observations on the genus Quadrula. *The Nautilus* 17(10):111–112.

Frierson, L. S. 1910. Description of a new species of Anodonta. *The Nautilus* 23(9):113–114, pl. 10.

Frierson, L. S. 1911. Remarks on Unio varicosus, cicatricosus and Unio compertus, new species. *The Nautilus* 25(5):51–54, pls. 2, 3.

Frierson, L. S. 1912. Unio (Obovaria) jacksonianus, new species. *The Nautilus* 26(2):23–24, pl. 3.

Frierson, L. S. 1914. Remarks on classification of the Unionidae. *The Nautilus* 28(1):6–8.

Frierson, L. S. 1927. *A classification and annotated check list of the North American naiades*. Baylor University Press, Waco, Texas. 111 pp.

Fuller, S. L. H. 1974. Clams and mussels (Mollusca: Bivalvia). Pp. 215–273, *In*: C. W. Hart, Jr., and S. L. H. Fuller, editors. *Pollution Ecology of Freshwater Invertebrates*. Academic Press, New York. 389 pp.

Fuller, S. L. H. 1978. *Fresh-water mussels (Mollusca: Bivalvia: Unionidae) of the Upper Mississippi River: Observations at selected sites within the 9-foot channel navigation project on behalf of the U.S. Army Corps of Engineers*. Academy of Natural Sciences of Philadelphia. 401 pp.

Geiser, S. W. 1911. The correct name for Tritogonia tuberculata. *The Academician* 1(1):15.

Gooch, C. H., W. J. Pardue, and D. C. Wade. 1979. Recent mollusk investigations on the Tennessee River: 1978. January 1979. Draft Report. Tennessee Valley Authority, Division of Environmental Planning, Water Quality and Ecology Branch, Muscle Shoals, Alabama, and Chattanooga, Tennessee. 126 pp.

Goodrich, C. 1913. Spring collecting in southwest Virginia. *The Nautilus* 27(7):81–82; 27(8):91–95.

Goodrich, C. 1931. *Pleurobema aldrichianum*, a new naiad. *Occasional Papers of the Museum of Zoology, University of Michigan* No. 229. 4 pp., 1 pl.

Goodrich, C. 1932. The Mollusca of Michigan. *Michigan Handbook Series No. 5*. University of Michigan Press, Ann Arbor, Michigan. 120 pp., 7 pls.

Goodrich, C., and H. van der Schalie. 1944. A revision of the Mollusca of Indiana. *The American Midland Naturalist* 32(2):257–326.

Gordon, M. E. 1995. *Venustaconcha sima* (Lea), an overlooked freshwater mussel (Bivalvia: Unionoidea) from the Cumberland River Basin of Central Tennessee. *The Nautilus* 108(3):55–60.

Gordon, M. E., and J. B. Layzer. 1993. Glochidial host of *Alasmidonta atropurpurea* (Bivalvia: Unionoidea, Unionidae). *Transactions of the American Microscopical Society* 112(2):145–150.

Gordon, M. E., J. B. Layzer, and L. M. Madison. 1994. Glochidial host of *Villosa taeniata* (Mollusca: Unionidae). *Malacological Review* 27:113–114.

Gordon, M. E., and D. G. Smith. 1990. Autumnal reproduction in *Cumberlandia monodonta* (Unionoidea: Margaritiferidae). *Transactions of the American Microscopical Society* 109(4):407–411.

Gould, A. A. 1841. *Report on the Invertebrata of Massachusetts, comprising the Mollusca, Crustacea, Annelida, and Radiata*. Folsom, Wells and Thurston, Cambridge. 373 pp., 213 figs.

Gould, A. A. 1855. New species of land and fresh-water shells from western (N.) America. (Cont.). *Proceedings of the Boston Society of Natural History* 5(15):228–229.

Gould, A. A. 1856. [Descriptions of shells]. *Proceedings of the Boston Society of Natural History* 5:228–229, 6:11–16.

Gray, J. E. 1847. A list of the genera of recent Mollusca, their synonyma and types. *Proceedings of the Zoological Society of London* 1847(15):129–219.

Green, J. 1827. Some remarks on the *Unios* of the United States, with a description of a new species. *Contributions of the Maclurian Lyceum to the Arts and Sciences* 1(2):41–47, pl. 3.

Grier, N. M. 1918. New varieties of naiades from Lake Erie. *The Nautilus* 32(1):9–12.

Grier, N. M. 1922. Final report on the study and appraisal of mussel resources in selected areas of the Upper Mississippi River. *The American Midland Naturalist* 8(1):1–33, 1 pl.

Grier, N. M., and J. F. Mueller. 1922. Notes on the naiad fauna of the Upper Mississippi River. II. The naiades of the Upper Mississippi drainage. *The Nautilus* 36(2):46–49; 36(3):96–103.

Grier, N. M., and J. F. Mueller. 1926. Further studies in correlation of shape and station in fresh water mussels. *Bulletin of the Wagner Free Institute of Science of Philadelphia* 1(2/3):11–28.

Guérin-Méneville, F. E. 1828–1844. *Iconographie du regne animal du G. Cuvier: ou Representation d'apres nature de l'une des especies les plus remarquables, et souvent non enore figurees, de chaque genre d'animaux. Avec un texte descriftif mis au courant de la science. Ouvarage pouvant sevir d'atlas a tous les traites de zoologie*. Paris, J. B. Baiaere. vol. 2

Haag, W. R., R. S. Butler, and P. D. Hartfield. 1995. An extraordinary reproductive strategy in freshwater bivalves: prey mimicry to facilitate larval dispersal. *Freshwater Biology* 34:471–476.

Haag, W. R., and M. L. Warren, Jr. 1996. Host fishes of some Mobile Basin unionids. *Triannual Unionid Report* 9:13.

Haag, W. R., and M. L. Warren, Jr. 1997. Host fishes and reproductive biology of 6 freshwater mussel species from the Mobile Basin, USA. *Journal of the North American Benthological Society* 16(3):576–585.

Haag, W. R., M. L. Warren, Jr., and M. Shillingsford. 1997. Identification of host fishes for *Lampsilis altilis* and *Villosa vibex*. *Triannual Unionid Report* 12:13.

Haas, F. 1930. Über nord- und mittelamerikanische Najaden. *Senckenbergiana Biologica* 12(6):317–330.

Haas, F. 1969a. Superfamilia Unionacea. *Das Tierreich* (Berlin) 88:x + 663 pp.

Haas, F. 1969b. Superfamily Unionacea. Pp. N:411–470. *In:* R. C. Moore, editor. *Treatise on Invertebrate Paleontology*. Geological Society of America and the University of Kansas. Part N, Vol. 1 (of 3), Mollusca. xxxviii + N1–N489.

Haldeman, S. S. 1842. Description of five new species of North American fresh-water shells. *Journal of the Academy of Natural Sciences of Philadelphia* 8:200–202.

Haldeman, S. S. 1846. Description of Unio abacoides, a new species. *Proceedings of the Academy of Natural Sciences of Philadelphia* 3(3):75. Also published verbatim in *American Journal of Science and Arts* Series 2, 52(5):274; *Annals of Natural History* 18:430.

Hanley, S. 1842a. *Testaceous Mollusca, Lamarck's species of shells.* 224 pp., 3 pls.

Hanley, S. 1842b–1856. *An illustrated and descriptive catalogue of recent bivalve shells, Appendix to Index Testaceologicus.* Williams and Norgate, London. i–xviii + 1–392, 25 pls.

Hannibal, H. 1912. A synopsis of the Recent and Tertiary freshwater Mollusca of the Californian Province, based upon an ontogenetic classification. *Proceedings of the Malacological Society of London* 10(2):112–166; 10(3):167–211.

Harris, J. L., and M. E. Gordon. 1990. *Arkansas mussels.* Arkansas Game and Fish Commission, Little Rock. 32 pp.

Hartfield, P. D. 1988. Mussel survey of the Amite River, Louisiana. 9–13 May 1988. Appendix D.5 *In:* Amite River Flood Control Study Environmental Report. State Project No. 575-99-30. Prepared by Espey, Huston & Associates. Metairie, Louisiana. 16 pp.

Hartfield, P. D., and E. Hartfield. 1996. Observations on the conglutinates of Ptychobranchus greeni (Conrad, 1834) (Mollusca: Bivalvia: Unionoidea). *The American Midland Naturalist* 135(2):370–375.

Heard, W. H., and R. H. Guckert. 1970. A re-evaluation of the recent Unionacea (Pelecypoda) of North America. *Malacologia* 10(2):333–355.

Herrmannsen, A. N. 1852. *Indicis Generum Malacozoorum. Supplementa et Corrigenda.* Cassellis, London. v + 140 pp.

Hickman, M. E. 1937. A contribution to the knowledge of the molluscan fauna of East Tennessee. Unpublished M.S. Thesis. University of Tennessee. Knoxville, Tennessee. 165 pp., 104 pls.

Hildreth, S. P. 1828. Observations on, and descriptions of the shells found in the waters of the Muskingum River, Little Muskingum, and Duck Creek, in the vicinity of Marietta, Ohio. *American Journal of Science and Arts* 14(2):276–291, 2 pls. Reprinted in *Sterkiana* 8(1962):39–48.

Hinkley, A. A. 1906. Some shells from Mississippi and Alabama. *The Nautilus* 20(3):34–36; 20(4):40–44; 20(5):52–55.

Hoeh, W. R. 1990. Phylogenetic relationships among Eastern North American *Anodonta* (Bivalvia: Unionidae). *Malacological Review* 23(1–2):63–82.

Hoff, C. C. 1943. Some records of sponges, branchiobdellids, and molluscs from the Reelfoot Lake Region. *Journal of the Tennessee Academy of Science* 18(3):223–227.

Hoggarth, M. A., and A. S. Gaunt. 1988. Mechanics of glochidial attachment (Mollusca: Bivalvia: Unionidae). *Journal of Morphology* 198:71–81.

Hove, M. C. 1995a. Early life history research on the squawfoot, *Strophitus undulatus*. *Triannual Unionid Report* 7:28–29.

Hove, M. C. 1995b. Host research on round pigtoe glochidia. *Triannual Unionid Report* 8:8.

Hove, M. C. 1995c. Suitable fish hosts of the lilliput, *Toxolasma parvus*. *Triannual Unionid Report* 8:9.

Hove, M. 1997. Ictalurids serve as suitable hosts for the purple wartyback. *Triannual Unionid Report* 11:4.

Hove, M. C., R. Engelking, E. Evers, M. Peteler, and E. Peterson. 1994a. *Ligumia recta* host fish suitability tests. *Triannual Unionid Report* 5:8.

Hove, M. C., R. Engelking, E. Evers, M. Peteler, and E. Peterson. 1994b. *Cyclonaias tuberculata* host suitability tests. *Triannual Unionid Report* 5:9.

Hove, M. C., R. Engelking, E. Long, M. Peteler, and L. Sovell. 1994. Life history research on *Cyclonaias tuberculata*, the purple wartyback. *Triannual Unionid Report* 3:20.

Hove, M. C., R. A. Engelking, M. E. Peteler, and E. M. Peterson, 1995. *Anodontoides ferussacianus* and *Anodonta imbecillis* host suitability tests. *Triannual Unionid Report* 6:22

Howard, A. D. 1915. Some exceptional cases of breeding among the Unionidae. *The Nautilus* 29(1):4–11.

Howard, A. D. 1951. A river mussel parasitic on a salamander. *The Chicago Academy of Sciences. Natural History Miscellanea* No. 77, 6 pp.

Howard, A. D., and B. J. Anson. 1922. Phases in the parasitism of the Unionidae. *Journal of Parasitology* 9(2):68–82, 2 pls.

Howells, R. G. 1996. Pistolgrip and Gulf Mapleleaf hosts. *Info-Mussel Newsletter* 4(3):3.

Howells, R. G., R. W. Neck, and H. D. Murray. 1996. *Freshwater mussels of Texas*. Texas Parks and Wildlife Department, Inland Fisheries Division. Austin, Texas. 218 pp.

Hubbs, D. W. 1995. 1993 Statewide commercial mussel report. *Fisheries Report, 95–15*. Tennessee Wildlife Resources Agency, Nashville, Tennessee. 63 pp.

Hurd, J. C. 1974. Systematics and zoogeography of the unionacean mollusks of the Coosa River drainage of Alabama, Georgia and Tennessee. Unpublished Ph.D. Dissertation. Auburn University. Auburn, Alabama. 240 pp.

Isom, B. G. 1968. The naiad fauna of Indian Creek, Madison County, Alabama. *The American Midland Naturalist* 79(2):514–516.

Isom, B. G. 1969. The mussel resource of the Tennessee River. *Malacologia* 7(2–3):397–425.

Isom, B. G. 1972. Mussels in the unique Nickajack Dam construction site, Tennessee River, 1965. *Malacological Review* 5(1):4–6.

Isom, B. G. 1974. Mussels of the Green River, Kentucky. *Transactions of the Kentucky Academy of Science* 35(1/2):55–57.

Isom, B. G., C. Gooch, and S. D. Dennis. 1979. Rediscovery of a presumed extinct river mussel, *Dysnomia sulcata* (Unionidae). *The Nautilus* 93(2–3):84.

Isom, B. G., and P. Yokley, Jr. 1968a. The mussel fauna of Duck River in Tennessee, 1965. *The American Midland Naturalist* 80(1):34–42.

Isom, B. G., and P. Yokley, Jr. 1968b. Mussels of Bear Creek watershed, Alabama and Mississippi, with a discussion of the area geology. *The American Midland Naturalist* 79(1):189–196.

Isom, B. G., P. Yokley, Jr., and C. H. Gooch. 1973. Mussels of the Elk River basin in Alabama and Tennessee—1965–1967. *The American Midland Naturalist* 89(2):437–442.

Jay, J. C. 1839. *A catalogue of the shells arranged according to the Lamarckian system; together with descriptions of new or rare species, contained in the collection of John C. Jay, M.D.* 3d ed. Wiley and Putnam, New York. 125 pp.

Jay, J. C. 1850. *A catalogue of the shells, arranged according to the Lamarckian System, with their authorities, synonymies, and references to the works where figured or described, contained in the collection of John C. Jay, M.D. Fourth Edition.* New York, New York. 460 pp.

Jenkinson, J. J., and F. L. Kokai. 1977. *Villosa lienosa* (Conrad, 1834) in Ohio. *Bulletin of the American Malacological Union, Inc., for 1977*, 82–83.

Johnson, R. I. 1967a. *Carunculina pulla* (Conrad), an overlooked Atlantic drainage unionid. *The Nautilus* 80(4):127–131.

Johnson, R. I. 1967b. Illustrations of all the mollusks described by Berlin Hart and Samuel Hart Wright. *Occasional Papers on Mollusks, Museum of Comparative Zoology, Harvard University* 3(35):1–35, pls. 1–13.

Johnson, R. I. 1969a. Further additions to the unionid fauna of the Gulf drainage of Alabama, Georgia and Florida. *The Nautilus* 83(1):34–35.

Johnson, R. I. 1969b. The Unionacea of William Irvin Utterback. *The Nautilus* 82(4):132–135.

Johnson, R. I. 1970. The systematics and zoogeography of the Unionidae (Mollusca: Bivalvia) of the southern Atlantic Slope Region. *Bulletin of the Museum of Comparative Zoology* 140(6):263–449.

Johnson, R. I. 1972. The Unionidae (Mollusca: Bivalvia) of Peninsular Florida. *Bulletin of the Florida State Museum, Biological Sciences* 16(4):181–249 + addendum.

Johnson, R. I. 1975. Simpson's unionid types and miscellaneous unionid types in the National Museum of Natural History. *Special Occasional Papers, Museum of Comparative Zoology, Harvard University* No. 4. 56 pp., 3 pls.

Johnson, R. I. 1977. Monograph of the genus *Medionidus* (Bivalvia: Unionidae) mostly from the Apalachicolan region, southeastern United States. *Occasional Papers on Mollusks, Museum of Comparative Zoology, Harvard University* 4(56):161–187.

Johnson, R. I. 1978. Systematics and zoogeography of *Plagiola* (=*Dysnomia* =*Epioblasma*), an almost extinct genus of freshwater mussels (Bivalvia: Unionidae) from Middle North America. *Bulletin of the Museum of Comparative Zoology* 148(6):239–320.

Johnson, R. I. 1980. Zoogeography of North American Unionacea (Mollusca: Bivalvia) north of the maximum Pleistocene glaciation. *Bulletin of the Museum of Comparative Zoology* 149(2):77–189.

Johnson, R. I., and H. B. Baker. 1973. The types of Unionacea (Mollusca: Bivalvia) in the Academy of Natural Sciences of Philadelphia. *Proceedings of the Academy of Natural Sciences of Philadelphia* 125(9):145–186, pls. 1–10.

Kelly, H. M. 1899. A statistical study of the parasites of the Unionidae. *Bulletin of the Illinois State Laboratory of Natural History* 5(8):399–418.

Kesler, D. H., and D. Manning. 1996. A new mussel record for Tennessee: Lampsilis siliquoidea (Mollusca: Unionidae) from the Wolf River. *Journal of the Tennessee Academy of Science* 71(4):90–94.

Kirtland, J. P. 1834. Observations on the sexual characters of the animals belonging to Lamarck's family of naiades. *American Journal of Science* 26(21):117–120.

Kitchel, H. E. 1985. Life history of the endangered shiny pigtoe mussel, *Fusconaia edgariana*, in the North Fork Holston River, Virginia. Unpublished M.S. Thesis. Virginia Polytechnic Institute and State University. Blacksburg, Virginia. 124 pp.

Klippel, W. E., and P. W. Parmalee. 1979. The naiad fauna of Lake Springfield, Illinois: an assessment after two decades. *The Nautilus* 93(4):189–197.

Knudsen, K. A., and M. C. Hove. 1997. Spectaclecase *(Cumberlandia monodonta)* conglutinates unique, host(s) elusive. *Triannual Unionid Report* 11:2.

Koch, L. M. 1989. *Mytilopsis leucophaeata* (Conrad, 1831) from the Upper Mississippi River (Bivalvia: Dreissenidae). *Malacology Data Net* 2(5–6):153–154.

Kraemer, L. R. 1970. The mantle flap in three species of *Lampsilis* (Pelecypoda: Unionidae). *Malacologia* 10(1):225–282.

Küster, H. C. 1852–1862, 1873. *In: Systematisches Conchylien-Cabinet von Martini und Chemnitz.* 2d ed.

La Rocque, A. 1953. Catalogue of the Recent Mollusca of Canada. *National Museum of Canada Bulletin* 129:ix + 1–406.

La Rocque, A. 1966–1970. Pleistocene Mollusca of Ohio. *Bulletin of the Geological Survey of Ohio* 1966, 62(1):iii + 1–111; 1967, 62(2):vii–xiv + 113–356, pls. 1–8; 1968, 62(3):xvii–xxiv + 357–553, pls. 9–14; 1970, 62(4):xxvii–xxxiv + 555–800, pls. 15–18.

La Rocque, A., and J. Oughton. 1937. A preliminary account of the Unionidae of Ontario. *Canadian Journal of Research* 15(8):147–155.

Lamarck, J. B. P. A. 1815–1822. *Histoire naturelle des Animaux sans Vertébres.* 8 volumes.

Lapham, I. A. 1852. Catalogue of the Mollusca of Wisconsin. *Transactions of the Wisconsin State Agricultural Society* 2:367–370.

Lapham, I. A. 1860. A list of the shells of the state of Wisconsin. *Proceedings of the Academy of Natural Sciences of Philadelphia* 12:154–156.

Latchford, F. R. 1882. Notes on the Unionidae found in the vicinity of Ottawa, Ont. *Transactions of the Ottawa Field-Naturalists' Club* 1(3):48–57. Reprinted in *Sterkiana* 8(1962):19–28.

Layzer, J. B., and R. M. Anderson. 1991. Fish hosts of the endangered Cumberland bean pearly mussel *(Villosa trabalis). NABS Bulletin* 8(1):110.

Layzer, J. B., and R. M. Anderson. 1992. Impacts of the coal industry on rare and endangered aquatic organisms of the upper Cumberland River Basin. Final Report to Kentucky Department of Fish and Wildlife Resources, Frankfurt, KY, and Tennessee Wild Resources Agency, Nashville, TN.

Lea, I. 1828. Description of six new species of the genus Unio, embracing the anatomy of the oviduct of one of them, together with some anatomical observations on the genus. *Transactions of the American Philosophical Society* 3[New Series]:259–273, pls. 3–6.

Lea, I. 1829. Description of a new genus of the family of naïades, including eight species, four of which are new; also the description of eleven new species of the genus Unio from the rivers of the United States: with observations on some of the characters of the naïades. *Transactions of the American Philosophical Society* 3[New Series]:403–457, pls. 7–14.

Lea, I. 1831. Observations on the naïades, and descriptions of new species of that and other families. *Transactions of the American Philosophical Society* 4[New Series]:63–121, pls. 3–18.

Lea, I. 1834. Observations on the naïades; and descriptions of new species of that, and other families. *Transactions of the American Philosophical Society* 5[New Series]:23–119, pls. 1–19.

Lea, I. 1836. *A synopsis of the family of naïades.* Philadelphia, Pennsylvania. Carey, Lea and Blanchard, Philadelphia. 1–59, 1 pl.

Lea, I. 1838a. *A synopsis of the family of naiades.* 2d ed. enlarged and improved. Carey, Lea and Blanchard, Philadelphia. 44 pp.

Lea, I. 1838b. Description of new freshwater and land shells. *Transactions of the American Philosophical Society* 6[New Series]:1–154, pls. 1–24.

Lea, I. 1840. Descriptions of new fresh water and land shells. *Proceedings of the American Philosophical Society* 1(13):284–289.

Lea, I. 1841a. Continuation of paper [On fresh water and land shells]. *Proceedings of the American Philosophical Society* 2:30–35.

Lea, I. 1841b. Continuation of Mr. Lea's Paper [On fresh water and land shells]. *Proceedings of the American Philosophical Society* 2:81–83.

Lea, I. 1842a. Continuation of paper [On new fresh water and land shells]. *Proceedings of the American Philosophical Society* 2:224–225.

Lea, I. 1842b. Description of new fresh water and land shells. *Transactions of the American Philosophical Society* 8[New Series](Part 2, Art. 12):163–250, pls. 5–27.

Lea, I. 1845. Descriptions of new fresh water and land shells. *Proceedings of the American Philosophical Society* 4(33):162–168.

Lea, I. 1848. Description of new fresh water and land shells. *Transactions of the American Philosophical Society* 10[New Series](Pt. I):67–101, pls. 1–9.

Lea, I. 1852a. Descriptions of New Species of the Family Unionidae. *Proceedings of the American Philosophical Society* 5:251–252.

Lea, I. 1852b. Descriptions of new species of the family Unionidae. *Transactions of the American Philosophical Society* 10[New Series](Pt. 2):253–294, pls. 12–29.

Lea, I. 1852c. *A synopsis of the family of naïades.* 3d ed. greatly enlarged and improved. Blanchard and Lea. Philadelphia, xx + pp. 17–88.

Lea, I. 1854. Rectification of Mr. T. A. Conrad's "Synopsis of the Family of *Naïades* of North America" published in the "Proceedings of the Academy of Natural Sciences of Philadelphia, February, 1853." *Proceedings of the Academy of Natural Sciences of Philadelphia* 7(6):236–249.

Lea, I. 1856a. Description of four new species of exotic uniones. *Proceedings of the Academy of Natural Sciences of Philadelphia* 8:103.

Lea, I. 1856b. Description of eleven new species of uniones, from Georgia. *Proceedings of the Academy of Natural Sciences of Philadelphia* 8:262–263.

Lea, I. 1857a. Description of thirteen new species of uniones, from Georgia. *Proceedings of the Academy of Natural Sciences of Philadelphia* 9:31–32.

Lea, I. 1857b. Description of six new species of uniones from Alabama. *Proceedings of the Academy of Natural Sciences of Philadelphia* 9:83.

Lea, I. 1857c. Description of eight new species of naïades from various parts of the United States. *Proceedings of the Academy of Natural Sciences of Philadelphia* 9:84.

Lea, I. 1857d. Description of twelve new species of naïades from North Carolina. *Proceedings of the Academy of Natural Sciences of Philadelphia* 9:85–86.

Lea, I. 1857e. Description of six new species of fresh water and land shells of Texas and Tamaulipas, from the collection of the Smithsonian Institution. *Proceedings of the Academy of Natural Sciences of Philadelphia* 9:101–102.

Lea, I. 1857f. Descriptions of twenty-seven new species of uniones from Georgia. *Proceedings of the Academy of Natural Sciences of Philadelphia* 9:169–172.

Lea, I. 1858a. Descriptions of new species of Unio, from Tennessee, Alabama and North Carolina. *Proceedings of the Academy of Natural Sciences of Philadelphia* 10:40–41.

Lea, I. 1858b. Descriptions of seven new species of Margaritanae, and four new species of Anodontae. *Proceedings of the Academy of Natural Sciences of Philadelphia* 10:138–139.

Lea, I. 1858c. Descriptions of twelve new species of uniones and other fresh-water shells of the United States. *Proceedings of the Academy of Natural Sciences of Philadelphia* 10:165–166.

Lea, I. 1858d. Descriptions of exotic genera and species of the family Unionidae. *Journal of the Academy of Natural Sciences of Philadelphia* 3[New Series](4):289–321, pls. 21–33.

Lea, I. 1858e. New Unionidae of the United States. *Journal of the Academy of Natural Sciences of Philadelphia* 4[New Series](1):51–95, pls. 6–20.

Lea, I. 1859a. Descriptions of eight new species of Unionidae, from Georgia, Mississippi and Texas. *Proceedings of the Academy of Natural Sciences of Philadelphia* 11:112–113.

Lea, I. 1859b. Descriptions of Two New Species of Uniones, from Georgia. *Proceedings of the Academy of Natural Sciences of Philadelphia* 11:154.

Lea, I. 1859c. Descriptions of seven new species of Uniones from South Carolina, Florida, Alabama and Texas. *Proceedings of the Academy of Natural Sciences of Philadelphia* 11:154–155.

Lea, I. 1859d. Descriptions of twelve new species of uniones, from Georgia. *Proceedings of the Academy of Natural Sciences of Philadelphia* 11:170–172.

Lea, I. 1859e. [Change of name of *Margaritana Etowahensis* to *M. Georgiana*.] *Proceedings of the Academy of Natural Sciences of Philadelphia* 11:280.

Lea, I. 1859f. New Unionidae of the United States. *Journal of the Academy of Natural Sciences of Philadelphia* 4[New Series](2):191–233, pls. 21–32.

Lea, I. 1860a. Descriptions of five new species of uniones from north Alabama. *Proceedings of the Academy of Natural Sciences of Philadelphia* 12:92–93.

Lea, I. 1860b. Descriptions of two new species of uniones from Georgia. *Proceedings of the Academy of Natural Sciences of Philadelphia* 12:305.

Lea, I. 1860c. Descriptions of seven new species of Unionidae from the United States. *Proceedings of the Academy of Natural Sciences of Philadelphia* 12:306–307.

Lea, I. 1860d. Descriptions of exotic Unionidae. *Journal of the Academy of Natural Sciences of Philadelphia* 4[New Series](2):235–273, pls. 33–45.

Lea, I. 1860e. New Unionidae of the United States and northern Mexico. *Journal of the Academy of Natural Sciences of Philadelphia* 4[New Series](4):327–374, pls. 51–66.

Lea, I. 1861a. Descriptions of twenty-five new species of Unionidae from Georgia, Alabama, Mississippi, Tennessee and Florida. *Proceedings of the Academy of Natural Sciences of Philadelphia* 13:38–41.

Lea, I. 1861b. Descriptions of two new species of *Anodontae*, from Arctic America. *Proceedings of the Academy of Natural Sciences of Philadelphia* 13:56.

Lea, I. 1861c. Descriptions of twelve new species of *Uniones*, from Alabama. *Proceedings of the Academy of Natural Sciences of Philadelphia* 13:59–60.

Lea, I. 1861d. Descriptions of eleven new species of the genus *Unio* from the United States. *Proceedings of the Academy of Natural Sciences of Philadelphia* 13:391–393.

Lea, I. 1862a. Descriptions of ten new species of Unionidae of the United States. *Proceedings of the Academy of Natural Sciences of Philadelphia* 14:168–169.

Lea, I. 1862b. New Unionidae of the United States. *Journal of the Academy of Natural Sciences of Philadelphia* 5[New Series](1):53–109, pls. 1–18.

Lea, I. 1862c. New Unionidae of the United States and Arctic America. *Journal of the Academy of Natural Sciences of Philadelphia* 5[New Series]:187–216, pls. 24–33.

Lea, I. 1863. Descriptions of twenty-four new species of Unionidae of the United States. *Proceedings of the Academy of Natural Sciences of Philadelphia* 15:191–194.

Lea, I. 1865. Descriptions of eight new species of *Unio* of the United States. *Proceedings of the Academy of Natural Sciences of Philadelphia* 17:88–89.

Lea, I. 1866. New Unionidae, Melanidae, etc., chiefly of the United States. *Journal of the Academy of Natural Sciences of Philadelphia* 6[New Series](1):5–65, pls. 1–21.

Lea, I. 1867. Descriptions of five new species of Unionidae and one *Paludina* of the United States. *Proceedings of the Academy of Natural Sciences of Philadelphia* 19:81.

Lea, I. 1868a. Description of sixteen new species of the genus *Unio* of the United States. *Proceedings of the Academy of Natural Sciences of Philadelphia* 20:143–145.

Lea, I. 1868b. New Unionidae, Melanidae, etc., chiefly of the United States. *Journal of the Academy of Natural Sciences of Philadelphia* 6[New Series](3):249–302, pls. 29–45.

Lea, I. 1868c. New Unionidae, Melanidae, etc., chiefly of the United States. *Journal of the Academy of Natural Sciences of Philadelphia* 6[New Series](4):303–343, pls. 46–54.

Lea, I. 1870. *A synopsis of the family Unionidae.* 4th ed., very greatly enlarged and improved. Henry C. Lea, Philadelphia, Pennsylvania. xxx + 184 pp.

Lea, I. 1871. Descriptions of twenty new species of uniones of the United states. *Proceedings of the Academy of Natural Sciences of Philadelphia* 23(3):189–193.

Lea, I. 1872a. *Rectification of T. A. Conrad's "Synopsis of the family of Naïades of North America," published in the "Proceedings of the Academy of Natural Sciences of Philadelphia, February, 1853."* New ed. Collins, Printer, 705 Jayne Street, Philadelphia. 45 pp.

Lea, I. 1872b. Descriptions of twenty-nine species of Unionidae from the United States. *Proceedings of the Academy of Natural Sciences of Philadelphia* 24(2):155–161.

Lea, I. 1874. Descriptions of fifty-two species of Unionidae. *Journal of the Academy of Natural Sciences of Philadelphia* 8[New Series](1):5–54, pls. 1–18.

Lefevre, G., and W. C. Curtis. 1910. Experiments in the artificial propagation of fresh-water mussels. Bulletin of the Bureau of Fisheries 28(1908):615–626. Issued separately as U.S. Bureau of Fisheries Document 671.

Lewis, J. 1870. On the shells of the Holston River. *American Journal of Conchology* 6(3):216–226.

Lightfoot, J. 1786. *A catalogue of the Portland Museum, lately the property of the duchess Dowager of Portland, deceased, which will be sold at auction by Mr. Skinner and Co.* London. vii + 194 pp.

Lister, M. 1685–1692. *Histoiriae sive synopsis methodicae conchyliorum quorum ommium picturae, ad vivum delineatae, exhibetur. Liber Primus, qui est de cochleis terrestribus,* 1st ed. Oxford. iv + 12 + 77 + 6 pp., 1059 + 22 pls.

Luo, M. 1993. Host fishes of four species of freshwater mussels and development of an immune response. Unpublished M.S. thesis in Biology, Tennessee Technological University, Cookeville. v + 32 pp.

Lydeard, C., M. Mulvey, and G. M. Davis. 1996. Molecular systematics and evolution of reproductive traits of North American freshwater unionacean mussels (Mollusca: bivalia) as inferred from 16S rRNA gene sequences. *Philosophical Transactions of the Royal Society, London B*, 351:1593–1603.

Manning, D. 1989. Freshwater mussels (Unionidae) of the Hatchie River, a tributary of the Mississippi River, in West Tennessee. *Sterkiana* 72:11–18.

Manzano, B. L. 1986. Faunal resources, butchering patterns, and seasonality at the Eastman Rockshelter (40SL34): An interpretation of function. Unpublished M.A. thesis, Department of Anthropology, University of Tennessee, Knoxville, 217 pp.

Marsh, P. 1885. *List of shells collected in central Tennessee by A. A. Hinkley and P. Marsh with notes on species.* Privately Printed. Aledo, Illinois. 10 pp.

Marsh, W. A. 1887–1889. Brief notes on the land and freshwater shells of Mercer County, Illinois. *Conchologists Exchange* 1887: 1(8):42–43; 1(9/10):50–51; 1(11):62–63; 1(12):74–75; 2(1):4–5; 2(2):20–21; 2(3):36–37; 2(4):48–50; 2(5):65–67; 2(6):80–81; 1888: 2(7):90–92; 2(8):103–104; 2(9):110–111; 1889: 3(2):23–24; 3(3):34–35.

Marsh, W. A. 1891. Description of two new species of Unio from Arkansas. *The Nautilus* 5(1):1–2.

Marsh, W. A. 1897. New American Unionidae. *The Nautilus* 10(9):103–104, 10(11):pl. 1.

Marsh, W. A. 1901. Description of a new Unio from Missouri. *The Nautilus* 15(7):74–75.

Marsh, W. A. 1902a. Description of a new Unio from Tennessee. *The Nautilus* 15(10):115–116.

Marsh, W. A. 1902b. Description of a new Unio from Tennessee. *The Nautilus* 16(1):7–8, pl. 1.

Marshall, W. B. 1916. A new genus and species of naiad from the James River at Huron, South Dakota. *The Nautilus* 29(12):133–135, pl. 4.

Mathiak, H. A. 1979. *A river survey of the unionid mussels of Wisconsin 1973–1977.* Sand Shell Press. Horicon, Wisconsin. 75 pp.

McGregor, M. A., and M. E. Gordon. 1992. *Commercial musseling in Tennessee.* 7 pp., color figs. Tennessee Wildlife Resources Agency, Nashville.

Menke, C. T. 1828. *Synopsis methodica molluscorum generum omnium et specierum earum quae in museo menkeano adservantur; cum synonymia critica et novarum specierum diagosibus.* 91 pp.

Miles, M. 1861. A catalogue of the mammals, birds, reptiles and mollusks, of Michigan. Pp. 219–241. *In:* A. Winchell, editor. *First biennial report of the progress of the Geological Survey of Michigan, embracing observations on the geology, zoölogy and botany of the Lower Peninsula.* Homer & Kerr, Lansing.

Modell, H. 1942. Das natüruliche system der Najaden. *Archiv für Molluskenkunde* 74(5/6):161–191.

Modell, H. 1949. Das natüruliche system der Najaden. 2. *Archiv für Molluskenkunde* 78(1/3):29–46.

Model, H. 1964. Das natüruliche system der Najaden. 3. *Archiv für Molluskenkunde* 93(3/4):71–126.

Morrison, J. P. E. 1942. Preliminary report on mollusks found in the shell mounds of the Pickwick Landing basin in the Tennessee River valley. *Bureau of American Ethnology Bulletin* 129:339–392.

Morrison, J. P. E. 1969. The earliest names for North American naiads. *Annual Report for 1969 of the American Malacological Union, Inc.* 36:22–24.

Morrison, J. P. E. 1976. Species of the genus *Uniomerus.* *Bulletin of the American Malacological Union, Inc.* 1976:10–11.

Murray, H. D., and A. B. Leonard. 1962. Handbook of unionid mussels in Kansas. *Miscellaneous Publication, Museum of Natural History, University of Kansas.* Lawrence, Kansas No. 28. 184 pp.

Murray, H. D., and E. C. Roy, Jr. 1968. Checklist of freshwater and land mollusks of Texas. *Sterkiana* 30:25–42.

Najarian, H. H. 1955. Notes on aspidogastrid trematodes and hydracarina from some Tennessee mussels. *Journal of the Tennessee Academy of Science* 30(1):11–14.

Neck, R. W. 1982. Preliminary analysis of the ecological zoogeography of the freshwater mussels of Texas. Pp. 33–42, *In:* J. R. Davis, editor. *Proceedings of the Symposium on Recent Benthological Investigations in Texas and Adjacent States.* Aquatic Science Section, Texas Academy of Science. 278 pp.

Neck, R. W. 1986. Freshwater bivalves of Lake Tawakoni, Sabine River, Texas. *Texas Journal of Science* 38(3):241–249.

Neel, J. K., and W. R. Allen. 1964. The mussel fauna of the Upper Cumberland Basin before its impoundment. *Malacologia* 1(3):427–459.

Nelson, D. 1982. Relocation of *Lampsilis higginsi* in the upper Mississippi River. Pp. 104–107. *In:* A. C. Miller, editor. *Report of freshwater mussels workshop.* U.S. Army Corps of Engineers Waterways Experiment Station, Vicksburg, Mississippi.

Neves, R. J. 1991. Mollusks. Pp. 251–320, *In:* K. Terwilliger, editor. *Virginia's Endangered Species. Proceedings of a Symposium.* Department of Game and Inland Fisheries, Commonwealth of Virginia. 672 pp.

Neves, R. J., A. E. Bogan, J. D. Williams, S. A. Ahlstedt, and P. W. Hartfield. 1997. Status of aquatic mollusks in the southeastern United States: A downward spiral of diversity. Pp. 43–85. *In:* G. W. Bense and D. E. Collins, editors. *Aquatic fauna in peril: The Southeastern perspective.* Special publication 1, Southeast Aquatic Research Institute, Lenz Design and Communications, Decatur, Georgia.

Nordstrom, G. R., W. L. Pflieger, K. C. Sadler, and W. H. Lewis. 1977. *Rare and endangered species of Missouri.* Missouri Department of Conservation. U.S. Department of Agriculture and Soil Conservation Service. 129 pp.

O'Hara, M. 1980. Mr. Boepple and his buttons. *The Iowan* Fall:46–51.

Oblad, B. R. 1980. An experiment in relocating endangered and rare naiad mollusks from a proposed bridge construction site at Sylvan Slough, Mississippi River near Moline Illinois. Pp. 211–222. *In:* J. L. Rasmussen, editor. *Proceedings of the UMRCC symposium on Upper Mississippi River bivalve mollusks.* Upper Mississippi River Conservation Committee, Rock Island.

Oesch, R. D. 1984. *Missouri naiades. A guide to the mussels of Missouri.* Missouri Department of Conservation. Jefferson City, Missouri. vii + 270 pp.

Ortmann, A. E. 1909a. The destruction of the fresh-water fauna in western Pennsylvania. *Proceedings of the American Philosophical Society* 48(191):90–110.

Ortmann, A. E. 1909b. The breeding season of Unionidae in Pennsylvania. *The Nautilus* 22(9):91–95; 22(10):99–103.

Ortmann, A. E. 1909c. Unionidae from an Indian garbage heap. *The Nautilus* 23(1):11–15.

Ortmann, A. E. 1910. A new system of the Unionidae. *The Nautilus* 23(9):114–120.

Ortmann, A. E. 1911. A monograph of the najades of Pennsylvania. Parts I and II. *Memoirs of the Carnegie Museum* 4(6):279–347, 4 pls.

Ortmann, A. E. 1912a. Notes upon the families and genera of the najades. *Annals of the Carnegie Museum* 8(2):222–365, pls. 18–20.

Ortmann, A. E. 1912b. Cumberlandia, a new genus of naiades. *The Nautilus* 26(2):13–14.

Ortmann, A. E. 1912c. Lampsilis ventricosa (Barnes) in the Upper Potomac drainage. *The Nautilus* 26(4):51–54.

Ortmann, A. E. 1913a. The Alleghenian Divide, and its influence upon the freshwater fauna. *Proceedings of the American Philosophical Society* 52(210):287–390, pls. 12–14.

Ortmann, A. E. 1913b. Studies in najades. *The Nautilus* 27(8):88–91.

Ortmann, A. E. 1914. Studies in najades (cont.). *The Nautilus* 28(2):20–22; 28(3):28–34; 28(4):41–47; 28(5[sic]):65–69.

Ortmann, A. E. 1915. Studies in najades (cont.). *The Nautilus* 28(9):106–108; 28(11):129–131; 28(12):141–143; 29(6):63–67.

Ortmann, A. E. 1916. The anatomical structure of Gonidea angulata (Lea). *The Nautilus* 30(5):50–53.

Ortmann, A. E. 1917. A new type of the nayad-genus Fusconaia. Group of F. barnesiana Lea. *The Nautilus* 31(2):58–64.

Ortmann, A. E. 1918. The nayades (freshwater mussels) of the Upper Tennessee drainage. With notes on synonymy and distribution. *Proceedings of the American Philosophical Society* 57:521–626.

Ortmann, A. E. 1919. A monograph of the naiades of Pennsylvania. Part III: Systematic account of the genera and species. *Memoirs of the Carnegie Museum* 8(1):xvi–384, 21 pls.

Ortmann, A. E. 1920. Correlation of shape and station in freshwater mussels (naiades). *Proceedings of the American Philosophical Society* 59(4):269–312.

Ortmann, A. E. 1921. The anatomy of certain mussels from the Upper Tennessee. *The Nautilus* 34(3):81–91.

Ortmann, A. E. 1923. Notes on the anatomy and taxonomy of certain Lampsilinae from the Gulf Drainage. *The Nautilus* 37(2):56–60.

Ortmann, A. E. 1924a. The naiad-fauna of Duck River in Tennessee. *The American Midland Naturalist* 9(1):18–62.

Ortmann, A. E. 1924b. Distributional features of naiades in tributaries of Lake Erie. *The American Midland Naturalist* 9(3):101–117.

Ortmann, A. E. 1925. The naiad-fauna of the Tennessee River system below Walden Gorge. *The American Midland Naturalist* 9(7):321–372.

Ortmann, A. E. 1926a. Unionidae from the Reelfoot Lake Region in west Tennessee. *The Nautilus* 39(3):87–94.

Ortmann, A. E. 1926b. The naiades of the Green River drainage in Kentucky. *Annals of the Carnegie Museum* 17(1):167–188, pl. 8.

Ortmann, A. E., and B. Walker. 1922. On the nomenclature of certain North American naiades. *Occasional Papers of the Museum of Zoology, University of Michigan* No. 112. 75 pp.

Paetel, F. 1887–1890. *Catalog der Conchylien-Sammlung von Fr. Paetel, mit hinzu Fügung der bis jetzt Publieirten Recenten arten, sowie der emittelten Synonyma.* Three volumes.

Parmalee, P. W. n.d.a. A molluscan assemblage from a Late Archaic shell midden along the Cumberland River at Ashland City, Cheatham County, Tennessee. Species list on file, Frank H. McClung Museum, University of Tennessee, Knoxville.

Parmalee, P. W. n.d.b. Freshwater mussels from the Lyons Site and Looney Island Site, Knoxville, Knox County, Tennessee. Species list on file, Frank H. McClung Museum, University of Tennessee, Knoxville.

Parmalee, P. W. n.d.c. Freshwater mussels from the McCrosky Island Site (40SV43), Sevier County, Tennessee. Species list on file, Frank H. McClung Museum, University of Tennessee, Knoxville.

Parmalee, P. W. n.d.d. Freshwater mussels from the Dallas component recovered at the McCrosky Site (40SV9), Sevier County, Tennessee. Species list on file, Frank H. McClung Museum, University of Tennessee, Knoxville.

Parmalee, P. W. n.d.e. Freshwater mussels from a Middle Woodland Site on Diamond Island (TRM196.0), Hardin County, Tennessee. Species list on file, Frank H. McClung Museum, University of Tennessee, Knoxville.

Parmalee, P. W. 1959. Use of mammalian skulls and mandibles by prehistoric Indians of Illinois. *Transactions of the Illinois State Academy of Science* 52(3–4):85–95.

Parmalee, P. W. 1967. The fresh-water mussels of Illinois. *Illinois State Museum Popular Science Series* 8:1–108.

Parmalee, P. W. 1988. A comparative study of late prehistoric and modern molluscan faunas of the Little Pigeon River system, Tennessee. *American Malacological Bulletin* 6(2):165–178.

Parmalee, P. W. 1990. Freshwater mollusk and vertebrate remains identified from 40LD208, Loudon County, Tennessee (Appendix 2, pp. 60–64), and Animal remains from the 40LD207 Site, Loudon County, Tennessee (Appendix 3, pp. 65–73). *In:* The Kimberly-Clark site (40LD208) and site 40LD207 by J. Chapman. *Tennessee Anthropological Association Miscellaneous Paper No. 14/ University of Tennessee, Department of Anthropology Report of Investigations No. 51/ Frank H. McClung Museum Occasional Paper No. 8.*

Parmalee, P. W. 1994. Freshwater mussels from Dust and Smith Bottom caves, Alabama. Pp. 135–162. *In:* N. S. Goldman-Finn and B. N. Driskell, editors. Preliminary archaeological papers on Dust Cave, Northwest Alabama. *Journal of Alabama Archaeology* 40(1–2):1–255.

Parmalee, P. W., and A. E. Bogan. n.d. A mollusk sample from the Eva site (40BN12), Benton County, Tennessee. Species list on file, Frank H. McClung Museum, University of Tennessee, Knoxville.

Parmalee, P. W., and A. E. Bogan. 1986. Molluscan remains from aboriginal middens at the Clinch River Breeder Reactor Plant Site, Roane County, Tennessee. *American Malacological Bulletin* 4(1):25–37.

Parmalee, P. W., and M. H. Hughes. n.d.a. Freshwater mussels from the Hogan Site (40SW24), a Mississippian cultural period shell midden along the Cumberland River, Stewart County, Tennessee. Species list on file, Frank H. McClung Museum, University of Tennessee, Knoxville.

Parmalee, P. W., and M. H. Hughes. n.d.b. Freshwater mussels from the Daniels Landing Site (40PY4), a Mississippian cultural period shell midden along the Tennessee River, Perry County, Tennessee. Species list on file, Frank H. McClung Museum, University of Tennessee, Knoxville.

Parmalee, P. W., and M. H. Hughes. 1993. Freshwater mussels (Mollusca: Pelecypoda: Unionidae) of Tellico Lake: Twelve years after impoundment of the Little Tennessee River. *Annals of Carnegie Museum* 62(1):81–93.

Parmalee, P. W., and M. H. Hughes. 1994. Freshwater mussels (Bivalvia: Unionidae) of the Hiwassee River in East Tennessee. *American Malacological Bulletin* 11(1):21–27.

Parmalee, P. W., and W. E. Klippel. n.d. Freshwater mussels from an aboriginal rockshelter midden along the Cumberland River (ca. CRM 335.5), Jackson County, Tennessee. Species list on file, Frank H. McClung Museum, University of Tennessee, Knoxville.

Parmalee, P. W., and W. E. Klippel. 1974. Freshwater mussels as a prehistoric food resource. *American Antiquity* 39(3):421–434.

Parmalee, P. W., and W. E. Klippel. 1982. A relic population of *Obovaria retusa* in the middle Cumberland River, Tennessee. *The Nautilus* 96(1):30–32.

Parmalee, P. W., and W. E. Klippel. 1984. The naiad fauna of the Tellico River, Monroe County, Tennessee. *American Malacological Bulletin* 3(1):41–45.

Parmalee, P. W., and W. E. Klippel. 1986. A prehistoric aboriginal freshwater mussel assemblage from the Duck River in Middle Tennessee. *The Nautilus* 100(4):134–140.

Parmalee, P. W., W. E. Klippel, and A. E. Bogan. 1980. Notes on the prehistoric and present status of the naiad fauna of the middle Cumberland River, Smith County, Tennessee. *The Nautilus* 94(3): 93–105.

Parmalee, P. W., W. E. Klippel, and A. E. Bogan. 1982. Aboriginal and modern freshwater mussel assemblages (Pelecypoda: Unionidae) from the Chickamauga Reservoir, Tennessee. *Brimleyana* 8: 75–90.

Parmalee, P. W., and J. B. Layzer. n.d. Freshwater mussels from a Middle Woodland site (40DR305) along the Tennessee River (Kentucky Lake), Decatur County, Tennessee. Species list on file, Frank H. McClung Museum, University of Tennessee, Knoxville.

Parmalee, P. W., and C. O'Hare. 1989. Snails and freshwater mussels from the Anderson Site. Chapter VI, pp. 37–42. *In:* The Anderson Site: Middle Archaic adaptation in Tennessee's Central Basin by John T. Dowd. *Tennessee Anthropological Association. Miscellaneous Paper No. 13.*

Patch, D. C. 1976. An analysis of the archaeological shell of freshwater mollusks from the Carlston Annis Shellmound West Central Kentucky. Unpublished Bachelor's Honors thesis. Washington University, St. Louis. 76 pp.

Pekkarinen, M., and I. Valovirta. 1996. Anatomy of the glochidia of the freshwater pearl mussel, *Margaritifera margaritifera* (L.). *Archiv für Hydrobiologie* 137(3):411–423.

Pepi, V. E., and M. C. Hove. 1997. Suitable fish hosts and mantle display behavior of *Tritogonia verrucosa*. *Triannual Unionid Report* 11:5.

Petit, G. 1984. Reelfoot—troubled waters. *Tennessee Wildlife*. January–February. 1984:6–10.

Pharris, G. L., C. C. Chandler, and J. B. Sickel. 1982. Range extension for *Plectomerus dombeyanus* (Bivalvia: Unionidae) into Kentucky (Abstract). *Transactions of the Kentucky Academy of Science* 43(1/2):95–96.

Pilsbry, H. A. 1892. New and unfigured Unionidae. *Proceedings of the Academy of Natural Sciences of Philadelphia* 44:131–132, pls. 7, 8.

Pilsbry, H. A., and S. N. Rhoads. 1896. Contributions to the zoology of Tennessee. No. 4, Mollusks. *Proceedings of the Academy of Natural Sciences of Philadelphia* 47:487–506.

Potiez, V. L. V., and A. L. G. Michaud. 1838, 1844. *Galereie des Mollusques ou Catalogue Méthodique, Descriptif et Raisonné des Mollusques et Coquilles du Muséum de Douai, 2.* Two volumes. Bailliè, Paris.

Poulson, C. A. 1832. *A monograph of the fluviatile bivalve shells of the river Ohio, containing twelve genera & sixty-eight species.* Translated from the French of C. S. Rafinesque, Prof. Bot. and Nat. Hist. in Transylvania University. J. Dobson, Philadelphia. Pp. [i–iii], iv–vi, [7–9], 10–70, frontispiece.

Pratt, W. H. 1876. Description of a Unio shell found on the south bank of the Mississippi River, opposite the Rock Island Arsenal, in 1870. *Proceedings of the Davenport Academy of Natural Sciences* 1:167–168, 1 pl.

Price, S. F. 1900. Mollusca of Southern Kentucky. *The Nautilus* 14(7):75–79.

Putnam, J. D. 1971. A taxonomic study of two forms of the *Lampsilis ovata* complex in the Ohio River drainage system (Mollusca: Bivalvia: Naiadoida). Unpublished M.S. Thesis. Ohio State University. Columbus, Ohio. i–v + 76 pp.

Rafinesque, C. S. 1818a. Art. 3. Museum of Natural History. Discoveries in natural history, made during a journey through the Western region of the United States, by Constantine Samuel Rafinesque, Esq. Addressed to Samuel L. Mitchill, President and the other members of the Lyceum of Natural History, in a letter dated at Louisville, Falls of Ohio, 20th July 1818. *The American Monthly Magazine and Critical Review* 3(1):354–356.

Rafinesque, C. S. 1818b. Art. 3. Museum of Natural History. General account of the discoveries made in the zoology of the Western States by C. S. Rafinesque, in 1818. *The American Monthly Magazine and Critical Review* 4(2):106–107.

Rafinesque, C. S. 1819a. Prodrome de 70 nouveaux Genres d'Animaux découverts dans l'intérieur des États-Unis d'Amérique, durant l'année 1818. *Journal de Physique, de Chimie, d'Histoire Naturelle et des Arts* 88:417–429.

Rafinesque, C. S. 1819b. Conchology [of Louisville, Kentucky]. Pp. 65–66. *In: Sketches of Louisville and its environs; including, among a great variety of miscellaneous matter, a Florula Louisvillensis; or a catalogue of nearly 400 genera and 600 species of plants that grow in the vicinity of the town, exhibiting their generic, specific and vulgar English names.* By H. M'Murtrie, 1st ed. Louisville.

Rafinesque, C. S. 1820. Monographie des coquilles bivalves fluviatiles de la Rivière Ohio, contenant douze genres et soixante-huit espèces. *Annales générales des sciences Physiques, a Bruxelles* 5(5):287–322, pls. 80–82.

Rafinesque, C. S. 1831. *Continuation of a monograph of the bivalve shells of the river Ohio, and other rivers of the western states. By Prof. C. S. Rafinesque. (Published at Brussels, September, 1820) Containing 46 species, from No. 76 to no. 121. Including an appendix on some bivalve shells of the rivers of Hindustan, with a supplement on the fossil bivalve shells of the Western states, and the Tulosites, a new genus of fossils.* Philadelphia, Pennsylvania. 8 pp.

Rafinesque, C. S. 1832. *Odatelia* N.G. of N. American bivalve fluviatile shell. *Atlantic Journal and Friend of Knowledge* 4:154.

Reeve, L. 1841–1842. *Conchologia Systematica or complete system of conchology.* 2 volumes. 522 pp., 300 colored pls.

Reeve, L. 1843–1878. *Conchologia Iconica: or Illustrations of the shells of molluscous animals.* L. Reeve and Company, London. 20 vols., 2,727 colored pls.

Richardson, R. E. 1928. The bottom fauna of the middle Illinois River, 1913–1925. Its distribution, abundance, valuation, and index value in the study of stream pollution. *Illinois State Natural History Survey Bulletin* 17(12):387–475.

Roback, S. S., D. J. Bereza, and M. F. Vidrine. 1980. Description of an *Ablabesmyia* [Diptera: Chironomidae: Tanypodinae] symbiont of unionid fresh-water mussels [Mollusca: Bivalvia: Unionacea], with notes on its biology and zoogeography. *Transactions of the American Entomology Society* 105(4):577–620.

Robertson, I. C. S., and C. L. Blakeslee. 1948. The Mollusca of the Niagara frontier region and adjacent territory. *Bulletin of the Buffalo Society of Natural Sciences* 19(3):xi–191, 1 map.

Robison, N. D. 1978. An analysis of the faunal remains. Pp. 179–200. *In:* Excavations at the Tomotley Site (40MR5), Monroe County, Tennessee 1973–1974. By Alfred K. Guthe and E. Marion Bistline. *Report of Investigations No. 24*, Department of Anthropology, University of Tennessee, Knoxville.

Robison, N. D. 1986. An analysis and interpretation of the faunal remains from eight late Middle Woodland Owl Hollow Phase sites in Coffee, Franklin and Bedford Counties, Tennessee. Unpublished Ph.D. Dissertation, University of Tennessee, Knoxville, Tennessee. 390 pp.

Robison, N. D., and A. E. Bogan. n.d. An analysis of a surface collection of faunal remains from 40MI69 and 1JA331. Species list on file, Frank H. McClung Museum, University of Tennessee, Knoxville.

Savazzi, E., and Y. Peiyi. 1992. Some morphological adaptations in freshwater bivalves. *Lethaia* 25:195–209.

Say, T. 1817. Article Conchology. *In:* W. Nicholson, editor. *American Edition of the British Encyclopedia or Dictionary of Arts and Sciences, Comprising an Accurate and Popular View of the Present Improved State of Human Knowledge.* Vol. 2. 1st ed. No pagination. Pls. 1–4. Samuel A. Mitchel and Horace Ames, Philadelphia, Pennsylvania.

Say, T. 1818a. Article Conchology. *In:* W. Nicholson, editor. *American Edition of the British Encyclopedia or Dictionary of Arts and Sciences* 4(1818), 4 pls., no pagination.

Say, T. 1818b. Description of a new genus of fresh water bivalve shells. *Journal of the Academy of Natural Sciences of Philadelphia* 1:459–460.

Say, T. 1819. Article Conchology. *In:* W. Nicholson, editor. *Third American Edition of the British Encyclopedia or Dictionary of Arts and Sciences, illustrated by upwards of 180 elegant engravings.* Mitchel, Ames, White, Philadelphia, Pennsylvania. No pagination.

Say, T. 1825. Descriptions of some new species of fresh water and land shells of the United States. *Journal of the Academy of Natural Sciences of Philadelphia* 5:119–131.

Say, T. 1829. Descriptions of some new terrestrial and fluviatile shells of North America. *The Disseminator of Useful Knowledge; containing hints to the youth of the United States, from the School of Industry,* New Harmony, Indiana 2(19):291–293, 23 September 1829; 2(20):308–310 7 October 1829; 2(21):323–325, 21 October 1829; 2(22):339–341, 4 November 1829; 2(23):355–356, 18 November 1829.

Say, T. 1830a–1834. *American Conchology, or descriptions of the shells of North America. Illustrated by colored figures from original drawings executed from nature.* School Press. New Harmony, Indiana, seven parts. Pt. 1: 1830; pt. 2: April 1831; pt. 3: September 1831; pt. 4: March 1832; pt. 5: August 1832; pt. 6: April 1834; pt. 7: [1834?] published after Say's death, edited by T. A. Conrad.

Say, T. 1830b–1831b. New terrestrial and fluviatile shells of North America (cont.). *The Disseminator. [2nd Series].* New Harmony, Indiana 1(27):no pagination, 28 December 1830; 1(29): no pagination, 15 January 1831; 1(31):no pagination, 29 January 1831.

Say, T. 1831c. Descriptions of several new species of shells and of a new species of *Lumbricus. Transylvania Journal of Medicine* 4:525–528.

Scammon, R. E. 1906. The Unionidae of Kansas. Part I. An illustrated catalogue of the Kansas Unionidae. *The Kansas University Science Bulletin* 3(9):279–373, 3(10):pls. 52–85.

Schmidt, J. E. 1982. *The freshwater mussels of the Stones River above J. Percy Priest Reservoir, Tennessee.* U.S. Army Corps of Engineers, Nashville District. 66 pp.

Scudder, N. P. 1885. The published writings of Isaac Lea, L.L.D. Bibliographies of American Naturalists. *United States National Museum Bulletin* 23: lix + 278 pp.

Sepkoski, J. J., Jr., and M. A. Rex. 1974. Distribution of freshwater mussels: Coastal rivers as biogeographic islands. *Systematic Zoology* 23(2):165–188.

Sheehan, R. J., R. J. Neves, and H. E. Kitchel. 1989. Fate of freshwater mussels transplanted to formerly polluted reaches of the Clinch and North Fork Holston rivers, Virginia. *Journal of Freshwater Ecology* 5:139–149.

Short, C. W., and H. H. Eaton. 1831. Notices on western botany and conchology. *Transylvania Journal of Medicine* 4(1):69–82.

Shoup, C. S., J. H. Peyton, and G. Gentry. 1941. A limited biological survey of the Obey River and adjacent streams in Tennessee. *Journal of the Tennessee Academy of Science* 16(1):48–76.

Simpson, C. T. 1896. The classification and geographical distribution of the pearly fresh-water mussels. *Proceedings of the United States National Museum* 18(1068):295–343, 1 map.

Simpson, C. T. 1900a. Synopsis of the naiades, or pearly fresh-water mussels. *Proceedings of the United States National Museum* 22(1205):501–1044.

Simpson, C. T. 1900b. New and unfigured Unionidae. *Proceedings of the Academy of Natural Sciences of Philadelphia* 52:74–86, pls. 1–5.

Simpson, C. T. 1914. *A descriptive catalogue of the naiades, or pearly fresh-water mussels.* Parts I–III. Bryant Walker, Detroit, Michigan, xii + 1540 pp.

Sinclair R. M., and B. G. Isom. 1961. A preliminary report on the introduced Asiatic Clam *Corbicula* in Tennessee. Tennessee Stream Pollution Control Board, Tennessee Department of Public Health. v +79 pp.

Smith, H. M. 1899. The mussel fishery and pearl-button industry of the Mississippi River. *Bulletin of the U.S. Fish Commission* 18(1898):289–314, pls. 65–85

Smith, D. G., and W. P. Wall. 1984. The Margaritiferidae reinstated: A reply to Davis and Fuller (1981), "Genetic relationships among recent Unioniacea (Bivalvia) of North America." *Occasional Papers on Mollusks, Museum of Comparative Zoology, Harvard University* 4(64):321–330.

Sowerby, G. B. 1839. *A conchological manual*. London. G. B. Sowerby. v + 130 pp., 24 plates.

Sowerby, G. B. 1842. *A conchological manual*. 2d ed. London.

Sowerby, G. B. 1864–1868. Monograph of the genus *Unio*. *In:* L. Reeve and G. B. Sowerby, editors. *Conchologia Iconica* 16:1–163, 96 pls.

Sowerby, G. B. 1867–1870. Monograph of the genus *Anodon*. *In:* L. Reeve and G. B. Sowerby, editors. *Conchologia Iconica* 17:1–57, 37 pls.

Sowerby, J. de C. 1836. Mollusca. *In:* Richardson's *Fauna Boreali Americana*. III.

Sowerby, J., and G. B. Sowerby. 1821–1834. *The genera of Recent and fossil shells, for the use of students in conchology and geology.* Two volumes, London, G. B. Sowerby, Regent Street. pls. 1–279.

Spamer, E. E. 1996. Academy helps save habitat of endangered snail. *Explore* (February/March):3.

Spengler, C. L. 1793. Beskrivelse over et nyt Slaegt of de toskallede Konkylier, forhen of mig Kakdet Chaena, saa og over det Linneiske Slaegt Mya, hvilket noiere bestemnes, og inddeles i tvende Slaegter. *Skrivter of Naturhistorie-Selbskebet Kjobenhavn*. Volume 3.

Stansbery, D. H. 1964. The Mussel (Muscle) Shoals of the Tennessee River revisited (Abstract). *American Malacological Union, Inc., Annual Reports for 1964.* 31:25–28.

Stansbery, D. H. 1966. Observations on the habitat distribution of the naiad *Cumberlandia monodonta* (Say, 1829) (Abstract). *American Malacological Union, Inc., Annual Reports for 1966.* 33:29–30.

Stansbery, D. H. 1969. Changes in the naiad fauna of the Cumberland River at Cumberland Falls in Eastern Kentucky (Abstract). *American Malacological Union, Inc., Annual Reports* 36(1969):16–17.

Stansbery, D. H. 1970. 2. Eastern freshwater mollusks. (I.) The Mississippi and St. Lawrence River systems. American Malacological Union Symposium on Rare and Endangered Mollusks. *Malacologia* 10(1):9–22.

Stansbery, D. H. 1971. Rare and endangered freshwater mollusks in eastern United States. Pp. 5–18f, 50 figs. *In:* S. E. Jorgensen and R. E. Sharp, editors. *Proceedings of a symposium on rare and endangered mollusks (naiads) of the U.S. Region 3, Bureau Sport Fisheries and Wildlife, U.S. Fish Wildlife Service.* Twin Cities, Minnesota. 79 pp.

Stansbery, D. H. 1972. The mollusk fauna of the North Fork Holston River at Saltville, Virginia. *Bulletin of the American Malacological Union, Inc., for 1971:* 45–46.

Stansbery, D. H. 1973. A preliminary report on the naiad fauna of the Clinch River in the southern Appalachian Mountains of Virginia and Tennessee (Mollusca: Bivalvia: Unionoida). *Bulletin of the American Malacological Union, Inc., for 1972:* 20–22.

Stansbery, D. H. 1976a. Naiad mollusks. Pp. 42–52. *In:* H. T. Boschung, editor. Endangered and Threatened Plants and Animals of Alabama. *Bulletin of the Alabama Museum of Natural History* 2:1–92.

Stansbery, D. H. 1976b. The status of endangered fluviatile mollusks in central North America. II. *Pegias fabula* (Lea, 1838). The Ohio State University Research Foundation, [report for the U.S. Department of the Interior, Fish and Wildlife Service]. Final No. 2, 6 pp., 1 map, 2 figs. Also listed as: The Ohio State University, Museum of Zoology Reports 1976(10):1–8, 1 pl., 1 map.

Stansbery, D. H. 1976c. The status of endangered fluviatile mollusks in central North America. IV. *Toxolasma cylindrellus* (Lea, 1868). The Ohio State University Research Foundation, [report for the U.S. Department of the Interior, Fish and Wildlife Service]. Final No. 4, 7 pp., 1 map, 1 fig. Also listed as: The Ohio State University, Museum of Zoology Reports 1976(12):1–9, 1 pl., 1 map.

Stansbery, D. H. 1976d. The status of endangered fluviatile mollusks in central North America. *Quadrula sparsa* (Lea, 1841). The Ohio State University Research Foundation, [report for the U.S. Department of the Interior, Fish and Wildlife Service]. Final No. 1, 5 pp., 1 map, 1 fig. Also listed as: The Ohio State University, Museum of Zoology Reports 1976(9):1–6, 1 pl., 1 map.

Stansbery, D. H. 1976e. The status of endangered fluviatile mollusks in central North America. *Epioblasma walkeri* (Wilson and Clark, 1914). The Ohio State University Research Foundation, [report for the U.S. Department of the Interior, Fish and Wildlife Service]. Final No. 6, 10 pp., 1 map, 1 fig. Also listed as: The Ohio State University, Museum of Zoology Reports 1976(14):1–10, 1 pl., 1 map.

Stansbery, D. H. 1976f. The status of endangered fluviatile mollusks in central North America. *Epioblasma turgidula* (Lea, 1858). The Ohio State University Research Foundation, [report for the U.S. Department of the Interior, Fish and Wildlife Service]. Final No. 3, 14 pp., 1 map, 1 fig. Also listed as: The Ohio State University, Museum of Zoology Reports 1976(11):1–14, 1 pl., 1 map.

Stansbery, D. H. 1976g. The status of endangered fluviatile mollusks in central North America. *Quadrula intermedia* (Conrad, 1836). The Ohio State University Research Foundation, [report for the U.S. Department of the Interior, Fish and Wildlife Service]. Final No. 5, 9 pp., 1 map, 1 fig. Also listed as: The Ohio State University, Museum of Zoology Reports 1976(13):1–9, 1 pl., 1 map.

Stansbery, D. H. 1979. The status of *Lemiox rimosus* (Rafinesque, 1831) (Mollusca: Bivalvia: Unionoida). Report of the Office of Endangered Species, Fish and Wildlife Service, U.S. Department of the Interior, Washington, D.C., 9 pp., 1 map, 2 figs.

Stark, J. 1828. *Elements of natural history, adapted to the present state of the science, containing the generic characters of nearly the whole animal kingdom, and the descriptions of the principal species.* Two volumes. Edinburgh, W. Blackwood, 8 pls.

Starnes, L. B., and A. E. Bogan. 1982. Unionid Mollusca (Bivalvia) from Little South Fork Cumberland River, with ecological and nomenclatural notes. *Brimleyana* 8:101–119.

Starnes, L. B., and A. E. Bogan. 1988. The mussels (Mollusca: Bivalvia: Unionidae) of Tennessee. *American Malacological Bulletin* 6(1):19–37.

Starobogatov, Ya. I. 1970. *Fauna mollyuskov i zoogeographicheskoe raionirovanie kontinental'nykh vodoemov zemnogo shara* [Mollusk fauna and zoogeographical partitioning of continental water reservoirs of the world]. Akademiya Nauk SSSR. Zoologischeskii Instituti Nauka. Leningrad. 372 pp., 39 figs., 12 tables [in Russian].

Starrett, W. C. 1971. A survey of the mussels (Unionacea) of the Illinois River: A polluted stream. *Illinois Natural History Survey Bulletin* 30(5):267–403.

Stein, C. B. 1963. Notes on the naiad fauna of the Olentangy River in central Ohio (Abstract). *American Malacological Union, Inc., Annual Reports for 1963* 30:19.

Stein, C. B. 1965. The naiad fauna of Little Darby Creek in central Ohio (Abstract). *American Malacological Union, Inc., Annual Reports for 1965* 32:22–23.

Stein, C. B. 1972. Population changes in the naiad mollusk fauna of the lower Olentangy River following channelization and highway construction. *Bulletin of the American Malacological Union, Inc., for 1971*, 47–48, 1 pl.

Sterki, V. 1898. Anodonta imbecillis, hermaphroditic. *The Nautilus* 12(8):87–88.

Sterki, V. 1907. A preliminary catalogue of the land and fresh-water Mollusca of Ohio. *Proceedings of the Ohio State Academy of Science* 4(8):367–402.

Sterki, V. 1914. Ohio mollusca. Additions and corrections. *Ohio Naturalist* 14(5):270–272.

Stern, E. M., and D. L. Felder. 1978. Identification of host fishes for four species of freshwater mussels (Bivalvia: Unionidae). *The American Midland Naturalist* 100(1):233–236.

Stimpson, W. 1851. *Shells of New England; a revision of the synonymy of the testaceous mollusks of New England, with notes on their structure and their geographical and bathymetrical distribution, with figures of new species.* Boston, Massachusetts. 58 pp. + 2 pls.

Strecker, J. K., Jr. 1931. The distribution of the naiades or pearly fresh-water mussels of Texas. *Baylor University Museum Special Bulletin* 2:1–71.

Surber, T. 1912. Identification of the glochidia of freshwater mussels. *Report and Special Papers of the U.S. Bureau of Fisheries* 1912:1–10, pls. 1–3. Issued separately as U.S. Bureau of Fisheries Document 771.

Surber, T. 1913. Notes on the natural hosts of fresh-water mussels. *Bulletin of the Bureau of Fisheries* 32(1912):101–116, pls. 29–31. Issued separately as U.S. Bureau of Fisheries Document 778.

Swainson, W. 1820–1823a. *Zoological illustrations or original figures and descriptions of rare, or interesting animals.* First series. London. 3 volumes.

Swainson, W. 1823b. Comments on *Iridina*, etc. *Philosophical Magazine* 1823:112–113.

Swainson, W. 1824. Description of two new remarkable freshwater shells, *Melania setosa* and *Unio gigas*. *Quarterly Journal of Science* 17:13–17.

Swainson, W. 1829–1833. *Zoological illustrations.* Second series. 6 volumes.

Swainson, W. 1840. *A treatise on Malacology or the natural classification of shells and shell-fish.* London. 419 pp.

Swainson, W. 1841. *Exotic conchology.* Edited by S. Hanley. 2d ed.

Tankersley, R. A. 1996. Multipurpose gills: Effect of larval brooding on the feeding physiology of freshwater unionid mussels. *Invertebrate Biology* 115(3):243–255.

Tankersley, R. A., and R. V. Dimock, Jr. 1992. Quantitative analysis of the structure and function of the marsupial gills of the freshwater mussel *Anodonta cataracta*. *Biological Bulletin* 182:145–154.

Tappan, B. 1839. Description of some new shells. *American Journal of Science and Arts* 35:268–270, pl. 3.

Tennessee Valley Authority. 1986. *Cumberlandian mollusk conservation program. Activity 3: Identification of fish hosts.* Tennessee Valley Authority, Knoxville, Tennessee: Office of Natural Resources and Economic Development. 57 pp.

Thiele, J. 1929–1935. *Handbuch der systematischen Weichtierkunde.* In four parts. 1154 pp. Gustav Fischer, Jena.

Thorp, J. H., and A. P. Covich, editors. 1991. *Ecology and classification of North American freshwater invertebrates.* Academic Press, New York. 911 pp.

Trdan, R. J., and W. R. Hoeh. 1982. Eurytopic host use by two congeneric species of freshwater mussel (Pelecypoda: Unionidae: Anodonta). *The American Midland Naturalist* 108:381–388.

Troost, G. 1846. Catalogue of the shells of Tennessee. Pp. 40–42. *Seventh Annual Report on the Geology of Tennessee.*

Troschel, F. H. 1847. Ueber die Brauchbarkeit der Mundlappen und Kiemen zur Familienunterscheidung und über die Familie der Najaden. *Archiv für Naturgeschichte* 13(1):257–274, pl. 6.

Tucker, M. E. 1927. Morphology of the glochidium and juvenile of the mussel *Anodonta imbecillis*. *Transactions of the American Microscopical Society* 46(4):286–293.

Turgeon, D. D., A. E. Bogan, E. V. Coan, W. K. Emerson, W. G. Lyons, W. L. Pratt, C. F. E. Roper, A. Scheltema, F. G. Thompson, and J. D. Williams. 1988. Common and scientific names of aquatic invertebrates from the United States and Canada: Mollusks. *American Fisheries Society, Special Publication* 16:viii + 277 pp., 12 pls.

Turgeon, D. D., J. F. Quinn, Jr., A. E. Bogan, E. V. Coan, F. G. Hochberg, W. G. Lyons, P. M. Mikkelsen, C. F. E. Roper, G. Rosenberg, B. Roth, A. Scheltema, M. J. Sweeney, F. G. Thompson, M. Vecchione, and J. D. Williams. n.d. Common and scientific names of aquatic invertebrates from the United States and Canada: Mollusks. *American Fisheries Society Special Publication*. 2d ed. [in press].

U.S. Fish and Wildlife Service. 1982. Cumberland monkeyface pearly mussel recovery plan. U.S. Fish and Wildlife Service, Atlanta, Georgia. 59 pp.

U.S. Fish and Wildlife Service. 1983a. Birdwing pearly mussel recovery plan. U.S. Fish and Wildlife Service, Atlanta, Georgia. 56 pp.

U.S. Fish and Wildlife Service. 1983b. Dromedary pearly mussel recovery plan. U.S. Fish and Wildlife Service, Atlanta, Georgia. 58 pp.

U.S. Fish and Wildlife Service. 1983c. Green-blossom pearly mussel recovery plan. U.S. Fish and Wildlife Service, Atlanta, Georgia. 50 pp.

U.S. Fish and Wildlife Service. 1983d. Shiny pigtoe pearly mussel recovery plan. U.S. Fish and Wildlife Service, Atlanta, Georgia. 67 pp.

U.S. Fish and Wildlife Service. 1983e. Appalachian monkey-face pearly mussel recovery plan. U.S. Fish and Wildlife Service, Atlanta, Georgia. 55 pp.

U.S. Fish and Wildlife Service. 1984a. Tan riffle shell pearly mussel recovery plan. U.S. Fish and Wildlife Service, Atlanta, Georgia. 59 pp.

U.S. Fish and Wildlife Service. 1984b. Fine-rayed pigtoe pearly mussel recovery plan. U.S. Fish and Wildlife Service, Atlanta, Georgia. 67 pp.

U.S. Fish and Wildlife Service. 1984c. Orange-footed pearly mussel recovery plan. U.S. Fish and Wildlife Service, Atlanta, Georgia. 46 pp.

U.S. Fish and Wildlife Service. 1984d. White warty-back pearly mussel recovery plan. U.S. Fish and Wildlife Service, Atlanta, Georgia. 43 pp.

U.S. Fish and Wildlife Service. 1984e. Rough pigtoe pearly mussel recovery plan. U.S. Fish and Wildlife Service, Atlanta, Georgia. 51 pp.

U.S. Fish and Wildlife Service. 1984f. Pale lilliput pearly mussel recovery plan. U.S. Fish and Wildlife Service, Atlanta, Georgia. 46 pp.

U.S. Fish and Wildlife Service. 1984g. Cumberland bean pearly mussel recovery plan. U.S. Fish and Wildlife Service, Atlanta, Georgia. 58 pp.

U.S. Fish and Wildlife Service. 1985a. Recovery plan for the tubercled-blossom pearly mussel *Epioblasma (Dysnomia) torulosa torulosa* (Rafinesque, 1820); Turgid-blossom pearly mussel *Epioblasma (Dysnomia) turgidula* (Lea, 1858) and yellow-blossom pearly mussel *Epioblasma (Dysnomia) florentina florentina*. U.S. Fish and Wildlife Service, Atlanta, Georgia. 42 pp.

U.S. Fish and Wildlife Service. 1985b. Recovery plan for the pink mucket pearly mussel (*Lampsilis orbiculata* (Hildreth, 1828)). U.S. Fish and Wildlife Service, Atlanta, Georgia. 47 pp.

U.S. Fish and Wildlife Service. 1985c. Alabama lamp pearly mussel recovery plan. U.S. Fish and Wildlife Service, Atlanta, Georgia. 41 pp.

U.S. Fish and Wildlife Service. 1986. Curtis' pearly mussel recovery plan. U.S. Fish and Wildlife Service, Twin Cities, Minnesota. 92 pp.

U.S. Fish and Wildlife Service. 1989a. Little-wing pearly mussel recovery plan. U.S. Fish and Wildlife Service, Atlanta, Georgia. 29 pp.

U.S. Fish and Wildlife Service. 1989b. Recovery plan watershed implementation schedules for fifteen mussels in Alabama, Illinois, Kentucky, Tennessee, and Virginia. U.S. Fish and Wildlife Service, Atlanta, Georgia. 83 pp.

U.S. Fish and Wildlife Service. 1990a. Purple cat's paw pearly mussel recovery plan. U.S. Fish and Wildlife Service, Atlanta, Georgia. 26 pp.

U.S. Fish and Wildlife Service. 1990b. Cracking pearly mussel *(Hemistena (=Lastena) lata)* recovery plan. U.S. Fish and Wildlife Service, Atlanta, Georgia. 25 pp.

U.S. Fish and Wildlife Service. 1991a. Fanshell *(Cyprogenia stegaria (=Cyprogenia irrorata))* recovery plan. U.S. Fish and Wildlife Service, Atlanta, Georgia. 37 pp.

U.S. Fish and Wildlife Service. 1991b. Ring pink pearly mussel recovery plan. U.S. Fish and Wildlife Service, Atlanta, Georgia. 24 pp.

U.S. Fish and Wildlife Service. 1991c. Cumberland pigtoe mussel *(Pleurobema gibberum)* recovery plan. U.S. Fish and Wildlife Service, Atlanta, Georgia. 20 pp.

U.S. Fish and Wildlife Service. 1993. Clubshell *(Pleurobema clava)* and northern riffleshell *(Epioblasma torulosa rangiana)* recovery plan. Technical/Agency Draft. Hadley, Massachusetts. 55 pp.

U.S. Fish and Wildlife Service. 1994. Technical/Agency Draft Mobile River Basin Ecosystem recovery plan. U.S. Fish and Wildlife Service, Jackson, Mississippi. 128 pp.

U.S. Fish and Wildlife Service. 1995. Endangered and threatened wildlife. *Code of Federal Regulations* 50CFR §17.11:94–129.

U.S. Fish and Wildlife Service. 1996. Recovery plan for the Appalachian elktoe (*Alasmidonta raveneliana* Lea). U.S. Fish and Wildlife Service, Atlanta, Georgia. 31 pp.

Utterback, W. I. 1915. The naiades of Missouri. *The American Midland Naturalist* 4(3):41–53; 4(4):97–152; 4(5):181–204, 4(6):244–273.

Utterback, W. I. 1916a. The naiades of Missouri. *The American Midland Naturalist* 4(7):311–327; 4(8):339–354; 4(9):387–400; 4(10):432–464, pls. 1–28.

Utterback, W. I. 1916b. *The naiades of Missouri*. University Press, Notre Dame, Indiana. 200 pp., 28 pls.

Utterback, W. I. 1916c. Breeding record of Missouri mussels. *The Nautilus* 30(2):13–21.

Valenciennes, A. 1827. Coquilles marines bivalves de l'Amerique Équinoxiale recueilles pendant le Voyage de M. M. de Humboldt et Bonpland. *In:* F. H. A. von Humboldt and A. J. A. Bonpland, *Voyage aux régions équinoxiales du Nouveau Continent*. Paris. 2(2):217–224, pl. 48–50.

Valentine, B. D., and D. H. Stansbery. 1971. An introduction to the naiades of the Lake Texoma region, Oklahoma, with notes on the Red River fauna (Mollusca: Unionidae). *Sterkiana* 42:1–40.

van der Schalie, H. 1938. The naiades (freshwater mussels) of the Cahaba River in northern Alabama. *Occasional Papers of the Museum of Zoology, University of Michigan* No. 392:1–29.

van der Schalie, H. 1939. *Medionidus mcglameriae,* a new naiad from the Tombigbee River, with notes on other naiads of that drainage. *Occasional Papers of the Museum of Zoology, University of Michigan* No. 407. 6 pp., 1 pl.

van der Schalie, H. 1966. Hermaphroditism among North American freshwater mussels. *Malacologia* 5(1):77–78.

van der Schalie, H. 1970. Hermaphroditism among North American freshwater mussels. *Malacologia* 10(1):93–112.

van der Schalie, H. 1973. The mollusks of the Duck River drainage in central Tennessee. *Sterkiana* 52:45–55.

van der Schalie, H. 1981. Past, present and future status of the Mollusca of the Upper Tombigbee River. *Sterkiana* 71:8–11.

van der Schalie, H., and A. van der Schalie. 1950. The mussels of the Mississippi River. *The American Midland Naturalist* 44(2):448–466.

Vanatta, E. G. 1910. Unionidae from southeastern Arkansas and N.E. Louisiana. *The Nautilus* 23(8):102–104.

Vanatta, E. G. 1915. Rafinesque's types of Unio. *Proceedings of the Academy of Natural Sciences of Philadelphia* 67(1915):549–559.

Vidrine, M. F. 1980. Systematics and coevolution of unionicolid water-mites and their unionid mussel hosts in the eastern United States. Unpublished Ph.D. Dissertation. Department of Biology, University of Southwestern Louisiana, Lafayette. xvii + 661 pp.

Vidrine, M. F. 1993. *The historical distributions of freshwater mussels in Louisiana*. Gail Q. Vidrine Collectibles. Eunice, Louisiana. 225 pp., 136 maps, 7 tables, 20 color pls.

Vidrine, M. F. 1996. *Najadicola and Unionicola: I. Diagnoses of genera and subgenera. II. Key. III. List of reported hosts*. Gail Q. Vidrine Collectibles. Eunice, Louisiana. 182 pp.

Vidrine, M. F., and J. L. Wilson. 1991. Parasitic mites (Acari: Unionicolidae) of fresh-water mussels (Bivalvia: Unionidae) in the Duck and Stones rivers in central Tennessee. *The Nautilus* 105(4):152–158.

von Ihering, H. 1893. Najaden von S. Paulo und die geographische Verbreitung der Süsswasser-Faunen von Südamerika. *Archiv für Naturgeschichte* 1893:45–140, pls. 3, 4.

Walker, B. 1910a. Description of a new species of Truncilla. *The Nautilus* 24(4):42–44, pl. 3.

Walker, B. 1910b. Notes on Truncilla, with a key to the species. *The Nautilus* 24(7):75–81.

Walker, B. 1911. Note on the distribution of Margaritana monodonta Say. *The Nautilus* 25(5):57–58.

Walker, B. 1913. The Unione fauna of the Great Lakes. *The Nautilus* 27(2):18–23; 27(3):29–34; 27(4):40–47; 27(5):56–59.

Walker, B. 1915. Pleurobema missouriensis Marsh. *The Nautilus* 28(12):140–141.

Walker, B. 1916. The Rafinesque-Poulson Unios. *The Nautilus* 30(4):43–47.

Walker, B. 1918a. A synopsis of the classification of the freshwater Mollusca of North America, North of Mexico, and a catalogue of the more recently described species, with notes. *Miscellaneous Publications, Museum of Zoology, University of Michigan* 6:1–213.

Walker, B. 1918b. The Mollusca. Pp. 957–1020. *In:* H. B. Ward and G. C. Whipple, editors. *Fresh-water Biology*. New York, 1st ed. 1111 pp.

Waller, D. L., L. E. Holland, L. G. Mitchell, and T. W. Kammer. 1985. Artificial infestation of largemouth bass and walleye with glochidia of *Lampsilis ventricosa* (Pelecypoda: Unionidae). *Freshwater Invertebrate Biology* 4(3):152–153.

Walling, R., L. Alexander, and E. Peacock. 1993. Archaeological data recovery, Jefferson Street (FAU-3258) Bridge; The East Nashville Mounds (40DV4) and French Lick/Sulphur Dell (40DV5) sites, Nashville, Davidson County, Tennessee. Draft report submitted to the Tennessee Department of Transportation by Panamerican Consultants, Inc., Tuscaloosa, Alabama. Two volumes, currently under revision for final publication.

Ward, F. 1985. The pearl. *National Geographic* 168(2):193–223.

Ward, H. B., and G. C. Whipple, editors. 1918. *Fresh-water Biology*. 1st ed. 1111 pp.

Warren, R. E. 1975. Prehistoric Unionacean (freshwater mussel) utilization at the Widows Creek site (1JA305), northeast Alabama. Unpublished M.A. Thesis. University of Nebraska. Lincoln, Nebraska. 245 pp.

Watkins, S. R. 1962. *"Co. Aytch," a side show of the big show.* Collier Books, Division of Macmillan Publishing Co., Inc., New York. 255 pp.

Watson, B. T., and R. J. Neves. 1996. Host fishes for two federally endangered species of mussels. *Triannual Unionid Report* 10:13.

Watters, G. T. 1993. *A guide to the freshwater mussels of Ohio.* Division of Wildlife, The Ohio Department of Natural Resources. Rev. ed. 106 pp.

Watters, G. T. 1994a. An annotated bibliography of the reproduction and propagation of the Unionoidea (Primarily of North America). *Ohio Biological Survey Miscellaneous Contributions* No. 1, 158 pp.

Watters, G. T. 1994b. Form and function of unionoidean shell sculpture and shape. *American Malacological Bulletin* 11:1–20.

Watters, G. T. 1995. New hosts for *Anodontoides ferussacianus* (Lea, 1834). *Triannual Unionid Report* 7:7–8.

Watters, G. T. 1996a. New hosts for *Lampsilis cardium. Triannual Unionid Report* 9:8.

Watters, G. T. 1996b. Hosts for the northern riffle shell (*Epioblasma torulosa rangiana*). *Triannual Unionid Report* 10:14.

Watters, G. T. 1996c. Small dams as barriers to freshwater mussels (Bivalvia, Unionoida) and their hosts. *Biological Conservation* 75:79–85.

Watters, G. T. 1997. Surrogate hosts: transformation on exotic and non-piscine hosts. *Triannual Unionid Report* 11:35.

Weaver, L. R., G. B. Pardue, and R. J. Neves. 1991. Reproductive biology and fish hosts of the Tennessee clubshell Pleurobema oviforme (Mollusca: Unionidae) in Virginia. *The American Midland Naturalist* 126:82–89.

Weiss, J. L., and J. B. Layzer. 1995. Infestations of glochidia on fishes in the Barren River, Kentucky. *American Malacological Bulletin* 11(2):153–159.

Western Academy of Natural Sciences of Cincinnati. 1849. *Catalogue of the Unios, Alasmodonta, and anodontas of the Ohio River and its northern tributaries, adopted by the Western Academy of Natural Sciences of Cincinnati, January 1849.* J. A. and U. P. James, Cincinnati. 19 pp.

Wheeler, H. E. 1914. The unione fauna of Cache River, with description of a new Fusconaia from Arkansas. *The Nautilus* 28(7):73–78, pl. 4.

Williams, J. D., A. E. Bogan, and R. J. Neves. n.d. Possibly extinct mollusks of North America. *In:* Turgeon et al., n.d. Common and scientific names of aquatic invertebrates from the United States and Canada: Mollusks. *American Fisheries Society Special Publication.* 2d ed. [in review].

Williams, J. D., M. L. Warren, Jr., K. S. Cummings, J. L. Harris, and R. J. Neves. 1993. Conservation status of the freshwater mussels of the United States and Canada. *Fisheries* 18(9):6–22.

Wilson, C. B. 1916. Copepod parasites of fresh-water fishes and their economic relations to mussel glochidia. *Bulletin of the Bureau of Fisheries* 34:331–374, pls. 60–74. Issued separately as *U.S. Bureau of Fisheries Document* 824.

Wilson, C. B., and H. W. Clark. 1912. The mussel fauna of the Kankakee basin. *Report and Special Papers of the U.S. Fish Commission* 1911:1–52, 1 map. Issued separately as U.S. Bureau of Fisheries Document 758.

Wilson, C. B., and H. W. Clark. 1914. The mussels of the Cumberland River and its tributaries. *U.S. Bureau of Fisheries Document* 781:1–63.

Wilson, K. A., and K. Ronald. 1967. Parasite fauna of the sea lamprey (*Petromyzon marinus* von Linné) in the Great Lakes region. *Canadian Journal of Zoology* 45:1083–1092.

Wood, W. 1828. *Supplement to the Index testaceologicus, or A catalogue of shells, British and foreign, illustrated with 480 figures.* London, printed for W. Wood. iv, I, 59 pp., 8 pls.

Wood, W. 1856. *Index testaceologicus, an illustrated catalogue of British and foreign shells, containing about 2,800 figures accurately coloured after nature, by W. Wood, a new and entirely revised edition, with ancient and modern appellations, synonyms, localities, etc. . . .* Edited by S. Hanley. London, Willis and Sotheran. xx + 234 pp., colored ill.

Wright, B. H. 1888a. Descriptions of new species of uniones from Florida. *Proceedings of the Academy of Natural Sciences of Philadelphia* 40:113–120, 5 pls.

Wright, B. H. 1888b. *Check list of North American Unionidae and other fresh water bivalves.* Dore and Cook, Portland. 8 pp.

Wright, B. H. 1896. New American Unionidae. *The Nautilus* 9(12):133–135, pl. 3.

Wright, B. H. 1898a. A new undulate Unio from Alabama. *The Nautilus* 11(9):101–102.

Wright, B. H. 1898b. New varieties of Unionidae. *The Nautilus* 11(11):123–124.

Wright, B. H. 1898c. New Unionidae. *The Nautilus* 12(1):5–6.

Wright, B. H. 1899. New southern Unios. *The Nautilus* 13(1):6–8; 13(2):22–23; 13(3):31; 13(4):42–43; 13(5):50–51; 13(6):69; 13(7):75–76; 13(8):89–90.

Wright, S. H. 1897. Contribution to a knowledge of United States Unionidae. *The Nautilus* 10(12):136–139; 11(1):4–5.

Yeager, B. L. 1987. Fish hosts for glochidia of *Epioblasma brevidens, E. capsaeformis,* and *E. triquetra* (Pelecypoda: Unionidae) from the upper Tennessee River drainage. Unpublished report on file with Office of Natural Resources and economic development, Tennessee Valley Authority, Norris, Tennessee.

Yeager, B. L., and R. J. Neves. 1986. Reproductive cycle and fish hosts of the rabbit's foot mussel, Quadrula cylindrica strigillata (Mollusca: Bivalvia: Unionidae) in the upper Tennessee River drainage. *The American Midland Naturalist* 116:329–340.

Yeager, B. L., and C. F. Saylor. 1995. Fish hosts for four species of freshwater mussels (Pelecypoda: Unionidae) in the Upper Tennessee River drainage. *The American Midland Naturalist* 133(1):1–6.

Yokley, P., Jr. 1972. Life history of *Pleurobema cordatum* (Rafinesque, 1820) (Bivalvia: Unionacea). *Malacologia* 11(2):351–364.

Zale, A. V., and R. J. Neves. 1982a. Fish hosts of four species of lampsiline mussels (Mollusca: Unionidae) in Big Moccasin Creek, Virginia. *Canadian Journal of Zoology* 60(11):2535–2542.

Zale, A. V., and R. J. Neves. 1982b. Reproductive biology of four freshwater mussel species (Mollusca: Unionidae) in Virginia. *Freshwater Invertebrate Biology* 1(1):17–28.

Zale, A. V., and R. J. Neves. 1982c. Identification of a fish host for Alasmidonta minor (Mollusca: Unionidae). *The American Midland Naturalist* 107(2):386–388.

Note: *Triannual Unionid Report* compiled and distributed by Endangered Species Field Office, U.S. Fish and Wildlife Service, Asheville, North Carolina.

Index

Pagination for primary treatment of Tennessee species appears in **boldface.**

306   *Index*

www.ingramcontent.com/pod-product-compliance
Lightning Source LLC
Chambersburg PA
CBHW040255100426
42811CB00011B/1269